Springer Texts in Statistics

For further volumes:
http://www.springer.com/series/417

René Carmona

Statistical Analysis of Financial Data in R

Second Edition

Springer

René Carmona
Department of Operations Research
and Financial Engineering
Princeton University
Princeton, NJ, USA

ISSN 1431-875X ISSN 2197-4136 (electronic)
ISBN 978-1-4939-3835-3 ISBN 978-1-4614-8788-3 (eBook)
DOI 10.1007/978-1-4614-8788-3
Springer New York Heidelberg Dordrecht London

Printed on acid-free paper

Springer is part of Springer Science+Business Media (www.springer.com)

To Chanel, Chelsea, and Stéphanie

Preface

Like its earlier incarnation in `S-Plus` written over 10 years ago, this book is a polished version of the lecture notes written for a one-semester junior statistics course offered to the undergraduate students majoring in the Department of Operations Research and Financial Engineering and a core course of the Master's program of the Bendheim Center for Finance at Princeton University.

The common goal of both courses is to introduce students to modern data analysis used in the financial industry. The prerequisites are minimal, though students are expected to have already taken a basic introductory statistics course. Elementary notions of random variables, expectation, and correlation are taken for granted, and earlier exposure to statistical inference (estimation, tests, and confidence intervals) is assumed. It is also expected that the students are familiar with a minimum of linear algebra as well as vector and matrix calculus. However, all the background concepts and results necessary for the comprehension of the material presented in the book (as well as the solutions of the homework problems) are recalled before they are used or needed.

By choice, the courses are both computational and mathematical in nature. Most problems considered are formulated in a rigorous manner. Mathematical facts are motivated by applications, stated precisely, justified at an intuitive level, but essentially never proven rigorously. The emphasis is more on the relevance of concepts and on the practical use of tools, rather than on their theoretical underpinnings.

I chose to illustrate concepts, manipulate data, build models, and implement estimation and prediction procedures in the `R` computer environment. For this reason the text is sprinkled with the `R` commands needed to perform the analyses and produce the plots. The first incarnation of this text was written for `S-Plus` on `Windows` platforms. The growing presence of Mac computers in the classrooms and the ease with which `Linux`, `Windows`, and `MacOS` versions of `R` can be downloaded and installed at no cost were influential in my decision to switch from `S-Plus` to `R`. To my surprise, the port was not as seamless as I originally expected. It took me several years to complete the transition, and in the process, an entire chapter, several sections, and a large number of examples and problems have been added to the original contents.

The text is divided into three parts. Part I, *Data Exploration, Estimation, and Simulation*, introduces heavy tail distributions and dependence concepts for multivariate data, with an emphasis on the practical applications of copulas. Part II, *Regression*, introduces the students to modern regression with an emphasis on robustness and nonparametric techniques. Part III, *Time Series and State Space Models*, is concerned with the theories of time series and state space models, including filtering applications.

CONTENTS

Part I comprises three chapters. Chapter 1 begins with a review of the classical probability distributions encountered throughout the book and presents the exploratory data analysis techniques (histograms, kernel density estimators, Q-Q plots, etc.) used to handle empirical samples. As a preparation for many analyses and problems based on random simulations, the chapter concludes with a discussion of Monte Carlo computations.

Chapter 2 is devoted to the detection, estimation, and simulation of *heavy tail* distributions already showcased in the first chapter. It contains more statements and discussions of theoretical results than most other chapters, the reason being the desire to provide insight in the estimation and simulation algorithms implemented in the `R` library `Rsafd` used in the practical applications. Illustrative examples are used to demonstrate the impact of the presence of heavy tails on the computations of measures of risk such as value at risk (also known as VaR).

The third chapter is concerned with multivariate distributions and the various concepts of dependence. We review the classical measures of correlation, demonstrate the shortcomings of the Pearson correlation coefficient, and study the notion of copula, and the important role it plays when the marginal distributions have heavy tails, both in the bivariate case and in the high dimensional case. We learn how to detect unusual dependencies, estimate and simulate them, and bring this expertise to bear on the analysis of large portfolios of financial instruments including stocks and credit derivatives. The chapter concludes with a complete discussion of principal component analysis and two applications to the fixed income markets.

Part II is concerned with regression, and it is naturally divided into two chapters: the first devoted to parametric methods and the second to nonparametric ones. Chapter 4 deals with linear models and their applications. The notion of robustness is introduced, and examples are used to illustrate the differences between least squares and least absolute deviations regressions. Applications of linear models include polynomial and more general nonlinear regressions. We use financial examples throughout, and we analyze the term structure of interest rates in detail. Chapter 5 is concerned with nonparametric regression. We compare the properties of data smoothers for univariate data, and we analyze in detail the multivariate kernel regression. For larger dimensions, we use projection pursuit. Examples of energy forward curves and intraday S&P 500 futures tick data are given. The last part of this chapter is devoted to the use of semi-parametric and nonparametric methods in option pricing. We demonstrate the implementation of modern regression techniques as pricing alternatives to the classical Black-Scholes pricing formula.

The first chapter of Part III is devoted to the classical linear models for time series and to the idiosyncrasies of the `R` objects and methods included in the library `Rsafd` for the sole purpose of their analyses. We discuss autoregressive and moving-average models, and we give examples of their use in practice. The main application is the analysis of temperature data. Even if it may not appear to be much of a financial application at first, we recast this analysis in the framework of financial risk management via a thorough discussion of the market of weather derivatives. We give practical examples to illustrate the use of the statistical techniques introduced in this chapter to the control of these financial instruments.

In the following two chapters, we turn to the analysis of partially observed state space systems. Chapter 7 deals with linear models and the classical Kalman filter. For illustration purposes, we study two financial applications, one related to an extension of the CAPM model and a second dealing with the analysis of quarterly company earnings. Chapter 8 is devoted to the analysis of nonlinear time series. We first consider the natural generalizations of the linear time series models, and we provide an extensive review of the theory and the practice of the famous ARCH and GARCH models. We also consider models from continuous time finance through their discretized forms. A special section is devoted to the use of scenarios for economic modeling. We concentrate on scenarios for a stock index and the short and long interest rates. These scenarios are of crucial importance in risk management where they are used as input to large stochastic optimization programs. Finally, we revisit the theory presented in the case of partially observed linear systems, and we extend the filtering paradigm to nonlinear systems with the help of recent advances in Monte Carlo techniques and the so-called particle filters. We give several applications of this material, including the estimation of stochastic volatility and commodity convenience yield.

Each chapter contains a problem section. Most practical problems are rooted in financial applications. Each problem is preceded by one or several symbols Ⓔ, Ⓢ, and/or Ⓣ intended as hints suggesting if it is of an empirical, simulation, and/or theoretical nature. Chapters end with Notes and Complements sections that include complements and bibliographic references for the readers interested in acquiring a deeper understanding of the topics of that chapter. The book ends with an appendix and a suite of indexes. The appendix contains the text of an introductory session to `R` intended to help the reader unfamiliar with `R` get started and to a crash course on Black-Scholes option pricing theory used in several chapters.

The code and the data sets used in the text and the problems are contained in the library `Rsafd` developed as a companion to the book. It can be downloaded from the URL:

`http://www.princeton.edu/~rcarmona`

This web page will be updated regularly, with corrections, complements, new data sets, code updates, etc.

ACKNOWLEDGMENTS

First and foremost, I want to thank all the students who suffered painfully through early versions of the course and primitive drafts of the lecture notes. Their patience and encouragements helped fine-tune the course.

My interest in computational statistics was sparked over 20 years ago by two dear friends: Anestis Antoniadis and Jacques Berruyer. Time and distance pulled us apart, but what their collaboration taught me will remain with me forever. The library implementation of heavy tail estimation and two-dimensional copulas is based on original code of Julia Morrison. It was a real pleasure to work with her on the development of the `S-Plus` library `EVANESCE`. I am very grateful for this experience.

As for the `S-Plus` version of the book, I want to thank my wife Debra for tolerating my insane working habits and my three wonderful daughters Stephanie, Chelsea, and Chanel for their limitless patience and unconditional love. I may not deserve it, but I am definitely proud of it.

Princeton, NJ, USA René Carmona

Contents

Part I

DATA EXPLORATION, ESTIMATION AND SIMULATION

1

UNIVARIATE DATA DISTRIBUTIONS

The first part of the chapter gives a quick review of the classical parametric families of probability distributions and the statistical estimation of their parameters. We also review nonparametric density estimation, but our interest in financial data and heavy tail distributions prompts us to focus on quantile comparison. We introduce Q-Q plots as the main graphical tool to detect the presence of heavy tails. Because random simulation will be used throughout the book, the last part of the chapter presents the basics of Monte Carlo computations. The first two fundamental theorem of the calculus of probability (the law of large numbers and the central limit theorem) are introduced as a justification for the numerical approximations provided by Monte Carlo computations.

1.1 Probability Distributions and Their Parameters

This first section is of probabilistic nature. Its purpose is to introduce some of the most commonly used parametric families of probability distributions. This part is included for the sake of completeness. The reader familiar with this material can skip it in a first reading, and use it whenever the needs to check the terminology and the notation arise.

1.1.1 Standard Probability Distribution Families

We review the most frequently used probability distributions. Because of the very nature of financial data, we are primarily interested in distributions with heavy tails, and as a consequence, we shall concentrate our efforts on understanding continuous distributions extending to plus or minus infinity.

R. Carmona, *Statistical Analysis of Financial Data in R*, Springer Texts in Statistics,
DOI 10.1007/978-1-4614-8788-3_1, © Springer Science+Business Media New York 2014

1.1.1.1 **The Uniform Distribution**

Despite the fact that its support is bounded, and hence does not have any tail to speak of, the uniform distribution is of crucial importance for random simulations and Monte Carlo computations.

The uniform distribution over an interval is the distribution of random numbers in this interval when they are equally likely to fall into different intervals as long as the lengths of those intervals are equal. It is also (hopefully) the distribution of the samples produced by the random number generators provided by the computing environment you are using. The density of the uniform distribution over the interval $[a, b]$ is given by the formula:

$$f_{a,b}(x) = \begin{cases} 0 & \text{if } x < a \text{ or } x > b \\ 1/(b-a) & \text{if } a \le x \le b. \end{cases} \tag{1.1}$$

The corresponding cumulative distribution function is given by:

$$F_{a,b}(x) = \begin{cases} 0 & \text{if } x \le a \\ (x-a)/(b-a) & \text{if } a \le x \le b, \\ 1 & \text{if } x > b \end{cases} \tag{1.2}$$

This probability distribution is denoted by $U(a, b)$. The uniform distribution over the unit interval $[0, 1]$ is most frequently used. It corresponds to the end points $a = 0$ and $b = 1$. Figure 1.1 gives a plot of the density and of the cumulative distribution function (cdf for short) of this uniform distribution. Values of $f_{a,b}(x)$ and $F_{a,b}(x)$

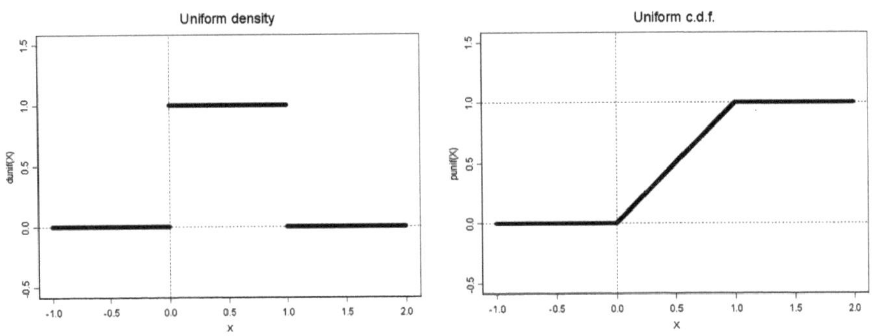

Fig. 1.1. Graphs of the density (*left*) and corresponding cdf (*right*) of the uniform distribution $U(0, 1)$ over the unit interval $[0, 1]$

can be computed with the `R` functions `dunif` and `punif`.

Remark. Formulae (1.1) and (1.2) are simple enough so that we should not need special commands for the computations of the values of the density and the cdf of the uniform distribution. Nevertheless, we mention the existence of these commands to emphasize the fact that their format is the same for all the common distribution families.

1.1.1.2 `R` *Convention*

`R` follows a very simple convention when it comes to computing densities, cumulative distribution functions, quantiles (which we consider in detail in Sect. 1.1.3 below), and random samples (which we study systematically in Sect. 1.3). To describe this convention, let us assume that `name` is the name or short name for a family of probability distributions. For example, `name` was `unif` in the case of the family of uniform distributions discussed above, and it will be `norm` in the case of the normal or Gaussian distribution discussed below. Then, provided with appropriate arguments, the `R` command

- `dname` gives values of the density function;
- `pname` gives values of the cumulative distribution function;
- `qname` gives values of the quantiles;
- `rname` produces random samples.

It is important to keep this convention in mind as it will be used throughout the book.

1.1.1.3 ***The Gaussian (Normal) Distribution***

The univariate normal distribution, also called the Gaussian distribution, is most often defined by means of its density function. It depends upon two parameters μ and σ^2, and is given by the formula:

$$\varphi_{\mu,\sigma^2}(x) = \frac{1}{\sqrt{2\pi\sigma^2}} e^{-(x-\mu)^2/2\sigma^2}, \qquad x \in \mathbb{R}. \tag{1.3}$$

The two parameters μ and σ^2 are the mean and the variance of the distribution respectively. Indeed, if X is a random variable with such a distribution (in which case we use the notation $X \sim N(\mu, \sigma^2)$) we have:

$$\mathbb{E}\{X\} = \mu, \qquad \text{and} \qquad \text{var}\{X\} = \sigma^2.$$

The corresponding cdf

$$\Phi_{\mu,\sigma^2}(x) = \int_{-\infty}^{x} f_{\mu,\sigma^2}(x')\, dx' \tag{1.4}$$

cannot be given by a formula in closed form involving standard functions. As a consequence, it will have to be evaluated numerically via approximation procedures. The `R` function `norm` is used throughout to compute values of this cdf. We drop the subscripts μ and σ^2 when $\mu = 0$ and $\sigma^2 = 1$. In this case, we call the distribution standard normal distribution, or standard Gaussian distribution, we denote it by $N(0,1)$, and the Greek letters φ and Φ are used for the density and the cdf. Figure 1.2 gives plots of three Gaussian densities. The density with the smallest variance has the highest central peak and it gets close to zero faster than the other two densities.

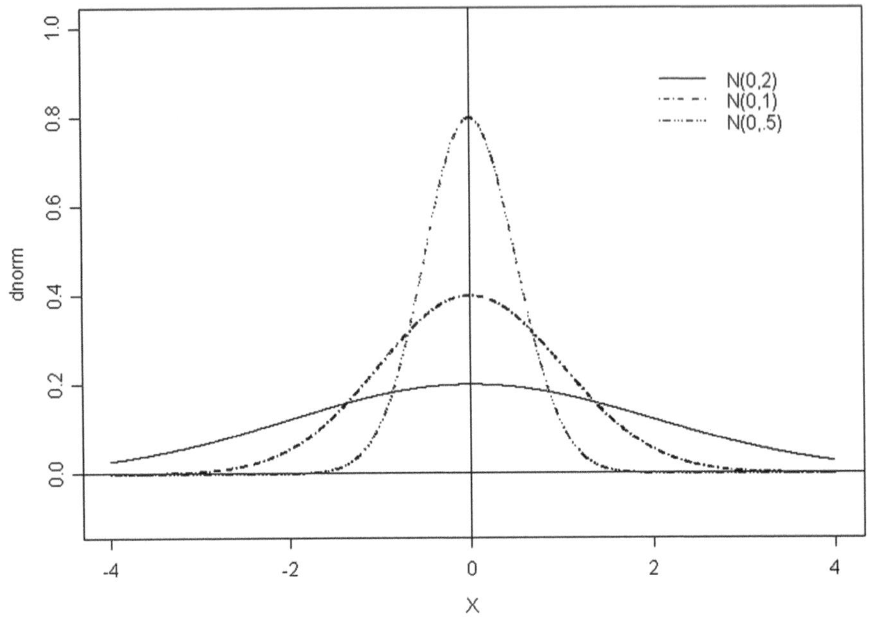

Fig. 1.2. Densities of the mean zero normal distributions $N(0, 0.5)$, $N(0, 1)$ and $N(0, 2)$ with variances 0.5, 1 and 2 respectively

The density with the largest variance has a flatter central bump, and it goes to zero later than the other ones. By shifting a general normal distribution we can center it around the origin, and in doing so, its mean becomes 0. By rescaling a mean zero normal random variable by its standard deviation, we turn it into a unit variance one. This qualitative statement can be turned into a rigorous mathematical fact in the following logical equivalence:

$$X \sim N(\mu, \sigma^2) \quad \Longleftrightarrow \quad \frac{X - \mu}{\sigma} \sim N(0, 1). \tag{1.5}$$

Because of this fact, most computations are done with the $N(0, 1)$ distribution only. More on this in the subsection *Effects of Affine Transformations* later in the chapter.

As mentioned earlier we shall compute values of the cumulative distribution function of a normal distribution using the command `pnorm` with arguments giving, respectively, the list of the values at which the computations are desired, the mean and the standard deviation. For example, the following command computes the probabilities that a standard normal variate is within one, two and three standard deviations of its mean.

```
> pnorm(c(1,2,3),mean=0,sd=1)-pnorm(c(-1,-2,-3),mean=0,sd=1)
[1] 0.6826895 0.9544997 0.9973002
```

The reader unfamiliar with the syntax of R can benefit from the following remarks. The seemingly simpler command

```
> pnorm(c(1,2,3))-pnorm(c(-1,-2,-3))
```

would produce the same output since the default values of the parameters `mean` and `sd` are 0 and 1 respectively. When parameters of R functions are not provided in a call, default values are used whenever these default values are available. The second remark concern the first argument which, in the two calls to the function `pnorm` were `c(1,2,3)` and `c(-1,-2,-3)` respectively. Both arguments are numeric vectors of lengths 3. Indeed the role of the R function `c` is to *concatenate* the objects passed as parameters into a single object. Hence `c(1,2,3)` can be understood as the vector `[1,2,3]` and `c(-1,-2,-3)` as the vector `[-1,-2,-3]`. The third remark is that, whenever a numeric vector is passed as a parameter to a numeric function originally intended for numeric parameters, the function returns a vector of the same length as the vector passed as parameter, and the entries of the output vector are the values of the function when evaluated at the entries of the input vector. In the present situation, `pnorm(c(1,2,3))` is the vector of length 3 with entries `pnorm(1)`, `pnorm(2)`, and `pnorm(3)`. Similarly, `pnorm(c(-1,-2,-3))` is the vector of length 3 with entries `pnorm(-1)`, `pnorm(-2)`, and `pnorm(-3)`, and since the difference of two numeric vectors of the same lengths is the vector of the differences of the respective entries, `pnorm(c(1,2,3))-pnorm(c(-1,-2,-3))` is the vector of length 3 with entries the numbers `pnorm(1)-pnorm(-1)`, `pnorm(2)-pnorm(-2)`, and finally `pnorm(3)-pnorm(-3)`. This is what we intended to compute.

Warning. The notation $N(\mu, \sigma^2)$ most frequently used in statistical textbooks uses the mean μ and the variance σ^2 as parameters. However, the R functions, `rnorm`, `dnorm`, `pnorm` and `qnorm` use the mean and the standard deviation σ (i.e. the square root of the variance) as parameters. So the probability that a normal random variable with mean 1 and variance 9 is not greater than 2 is given by the command

```
> pnorm(2,mean=1,sd=3)s
```

Also, recall that if $X \sim N(\mu, \sigma^2)$, then the scaling property (1.5) implies that $Z = (X - \mu)/\sigma \sim N(0, 1)$ and the results of the computations done above with the R function `pnorm` can be restated as:

$$\begin{aligned}
\mathbb{P}\{-\sigma \le X - \mu \le \sigma\} &= \mathbb{P}\{-1 \le Z \le 1\} = \Phi(1) - \Phi(-1) = 0.683 \\
\mathbb{P}\{-2\sigma \le X - \mu \le 2\sigma\} &= \mathbb{P}\{-2 \le Z \le 2\} = \Phi(2) - \Phi(-2) = 0.955 \\
\mathbb{P}\{-3\sigma \le X - \mu \le 3\sigma\} &= \mathbb{P}\{-3 \le Z \le 3\} = \Phi(3) - \Phi(-3) = 0.997.
\end{aligned}$$

These facts can be restated in words as:

- The probability that a normal r.v. is one standard deviation, or less, away from its mean is 0.683;
- The probability that a normal r.v. is two standard deviations, or less, away from its mean is 0.955;
- The probability that a normal r.v. is three standard deviations, or less, away from its mean is 0.997.

In other words, Gaussian variates are most frequently found within two standard deviations of their means, essentially always within three standard deviations.

We close this discussion of the Gaussian distribution with the derivation of a useful formula.

$$X \sim N(\mu, \sigma^2) \quad \Longrightarrow \quad \mathbb{E}\{e^X\} = e^{\mu+\sigma^2/2} \tag{1.6}$$

To start with, notice that if $Z \sim N(0,1)$, direct calculations (by a trick going under the name of *completing the square*) give:

$$\mathbb{E}\{e^{\sigma Z}\} = \frac{1}{\sqrt{2\pi}}\int_{-\infty}^{+\infty} e^{\sigma z} e^{-z^2/2} dz = \frac{e^{\sigma^2/2}}{\sqrt{2\pi}}\int_{-\infty}^{+\infty} e^{-(z-\sigma)^2/2} dz = e^{\sigma^2/2}.$$

Now in general, if $X \sim N(\mu, \sigma^2)$, then as we saw, $X = \mu + \sigma Z$ with $Z \sim N(0,1)$, so that:

$$\mathbb{E}\{e^X\} = \mathbb{E}\{e^{\mu+\sigma Z}\} = e^{\mu}\mathbb{E}\{e^{\sigma Z}\} = e^{\mu} e^{\sigma^2/2}$$

which proves the desired formula (1.6) in full generality.

We now introduce three distribution families derived from the normal family.

1.1.1.4 **The Log-Normal Distribution**

The log-normal distribution is the major building block of the mathematical theory of continuous time finance, and it plays a central role in the Samuelson's model for stock prices dynamics and the Black-Scholes pricing theory. A random variable is said to be log-normal if it is the exponential of a Gaussian random variable, or in other words, if it is positive and if its logarithm is a Gaussian random variable. This definition has a clear consequence at the level of random samples, i.e. realizations of independent random variables with the same distribution. Indeed, the definition of the log-normal distribution can be restated as saying that

$$x_1, x_2, \cdots\cdots, x_n$$

is a sample from a log-normal distribution if and only if

$$x_1 = e^{y_1}, x_2 = e^{y_2}, \cdots\cdots, x_n = e^{y_n}$$

where $y_1, y_2, \cdots, y_n$ is a sample from a normal distribution.

Things are not as straightforward at the level of the densities and cdf's. Indeed, if $X = e^Y$ with $Y \sim N(\mu, \sigma^2)$, then if we denote by f_X and F_X the density and the cdf of X, we have:

$$F_X(x) = \Phi_{\mu,\sigma^2}(\log x) = \Phi\left(\frac{\log x - \mu}{\sigma}\right) \tag{1.7}$$

since $\mathbb{P}\{X \le x\} = \mathbb{P}\{Y = \log X \le \log x\}$, and using the fact that the density $f_X(x)$ can be computed as the derivative $F'_X(x)$ of the cdf whenever this derivative exists, we get

$$f_X(x) = \frac{1}{x}\varphi_{\mu,\sigma^2}(\log x) = \frac{1}{\sigma x}\varphi\left(\frac{\log x - \mu}{\sigma}\right) = \frac{1}{x\sigma\sqrt{2\pi}}\exp\left[-\frac{(\log x - \mu)^2}{2\sigma_2}\right]. \tag{1.8}$$

In accordance with the R-convention explained earlier, values of the density and the cdf of a log-normal distribution can be computed in R by means of the commands `dlnorm` and `plnorm`. In other words, we use `lnorm` for `name`. We used them to produce the plots of Fig. 1.3 which give the densities of the log-normal distributions with mean zero and variances 0.2, 1 and 3 respectively. As in the case of the exponential distribution which is considered later on, we plot the graphs only over the positive part of the x-axis because these densities vanish on the negative part of the x-axis.

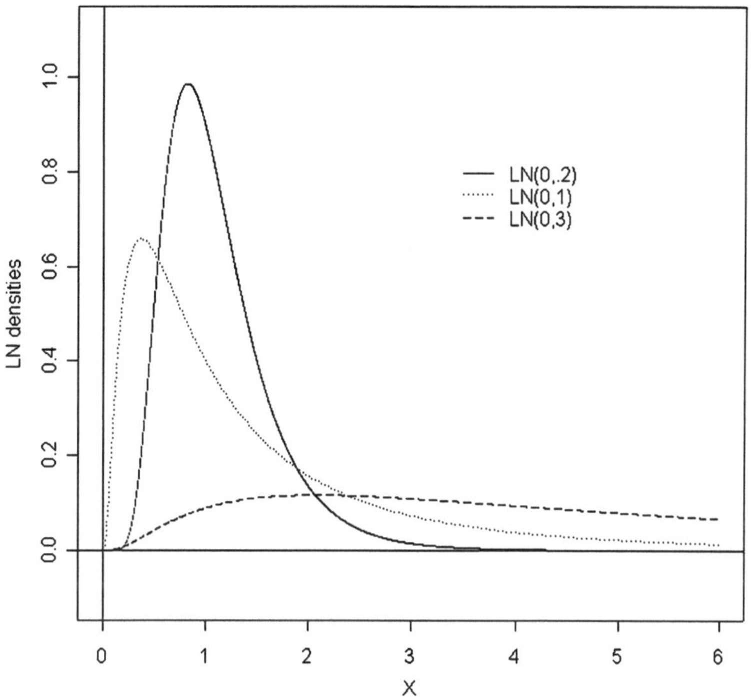

Fig. 1.3. Graphs of the densities of the log-normal distributions with mean zero and variances 0.2, 1 and 3

Warning. One often talks about a log-normal distribution with mean μ and variance σ^2 to mean that the corresponding normal distribution has mean μ and variance σ^2. This abuse of the terminology is very frequent despite the fact that it is misleading. We shall do our best to make clear what we mean when we mention the parameters of a log-normal distribution. In other words, saying that X is has a log-normal distribution with mean μ and variance σ^2 actually means that $\log X$ is Gaussian, μ is the mean of $\log X$ and σ^2 is the variance of $\log X$. This is emphasized by the way `R` names the parameters of the functions `rlnorm`, `dlnorm`, ... by calling them `meanlog` and `sdlog`.

So for example, the command `dlnorm(X, meanlog = 0, sdlog = 1)` will compute the density of the log-normal distribution with mean zero and variance one at `X`, or the entries of the array `X` if `X` is an array.

We conclude our discussion of the lognormal distribution with the computation of the mean and the variance. Let us assume that X is lognormal with mean μ and variance σ^2 and let us compute $\mu_X = \mathbb{E}\{X\}$ and $\sigma_X^2 = \text{var}\{X\}$. By definition, $X = e^Y$ with $Y \sim N(\mu, \sigma^2)$. In turn, this means that $Y = \mu + \sigma Z$ for some random variable $Z \sim N(0, 1)$. Consequently, using repeatedly formula (1.6) proved earlier:

$$\mathbb{E}\{X\} = \mathbb{E}\{e^Y\} = \mathbb{E}\{e^{\mu+\sigma Z}\} = e^{\mu}\mathbb{E}\{e^{\sigma Z}\} = e^{\mu+\sigma^2/2}.$$

Similarly:

$$\mathbb{E}\{X^2\} = \mathbb{E}\{e^{2Y}\} = \mathbb{E}\{e^{2\mu+2\sigma Z}\} = e^{2\mu}\mathbb{E}\{e^{2\sigma Z}\} = e^{2\mu+2\sigma^2}.$$

And finally:

$$\text{var}\{X\} = \mathbb{E}\{X^2\} - \mathbb{E}\{X\}^2 = e^{\mu+\sigma^2/2} - e^{2\mu+2\sigma^2} = e^{2\mu+\sigma^2}[e^{\sigma^2} - 1].$$

In summary, if $X \sim LN(\mu, \sigma^2)$, then

$$\mu_X = \mathbb{E}\{X\} = e^{\mu+\sigma^2/2} \quad \text{and} \quad \sigma_X^2 = \text{var}\{X\} = e^{2\mu+\sigma^2}[e^{\sigma^2} - 1]. \tag{1.9}$$

1.1.1.5 The Chi-Square Distribution

The χ^2-distribution, in words *chi square distribution*, is of crucial importance in statistical inference and hypothesis testing. Its role in this book will be quite marginal, its contribution being limited to the theoretical definition of the Student t distribution which we introduce and analyze in the following subsection. For the record we mention that whenever k is an integer, the χ^2-distribution with k degrees of freedom is the distribution of the sum of the squares of k independent standard Gaussian random variables. This distribution will be denoted $\chi^2(k)$. In other words, if $X_1, \cdots,$ X_k are independent $N(0, 1)$ random variables, then

$$X_1^2 + \cdots + X_k^2 \sim \chi^2(k).$$

The `R` commands producing values of the quantiles, the density, and the cdf of the χ^2-distribution are `qchisq`, `dchisq`, and `pchisq` while random samples are generated with the command `rchisq`. See the help files for details.

1.1.1.6 The Student t Distribution

For each integer $k \geq 1$, the Student distribution – also called the t distribution – with k degrees of freedom will be denoted by $t(k)$. Intuitively, it should be thought of as the distribution of a Gaussian random variable with a random variance which is independent and χ^2-distributed. More precisely, the t distribution with k degrees of freedom is the distribution of a random variable X of the form

$$X = \frac{\xi}{\sqrt{\chi/k}} \tag{1.10}$$

where $\xi \sim N(0,1)$ and $\chi \sim \chi^2(k)$ are independent. One can use the definition formula (1.10) to derive the density function $f_k^{(t)}(x)$ of the $t(k)$ distribution. It is given by the formula:

$$f_k^{(t)}(x) = \frac{\Gamma((k+1)/2)}{\sqrt{k\pi}\Gamma(k/2)}(1 + x^2/k)^{-(k+1)/2}, \qquad x \in \mathbb{R}, \tag{1.11}$$

where Γ denotes the classical *gamma function* defined by

$$\Gamma(\lambda) = \int_0^\infty x^{\lambda-1}e^{-x}dx, \qquad \lambda > 0. \tag{1.12}$$

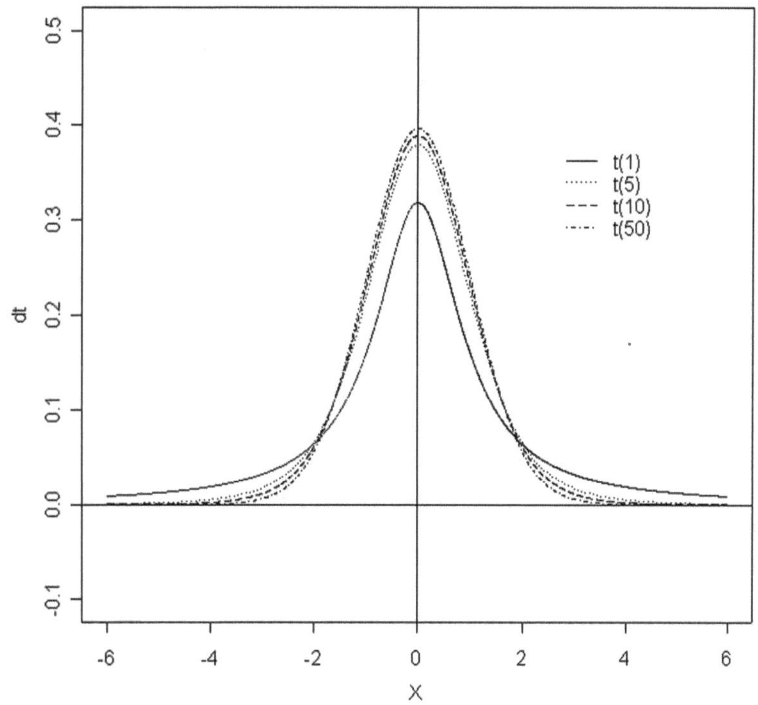

Fig. 1.4. Graphs of the densities of the t distributions with degrees of freedom 1, 5, 10 and 50 respectively

One of the nice properties of the gamma function is that it offers a generalization of the factorial function to real numbers. Indeed, if $\lambda = n + 1$ for some integer $n \geq 0$, then repeated integrations by parts show that $\Gamma(\lambda) = \Gamma(n+1) = n!$. The cdf of the t-distribution with k degrees of freedom will be denoted by $F_k^{(t)}(x)$.

Like the Gaussian distribution, the t-distribution is unimodal in the sense that the graph of the density has a unique maximum. This maximum is located at the origin, though it is not difficult to imagine that t-distributions centered around other values can be obtained by mere shifts. Figure 1.4 gives plots of four t densities. Values of the density and cdf of the Student t distribution can be computed in `R` with the functions `dt` and `pt` both of which take the number of degrees of freedom `df` as parameter. There is a non-central form of the t distribution. It needs an extra parameter, the so-called non centrality parameter `ncp`, but we shall not need it in this book. The distribution with one degree of freedom `df=1`, has the lowest central peak, and it tails off slower that the three other densities. We shall see later in this chapter the similarities between this distribution and the Cauchy distribution which we introduce next. The graphs of the three other densities are very similar, the higher the number of degrees of freedom, the higher the central bump and the faster the decay of the density at plus and minus infinity, i.e. the thinner the tails. This similarity with the role played by the standard deviation in the case of the normal family may be deceiving and it is important to emphasize that the role of the number of degrees of freedom is very different from a mere scale parameter. A Gaussian random variable has moments of all order. However, this is not the case for random variables with the t-distribution. Indeed, the number of degrees of freedom `df` determines how many moments are finite. To be more specific, if X is a random variable with a t-distribution with `df` degrees of freedom,

$$\mathbb{E}\{|X|^k\} < \infty \quad \Leftrightarrow \quad k < \texttt{df}$$

as we can see from the expression of the density of X given above. So the t-distribution is a distribution with heavy tails in a sense we will make precise later, and this should be of practical relevance, especially if the number of degrees of freedom is small. At the other end of the spectrum, namely when the number of degrees of freedom becomes large without bound, the number of finite moments also increases without bound, and the tail of the distribution become thinner. In fact, the t-distribution converges (in a mathematical sense which we shall not attempt to make precise here) toward the normal distribution. This theoretical result is what is known as the normal approximation to the t-distribution. It is used quite frequently in the computation of p-values and significance levels of statistical tests when the number of degrees of freedom is in the range of 50, and often quite smaller. This fact is illustrated in Fig. 1.5 which gives the plot of the standard normal density $N(0, 1)$ together with the t-densities with 5, 10 and 50 degrees of freedom.

1.1.1.7 ***The Fisher F Distribution***

The Fisher or F-distribution is another distribution derived from the Gaussian with important applications in statistical testing of hypotheses. We do not discuss it here because we shall not use it in this textbook. For the record we mention the `R` commands for dealing with the F-distribution. Not surprisingly they are `rf`, `df`, `pf`, and `qf` for the quantile function which we define later in this chapter.

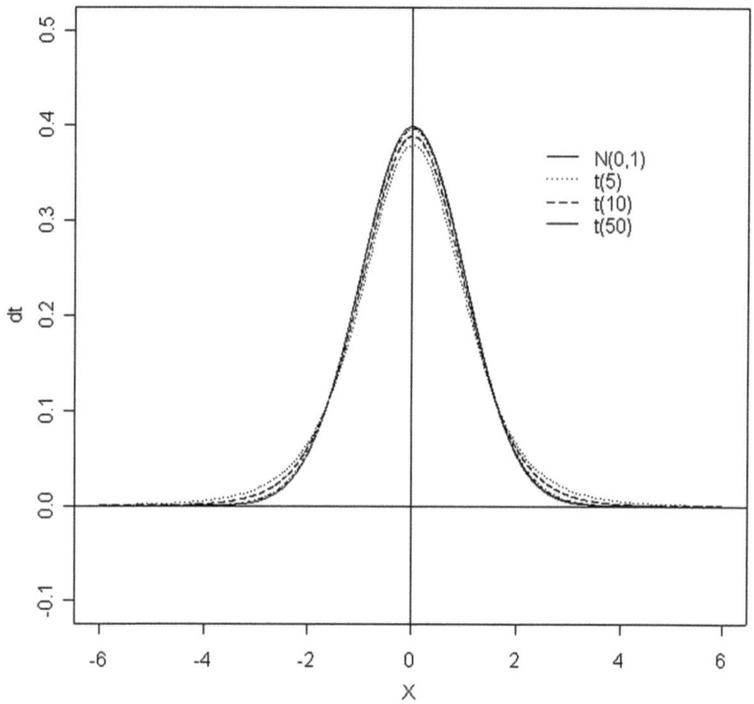

Fig. 1.5. Graphical comparison of the densities of the Gaussian distribution $N(0, 1)$, and the t distributions with degrees of freedom 5, 10 and 50

1.1.1.8 ***The Cauchy Distribution***

Among the probability distributions introduced in this section, the Cauchy distribution is the least known, presumably because of its lack of applications to statistical testing. It is usually introduced as a particular element of the class of *stable* distributions which are not discussed in this book, and of the class of Generalized Pareto Distributions (GDP for short) which we study in detail in the second chapter of the book. These distributions play an important role because of the thickness of their tails. The Cauchy distribution is of great pedagogical (and possibly practical) interest because it is one of the rare distribution from this class with explicit closed formulae for the density, cdf, quantile function, etc. Like the Gaussian distribution, it depends upon two parameters: a location parameter, say m, and a scale parameter, say λ. It can be defined from its density function $f_{m,\lambda}(x)$ by the formula:

$$f^{(C)}_{m,\lambda}(x) = \frac{1}{\pi}\frac{\lambda}{\lambda^2 + (x-m)^2}, \qquad x \in \mathbb{R}. \tag{1.13}$$

This distribution is denoted by $C(m, \lambda)$. The computation of its cdf $F^{(C)}_{m,\lambda}(x)$ leads to a simple formula. Indeed:

$$\begin{aligned}
F^{(C)}_{m,\lambda}(x) &= \int_{-\infty}^{x} f_{m,\lambda}(y)dy \\
&= \frac{1}{\pi}\int_{-\infty}^{x} \frac{1}{1+[(y-m)/\lambda]^2}\frac{dy}{\lambda} \\
&= \frac{1}{\pi}\int_{-\infty}^{(x-m)/\lambda} \frac{1}{1+z^2}dz \\
&= \frac{1}{\pi}[\tan^{-1}\frac{x-m}{\lambda} - \tan^{-1}(-\infty)] \\
&= \frac{1}{\pi}\tan^{-1}\frac{x-m}{\lambda} + \frac{1}{2},
\end{aligned} \tag{1.14}$$

where we used the substitution $z = (y-m)/\lambda$ to compute the indefinite integral. Like the Gaussian and the Student distributions, the Cauchy distribution is unimodal.

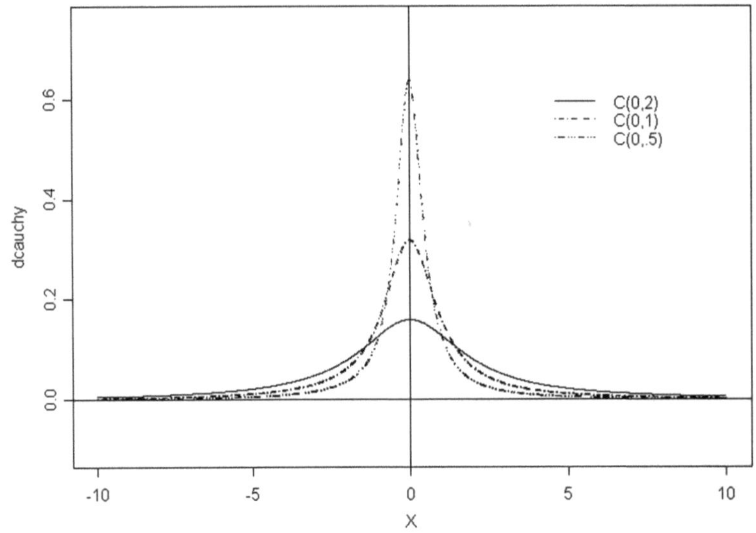

Fig. 1.6. Graphs of the densities of the Cauchy distributions $C(0, 0.5)$, $C(0, 1)$ and $C(0, 2)$ located around 0 and with scales 0.5, 1 and 2 respectively

The maximum of the central bump of the distribution is located at m. Figure 1.6 gives plots of three Cauchy densities with the same location parameter $m = 0$. The distribution with the smallest scale parameter λ has the highest central peak, and it tails off faster that the two other densities. The distribution with the largest scale parameter has a wider central bump, and as a consequence, it goes to zero later than the other ones. This figure seems to be very similar to Fig. 1.2 which shows graphs of normal densities, or even to Fig. 1.4 which shows graphs of t densities. Indeed, both Gaussian, t, and Cauchy distributions are unimodal in the sense that the graph of the density has a unique maximum. This maximum is located at the mean in the case of the normal distribution, and at the value of the location parameter m in the case of the Cauchy distribution. Moreover, if we associate the standard deviation of the normal distribution to the scale parameter of the Cauchy distribution, then the discussion of the qualitative features of the graphs in Fig. 1.2 also applies to those in Fig. 1.6. Nevertheless, major differences exist between these two families of distributions. Indeed, as we can see from Fig. 1.7, where the graphs of densities from both families are superimposed on the same plot, the tails of the normal distribution are much thinner than those of the Cauchy distribution. What we mean here is not so much that the density of the normal distribution approaches zero earlier than the density of the Cauchy distribution, but that it does so at a much faster rate. This is because the decay toward zero away from the center of the density is exponential in the negative of the square distance to the center, instead of being merely an inverse polynomial in this distance. These rates of convergence to zero are very different, and one should not be mislead by the apparent similarities between the two unimodal density graphs. As explained earlier, because of its lack of moments, the t distribution with a small

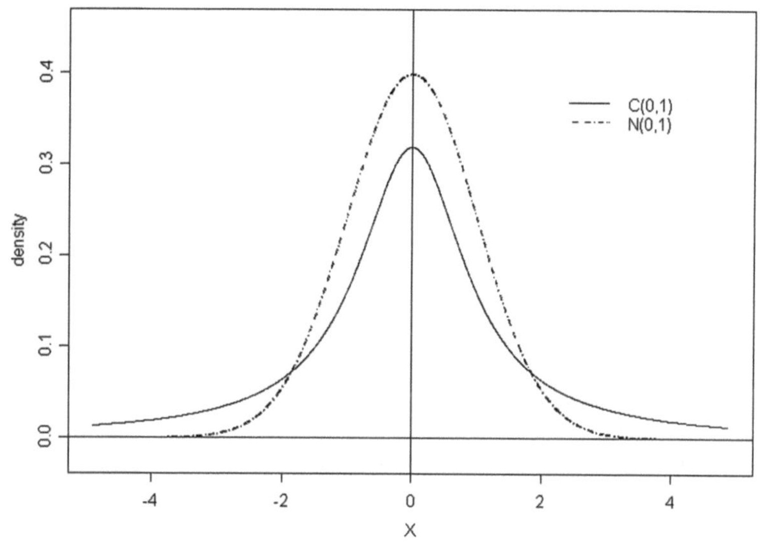

Fig. 1.7. Graphical comparison of the Cauchy distribution $C(0, 1)$ and the Gaussian distribution $N(0, 1)$

number of degrees of freedom bears to the Cauchy distribution, more similarity than to the Gaussian distribution. We illustrate this fact in Fig. 1.8 by plotting together the graphs of the $C(0, 1)$ Cauchy distribution and of the t distributions with 1 and 5 degrees of freedom. On this plot, it is impossible to distinguish the graph of the $C(0, 1)$ density from the graph of the $t(1)$ density. The standard Cauchy distribution

$C(0, 1)$ and the t distribution with one degree of freedom look very similar. Values of the density and cumulative distribution functions of the Cauchy distribution can be computed in R with the functions dcauchy and pcauchy both of which take the location and the scale of the distribution as parameters.

It is quite clear that density plots as those given in this section do not make it easy to distinguish distribution families according to the thickness of their tails.

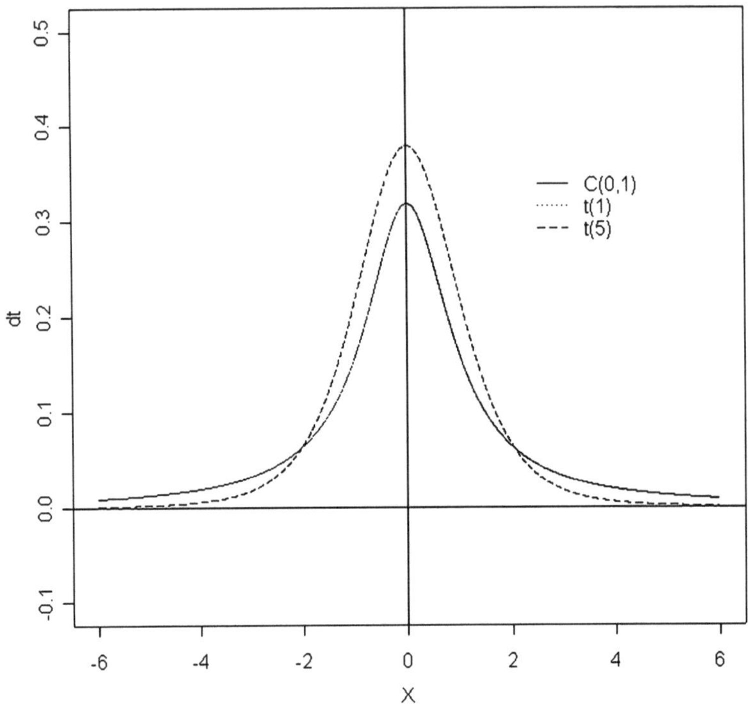

Fig. 1.8. Graphical comparison of the Cauchy distribution $C(0, 1)$ and the t distributions with df=1 and df=5 degrees of freedom

Figure 1.24 below shows clearly the effects of these differences in tail thickness on random samples. We will use Q-Q plots to emphasize and capture these differences in tail behavior.

A review of the classical probability-distribution families would not be complete without a discussion of the exponential distribution.

1.1.1.9 ***The Exponential Distribution***

The exponential distribution is one of the rare distributions for which many computations can be carried out explicitly because of the simplicity of its definition. It is extremely useful in modeling the length of time-intervals separating successive arrivals of events captured in a stochastic process model. The problems analyzed with these models include internet traffic, insurance, catastrophe and rainfall modeling, failure and reliability problems, queues, ... In financial applications, the exponential distribution is a building block for models of the time separating economic regime changes, jumps in prices, credit migrations, defaults, postings and cancellations of market and limit orders on an electronic exchange, ...

The exponential distribution shares with the log-normal distribution the fact that its support is the half line $\mathbb{R}_+ = [0, \infty)$. Random samples from the exponential distribution are positive numbers: the density is non-zero on the positive axis only. In particular, the tail or extreme values are only on the positive side, and the distribution has only one tail at $+\infty$. This distribution depends upon a parameter $r > 0$, called the rate of the distribution, and it is denoted by $E(r)$. It can be defined from its density function $f_r(x)$ which is given by the formula:

$$f_r(x) = \begin{cases} 0 & \text{if } x < 0 \\ re^{-rx} & \text{if } x \geq 0. \end{cases} \tag{1.15}$$

The positive number $\lambda = 1/r$ is called the scale of the distribution. The reason why λ is called the scale of the distribution is because of the following easily checked fact: if X is an exponential random variable with rate $r = 1$ (in which case the parameter λ is also equal to 1), then the random variable λX is also an exponential random variable, its rate r is equal to $1/\lambda$ and consequently, its scale parameter is equal to λ. The mean of this distribution is the inverse of the rate, in other words, the scale of the distribution. Indeed, if $X \sim E(r)$ then

$$\mathbb{E}\{X\} = \int x f_r(x)\, dx = r \int_0^\infty x e^{-rx}\, dx = \frac{1}{r} = \lambda.$$

In this respect, this property of the exponential distribution is quite unusual. Indeed, the mean is typically used as a measure of the *location* of the distribution while the scale is used as a measure of the spread of the distribution about its central location, and these two characteristics are different in general. The cdf of the exponential distribution is given by the following formula:

$$F_r(x) = \begin{cases} 0 & \text{if } x < 0 \\ 1 - e^{-rx} & \text{if } x \geq 0. \end{cases} \tag{1.16}$$

Also, note that the second moment $\mathbb{E}\{X^2\}$ can be computed by a simple integration by parts. We get;

$$\mathbb{E}\{X^2\} = \int x^2 f_r(x)\, dx = r \int_0^\infty x^2 e^{-rx}\, dx = \frac{2}{r^2} = 2\lambda^2$$

and this implies that

$$\text{var}\{X\} = \mathbb{E}\{X^2\} - \mathbb{E}\{X\}^2 = \frac{1}{r^2} = \lambda^2.$$

So in the case of the exponential distribution, the standard deviation is equal to the mean. This definitely blurs the interpretation of the parameter λ which could be a scale as well as a *location* parameter. Figure 1.9 gives plots of three exponential densities. The distribution with the highest rate has the highest starting point on the y-axis, and it tails off faster that the other two densities. The distribution with the lowest rate starts lower on the y-axis, and does not decay as fast. Values of the density and the cdf of the exponential distribution can be computed in R with the functions `dexp` and `pexp`, both of which take the rate of the distribution as parameter.

1.1.2 **Estimation from Empirical Data**

We now consider the challenging problem of the statistical estimation of the parameters which characterize the probability distributions from the families discussed

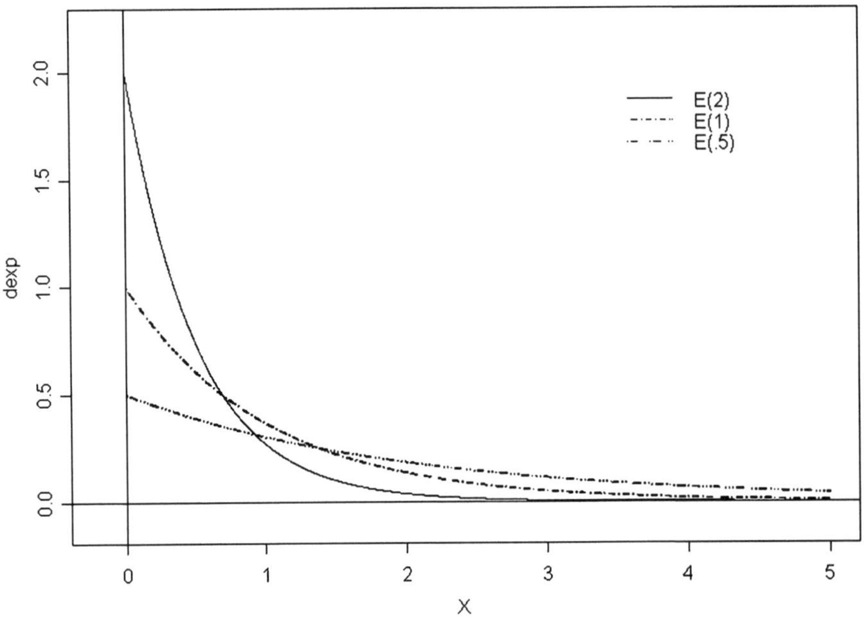

Fig. 1.9. Graphs of the densities of the exponential distributions $E(0.5)$, $E(1)$ and $E(2)$ with rates 0.5, 1 and 2. We plot the graphs only over the positive part of the x-axis because these densities vanish on the negative part of the x-axis

above. This kind of statistical inference is based on the analysis of sample observations, say:

$$x_1, x_2, \ldots\ldots, x_n$$

which we assume to be realizations of independent identically distributed (i.i.d. for short) random variables $X_1, X_2, \ldots, X_n$ with common cdf F and/or density function f, and the challenge is to estimate F from the data. The premise of parametric statistics is to assume that the unknown distribution belongs to a given family, and the whole estimation issue reduces to the estimation of the parameter(s) characterizing the specific elements of the family in question.

A time honored method of parameter estimation is based on the maximization of the likelihood of the observations. We describe this method briefly, and for the sake of completeness, we also say a few words of the classical method of moments.

1.1.2.1 ***Maximum Likelihood Estimation (MLE)***

The likelihood of a set $x_1, x_2, \cdots, x_n$ of observations from random variables X_1, X_2, $\cdots$, X_n is the value of the joint density $f_{(X_1,X_2,\cdots,X_n)}$ evaluated at the observations x_1, x_2, $\cdots$, x_n. In other words, the likelihood for this set of observations is the number $f_{(X_1,\cdots,X_n)}(x_1, \cdots, x_n)$. In the case of discrete random variables, the likelihood is in fact the probability that $X_1 = x_1$, at the same time as $X_2 = x_2$, ..., at the same time as $X_n = x_n$, which is clearly what one should expect from the likelihood of $(x_1, x_2, \cdots, x_n)$. We are interested in the case when this joint distribution depends upon a parameter θ. This parameter can be multidimensional, e.g. $\theta = (\mu, \sigma^2)$ in the case of the simultaneous estimation of the mean and the variance of a Gaussian distribution from a sample of observations. In any case, for each given set $x_1, x_2, \cdots, x_n$ of observations, we can look at the likelihood as a function of the parameter θ, and the so-called maximum likelihood estimate (MLE for short) of the parameter θ is the value of θ, say $\hat{\theta}_{MLE}$, which maximizes this function of θ. We usually use the notation

$$\theta \hookrightarrow L(\theta|x_1, x_2, \cdots, x_n)$$

for the likelihood function, and with this notation the above definition can be stated as:

$$\hat{\theta}_{MLE} = \arg \sup_{\theta} L(\theta|x_1, x_2, \cdots, x_n).$$

We should keep in mind the fact that this estimate is a function of the observations, i.e. $\hat{\theta}_{MLE} = \hat{\theta}_{MLE}(x_1, x_2, \cdots, x_n)$, and if we replace the values of the observations by the actual random variables, the MLE $\hat{\theta}_{MLE} = \hat{\theta}_{MLE}(X_1, X_2, \cdots, X_n)$ becomes a random variable. We will use these facts freely even though we shall most often drop the dependence on the observations and the random variables from our notation.

Most of the analyses conducted in this book deal with observations $x_1, x_2, \cdots, x_n$ from independent random variables $X_1, X_2, \cdots, X_n$ in which case

$$f_{(X_1,X_2,\cdots,X_n)}(x_1, x_2, \cdots, x_n) = f_{X_1}(x_1) f_{X_2}(x_2) \cdots f_{X_n}(x_n).$$

Moreover, when the random variables $X_1, X_2, \cdots, X_n$ are identically distributed, all the densities $f_{X_i}(x)$ are the same, say $f_{X_i}(x) = f(x)$, and

$$f_{(X_1,X_2,\cdots,X_n)}(x_1, x_2, \cdots, x_n) = f(x_1) f(x_2) \cdots f(x_n).$$

As explained above, in the parametric case, this common density f depends upon a parameter, say $f(x) = f_\theta(x)$, and the likelihood function appears as the product

$$L(\theta|x_1, \ldots, x_n) = f_\theta(x_1) f_\theta(x_2) \cdots f_\theta(x_n).$$

Now, since the logarithm function is monotone increasing, the maximum (if any) of the likelihood function is attained for the same values of θ as the maximum of the logarithm of the likelihood. So because of its product structure, the maximization of the likelihood function L is often replaced by the maximization of its logarithm which we denote by $\mathcal{L}$

$$\mathcal{L}(\theta|x_1, \ldots, x_n) = \log L(\theta|x_1, \ldots, x_n) = \log f_\theta(x_1) + \cdots + \log f_\theta(x_n)$$

and the MLE is computed as

$$\hat{\theta}_{MLE} = \arg \sup_\theta \mathcal{L}(\theta|x_1, \ldots, x_n).$$

We summarize the main properties of MLE's (requiring minimal assumptions on the nature of the common density $f_\theta(x)$ and the form of its dependence upon θ which will not be discussed here) in the following bullet points;

- $\hat{\theta}_{MLE}$ is consistent for large samples, in the sense that it converges in probability when the sample size grows without bound toward the true value of the parameter θ;
- $\hat{\theta}_{MLE}$ is asymptotically normal, fact from which one can derive approximate confidence intervals;
- Maximum likelihood estimation is covariant in the sense that the maximum likelihood estimate of a function of a parameter is that very function of the maximum likelihood estimate of this parameter. For example, in the case of an exponential family with rate r and mean $\lambda = 1/r$, this result implies that the maximum likelihood of the mean of the distribution is the inverse of the maximum likelihood of the rate, i.e. $\hat{\lambda}_{MLE} = 1/\hat{r}_{MLE}$.

The most commonly studied parameters of a distribution are the mean and median which are measures of location of the distribution, together with other parameters such as the standard deviation or the scale, which quantify the spread of the distribution around its location. All the distribution families reviewed earlier depend upon parameters which can be estimated by maximum likelihood methods. Not to distract from the objective of this book, we refrain from presenting examples here. They can be found in essentially any introductory statistics textbook. We shall implement the maximum likelihood estimation procedure in the case of GEV (General Extreme Values) distributions and GPDs (General Pareto Distributions) of crucial importance to us later in Chap. 2.

1.1.2.2 ***Classical Method of Moments***

Let us assume that the parameter θ is a scalar (think for example of the rate of an exponential distribution), and that the mean of the distribution,

$$\mu(\theta) = \mathbb{E}\{X\} = \int x f_\theta(x)\, dx$$

has a simple expression as a function of θ. In this case, given a sample $x_1, x_2, \cdots, x_n$ of numeric observations from independent random variables $X_1, X_2, \cdots, X_n$ with the same density $f_\theta(x)$ we can try to solve the equation

$$\mu(\theta) = \overline{x}$$

for θ, and call the solution, say $\hat{\theta}_{MOM}$, the method of moments estimator of θ. Here and throughout the book, we use the notation $\overline{x}$ for the sample mean $(x_1 + \cdots + x_n)/n$. For example, if $x_1, x_2, \ldots\ldots, x_n$ is a sample of independent observations from $E(r)$, and if the goal is to estimate the rate of the distribution, i.e. $\theta = r$, then since $\mu(\theta) = \mathbb{E}\{X\} = r^{-1}$, solving the equation $\mu(r) = \overline{x}$ for r gives the Method Of Moments (MOM for short) estimate

$$\hat{r}_{MOM} = \frac{1}{\overline{x}}.$$

When θ is not a scalar, we can use as many equations and as many empirical moments, e.g. sample mean, sample variance, ..., as we have parameters to estimate, and solve a system of equations to find the method of moment estimators. For example, if $x_1, x_2, \ldots\ldots, x_n$ is a sample of independent observations from $N(\mu, \sigma^2)$, and if the goal is to estimate the mean and the variance of the distribution, i.e. $\theta = (\mu, \sigma^2)$, and since $\mathbb{E}\{X\} = \mu$, which is the first component of the vector parameter θ, and the second moment $\mathbb{E}\{X^2\} = \sigma^2 + \mu^2$ has also a simple expression in terms of the components of the parameter θ, replacing the theoretical moments $\mathbb{E}\{X\}$ and $\mathbb{E}\{X^2\}$ by the empirical moments $\overline{x}$ and $\overline{x^2}$ defined by

$$\overline{x^2} = \frac{x_1^2 + x_2^2 + \cdots + x_n^2}{n}$$

and solving for μ and σ^2 we find the moment estimates

$$\hat{\mu}_{MOM} = \overline{x} \qquad \text{and} \qquad \hat{\sigma^2}_{MOM} = \overline{x^2} - \overline{x}^2.$$

Notice that in the case of the Gaussian distribution, if one uses the first two empirical moments $\overline{x}$ and $\overline{x^2}$ computed from sample data, the MOM estimators of the mean and the variance are uniquely defined. It is also interesting to notice that in this case, they coincide with the MLE estimates of the mean and the variance.

Clearly, estimating parameters with the method of moments is even more intuitive than with the maximum likelihood estimation procedure. However, its realm of applicability is much more limited and the MOM estimators do not share the desirable theoretical properties of the ME estimators listed above. Worse, in many simple cases, it is not well defined and can lead to inconsistent results. Indeed, for a given parameter, one can get different estimates depending on which moment equation we use! Examples of some of these facts are given in problems at the end of the chapter.

We discussed MOM estimators here because of the implementation in the library `Rsafd` of the so-called method of L-moments used to estimate the parameters of generalized Pareto distributions in Chap. 2.

1.1.3 **Quantiles and Q-Q Plots**

So far, the graphical tools used to compare two distributions were basically limited to the comparison of the graphs of the densities. As we saw, this can give a clear picture of the differences in the center of the distributions in some cases. However, these graphical comparisons are not informative for the comparison of the distribution tails. This subsection is devoted to the introduction of a more efficient way to quantify the thickness of a tail, and as a by-product, we will get tools allowing us to compare the relative sizes of the tails to two distributions.

1.1.3.1 Quantiles of a Distribution

Given a (theoretical) distribution function F, and a number $p \in [0, 1]$, the p-quantile, or the $100p$th percentile of the distribution, is the number $\pi_p = \pi_p(F)$ satisfying $F(\pi_p) = p$. If we think of F as the distribution function of a random variable X, then the quantile definition can be restated as:

$$F(\pi_p) = \mathbb{P}\{X \le \pi_p\} = p. \tag{1.17}$$

In words, the $100p$th percentile, is the number π_p such that the probability that X is not greater than π_p is exactly equal to p.

Remark. Even though statement (1.17) is very intuitive, it cannot be a non-ambiguous definition. Indeed, there may not be any real number x satisfying $F(x) = \mathbb{P}\{X \le x\} = p$. Indeed, this can be the case for some values of p when the distribution function F has jumps, i.e. when the random variable X can take discrete values with positive probabilities. When such jumps occur, there may be plenty of possible choices. In fact, all the real numbers x satisfying:

$$\mathbb{P}\{X < x\} \le p \le \mathbb{P}\{X \le x\} = F(x)$$

can be regarded as reasonable candidates for the p-quantile of the distribution. A more precise definition would state that the set of p-quantiles is the closed interval $[x_p^-, x_p^+]$ where:

$$x_p^- = \inf\{x;\ F(x) \ge p\} \qquad \text{and} \qquad x_p^+ = \inf\{x;\ F(x) > p\}.$$

In any case, we get a uniquely defined quantile $\pi_p(F)$ (i.e. we have $x_p^- = x_p^+$) except for at most countably many p's in $[0, 1]$. For the sake of definiteness, for any of these countably many values of p, we shall use the left endpoint x_p^- of the interval as our definition of the percentile.

Most of the cdf's F used in this book are invertible. When it exists, the inverse function F^{-1} is called the quantile function because (1.17) can be rewritten as:

$$\pi_p = F^{-1}(p). \tag{1.18}$$

1.1.3.2 ***Examples***

As an illustration, let us consider the problem of the computation of the quantiles of the classical distributions introduced earlier in the chapter. The results of these computations will come handy when we discuss random number generators later on.

The quantiles of the uniform distribution are very easy to compute. However, they are rarely needed. The quantiles of the Gaussian distribution cannot be given in closed form: we cannot compute the Gaussian cdf in closed form, neither can we compute its inverse in closed form, and we need to rely on numerical approximation schemes to compute the quantiles of the Gaussian distributions. We give examples of these computations in the next subsection below. For the exponential distribution, the percentiles are easily computed from formula (1.16) which gives a simple expression for the cdf $F_r(x)$. One finds:

$$\pi_p = \frac{1}{r}\log\frac{1}{1-p}. \tag{1.19}$$

Finally, in the case of the Cauchy distribution, the explicit form of the cdf can also be inverted, and from trigonometry we find that the quantile function is given by:

$$\pi_p = F_{m,\lambda}^{-1}(p) = m + \lambda\tan\left(p\pi - \frac{\pi}{2}\right) \tag{1.20}$$

Quantiles and percentiles are numbers dividing the real line into intervals in which a prescribed proportion of the probability distribution lives. For example, if we compute the quantiles π_p for a sequence of regularly spaced probability levels p, patterns in the distribution of this set of percentiles can be interpreted as properties of the probability distribution. This remark is particularly useful when it comes to comparing several distributions. We develop this idea later on in this section with the introduction of the concept of Q-Q plot which we discuss in great detail as it is the cornerstone of the graphical analysis of heavy tail distributions.

1.1.3.3 ***Value at Risk (VaR)***

For better or worse, Value at Risk (VaR for short) is nowadays a crucial component of most risk management systems in the financial and insurance industries. Whether its computation is imposed by regulators, or done on a voluntary basis by portfolio managers, is irrelevant here. For the purpose of this subsection, we merely attempt to understand the rationale behind this measure of risk.

We introduce the concept of risk measure at an intuitive level, relying on the notion of required capital needed to make a financial position acceptable. We start with the discussion with a simple example.

Let us imagine that we need to track the performance of a portfolio. Typical portfolios comprise a large number of instruments, and a common assumption is to assume that the (log) returns on these portfolios are normally distributed. Note that in this book, we shall do our best *not to make this assumption* whenever we can help it. We denote by P_t the value of the portfolio at time t, we choose a specific level of confidence, say $p = 2\,\%$, and a time horizon Δt, say 1 year. Notice that the period $[t, t + \Delta t]$ is fixed. For that reason risk measures such as VaR are considered to be *static* risk measures. See the Notes & Complements at the end of the chapter for references.

Under these conditions, the associated required capital RC_t is defined as the capital needed to guarantee that the book will be in the red at time $t + \Delta t$ with probability no greater than p. In other words, RC_t, is defined by the identity:

$$\mathbb{P}\{P_{t+\Delta t} + RC_t < 0\} = p.$$

The Value at Risk at time t is then defined as the sum of the *current endowment* and the *required capital*

$$VaR = P_t + RC_t.$$

Notice that $\mathbb{P}\{P_{t+\Delta t} - P_t + VaR_t < 0\} = p$, which says that $-VaR_t$ is the p-quantile of the distribution of the change $P_{t+\Delta t} - P_t$ in the value of the portfolio over the given horizon. It is often more convenient to express VaR_t in units of P_t, i.e. to set $VaR_t = \widetilde{VaR}_t * P_t$. The definition

$$\mathbb{P}\{P_{t+\Delta t} - P_t + VaR_t < 0\} = p$$

of VaR can then be rewritten as:

$$\mathbb{P}\left\{\frac{P_{t+\Delta t} - P_t}{P_t} + \widetilde{VaR}_t < 0\right\} = p$$

which shows that, expressed in units of P_t, the negative of the value at risk is nothing more than the p-quantile of the distribution of the raw return over the period in question. As we will stress during our comparison of raw returns and log returns later in this section, they are very close to each other for small relative changes and we have:

$$\frac{P_{t+\Delta t} - P_t}{P_t} \sim \log \frac{P_{t+\Delta t}}{P_t}$$

which justifies the fact that we will often call value at risk at the level p for the horizon Δt, the negative of the p-quantile of the distribution of the log-return over a typical period of length Δt.

We now recast the above rather informal discussion in the framework we use throughout the book, and articulate a precise definition of Value at Risk. First, we slightly change the convention: in order to get rid of the annoying negative sign which we had to deal with above, it is convenient to think of a distribution function F as modeling the loss, i.e. the down side of the *profit and loss* (P&L for short) distribution in the standard financial jargon, associated to a specific financial position. A random variable X with cdf F could be for example the cumulative catastrophic insurance losses in a reporting period, or the credit losses of a bank or a credit card company, or the daily negative returns on a financial portfolio. Note that the length

Δt of the time horizon can, and will change from one application to another. We will consider many examples in which the frequency of the data will be different from 1 day. We will consider examples of minute by minute quotes used by day traders, and weekly and monthly quotes used by fund managers.

For a given level p, the value at risk VaR_p is defined as the $100p$-th percentile of the loss distribution. For the sake of definiteness, we shall choose the left hand point of the interval of possible percentiles when the loss distribution is not continuous. Value at Risk is widely used in the financial industry as a risk measure. Indeed, according to the first Basel agreement, financial institutions have to control the VaR of their exposures. Despite all that, VaR is not a satisfactory measure of risk for several reasons. The first one is quite obvious: it does not involve the actual size of the losses. The second one is less straightforward: it does not encourage diversification. This property is more subtle and more difficult to prove mathematically, and we refer the interested reader to the references given in the Notes & Complements at the end of the chapter. For these reasons we shall introduce another way to quantify the risk of a financial position modeled by a random variable X.

1.1.3.4 ***Effect of Affine Transformations***

As we already witnessed, shifting the location and changing the scale of a distribution are common practices in statistical data analysis, and controlling the effect of these transformations on the quantiles of a distribution is important. The results of the following discussion will be used in the interpretation of the Q-Q plots introduced and discussed below.

Let X be a random variable with cdf F_X, and let us denote by $\pi_p^{(X)}$ its quantiles. Let us now consider the random variable $Z = (X - m)/\lambda$ obtained by shifting the distribution of X by a real number m and scaling it by a positive number λ. In most applications we will use a location parameter for m (for example the mean of the distribution of X when it exists) and a scale parameter for λ (for example the standard deviation of the distribution when the variance exists). In particular, Z would be $N(0, 1)$ if we were to start with X Gaussian. We now show that the affine relationship

$$Z = \frac{X - m}{\lambda} \qquad \text{or equivalently} \qquad X = \lambda Z + m$$

carries over to the quantiles $\pi_p^{(X)}$ and $\pi_p^{(Z)}$ of X and Z respectively. Indeed, for any $p \in [0, 1]$ the quantile $\pi_p^{(X)}$ is characterized as the number π satisfying $F_X(\pi) = p$ and since

$$\begin{aligned} F_X(\pi) = \mathbb{P}\{X \leq \pi\} &= \mathbb{P}\{\lambda Z + m \leq \pi\} \\ &= \mathbb{P}\{Z \leq \frac{\pi - m}{\lambda}\} = F_Z(\frac{\pi - m}{\lambda}), \end{aligned}$$

(recall that $\lambda > 0$ which justifies our manipulations of the above inequalities) the number π satisfying $F_X(\pi) = p$ should satisfy $F_Z(\frac{\pi - m}{\lambda}) = p$ and hence we have

$$\frac{\pi - m}{\lambda} = \pi_p^{(Z)}$$

which give the desired equivalence

$$\frac{\pi^{(X)} - m}{\lambda} = \pi_p^{(Z)} \qquad \text{or equivalently} \qquad \pi_p^{(X)} = \lambda\pi_p^{(Z)} + m \tag{1.21}$$

between the quantiles of X and Z.

1.1.3.5 Comparing Quantiles

We motivate the introduction of the theoretical Q-Q plots with a simple experiment intended to illustrate the sizes of the quantiles of a distribution as they relate to the size (e.g. thickness) of its tail.

	$X \sim N(0,1)$	$X \sim C(0,1)$
$\pi_{0.8} = F_X^{-1}(0.8)$	0.842	1.376
$\pi_{0.85} = F_X^{-1}(0.85)$	1.036	1.963
$\pi_{0.9} = F_X^{-1}(0.9)$	1.282	3.078
$\pi_{0.95} = F_X^{-1}(0.95)$	1.645	6.314
$\pi_{0.975} = F_X^{-1}(0.975)$	1.960	12.706
$\pi_{0.99} = F_X^{-1}(0.99)$	2.326	31.821

Table 1.1. Comparison of the quantiles of the standard Gaussian and Cauchy distributions

We already emphasized how thin the tails of the Gaussian distribution are by quantifying the concentration of the probability mass around the mean of the distribution. We now compute quantiles of the normal distribution with the command `qnorm`, and arguments giving respectively the list of quantiles we want, the mean, and the standard deviation of the distribution. Because of the symmetry of the distribution, we restrict ourselves to the upper tail.

```
> qnorm(c(.8,.85,.9,.95,.975,.99),mean=0,sd=1)
[1] 0.8416 1.0364 1.2816 1.6449 1.9599 2.3263
```

In words, these numbers tell us that 80 % of the probability mass is to the left of $x = 0.8416$, 85 % is to the left of $x = 1.0364$, ... The computation of the corresponding quantiles for the Cauchy distribution gives:

```
> qcauchy(c(.8,.85,.9,.95,.975,.99),location=0,scale=1)
[1]  1.376  1.963  3.078  6.314 12.706 31.821
```

We display these results in tabular form in Table 1.1.

We see that in the case of the Cauchy distribution, in order to have 80 % of the probability mass to its left, a quantile candidate has to be as large as $x = 1.376$, which is greater than $x = 0.842$ found for the normal distribution. Obviously, the same is true for the other quantiles in the above lists. This pattern may be visualized by plotting the quantiles of the Cauchy distribution against the corresponding quantiles of the normal distribution. We would have to plot the points

$$(0.8416212, 1.376382), (1.0364334, 1.962611), (1.2815516, 3.077684),$$
$$(1.6448536, 6.313752), (1.9599640, 12.706205), (2.3263479, 31.820516).$$

Note that all these points are above the diagonal $y = x$, and in fact they drift further and further away above this diagonal. This fact is at the core of the interpretation of a Q-Q plots which we introduce formally below: points above the diagonal in the rightmost part of the plot indicate that the upper tail of the first distribution (whose quantiles are on the horizontal axis) is thinner than the tail of the distribution whose quantiles are on the vertical axis. This phenomenon appears as well in the case of the empirical Q-Q plots introduced later and illustrated for example in Figs. 1.23 and 1.25.

1.1.3.6 ***Theoretical Q-Q Plots***

The above discussion seems to indicate that comparing the quantiles of two distributions carries powerful information on the relative properties of the tails of these distributions.

A quanrile-quantile plot (Q-Q plot for short) of a distribution $F^{(2)}$ against a distribution $F^{(1)}$, is the plot of the curve formed by the couples $(\pi_p^{(1)}, \pi_p^{(2)})$ when the real number p varies over $[0, 1]$. Obviously, we use the notation $\pi_p^{(1)}$ for the quantiles of the distribution $F^{(1)}$, and $\pi_p^{(2)}$ for the quantiles of the distribution $F^{(2)}$.

The interpretation of these plots is based on the following simple remarks.

- The Q-Q plot should be on the diagonal if the two distributions are equal;
- If both distributions extend to $+\infty$, and if the tail of $F^{(1)}$ is heavier than the tail of $F^{(2)}$, then when p increases toward 1, the quantile $\pi_p^{(1)}$ of $F^{(1)}$ should grow faster that the quantile $\pi_p^{(2)}$ of $F^{(2)}$, and consequently, the Q-Q plot should be below the diagonal and curve downward below the diagonal;
- Similarly, if both distributions have right tails extending to $+\infty$, and if the tail of $F^{(1)}$ is thinner than the tail of $F^{(2)}$, then as p increases toward 1, the quantile $\pi_p^{(1)}$ of $F^{(1)}$ should grow slower that the quantile $\pi_p^{(2)}$ of $F^{(2)}$, and consequently, the Q-Q plot should curve upward above the diagonal;
- If both distributions extend to $-\infty$, and if the tail of $F^{(1)}$ is heavier than the tail of $F^{(2)}$, then when p decreases toward 0, the quantile $\pi_p^{(1)}$ of $F^{(1)}$ should go to $-\infty$ faster that the quantile $\pi_p^{(2)}$ of $F^{(2)}$, and consequently, the Q-Q plot should curve upward above the diagonal;
- Finally, if both distributions have a left tails extending to $-\infty$, and if the tail of $F^{(1)}$ is thinner than the tail of $F^{(2)}$, then as p decreases toward 0, the quantile $\pi_p^{(1)}$ of $F^{(1)}$ should go to $-\infty$ slower that the quantile $\pi_p^{(2)}$ of $F^{(2)}$, and consequently, the Q-Q plot should curve downward below the diagonal.

We illustrate these general qualitative statements with specific examples of Q-Q plots of distributions from some of the families introduced in the first part of the chapter.

1.1.3.7 Examples of Theoretical Q-Q Plots

We close this section with a discussion of specific examples of theoretical Q-Q plots.

One Tailed Distributions. We first consider two distributions on the half line $\mathbb{R}_+ = [0, \infty)$, and we compare their tails at $+\infty$. We choose the exponential distribution with rate one, and the log-normal distribution with mean zero and variance one. We implement the definition of the theoretical Q-Q plot of these distributions with the R commands

```
> P <- seq(from=0,to=1,length=1025)
> plot(qexp(P,1),qlnorm(P,meanlog=0,sdlog=1),type="l",
      xlab="E(1) quantiles",ylab="LogNormal quantiles")
> abline(0,1)
```

which also superimpose the diagonal (which helps the interpretation of the result, especially when the scales on the vertical and horizontal axes differ). The result is given in Fig. 1.10. The first R command creates a vector P of length 1025 with

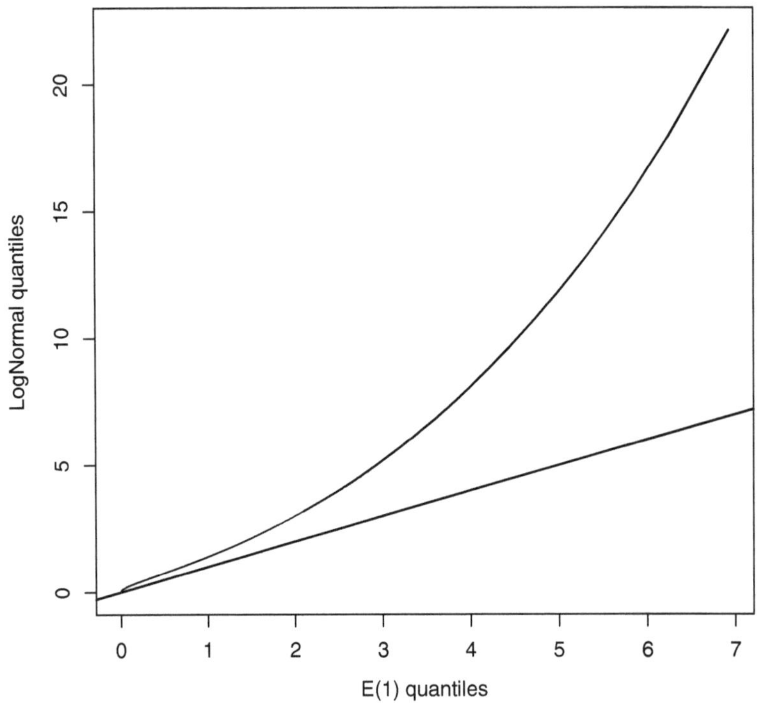

Fig. 1.10. Theoretical Q-Q plot of the log-normal distribution with mean zero and variance one, against the exponential distribution with unit rate

entries forming a regular grid of real numbers from 0 to 1. The first two arguments of the function plot give the x and y coordinates of the points to plot on the graph, the parameter `type="l"` implying that straight lines will be used to connect the points. Applying a function to a vector creates a vector of the same length with entries equal to the values of the function computed for the entries of the original vector. So `qexp(P,1)` is a vector of length `1025` , the entries of which are the quantiles of the exponential distribution with rate one (the second parameter of the function `qexp`) computed at the values found in the vector `P`. Similarly, `qlnorm(P,meanlog=0,sdlog=1)` is a vector of length `1025` whose entries are the quantiles of the log-normal distribution with mean zero and variance one computed at the values found in the vector `P`. The roles of the parameters `xlab` and `ylab` in setting the labels of the axes are self explanatory. Finally, the command `abline(0,1)` adds a line with intercept 0 and slope 1 (i.e. the first diagonal when the units on the two axes are the same) to the plot.

The interpretation of the plot is plain: the upward bend in the right hand side of Fig. 1.10, together with its convexity show that the log-normal distribution has a heavier tail than the exponential distribution.

Two Tailed Distributions. Next, we consider theoretical Q-Q plots of two distributions having tails extending to plus and minus infinity. The plots appearing in Fig. 1.11 epitomize the most important features which can be identified, and we discuss them one by one in the next bullet points.

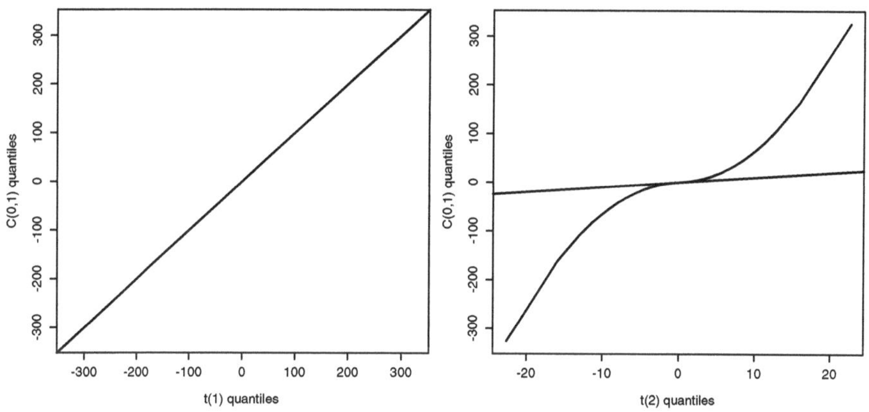

Fig. 1.11. Theoretical Q-Q plots of the Cauchy distribution $C(0, 1)$ against the t distribution with 1 (*left*) and 2 (*right*) degrees of freedom

- The left pane of Fig. 1.11 shows the theoretical Q-Q plot of the Cauchy distribution with location zero and scale one against the t distribution with one degree of freedom. The plot seems to align perfectly with the diagonal which shows that there is no significant numerical difference between the sizes of the tails as captured by this type of plot. A quick look at the plots given earlier of the densities of the two distributions is enough to explain why this had to be expected.
- The right pane of Fig. 1.11 shows the theoretical Q-Q plot of the same Cauchy distribution against the t distribution with two degrees of freedom. The plot is

obviously very different. First, we should notice that the scales on the two axes are very different (see the tick labels on these axes), as evidenced among other things, by the diagonal which is almost horizontal. Next the curvatures found on both ends show that both tails of the t distribution are thinner than the tail of the Cauchy distribution. Indeed, the fact that the Q-Q plot is above the diagonal on the right of the plot is due to the fact that the upper quantiles of the $t(2)$ distribution grow to $+\infty$ slower than those of the $C(0,1)$ distribution, implying that the right tail of the $t(2)$ distribution is lighter (we also say thinner) than the right tail of the Cauchy distribution $C(0,1)$. Similarly, fact that the Q-Q plot is below the diagonal on the left of the plot is an indication that the lower quantiles of the $t(2)$ distribution decrease to $-\infty$ slower than those of the $N(0,1)$ distribution, implying as before that the left tail of the $t(2)$ distribution is thinner than the left tail of the $C(0,1)$ distribution. The argument given for the upper tails was repeated mutatis mutandis for the lower tails for pedagogical reasons. We could have used the fact that both distributions are *symmetric* and as a consequence, have same size at $+\infty$ and $-\infty$.

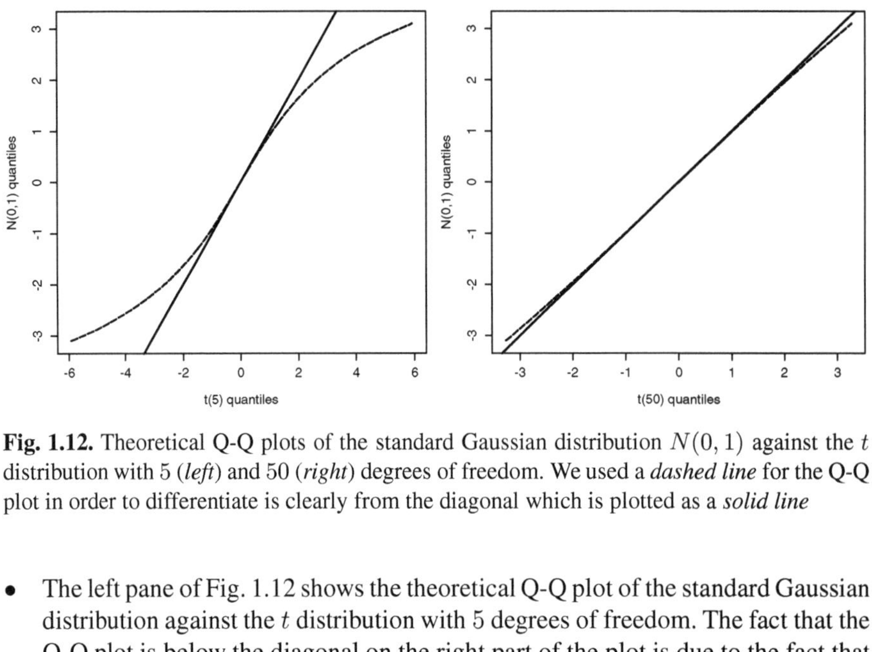

Fig. 1.12. Theoretical Q-Q plots of the standard Gaussian distribution $N(0,1)$ against the t distribution with 5 (*left*) and 50 (*right*) degrees of freedom. We used a *dashed line* for the Q-Q plot in order to differentiate is clearly from the diagonal which is plotted as a *solid line*

- The left pane of Fig. 1.12 shows the theoretical Q-Q plot of the standard Gaussian distribution against the t distribution with 5 degrees of freedom. The fact that the Q-Q plot is below the diagonal on the right part of the plot is due to the fact that the upper quantiles of the $t(5)$ distribution grow to $+\infty$ faster than those of the $N(0,1)$ distribution, implying that the right tail of the $t(5)$ distribution is heavier (we also say thicker) than the right tail of the standard Gaussian distribution. Similarly, the fact that the Q-Q plot is above the diagonal on the left of the plot is an indication that the lower quantiles of the $t(5)$ distribution decrease to $-\infty$ faster than those of the $N(0,1)$ distribution, implying that the left tail of the $t(5)$

distribution is heavier than the left tail of the $N(0,1)$ distribution. As before, a symmetry argument can be used to reduce the comparison of the tails to the right tails.

- However, as evidenced by the plot on the right pane of Fig. 1.12, the t distribution with 50 degrees of freedom has tails of pretty much the same sizes as those of the normal distribution. The tails are still heavier than the tails of the $N(0,1)$ distribution since the Q-Q plot is on the same side of the diagonal as before, but the difference is much smaller than earlier. This fact is a consequence of the normal approximation result which we mentioned earlier in the text.

1.2 Observations and Nonparametric Density Estimation

This section reviews some of the most basic nonparametric techniques of density estimation. Our stand point is more practical than theoretical, and we emphasize implementation in R.

1.2.1 **Sample Data**

The data used in this chapter come in the form of a sample:

$$x_1, x_2, \ldots\ldots, x_n$$

where the x_j are real numbers. They are analyzed with statistical tools based on concepts of statistical data analysis which we review in this chapter. Our presentation is sprinkled with numerical illustrations anchored on a small number of specific examples. We proceed to the introduction of the first of these examples.

1.2.1.1 ***The PCS Data***

The PCS Index is the year-to-date aggregate amount of total damage reported in the United States to the insurance industry. PCS stands for Property Claim Services. It is a division of ISO Inc (Insurance Services Office). Regional indexes also exist, and different regions have unique indexes, e.g. California, Florida, Texas, ..., but we only consider the national index in this example. Each index value represents \$100 million worth of damage. For example, a value of 72.4 for the national index in 1966, means that $\$(72.4 \times 100)$ million (i.e. \$7.24 billion) in damage were recorded on that year.

The Chicago Board of Trade began trading options on the PCS Index in 1996. Options and futures contracts on the PCS Index offer a possibility to securitize insurance catastrophe risk in a standardized fashion. These financial products were seen in the mid 1990s by the insurance industry as a way to tap into the inverstors enormous appetite for risk.

For the purpose of our analysis, we do not use the index values. Instead, we use some of the individual claim reports used to compute the final values of the index. The data we use come in the form of a matrix with 2 columns and 380 rows:

```
> head(PCS)
  Col1  Col2
1   13  4.00
2   16  0.07
3   46  0.35
4   60  0.25
5   87  0.36
6   95  1.00
```

where the first column contains time stamps $13, 16, \cdots$ (i.e. codes for the dates of the catastrophic events), and the second column contains the aggregate amounts $4.00, 0.07, \cdots$ (again in \$100 million) of all the claims reported after each event. These data are contained in an `R` matrix called `PCS`. The command `head(PCS)` was used to print the first six rows of the data set. Notice that, since the object `PCS` does not have column names and row names, `R` uses default names `Col1` and `Col2` for columns and the integers `1,2, ...` for row names. The plot of the data is given in Fig. 1.13. In a first analysis, we do not use the information of the timing of

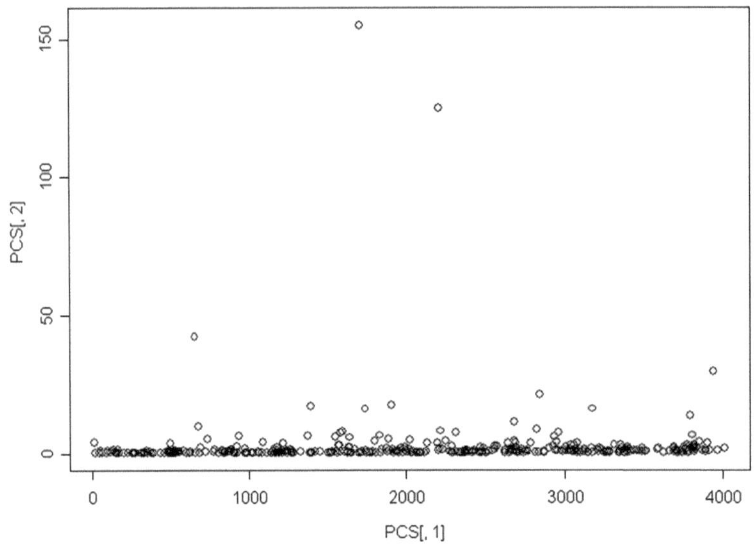

Fig. 1.13. Approximately 10 years worth of individual catastrophe costs used to compute the PCS index. For each catastrophic event included in the data set, the time stamp is reported on the *horizontal axis*, and the aggregate dollar amount of the claims attributed to this catastrophic event is reported on the *vertical axis*

the catastrophes. In other words, we first work with the second column of the data set, i.e. with the dollar amounts only. So, at least in this chapter, the data of interest to us will be:

$$x_1 = 4.00,\ x_2 = 0,07,\ x_3 = 0.35,\ x_4 = 0.25, \ldots\ldots$$

and we encapsulate this data set in the `R` object `PCS.index` (a numeric vector in this case) with the command:

```
> PCS.index <- PCS[,2]
```

which extracts the second column of the matrix `PCS`, and renames it `PCS.index`. From now on, we work with these values, ignoring the timing of the actual catastrophes. The analysis of the time dependence is the subject of time series analysis which we will tackle in the third part of the book.

Again, we refer to the `R` Tutorial contained in the first Appendix at the end of the book for details on the basic commands used in this chapter.

1.2.1.2 ***The S&P 500 Index and Financial Returns***

For this second example, the data give the weekly closing values of the S&P 500 index. They come in the form:

```
1950-01-03  16.98
1950-01-09  16.67
1950-01-16  16.90
1950-01-23  16.82
1950-01-30  17.29
1950-02-06  17.24
..................... ..........
```

These data are contained in the `R` object `WSP.ts` of class `timeSeries` and plotted in Fig. 1.14. `timeSeries` objects will be studied in Chap. 6. In the present chapter, we do not make use of the time stamps appearing in the first entry of each row. We concentrate on the data appearing in the second column. For the sake of convenience, we organized these data in an `R` numeric vector which we called `WSP`. We are not really interested in the raw values of the index. For reasons which will become clear later, we would rather analyze the returns, so we transform the data for the purposes of our analysis. However, before doing so we define the two notions of *return* used in finance. For the sake of simplicity, we ignore the possibility of dividend payments. Given the value of an index at time t, say S_t, and its value after a period of length Δt, say $S_{t+\Delta t}$, the raw-return over that period is defined as:

$$RR_t = \frac{S_{t+\Delta t} - S_t}{S_t} = \frac{S_{t+\Delta t}}{S_t} - 1 \tag{1.22}$$

while the log-return over the same period is defined by the formula:

$$LR_t = \log \frac{S_{t+\Delta t}}{S_t}. \tag{1.23}$$

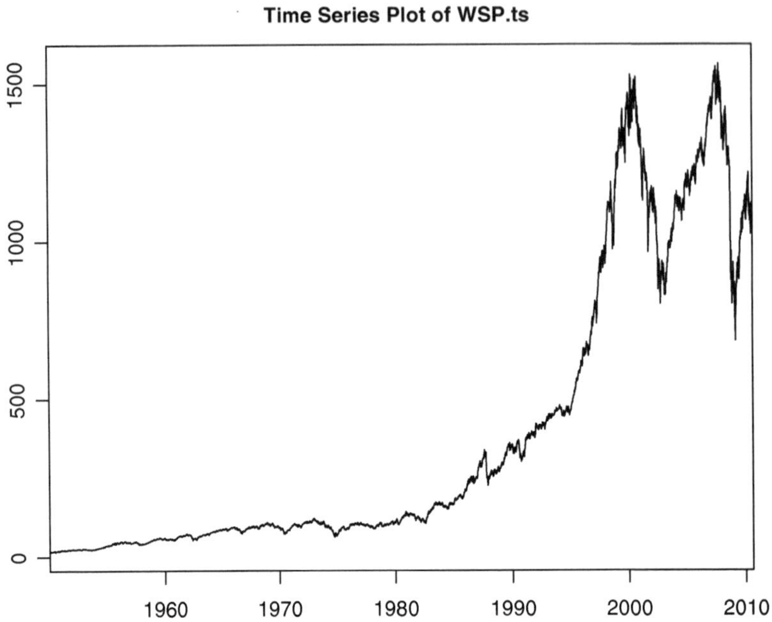

Fig. 1.14. Weekly values of the S&P index

Log-returns are natural in the context of continuous time discounting, while raw-returns are more natural when discounting is done at discrete time intervals. Notice that since $x \sim \log(1+x)$ when x is small, we have:

$$LR_t = \log\left(1 + \frac{S_{t+\Delta t} - S_t}{S_t}\right) \sim \frac{S_{t+\Delta t} - S_t}{S_t} = RR_t$$

whenever the ratio $S_{t+\Delta t}/S_t$ is close to 1. So we expect that the two methods of computing returns will give essentially the same results when Δt is small, or when the value of the index does not change much over a period of length Δt. Notice that both notions depend upon the time period over which the returns are computed. Δt is 1 week in the present situation. It will be 1 month in some examples, or even one quarter in others. However, it will be 1 day in most of the examples considered in the book.

The practical computation of the log-returns proceeds as follows. Except for the first value, we divide each closing value by its value the previous week (computing in this way the weekly return), and we then compute the logarithm of this ratio, obtaining in this way the weekly log-return. Since the log of a ratio is the difference of the logarithms of the numerator and denominator, we use the following R command

```
> WSPLRet <- diff(log(WSP))
```

to compute the vector of log-returns. When applied to a vector, the R function `log` produces a vector of the same length, whose entries are given by the logarithms of the entries of the original vector. The function `diff` gives a vector that is one element shorter than the original vector, and whose entries are given by the differences of two successive terms of that sequence. The first entries of the sample of log-returns are:

$$x_1 = -0.018425484,\ x_2 = 0.013702925,\ x_3 = -0.004744967,\ x_4 = 0.027559645,\ \ldots$$

After this transformation of the original data, the resulting sample appears as if the entries formed a set of observations of independent random variables with the same distribution. In this chapter, we concentrate on the analysis of log-return values instead of the original values of the index, and we try to infer statistical properties of the common distribution of these log-return observations. We stress that in what follows, the results of the analysis do not depend upon the order of the observations. In other words, were we to shuffle these numbers and change the order in which they appear, the results of our analysis would remain unchanged.

1.2.2 Nonparametric Estimation

1.2.2.1 Statistical Estimation

When working with sample observations from an unknown cdf F, if the quantities we want to estimate are numerical characteristics which can be computed from the cdf F, our preferred strategy is to use the sample observations to produce an estimate, say $\hat{F}$, of the cdf F, and then to use the corresponding characteristics computed from the estimate $\hat{F}$ as estimates of the desired characteristics of F. For example, the mean of $\hat{F}$ will be used to estimate the mean of F, the variance of $\hat{F}$ for the variance of F, etc. Often, we assume that the unknown distribution has a density $f = F'$, and we try to compute the characteristics as integrals involving f.

A good part of classical parametric estimation theory can be recast in the framework of density estimation: for instance, estimating the mean and the variance of a normal population is nothing but estimating the density of a normal population. Indeed, a Gaussian distribution is entirely determined by its first two moments, and knowing its mean and variance is enough to determine the entire distribution. Similarly, estimating the mean of an exponential population is the same as estimating the density of the population since the exponential distribution is determined by its rate parameter, which in turn is determined by the mean of the distribution. More generally, if the random mechanism governing the generation of the data is known to be producing samples from a given parametric family, estimating the distribution reduces to estimating the parameters of the distribution, and we can use classical estimation methods, such as maximum likelihood, method of moments, … reviewed in the previous section. We can then plug the estimated values into the formulae defining the parametric family and compute the desired characteristics accordingly. However, if the unknown characteristics of the random mechanism cannot be captured by a small set of parameters, we need to use a nonparametric approach.

1.2.2.2 **The Empirical Cumulative Distribution Function (cdf)**

When fitting a distribution to a data sample, different methods have to be brought to bear when no rationale exists for the choice of a specific parametric family of distributions. In the absence of specific information about the type of statistical distribution involved, the only recourse left is very often the so-called empirical distribution function $\hat{F}_n$ which we now define. Given a sample $x_1, x_2, \cdots, x_n$ from an unknown cdf F, the empirical distribution function $\hat{F}_n$ of this sample is the piecewise constant function defined by:

$$\hat{F}_n(x) = \frac{1}{n}\#\{i;\ 1 \le i \le n,\ x_i \le x\}. \tag{1.24}$$

$\hat{F}_n(x)$ represents the proportion of observations not greater than x. The function $\hat{F}_n(x)$ is piecewise constant, since it is equal to j/n for x in between the j-th and the $(j+1)$-th observations. Its graph is obtained by ordering the observations x_j's, and then plotting the flat plateaus in between the ordered observations.

The rationale behind this estimation procedure is the fact that the empirical cdf $\hat{F}_n(x)$ converges uniformly in $x \in \mathbb{R}$ toward the unknown cdf $F(x)$. This is known as the Glivenko-Cantelli theorem (or the fundamental theorem of statistics). Its proof is based on the Law of Large Numbers (LLN for short) which we state in Sect. 1.3 in the context of Monte Carlo computations. It can be found in most introductory texts in statistics.

According to the estimation strategy articulated earlier, once the empirical distribution function has been chosen as *the estimate* of the (theoretical) unknown cdf, characteristics of the unknown cdf can be estimated by the corresponding characteristics of the empirical distribution function $\hat{F}_n$. In particular, the percentiles can be estimated by the percentiles computed from $\hat{F}_n$. So, according to this estimation strategy, the $100p$-th percentile $\pi_p = \pi_p(F)$ is estimated by the empirical percentile $\pi_p(\hat{F}_n)$. This procedure has obvious applications to the computation of risk measures based on quantiles, such as the value at risk VaR_p.

1.2.2.3 **Order Statistics**

The actual construction of the empirical cdf given above emphasized the special role played by the ordered observations:

$$\min\{x_1, \ldots, x_n\} = x_{(1),n} \le x_{(2),n} \le \cdots \le x_{(n),n} = \max\{x_1, \ldots, x_n\}$$

Warning. The ordered observations $x_{(k),n}$ are not defined uniquely because of possible ties, i.e. cases when $x_i = x_j$ for different indices i and j. Ties cannot occur (to be more specific, we should say that the probability that they do occur is zero) when the distribution function F is continuous. The empirical cdf's are never continuous, but most of the cdf's occurring in practice are continuous, so since such an assumption does not restrict the scope of our analysis, we shall subsequently assume that there are no ties in the sample observations.

The ordered observations $x_{(k),n}$ can be viewed as realizations of random variables $X_{(1),n}, X_{(2),n}, \ldots X_{(n),n}$. Note that the latter are **neither independent nor identically distributed** any more. The random variable $X_{(k),n}$ is called the k-th order statistic. Our definition of the percentiles of a distribution with jumps, and of the left inverse of a cdf imply that for each probability level $p \in (0, 1)$:

$$x_{(k),n} = \hat{\pi}_p = \pi_p(\hat{F}_n) \qquad \text{for} \qquad \frac{k-1}{n} < p \leq \frac{k}{n}.$$

The following will have far-reaching consequences in the analysis of distribution tails.

Let us assume that $x_1, x_2, \ldots\ldots, x_n$ form a sample from an unknown cdf F, and let us assume that we are interested in the estimation of a quantile $\pi_p = \pi_p(F)$. As we already emphasized, one important financial application is the computation of VaR (value at risk). The probability that exactly k of the sample observations x_j fall in between π_p and 1 is the same as the probability that

$$X_{(n-k),n} \leq \pi_p < X_{(n-k+1),n}. \tag{1.25}$$

Since the cdf F is monotone increasing, the inequalities (1.25) can be equivalently rewritten as:

$$F(X_{(n-k),n}) \leq p < F(X_{(n-k+1),n}) \tag{1.26}$$

and the probability that this happens is equal to the probability that exactly k of the numbers $F(X_j)$ fall in the interval $[p, 1]$. Since the $F(X_j)$ are i.i.d. random variables, this probability is equal to the binomial probability:

$$\binom{n}{k} \overline{p}^k (1 - \overline{p})^{n-k}.$$

where $\overline{p}$ denotes the probability that a number $F(X_j)$ belongs to the interval $[p, 1]$. But according to the results presented in Sect. 1.3, and especially Fact 1, this probability is equal to $(1-p)$. Consequently, the probability that (1.26) occurs is given by:

$$\binom{n}{k} p^{n-k} (1 - p)^k. \tag{1.27}$$

This important result is used in practice to derive confidence intervals for the empirical quantiles of a distribution.

1.2.2.4 R ***Implementation***

For the purpose of illustration we work with the Calpine stock price data downloaded from the internet and imported in R as explained in the appendix. We first compute the daily log-returns. On any given day, the log return is the logarithm of the ratio of the price on that day divided by the price the day before. Given the properties of the logarithm function, this is also the difference between the logarithm of the price on that day and the logarithm of the price the day before. We create a vector `CPNLRet` containing the daily log returns with the command:

```
> CPNLRet <- diff(log(CPN))
```

As explained in our discussion of the weekly log returns of the S&P 500, the command `log(CPN)` creates a vector with the same length as `CPN`, each entry being the logarithm of the corresponding entry in `CPN`. The function `diff` creates a vector with one less element, each entry being equal to the difference between the corresponding entry in the argument vector, and its preceding entry (hence the shorter length). Now we assume that the entries of `CPNLRet` form a sample from an unknown distribution (the distribution of the daily log returns on Calpine stock) and we proceed to the computation of the empirical quantiles. Remember that, according to the above discussion, the latter can be viewed as estimates of the quantiles of the unknown distribution.

```
> quantile(CPNLRet,c(.01,.05,.25,.5,.75,.95,.99))
      1%        5%        25%       50%       75%       95%        99%
-0.1263   -0.0658   -0.0201   0.0000   0.0204   0.06353   0.1315
```

We use the `R` function `quantile`. Its first argument needs to be the numeric vector of the sample values, `CPNLRet` in our case, while its second argument should be the vector of probability levels at which we want to compute the quantiles. In the present case, we chose to compute the empirical p-quantiles for the values 0.01, 0.05, 0.25, 0.5, 0.75, 0.95, 0.99 of p which we encapsulated in a vector with the concatenation function `c`. The function `quantile` returns a vector of the same length as the vector of probability levels. Its entries are the desired quantiles.

Even though we should not try to re-invent the wheel, for pedagogical reasons, we think that it is important to use the tools offered by `R` to compute from scratch (i.e. without using the function `quantile`) the ordered statistics of a sample. The `R` function `order` gives the ranks of the entries of a numeric vector. Using the ranks as new indexes, we can construct the vector of ordered entries. The following commands illustrate these facts.

```
> RKS <- order(CPNLRet)
> OCPNLRet <- CPNLRet[RKS]
> par(mfrow=c(2,1))
> plot(CPNLRet,type="l")
> plot(OCPNLRet,type="l")
> par(mfrow=c(1,1))
```

The results are reproduced in Fig. 1.15. The left plot is typical of stock log returns. The only reason we give the left plot is to check that the observations have been ordered correctly. We can now recover the empirical quantiles computed above with the *pedestrian* approach based on the definition of these quantiles (e.g. the 1 – empirical – percentile is the value of the entry which has $1\,\%$ of the entries below)

```
> L <- length(OCPNLRet)
> P <- 0.01
> OCPNLRet[ceiling(P*L)]
[1] -0.1267517
> P <- 0.75
> OCPNLRet[ceiling(P*L)]
[1] 0.02040887
```

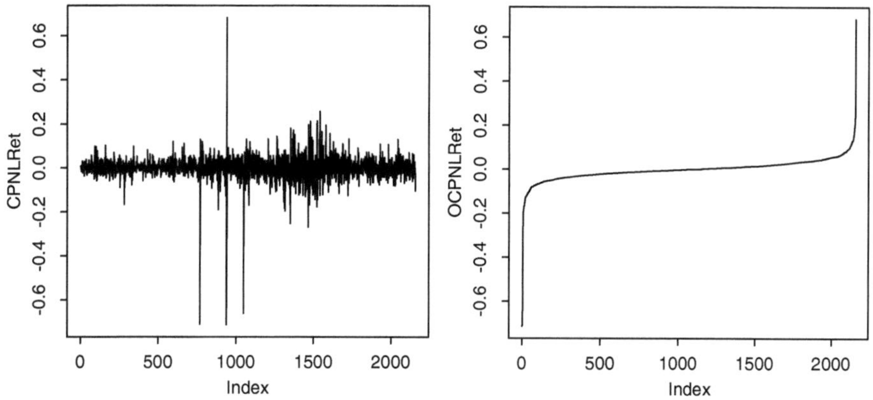

Fig. 1.15. Sequential plot of the daily log returns on Calpine stock before (*left*) and after (*right*) ordering the observations

1.2.3 **Histograms**

Together with pie charts, histograms are presumably the most frequently used graphical representations of data. In this book, we are interested in the analysis of data from continuous probability distributions, and we view histograms as estimators of the densities of the corresponding distributions. Since the construction of a histogram does not assume that the distribution is an element of a specific family of distributions, it can be viewed as a nonparametric estimation procedure. However, it is not a panacea, and like most nonparametric functional estimation procedures, the histogram relies on the specification of a few constants, two in most cases. For example, in order to produce a histogram, one chooses the width of the bins, and the origin from which the bins are lined up.

Recall that once the range of the observation sample vector is covered by a set of bins, plotting the histogram is plotting a rectangle above each bin, the height of the rectangle being the absolute (or relative) frequency of the data entries falling in the bin.

1.2.3.1 ***Implementation in*** `R`

Histograms are produced in `R` with the command `hist`. The left pane of Fig. 1.16 is the result of the command `hist(CPNLRet)`. If the only argument passed to the function `hist` is a numeric vector, `R` chooses automatically default values for all the other arguments. We specified a few of these arguments to produce the histogram appearing in the right pane of Fig. 1.16. We chose to force the histogram to have `25` bins by including `breaks=25`. In fact the parameter `breaks` can be used to specify not only the number of bins, but the end points of these bins. Indeed, when `breaks` is a numeric vector, its entries are interpreted as the end points of the bins. We chose to plot the bars over the bins in blue by setting the parameter `col` appropriately, namely by passing `col="blue"` to the function `hist`. Finally, we set

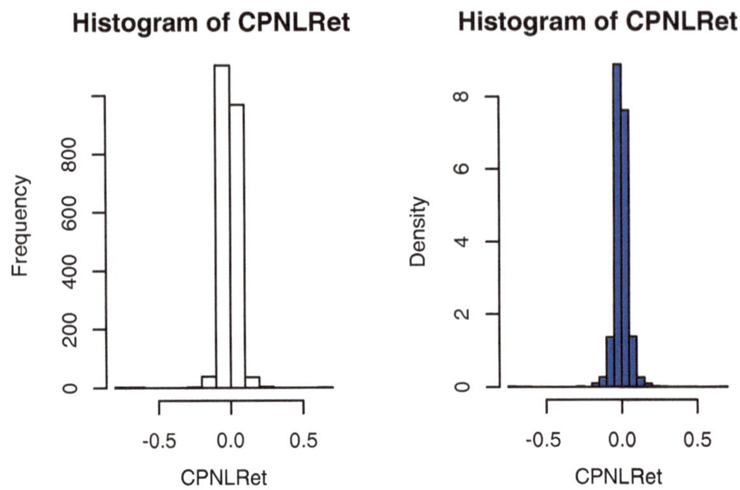

Fig. 1.16. Histograms of the daily log returns on the Calpine stock

the parameter `freq` to `FALSE` in order for *relative frequencies* to be used as labels on the vertical axis. This does not change the look of the histogram, but the numerical values correspond now to the values of a probability density, instead of raw frequency counts. We give the `R` commands used to produce Fig. 1.16 for the sake of definiteness.

```
> par(mfrow=c(1,2))
> hist(CPNLRet)
> hist(CPNLRet,breaks=25,col="blue",freq=F)
> par(mfrow=c(1,1))
```

Both histograms give a good rendering of the central bump in the density. However, it is difficult to gauge the relative importance of the bins in the right and left most parts of the graph. It seems that most of the bins are empty, except possibly for some extreme bins which justify the choice of the range of x-values over which the histogram is constructed.

1.2.3.2 ***More Shortcomings and Extensions***

The dependence of the histogram upon the choice of the origin is an undesirable artifact of the method. In order to circumvent this shortcoming, the notion of averaged histogram was introduced: one histogram is computed for each of a certain number of choices of the origin, and all these histograms are averaged out to produce a smoother curve expected to be robust to shifts in the origin. This estimate is called the `ASH` estimate of the density of the population, the three initials A, S and H standing for "average shifted histogram".

Even though ASH estimates are free of the artificial dependence upon the choice of the origin, they are still dependent on the particular choice of the bin width, the latter being responsible for the look of the final product: ragged curves due to a choice of small bin widths, and smoother looking blocks if the bins have larger widths. The decisive influence of this parameter should be kept in mind as we inch our way toward the introduction of our favorite density estimation procedure.

For the time being, we limit ourselves to the following remark: building a histogram is done by piling up rectangles. Given the choice of the subdivision of the range of the data into intervals of equal lengths (the so-called bins), the contribution of any given observation is a rectangle of height $1/(nb)$ (where n is the population size and b is the bin width), the rectangle being set on top of the bin in which the observation falls. In particular, if many observations fall near the boundary of a bin, the piling up of these rectangles will create an undesirable effect which we illustrate with one possible instance of this shortcoming. Let us assume that the bins have been chosen to be the unit intervals $[0, 1)$, $[1, 2)$, $[2, 3)$, ..., $[5, 6)$, and let us assume that the data is comprised of $6 \times 8 = 48$ points grouped in 8's around each of the integers $1, 2, 3, 4, 5$ and 6, in such a way that for each of these integers, four of the data points are smaller than (and very close to), the other four (still very close) being greater than the integer in question. Obviously, the distribution of the points shows a periodic regularity, and one would want the density estimator to account for it: the high concentration of points near the integers should be reflected in the presence of high values for the density, while this same density should vanish in the middle of the inter-integer intervals which are empty of data points. Unfortunately, our histogram will completely miss this pattern. Indeed, the bins having been chosen as they were, the histogram is flat throughout the interval $[0, 6)$ leading us (misleading us should I say!) to believe that the distribution is uniform over that interval.

One possible way out of this problem is to center the bins around the observation values. Doing so is just computing the kernel density estimator proposed below with the particular choice of the `box` kernel function !!!

1.2.4 Kernel Density Estimation

Given a sample $x_1, \ldots, x_n$ from a distribution with (unknown) density $f(x)$, the formal definition of a kernel density estimator of f is the function $\hat{f}_b$ defined by:

$$\hat{f}_b(x) = \frac{1}{nb} \sum_{i=1}^{n} K\left(\frac{x - x_i}{b}\right) \tag{1.28}$$

where the function K is a given non-negative function which integrates to one (i.e. a probability density function) which we call the kernel, and $b > 0$ is a positive number which we call the bandwidth. The interpretation of formula (1.28) is simple. Over each point x_i of the sample, we center a scaled copy of the kernel function K, and the final density estimate is the superposition of all these "bumps". The division by nb guarantees that the total mass is one (i.e. the integral of $\hat{f}_b(x)$ is one.)

1.2.4.1 Univariate Kernel Density Estimation in R

The function provided by R for univariate kernel density estimation is `density`. As in the case of the function `hist`, most of the parameters are given default values by the program. However, mostly for the sake of completeness, we mention the meaning and the role of some of the most important parameters. The bandwidth parameter b of formula (1.28) is given by the product of the parameters `bw` and `adjust`. The latter is equal to `1` by default while the default value of the former is specified as the value of an asymptotically optimal formula. However, the user can choose whatever values he or she pleases for these two parameters. The kernel estimate $f_b(x)$ given by (1.28) is computed at `n` equally spaced values of x. The program chooses `n=512` by default, but the parameter `n` can be user specified. Also, the values of x at which the density is computed can be chosen by specifying the parameters `from` and `to` which determine (together with `n`) the entire grid of x-values where the density is computed. The function `density` returns a list whose components include the vector `x` of points where the density function has been computed, the vector `y` of the values on the estimate of the density function. See the help file for details of the components returned by this function. For the purpose of illustration and comparison with the earlier histogram computations, we produce density estimates for the Calpine daily log returns. We use the Gaussian kernel and two different bandwidths.

```
> par(mfrow=c(1,2))
> DENS <- density(CPNLRet,bw=.1)
> plot(DENS$x,DENS$y,type="l",main="KDE with bw=.1")
> DENS <- density(CPNLRet,bw=5)
> plot(DENS$x,DENS$y,type="l",main="KDE with bw=5")
> par(mfrow=c(1,1))
```

each call to the function `density` returns a list which we call `DENS`, and in each case we extract the `x` and `y` components with the dollar sign and we plot them with the command `plot`. The results are given in Fig. 1.17. They are strikingly different. Indeed the scales on the horizontal axes as well as on the vertical axes are very different, and any attempt to super-impose these two graphs on the same plot would flatten one of them. Notice that, a small value of the bandwidth gives a sharp peak around the bulk of the data, while a larger bandwidth gives a smoother curve and a larger spread for the mass distribution. For the sake of comparison, we plot a histogram and a kernel density estimate of the same daily log return data `CPNLRet` on the same graph. We do this with the following commands.

```
> DENS <- density(CPNLRet)
> hist(CPNLRet,breaks=25,col="blue",freq=F,ylim=c(0,15),
  main="Histogram and Kernel Density Estimator of CLPLRet")
> lines(DENS$x,DENS$y)
```

The results are given in Fig. 1.18. It is important to choose the option `freq=F` to ensure that the histogram is normalized as a probability density. This guarantees that the two estimates are on the same vertical scale. We also forced the limits of the

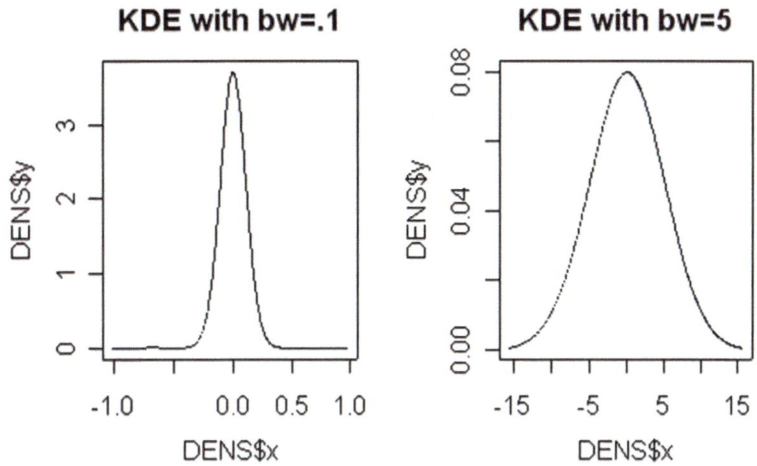

Fig. 1.17. Kernel density estimates of the daily log returns on the Calpine stock with Gaussian kernel and bandwidths equal to 0.1 (*left*) and 5 (*right*)

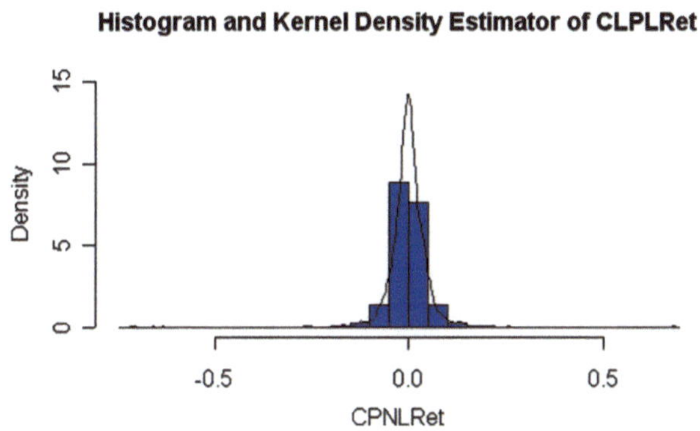

Fig. 1.18. Histogram and kernel density estimate of the daily log returns on the Calpine stock

vertical axis by setting the parameter `ylim` to guarantee that the graph of the kernel density estimate will not leak out of the plot.

Figure 1.18 illustrates perfectly the advantages and the limitations of the kernel density estimator. First and most importantly, it produces a smooth curve which has a more pleasing look than the histogram. However, it has the same problem with the extreme observations: they are poorly accounted for, especially if the bandwidth is too large. Indeed, too large a bandwidth can smooth features out of the picture!

The kernel functions $K(x)$ used in the kernel density estimation function `density` is specified by a string of characters assigned to the parameter `kernel`. There are seven possible choices. We give the plots of four of these possible kernel functions in Fig. 1.19. The interested reader is invited to look at the help file of the function `density` to find out about the other three options.

1.2.4.2 Comparison with the Histogram

Both histogram and kernel estimates start from a data sample $x_1, x_2, \ldots, x_n$. In order to construct a histogram, we choose an origin and n bins, and we define the histogram as the graph of the function:

$$x \hookrightarrow \text{Hist}(x) = \frac{1}{n} \sum_{i=1}^{n} \theta(x, x_i) \tag{1.29}$$

where $\theta(x, x_i) = 1/b$ if x and x_i belong to the same bin, and 0 otherwise (remember that we use the notation b for the width of the bins.) Notice that definition (1.28) of

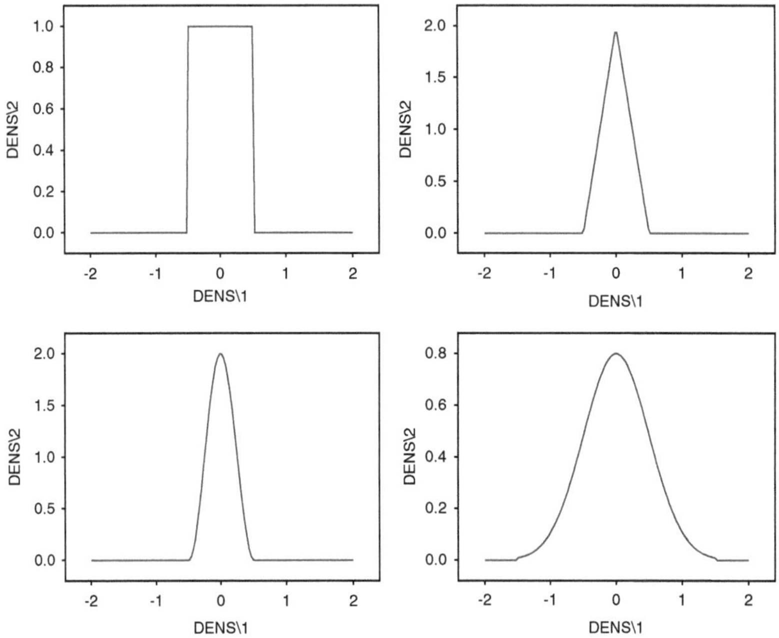

Fig. 1.19. Graphs of four of the kernel functions used by the R function `density`. *Top left*: `rectangular` kernel. *Top right*: `triangular` kernel. *Bottom left*: `cosine` kernel. *Bottom right*: `gaussian` kernel

the kernel density estimator has exactly the same form as the re-formulation (1.29) of the definition of the histogram, provided we re-define the function $\theta(x, x_i)$ by:

$$\theta(x, x_i) = \frac{1}{b} K \left(\frac{x - x_i}{b} \right).$$

The similarity is striking. Nevertheless there are fundamental differences between these two nonparametric density estimation methods.

- *First, at the level of the "bumps":* they are rectangular for the histogram, while their shape is given by the graph of the kernel function $K(x)$ rescaled by b in the case of the kernel estimate;
- *At the level of the locations of the bumps:* these locations are fixed more or less independently of the data for the histogram, while they are centered around the data points in the case of the kernel estimate;
- *At the level of the smoothness of the resulting density estimate:* It is determined by the number (and size) of the bins in the case of the histogram while it is determined by the value of the bandwidth $b > 0$ in the case of the kernel estimate.

1.2.4.3 ***Still Another Example***

The goal of this subsection is to review the two methods of nonparametric density estimation discussed in this chapter by testing them on another data set. We choose the S&P 500 index for purpose of illustration. The data comprise the series of daily closing values of the S&P 500 index over the period ranging from January 3, 1950 to August 20, 2010. Figure 1.20 gives a sequential plot of these values, together with the

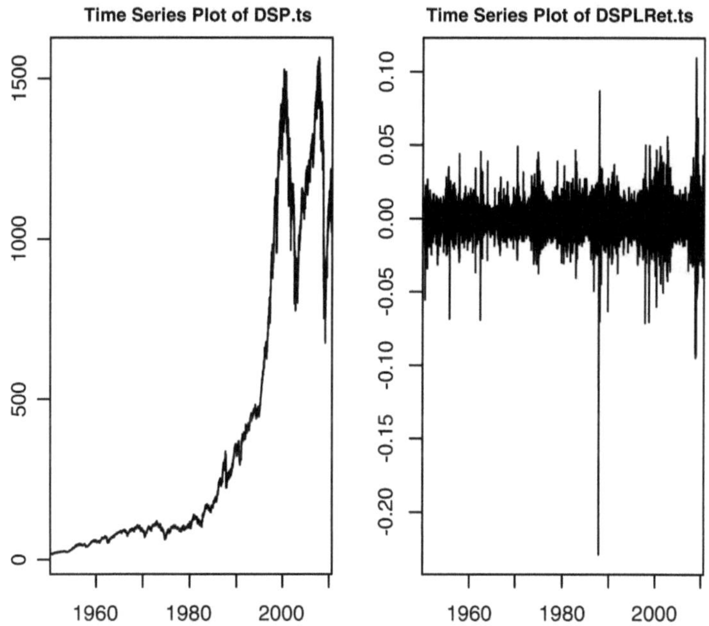

Fig. 1.20. Daily S&P 500 closing prices as imported from the Web (*left*) and daily log-returns over the same period (*right*)

series of the daily log-returns. Notice that in computing the returns, we ignored the fact that the returns are actually computed over periods of different lengths. Indeed we ignore the presence of weekends, typically when Monday's close follows the preceding Friday's close, and we use the quotes as if they were from consecutive days. We also ignored holidays and days the market was closed, like the week of September 11, 2001 of the terrorist attack on New York.

The other important point which needs to be made at this stage is that we are not interested in the time evolution of the values of the series, but instead, in the distribution of these values on a given day. In other words, instead of worrying about how the value on a given day depends upon the values on previous days, we care about statistics which would not change should we modify the order of the entries of the data vector. This seems to be quite a reasonable requirement in the case of the log-returns of the right pane of Fig. 1.20. We use the commands:

```
> hist(DSPLRet,col="blue",freq=F,ylim=c(0,30))
> DENS <- density(DSPLRet, adjust=12)
> lines(DENS)
```

to produce Fig. 1.21. As before, we use the option `freq=F` to force the area of the histogram to be one, so it will be on the same vertical scale as the kernel density estimate computed next. Also, we set the parameter `ylim` to force the limits on the vertical axis to be 0 and 30 so as to make sure that the kernel density estimate will be plotted inside the plot area. The parameter `adjust`, whose default value is 1, is a multiplicative factor which is used to modify the bandwidth. The bandwidth used by the function `density` is actually `adjust` times the default bandwidth. We set it to 12 because the result of the density estimation with the default bandwidth was producing a curve too peaked around 0. Notice also that we

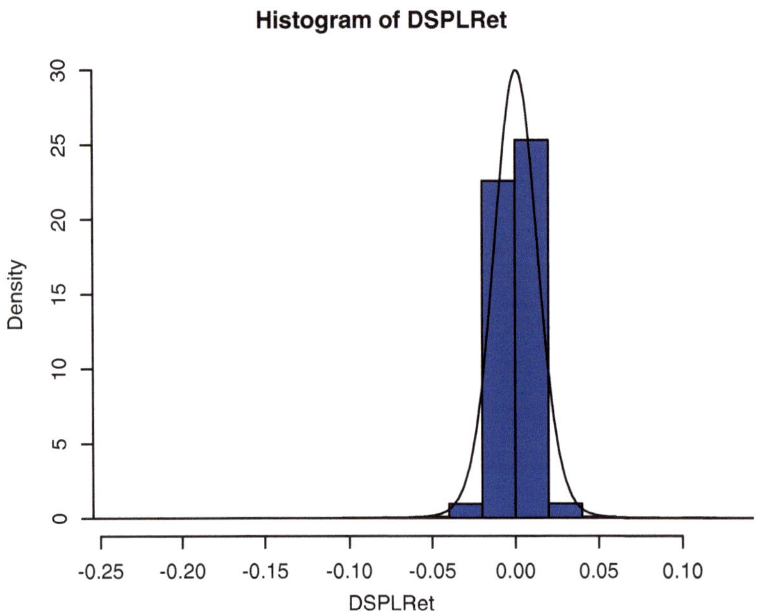

Fig. 1.21. Histogram and kernel density estimates of the daily log-returns of the S&P 500

used the command `lines(DENS)` instead of `lines(DENS$x,DENS$y)` which we used before.They are equivalent because `DENS` has two components, `DEN$x` and `DENS$y`.

The histogram reproduced in Fig. 1.21 is not satisfactory for two related reasons. First it gives very little information about the center of the distribution, the two central bars capturing too much of the information. Second, the extreme values to the left and to the right of the distribution do not appear because they are in such small numbers that the heights of the bars they create are too small compared to the heights of the central bars, for the former to be visible. Because of this shortcoming, we decided to use the kernel method as an alternative to the histogram, but despite a smoother look, the same problem plagues the estimate: the contributions of both ends of the distribution (which we call tails) to the graph are overwhelmed by the central part of the distribution.

1.2.4.4 ***Importance of the Choice of the Bandwidth***

The choice of bandwidth can have drastic consequences on the final look of the density estimate. Figure 1.22 shows two kernel density estimates for the same sample of daily log-returns of the S&P 500. The left pane was obtained with a bandwidth equal to 1 while the right pane was produced with a bandwidth equal to 0.1. The results look very different. Notice that in order to allow for a meaningful comparison, we forced identical scales on the vertical axes of the two panes. This ensures that the observed differences are not due to an artifact from the choice of axis scales. Indeed, the plots of these two density estimates would look almost identical if we were to let the program adjust the limits of the axes to the actual values of the functions. The strong influence of the value of the bandwidth will also be emphasized in Problem 1.2.

In any case, both histograms and kernel density estimators are unable to give a good account of the attributes of the tails. We appeal to other graphical tools to exhibit the tail properties in a more suggestive way.

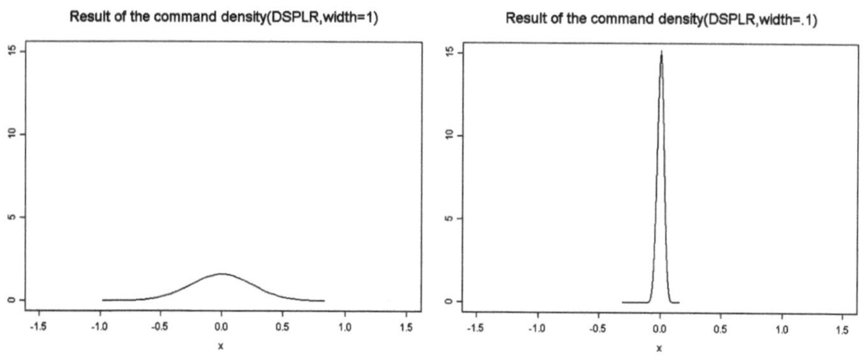

Fig. 1.22. Kernel density estimates of the daily log-returns of the S&P 500 produced by the R function `density` with bandwidths computed by setting the parameter `window` to 1 (*left*) and 0.1 (*right*)

1.2.5 Empirical Q-Q Plots

The purpose of this subsection is to show how the properties of the theoretical Q-Q plots introduced earlier can be used to infer tail properties of the distributions from which we have sample observations. This is done by producing a Q-Q plot of the empirical distribution of the data sample against a given theoretical distribution. Figure 1.23 shows two such Q-Q plots. It was produced with the commands:

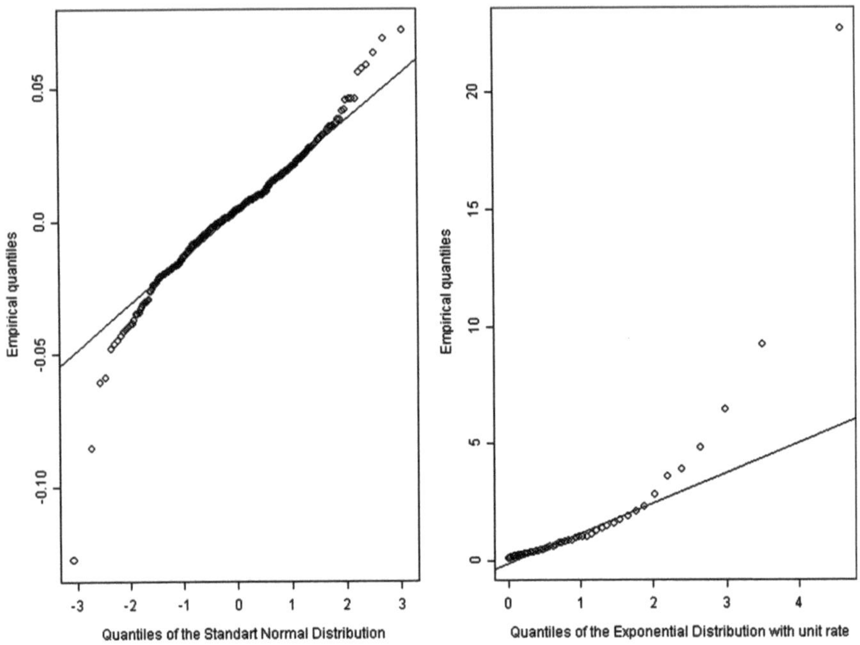

Fig. 1.23. Examples of Q-Q plots: normal Q-Q plot of the weekly log-returns of the S&P 500 (*left*) and exponential Q-Q plot of the PCS index (*right*)

```
> par(mfrow=c(1,2))
> qqnorm(WSPLRet)
> qqexp(PCS.index)
> par(mfrow=c(1,1))
```

The plot on the right shows the empirical percentiles of the PCS index data against the theoretical percentiles of the exponential distribution with unit rate. This plot was produced by the function `qqexp` which we wrote to this effect. On top of the points whose coordinates are the respective percentiles, it shows a line on which all these points would be found should the distribution of the data sample be exponential. The slope of this line should be the mean of the distribution (i.e. the inverse of the rate). Indeed, one sees from formula (1.19) that

$$\pi_q^{(r)} = \frac{1}{r}\pi_q^{(1)}$$

if we denote by $\pi_q^{(r)}$ the q-percentile of the exponential distribution $E(r)$ with rate r. So if the quantiles on the vertical axis are the quantiles of the distribution $E(r)$ and the quantiles on the horizontal axis are the quantiles of the distribution $E(1)$ with unit rate, the corresponding points in the plane should be on the straight line with slope $1/r$ which is equal to the mean of the distribution $E(r)$. So, the fact that the points in the rightmost part of the figure are above the line indicates that the right tail of the distribution is *thicker* than the tail of the exponential distribution. The plot on the left is for the weekly log-returns on the S&P 500 index. It uses the function `qqnorm` which we modified to add to the standard `R` function `qqnorm`, the plot of the line whose slope and intercept determine the parameters of the Gaussian distribution, should all the points sit on this line. This plot shows that both tails are heavier than the tails of the normal distribution.

Q-Q plots are very useful to detect heavy tails. We shall make an extensive use of Q-Q plots in our study of heavy tail distributions.

1.3 Monte Carlo Computations

This final section gives an introduction to the general principles of random number generation, and the basics of Monte Carlo computations of probabilities and expectations. This *crash course* is necessary because of the special emphasis of the book on the many modern data analysis computational techniques which rely on random simulations.

1.3.1 Generating Random Samples in R

Recall that by a random sample of size n we mean a set of n realizations $x_1, \cdots, x_n$ of independent random variables $X_1, \cdots, X_n$ with the same distribution. Notice also the fact that we use upper cases for random variables and lower cases for actual realizations. Finally, we often talk about a sample *from* a distribution, to specify the common distribution of the random variables, and we call it a white noise when this common distribution has mean 0.

Examples of `R` commands producing samples from some of the classical probability distribution families introduced earlier are given in the second column of Table 1.2. We refer the reader to the examples used in the text and to the help files of these `R` functions for details on the names and the meanings of the parameters of these functions. For the sake of convenience, we also add columns for the `R` commands to evaluate their densities, their cdf's and their quantiles. As we already emphasized, the rationale behind this terminology is easy to remember: a short name for the distribution family follows a r for random samples, a d for density, a p for probability (giving the values of the cdf) and a q for quantile.

Distribution	Random samples	Density	cdf	Quantiles
Uniform distribution	`runif`	`dunif`	`punif`	`qunif`
Univariate normal distribution	`rnorm`	`dnorm`	`pnorm`	`qnorm`
Exponential distribution	`rexp`	`dexp`	`pexp`	`qexp`
Log-normal distribution	`rlnorm`	`dlnorm`	`plnorm`	`qlnorm`
Student t distribution	`rt`	`dt`	`pt`	`qt`
Cauchy distribution	`rcauchy`	`dcauchy`	`pcauchy`	`qcauchy`

Table 1.2. `R` commands for the manipulation of the classical probability distributions

We already used some of these functions when we discussed white noise vectors in the introduction to `R` given in the appendix, and showed how to produce plots of density functions or compare distributions from their theoretical Q-Q plots. As another illustration, we now generate and plot samples from the Gaussian and Cauchy distributions. We use the commands:

```
> GWN <- rnorm(1024)
> CWN <- rcauchy(1024)
> par(mfrow=c(2,1))
> plot(GWN, type="l")
> title("Sequential Plot of a Standard Gaussian Sample")
> plot(CWN, type="l")
> title("Sequential Plot of a Standard Cauchy Sample")
> par(mfrow=c(1,1))
```

The command `rnorm(1024)` creates a vector of length 1,024 whose entries form a sample of size $n = 1{,}024$ from the standard normal distribution. Indeed, since we did not specify the parameters `mean` and `sd`, the function `rnorm` used the default values which are 0 and 1 respectively. Similarly, the command `rcauchy(1024)` creates a vector of length 1,024 whose entries form a sample of size $n = 1{,}024$ from the standard Cauchy distribution. Indeed, since we did not specify the parameters `location` and `scale`, the function `rcauchy` used the default values of 0 and 1. The corresponding plot is given in Fig. 1.24. These two samples look very different. Indeed the Cauchy distribution produces positive and negative numbers with very large absolute values while the values in the Gaussian sample seem to remain between -3 and $+3$ in line with our earlier discussion. To better understand the huge differences between these two plots, we first need to notice the differences in scales on the vertical axes. The relative size of the extreme values in the Cauchy sample forces the bulk of the other points to be crammed together, giving the false impression that they are trying to line up along the horizontal axis!!

Since we claim that the most significant differences are in the tails of the distributions, we use a Q-Q plot of the empirical distributions of the two samples in order to make our point. It is given in Fig. 1.25. The discrepancy between the ranges of the two samples is reflected by the fact that the line representing the diagonal is essentially horizontal. Moreover, the points on the right and left most parts of the plot depart from this line, showing that the tails of the distribution implied by the

Cauchy sample are much thicker than the tails of the distribution implied by the Gaussian sample, confirming the thrust of our discussion of the Q-Q plots of these distributions.

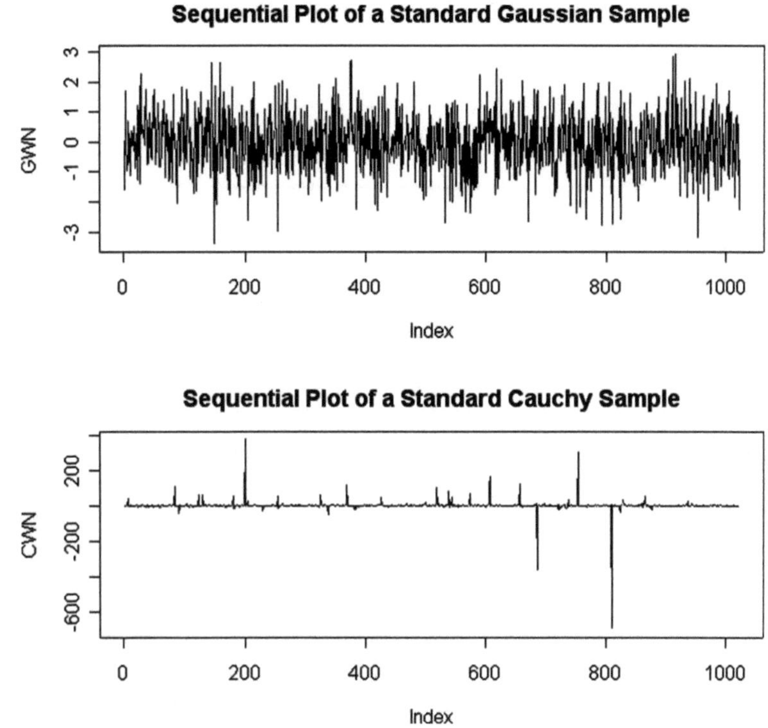

Fig. 1.24. Sequential plot of 1,024 i.i.d. samples from the Gaussian distribution $N(0,1)$ (*top*) and the Cauchy distribution $C(0,1)$ (*bottom*)

Remarks. Sequences of mean zero independent and identically distributed (i.i.d. for short) random variables are used extensively in the context of time series where they are called white noise sequences. This terminology explains our use of the names `GWN` and `CWN` for Gaussian white noise and Cauchy white noise respectively.

1.3.2 Limit Theorems and Monte Carlo Computations

In this section, we explain the relevance of the two most fundamental limit theorems of the calculus of probability to Monte Carlo computations.

1.3.2.1 The Law of Large Numbers (LLN)

The law of large numbers states that, if $Y, Y_1, Y_2, \cdots\cdots$ is a sequence of independent random variables of order one with the same probability distribution, then for almost all realizations, the following limit holds true

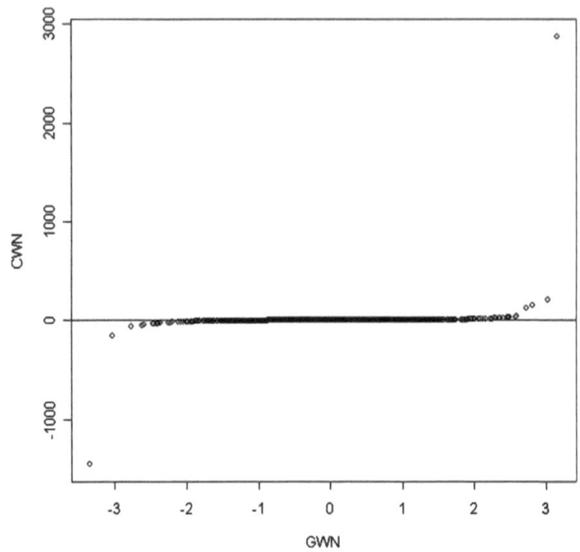

Fig. 1.25. Empirical Q-Q plot of the sample of size 1,024 from the Cauchy distribution $C(0,1)$ against the sample of size 1,024 from the Gaussian distribution $N(0,1)$ which were plotted sequentially in Fig. 1.24

$$\lim_{n\to\infty} \frac{1}{n}[Y_1 + \cdots + Y_n] = \mathbb{E}\{Y\}. \tag{1.30}$$

Recall that a random variable Y is said to be of order 1 if $\mathbb{E}\{|Y|\} < \infty$, in which case the expectation $\mathbb{E}\{Y\}$ exists. Remember also that even though this is the case for essentially all the distributions we encounter in this book, it is not the case of the Cauchy distribution. Consequently, if

$$y_1, y_2, \cdots, y_n$$

is a sample from a probability distribution with cdf $F(y)$, and if this distribution has a first moment in the sense that

$$\int |y|\, dF(y) < \infty$$

then the sample average

$$\overline{y} = \frac{y_1 + y_2 + \cdots + y_n}{n} = \frac{1}{n}\sum_{j=1}^{n} y_j$$

can be used as an approximation of the mean μ of the distribution

$$\mu = \int y\, dF(y),$$

and the larger the sample size n, the better the approximation. We can think of $y_1, y_2, \cdots, y_n$ as realizations of the first n random variables of an infinite sequence $Y_1, Y_2, \cdots, Y_n, \cdots$, of independent random variables with a common cdf F, μ being its expectation. Then the theorem explained in words can be stated mathematically as:

$$\lim_{n\to\infty} \frac{Y_1 + Y_2 + \cdots + Y_n}{n} = \mu,$$

the only thing missing being a formal definition for the convergence of the random variables in the left hand side toward the number appearing in the right hand side.

1.3.2.2 Computations of Probabilities and Expectations

One of the most pervasive use of the law of large numbers is the following. Let us imagine that we need to produce a numerical value for an expectation $\mathbb{E}\{\psi(X)\}$, or a probability of the form $\mathbb{P}\{\psi(X) \geq \alpha\}$ for a given number α, where X is a random variable (possibly multivariate) and ψ is a real valued function defined on the range of X. In financial applications, numerical values of expectations are often needed in pricing problems, and as we saw earlier in the chapter, probabilities enter measures of risk computations. Notice that, if one uses the fact that the probability of an event is equal to the expectation of the random variable which is equal to 1 when the event occurs and 0 otherwise, replacing the function ψ by another, one easily reduces the computation of such a probability to the computation of an expectation. Indeed, if we define the function $\tilde{\psi}$ by

$$\tilde{\psi}(x) = \begin{cases} 1 & \text{if } \psi(x) \geq \alpha \\ 0 & \text{if } \psi(x) < \alpha \end{cases}$$

then clearly, $\mathbb{P}\{\psi(X) \leq \alpha\} = \mathbb{E}\{\tilde{\psi}(X)\}$. So at the theoretical level, we only discuss Monte Carlo approximations of expected values, even though we apply the theoretical results to the computations of probabilities as well as expectations. Given a sample $x_1, x_2, \cdots, x_n$ from the distribution of X, one obtains a sample from the distribution of the random variable $Y = \psi(X)$ by evaluating the function ψ on the X-sample values:

$$y_1 = \psi(x_1), y_2 = \psi(x_2), \cdots, y_n = \psi(x_n)$$

and the Law of Large Numbers tells us that the number

$$\frac{\psi(x_1) + \psi(x_2) + \cdots + \psi(x_n)}{n} = \frac{1}{n}\sum_{j=1}^{n} y_j$$

is an approximation of the desired expectation $\mathbb{E}\{\psi(X)\}$, the larger the sample size n, the better the approximation. Applying this result to the function $\tilde{\psi}$ introduced above, we see that

$$\frac{\tilde{\psi}(x_1) + \tilde{\psi}(x_2) + \cdots + \tilde{\psi}(x_n)}{n}$$

provides an approximation of the probability $\mathbb{P}\{\psi(X) \le \alpha\}$, but since $\tilde{\psi}(x_j)$ is either 0 or 1, the sum in the above numerator is merely equal to the number of ones in the sum, and by definition of $\tilde{\psi}$, this is the number of observations x_j's for which $\psi(x_j) \ge \alpha$. So the Monte Carlo approximation of the probability $\mathbb{P}\{\psi(X) \le \alpha\}$ is given by the proportion of the sample $x_1, x_2, \cdots, x_n$ for which the condition $\psi(x_j) \ge \alpha$ is satisfied. In other words, the Monte Carlo approximation of the probability of an event is given by the relative frequency with which the event occurs in the sample.

1.3.2.3 ***Pricing a Call Option***

Let us assume that X represents the value at a future date T of an asset, and let us assume that we need to value a European call option with maturity T and strike K written on this underlying asset. Then, ignoring discounting issues for the sake of simplicity (say that the interest rates are 0 for the purpose of this illustration), according to Black-Scholes theory, the price of the option is given by the expectation $\mathbb{E}\{\psi(X)\}$ for the particular function ψ defined by $\psi(x) = \max\{x - K, 0\}$ (namely the payoff function of the option) provided we use a risk neutral, or risk adjusted expectation. Hence the price $C_{T,K}$ of the call option is given by the integral

$$C_{T,K} = \int \max\{x - K, 0\}\, f_X^{(rn)}(x)\, dx \tag{1.31}$$

where we use the notation $f_X^{(rn)}(x)$ for the risk neutral density of the value of the asset at maturity T. The classical Black-Scholes formula derived in the appendix at the end of the book is nothing but an expression for the value of this integral when $f_X^{(rn)}$ is the density of a log-normal distribution. The lognormal case is one of a handful of models for which one can derive an explicit formula for option prices, hence its popularity. Unfortunately, this is not the case for most of the financial models we would like to use. However, most financial models are amenable to Monte Carlo computations when generating samples from the risk neutral density is feasible. For this reason, the option price is often inferred from the Monte Carlo approximation

$$C_{T,K} \approx \frac{1}{n}[\max\{x_1 - K, 0\} + \cdots + \max\{x_n - K, 0\}] \tag{1.32}$$

computed from a random sample $x_1, x_2, \cdots, x_n$ from the risk neutral distribution of the asset price underlying the option. We shall use this Monte Carlo approximation many times in the sequel.

1.3.2.4 ***A Numerical Example in*** `R`

We now implement the computation of the Monte Carlo approximation described above in `R`. On Thursday January 27, 2005 the value of Calpine stock was $S = 3.36$

USD and the short interest rate was $r = 2.4\,\%$. What was the price on that day of a 2 weeks at-the-money European call option priced at $\sigma = 60\,\%$ volatility? An option is at the money when its strike is equal to the current value of the underlying stock, in other words when $S = K$. We have all the information needed to use Black-Scholes pricing formula. For the sake of comparison, we use the *home-grown* R function `bscall` whose code is explained in the appendix at the end of the book, to compute the price given by the Black-Scholes formula.

```
> C <- bscall(TAU=0.04, K=3.36, S=3.36, R=0.024, SIG=.6)
> C
[1] 0.1622971
```

We now use the algorithm described above to compute a Monte Carlo approximation to this exact Black-Scholes price. If we denote by X the value of the underlying stock at maturity (S_T in the notation used in the appendix at the end of the book), according to the theory reviewed in the appendix, the risk neutral distribution of X is the log-normal distribution with mean $\log S+(r-\sigma^2/2)\tau$ and variance $\tau\sigma^2$, and the price of the option is given by the risk neutral expectation of the discounted expected pay-off $e^{-r\tau}\max\{S-K,0\}$.

```
> TAU <- 0.04; K <- 3.36; S <- 3.36; R <- 0.024; SIG <- 0.6
> N <- 10000
> ML <- log(S) + TAU*(R-SIG^2/2)
> SL <- SIG*sqrt(TAU)
> XX <- rlnorm(N, meanlog=ML, sdlog=SL)
> PSIX <- pmax(XX-K,0)
> MCCall <- exp(-R*TAU)*mean(PSIX)
> MCCall
[1] 0.1623504
```

which is a reasonable approximation since the relative error is only of the order of 0.009.

1.3.2.5 Probability That an Option Will Eventually Be Exercised

Let us consider once more the case of a European call with strike K, but instead of worrying about its Black-Scholes price, let us compute the probability that the option will be exercised at maturity. Assuming that the owner of the option acts rationally, the option is exercised when and only when the value at maturity of the underlying asset, say X, is greater than the strike K. Hence, we need to evaluate the probability $\mathbb{P}\{X > K\}$ as we described earlier.

NB: A very important remark is in order at this stage: because we are not trying to price a derivative, we do not need to use risk neutral or risk adjusted probabilities. Instead, we need to work with real world probabilities also called *objective* probabilities. So when we generate the sample $x_1, \cdots, x_n$, we need to reproduce the historical statistics of the value of X. In other words, we need to generate a sample from the historical distribution of X.

1.3.2.6 **A Numerical Example in** R

We implement in R the second of the two applications of the Monte Carlo principle described above, namely the computation of the probability that an European call option will be exercised at maturity.

We work in the framework of Samuelson's model in which stocks dynamics are given by geometric Brownian motions. Black-Scholes option pricing theory was developed in this context. According to this model, at any time, the value of the stock is log-normally distributed, i.e. its logarithm has a normal distribution. More precisely, the distribution of its logarithm is

$$N\left(\log S + \tau(\mu - \frac{1}{2}\sigma^2), \sigma^2\tau\right)$$

where μ denotes the rate of growth of the stock, and σ the historical return volatility, i.e. the standard deviation of the log-returns. For the sake of comparison, we work with the same example as before. We compute the probability on Thursday January 27, 2005, when Calpine's stock was valued at $S = 3.36$ USD, that the option will be exercised when it matures, and we consider the same at-the-money European call option (i.e. with striket $K = S$) with time to maturity $\tau = 0.04$. In the present situation, we do not need to know the value the short interest rate, but we need to know the historical rate of return μ and the historical volatility σ. According to Samuelson's model, Calpine's stock value at maturity is a random variable say X, with a log-normal distribution with mean $\log S + (\mu - \sigma^2/2)\tau$ and variance $\sigma^2\tau$, and if Z is any standard normal random variable, the desired probability is given by:

$$\begin{aligned}\mathbb{P}\{X > K\} &= \mathbb{P}\{\log X > \log K\} \\ &= \mathbb{P}\{\log S + (\mu - \sigma^2/2)\tau + \sigma\sqrt{\tau}Z \ > \ \log K\} \\ &= \Phi\left(\frac{\log(S/K) + (\mu - \sigma^2/2)\tau}{\sigma\sqrt{\tau}}\right).\end{aligned} \tag{1.33}$$

Whatever numerical method we choose for the computation of this probability (direct evaluation using the cdf Φ or Monte Carlo approximation), we need estimates of the rate of growth μ and the historical volatility σ in order to get a numerical result from the above formula.

Estimation of the Parameters. In theory, the observation of a single trajectory (realization) of the price should be enough to completely determine the value of the volatility σ. This would be true if the price process S_t could be observed continuously! Unfortunately, this cannot be the case in practice and we have to settle for an approximation. Given observations S_{t_j} of past values of the risky asset (usually the times t_j are of the form $t_j = t - j\delta t$), we use the fact that in Samuelson's model the random variables $\log(S_{t_j}/S_{t_{j+1}})$ are independent and normally distributed with mean $(\mu - \sigma^2/2)\delta t$ and variance $\sigma^2\delta t$. Consequently, the volatility can be estimated by the formula:

$$\hat{\sigma} = \sqrt{\frac{1}{(N-1)\delta t}\sum_{j=0}^{N-1}[\log\frac{S_{t_j}}{S_{t_{j+1}}} - \overline{LS}]^2}, \tag{1.34}$$

where $\overline{LS}$ is the sample mean daily log-return

$$\overline{LS} = \frac{1}{N}\sum_{j=0}^{N-1}\log\frac{S_{t_j}}{S_{t_{j+1}}}.$$

The volatility estimate provided by formula (1.34) is called the historical volatility. Even though the corresponding estimate of the variance is unbiased, $\hat{\sigma}$ has a bias due to the nonlinearity of the square root function. $\overline{LS}$ provides an estimate of the rate of growth μ. It reads:

$$\hat{\mu} = \frac{1}{\delta t}\overline{LS} + \frac{\hat{\sigma}^2}{2}. \tag{1.35}$$

The Equity Premium Puzzle. The above discussion can be very misleading as it seems to imply that the parameters μ and σ can be easily estimated in a very natural manner. However, estimates of the volatility σ are usually much better than estimates of the rate of return μ. This fact has been amply documented in the econometrics literature, and is known under the name of the Equity Premium Puzzle (EPP for short).

We shall momentarily ignore this important issue as the main concern of this chapter is Monte Carlo computations. For the sake of illustration, we use the data of the Calpine daily log returns already massaged earlier. We get:

```
> LN <- mean(CPNLRet)
> DELTAT <- 1/252
> SIGHAT <- sqrt(var(CPNLRet)/DELTAT)
> MUHAT <- LN/DELTAT + SIGHAT^2/2
> MUHAT
[1] 0.1408325
> SIGHAT
[1] 0.853481
```

In order to compute the probability that the option is going to be exercised, we first consider the direct method of computation based on an implementation of the computation of the Gaussian cdf Φ. Plugging these estimates of μ and σ into formula (1.33) we get:

```
> TAU <- 0.04
> S <- 3.36
> K <- S
> MN <- log(S/K)
> PROBA <- pnorm((MN + TAU*LN/DELTAT)/(SIGHAT*sqrt(TAU)))
> PROBA
[1] 0.4791264
```

So the probability that the owner of the option ends up exercising her option at maturity is essentially $48\,\%$. This should not be too surprising for an option at the money and relatively short time to maturity. We should expect a small (resp. large) probability when the option is out of (resp. in) the money. A call option is said to be out of the money (resp. in the money) if its strike is higher (resp. smaller) than the current value of the underlying stock, in other words when $S < K$ (resp. $S > K$).

Monte Carlo Computation of the same Probability. The parametric model used in the Samuelson approach leads to formula (1.33) for the expression of the desired probability. In this case, a Monte Carlo computation is not really needed. We nevertheless explain how to compute a Monte Carlo approximation of this probability in order to illustrate the procedure as it can be used in many instances where one does not have the luxury of a closed form formula. We first produce a sample from the distribution of the stock at maturity.

```
> N <- 10000
> ML <- log(S) + TAU*LN/DELTAT
> SL <- SIGHAT*sqrt(TAU)
> X <- rlnorm(N, meanlog=ML, sdlog=SL)
> Y <- X>K
> MCPROBA <- mean(Y)
> MCPROBA
[1] 0.4821
```

The first command chooses the number `N` of Monte Carlo samples we plan to use in order to compute the Monte Carlo approximation. The next two commands set the mean `ML` and the standard deviation `SL` of the log-normal distribution. The next command produces a numerical vector `X` of length `N` containing the sample realizations of the log-normal random variable modeling the stock price at maturity. The value of `X>K` is a boolean vector with the same length as `X`. Its entries are equal to `TRUE` (which is automatically coerced one to the number 1 when needed) when the condition `X>K` is satisfied, and to `FALSE` (automatically coerced to 0) otherwise. Finally we compute the desired probability as the mean of this boolean vector, i.e. the proportion of entries of `X` satisfying the boolean condition. The relative error of the Monte Carlo approximation obtained with this run is $0.6\,\%$.

The following computations illustrate in the present situation, the convergence of the Monte Carlo approximations as predicted by the law of large numbers.

```
> NMAX <- 100000
> X <- rlnorm(NMAX, meanlog=ML, sdlog=SL)
> Y <- X>K
> CS <- cumsum(Y)
> MCAPPROXS <- CS/(1:NMAX)
> plot(MCAPPROXS,type="l")
> abline(h=PROBA)
> title("Convergence of Monte Carlo Approximations")
```

We construct a large random sample of log-normal random variables as before. We chose to have 100,000 sample realizations for this experiment. We also construct the boolean vector Y as before. The difference is that we now try to compute the Monte Carlo approximation produced by the first n Monte Carlo samples, for each $n \leq 100{,}000$, In order to do so, we create a vector CS which has the same length as Y, and whose n-th entry is the sum of the first n entries of the vector Y. This is exactly what the R function cumsum does (cumsum is short for cumulative sum). The remaining task is to divide the n-th entry of this vector CS by n in order to get the average of the first n entries of Y, providing in this way, the Monte Carlo approximation based on the first n samples. This is done by dividing the vector CS by the vector 1:NMAX of the first NMAX integers, entry by entry. Finally, note that the command abline is used to add a horizontal line to emphasize the true value of the probability: convergence of the approximations means that this horizontal line should be an asymptote.

The results are given in Fig. 1.26. The plot confirms the convergence at the same time that it shows that a residual error can persist for quite a long time.

NB. A Monte Carlo approximation is only an approximation of the true value we are trying to compute. So it is important to have an idea of the error in question. But contrary to most of the approximations offered by standard numerical algorithms, Monte Carlo approximations are random. As such, they are different each time we compute them: they depend upon the seed of the random number generator used to produce the random samples. Hence, quantifying the error will have to be done in a statistical sense. We tackle this problem next.

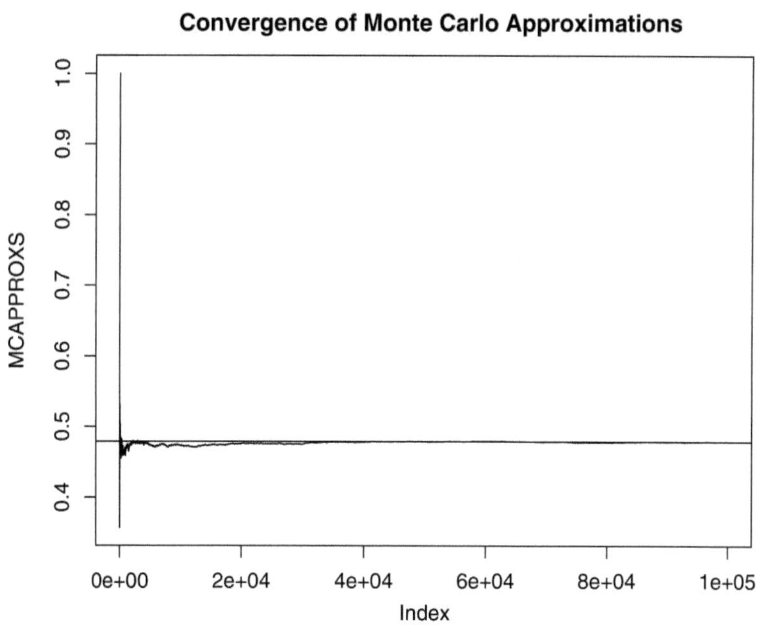

Fig. 1.26. Sequential plot of the Monte Carlo approximations of the probability that an option is exercised as a function of the number of Monte Carlo samples used. The *horizontal line* gives the true value of the probability

1.3.2.7 **The Central Limit Theorem (CLT)**

The law of large numbers tells us that the sample mean converges toward the true mean. The central limit theorem can be viewed as a successful attempt to quantify the rate of convergence. Indeed, the central limit theorem identifies (at least asymptotically) the statistical distribution of the fluctuations around the true mean. In order to describe this result precisely, we use the set up and the notation of Sect. 1.3.2.1 above, and we assume further that the random variables Y_j have a moment of order 2, namely that $\mathbb{E}\{Y_j^2\} < \infty$ and hence that their variances exist. We denote by $\mu = \mathbb{E}\{Y\}$ and $\sigma^2 = \mathbb{E}\{(Y-\mu)^2\}$ the common mean and variance of the variables $Y_1, Y_2. \cdots, Y_n$ in the sample. The central limit theorem states that the distribution of the random variable

$$\frac{Y_1 + \cdots + Y_n - n\mu}{\sigma\sqrt{n}} \tag{1.36}$$

converges toward the standard Gaussian distribution. In order to understand this result as information on the rate of convergence, we rewrite its conclusion in the more intuitive form:

$$\frac{Y_1 + \cdots + Y_n}{n} - \mu \approx \frac{\sigma}{\sqrt{n}} Z$$

where Z is a standard Gaussian random variable, i.e. $Z \sim N(0,1)$. Stated in this form, the relevance of the central limit theorem to Monte Carlo computations is clear. As predicted by the strong law of large numbers, the error of the Monte Carlo approximation goes to zero when the sample size n increases without bound. However, we can derive much more information from the above computation. Indeed, we learn that the rate of convergence is of the order $n^{-1/2}$. So the size of the error can be controlled by the choice of a large sample size. Furthermore, even though the size of Z cannot be predicted, after all it is a random variable, we know that it will be typically in the range $[-3, +3]$, and the only remaining way to control the error is to lower the variance of Y. Because of the increasing importance of Monte Carlo computations in modern scientific computing, entire research programs are devoted to variance reduction algorithms. We list the names of some of the most commonly used techniques.

- Stratified sampling
- Antithetic variables
- Control variates
- Importance sampling

Discussing any of these topics would take us far beyond the scope of this book. The interested reader should consult the Notes & Complements at the end of the chapter for bibliographical references on these variance reduction techniques.

1.3.3 **Home Grown Random Samples**

All scientific computing environments (`R` is no exception) come equipped with a random number generator for the uniform distribution $U(0,1)$. We shall use this generator as a building block. In fact, R comes with commands for the generation of samples from the most common parametric families of distributions. We reviewed some of them earlier. However, realistic applications require Monte Carlo simulations of quantities with more general (and less generic) statistical properties. We now review the elementary facts of the theory of probability which underpin the construction of random number generators, setting the stage for the development of more sophisticated tools needed for Monte Carlo computations. The following simple mathematical facts are fundamental to the discussion of the section.

Fact 1 Given a random variable X with cdf F_X, the random variable $U = F_X(X)$ is uniformly distributed in the interval $[0, 1]$

This is consistent with the fact that the cdf F_X can only take values between 0 and 1. The fact that the values of $F_X(X)$ are uniformly distributed is due precisely to the definition of the cdf. Indeed, if $0 \leq u \leq 1$, we have:

$$\mathbb{P}\{U \leq u\} = \mathbb{P}\{F_X(X) \leq u\} = \mathbb{P}\{X \leq F_X^{-1}(u)\} = F_X(F_X^{-1}(u)) = u. \quad (1.37)$$

This argument is perfectly rigorous when the cdf F_X is strictly increasing (and hence invertible). A little *song and dance* is needed to make the argument mathematically sound in the general case, but we shall not worry about such details here. So we proved that:

$$F_X \text{ cdf of } X \Longrightarrow U = F_X(X) \sim U(0, 1). \quad (1.38)$$

This result has a very important consequence in terms of random samples. Indeed, it implies that, if $x_1, \cdots, x_n$ is a sample from the cdf F, then, the sample $u_1, \cdots, u_n$ defined by

$$u_1 = F(x_1), \cdots, x_n = F(x_n)$$

is a sample from the uniform distribution $U(0, 1)$! We will use this fact repeatedly in the sequel.
The above simple mathematical result stated as **Fact 1** has a far-reaching converse. Indeed, reading (1.37) from right to left we get:

Fact 2 If $U \sim U(0, 1)$ and F is a cdf, then if we define the random variable X by $X = F^{-1}(U)$ we necessarily have $F_X = F$.

Indeed:

$$\mathbb{P}\{X \leq x\} = \mathbb{P}\{F^{-1}(U) \leq x\} = \mathbb{P}\{U \leq F(x)\} = F(x). \quad (1.39)$$

Consequently, if we want to generate a sample of size n from a distribution for which we do not have a random number generator, but for which we can compute the inverse F^{-1} of the cdf F, we only need to generate a sample from the uniform distribution, say

$$u_1, u_2, \ldots\ldots, u_n$$

and then compute the inverse of the target cdf for each of these outcomes. Because of Fact 2, the sample

$$x_1 = F^{-1}(u_1), x_2 = F^{-1}(u_2), \ldots\ldots, x_n = F^{-1}(u_n),$$

is a typical sample from the distribution determined by the cdf F.

This very elementary fact is extremely useful, and we shall use it many times in this book, especially to simulate extreme events. Its only limitation lies in the actual numerical computation of the inverse cumulative distribution function F^{-1} also known as the quantile function. This method is not used to generate samples from the Gaussian distribution because the inversion of the cdf Φ is numerically too costly as compared to other existing methods which we will not discuss here. On the other hand, it is routinely used for the generation of exponential samples. See Problem 1.9 for details. We illustrate the details of this random generation procedure on the example of the Cauchy distribution.

Example. Recall that a random variable X is said to have a Cauchy distribution (with location parameter m and scale parameter λ) if X has density:

$$f_{m,\lambda}(x) = \frac{1}{\pi} \frac{\lambda}{\lambda^2 + (x-m)^2}, \qquad x \in \mathbb{R}$$

which was defined in (1.13). Its cdf

$$F_{m,\lambda}(x) = \frac{1}{\pi}\left[\tan^{-1}\frac{x-m}{\lambda} + \frac{1}{2}\right]$$

was already computed in formula (1.14). From this expression of the cdf, we compute the quantile function:

$$F^{-1}_{m,\lambda}(p) = m + \lambda \tan\left(p\pi - \frac{\pi}{2}\right)$$

already defined in (1.20), and consequently, the R command:

```
> CAUCHY <- M +LAMBDA*tan(PI*(runif(N)-.5))
```

will produce a vector CAUCHY of length N of Cauchy variates with location M and scale parameter LAMBDA. We shall not use this command in the sequel because as we explained earlier, R has a special command rcauchy for the generation of Cauchy random samples. The above exercise was motivated by our desire to *open the box* on how many random number generators are designed.

PROBLEMS

Ⓣ **Problem 1.1** *Let us assume that $F_1(x)$ and $F_2(x)$ are two cdf satisfying*

$$F_1(x) \le F_2(x) \qquad \textit{for all values of } x.$$

1. *Which of these two distributions has the heavier lower tail? Explain.*
2. *Which of these two distributions has the heavier upper tail? Explain.*

3. *If these two distributions are proposed as models for the returns of a given portfolio over the next month, and if you are asked to compute* $VaR_{0.01}$ *for this portfolio over that period, which of these two distributions will give the larger value at risk?*

Ⓔ **Problem 1.2** 1. *In* R*, generate a sample of size* $N = 1{,}024$ *from the exponential distribution with rate parameter* $r = 0.2$*. Call* X *the vector containing the sample values.*

2. *Plot on the same graph, the exact (theoretical) density of the distribution of* X*, and a histogram of* X*. It is recommended to try several values for the numbers of bins, and to report only the result found most satisfactory.*
3. *Plot on the same graph, the same theoretical density as before, together with a kernel density estimate of the distribution of* X*. Again, it is recommended to try several values of the bandwidth, and to report only the result found most satisfactory.*
4. *Compare the two plots and explain the reasons for the differences. Say which estimate of the density you prefer, and explain why.*

Ⓔ **Problem 1.3** *Give an interpretation to each of the following four Q-Q plots of Fig. 1.27.*

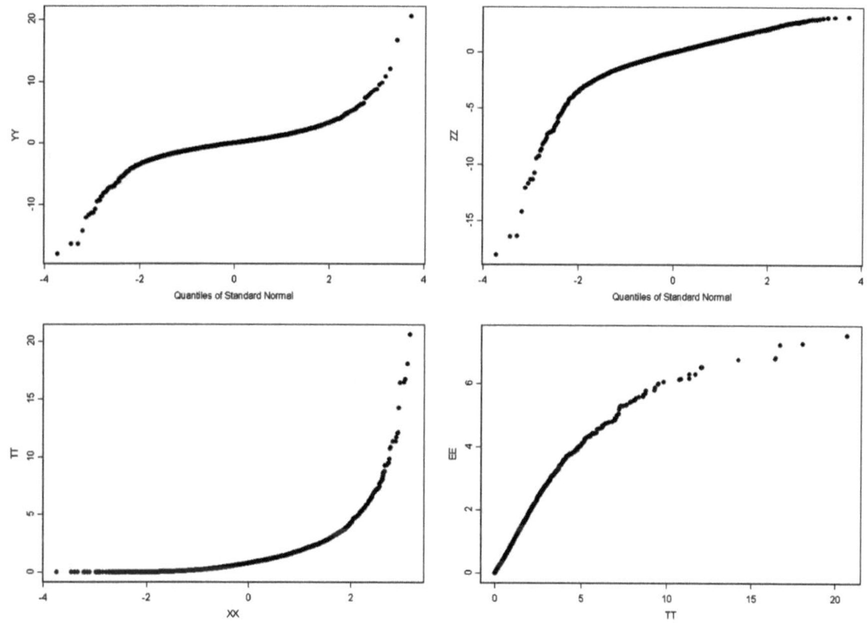

Fig. 1.27. The two plots of the *top row* were produced with the R command qqnorm while the last ones (*bottom row*) were produced with the command qqplot

Ⓔ **Problem 1.4** 1. *As explained in the caption, the plots of Fig. 1.28 were produced with the* R *command* qqnorm*. Articulate properties of the distributions of* XX *and* YY *which you can infer from these plots.*

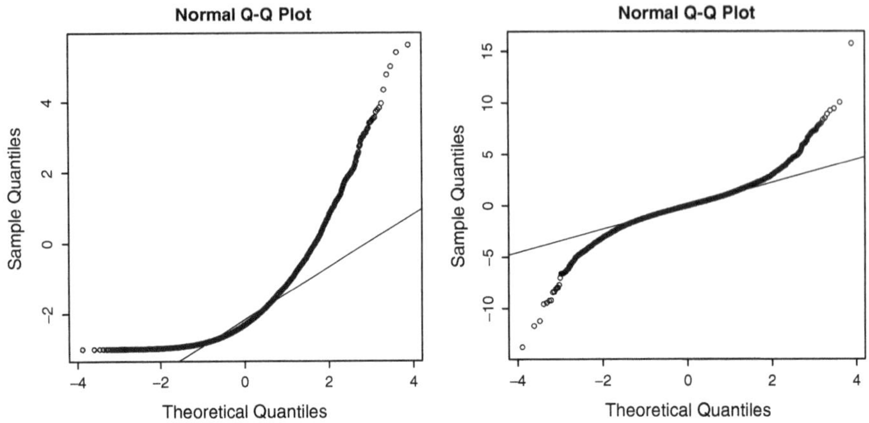

Fig. 1.28. Plots produced by the commands `qqnorm(XX)` (*left*) and `qqnorm(YY)` (*right*)

2. *We now assume that*

$$x_1, x_2, \cdots\cdots, x_m \quad \textit{and} \quad y_1, y_2, \cdots\cdots, y_n$$

are two univariate samples from possibly different distributions. In each of the following situations, sketch the Q-Q plot of `Y` *against* `X` *as given by the* `R` *command* `qqplot(X,Y)` *when:*

2.1. $x_1, x_2, \cdots x_m$ *is a sample from the log-normal distribution* $LN(0,1)$*, and* y_1, $y_2, \cdots y_n$ *is a sample from the Cauchy distribution* $C(0,1)$.

2.2. $x_1, x_2, \cdots x_m$ *is a sample from the log-normal distribution* $LN(0,1)$*, and* y_1, $y_2, \cdots y_n$ *is a sample from the Gaussian distribution* $N(0,1)$

Explain your answers, label the horizontal and vertical axes of your plots, and mark them with numerical values to specify the range of your plots.

Ⓔ **Problem 1.5** *1. As explained in the caption, the plots of Fig. 1.29 were produced with the* `R` *commands* `qqnorm` *and* `qqexp` *which give Q-Q plots of the empirical quantiles of the data against the theoretical quantiles of the normal and exponential distributions respectively. Articulate properties of the distributions of* `XX` *and* `YY` *which you can infer from these plots.*

2. *We now assume that*

$$x_1, x_2, \cdots\cdots, x_m \quad \textit{and} \quad y_1, y_2, \cdots\cdots, y_n$$

are two univariate samples from two possibly different distributions. In each of the following situations, sketch the Q-Q plot of `Y` *against* `X` *(as produced for example by the* `R` *command* `qqplot(X,Y)`*) when:*

2.1. $x_1, x_2, \cdots x_m$ *is a sample from the ordinary Pareto distribution with shape parameter* $\xi = 0.1$*, and* $y_1, y_2, \cdots y_n$ *is a sample from the Cauchy distribution with location parameter* $m = 0$ *and scale parameter* $\lambda = 1$.

2.2. $x_1, x_2, \cdots x_m$ *is a sample from the ordinary Pareto distribution with shape parameter* $\xi = 0.1$*, and* $y_1, y_2, \cdots y_n$ *is a sample from the exponential distribution with rate* $r = 1$.

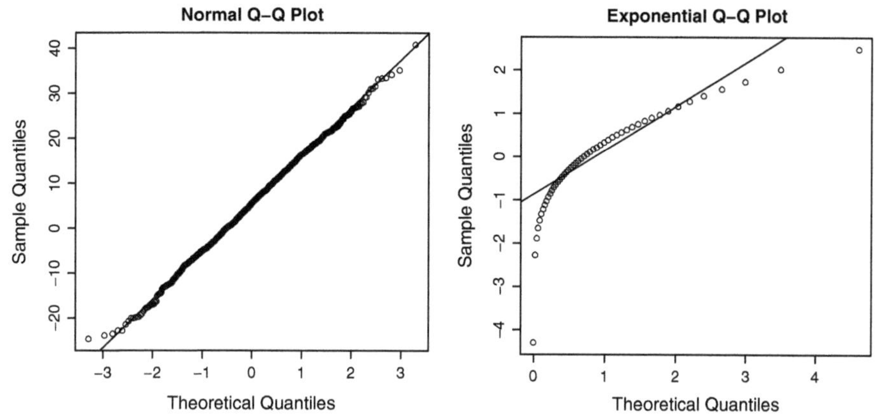

Fig. 1.29. Plots produced by the commands `qqnorm(XX)` (*left*) and `qqexp(YY)` (*right*)

The ordinary Pareto distribution is defined by its density given by formula (2.1) in Chap. 2. Explain your answers, label the horizontal and vertical axes of your plots, and mark them with numerical values to specify the range of your plots.

Ⓔ **Problem 1.6** *1. The plots of Fig. 1.30 were produced with the* R *command* `qqnorm`*. In each case, infer properties of the distribution of the data from the interpretations of these plots.*

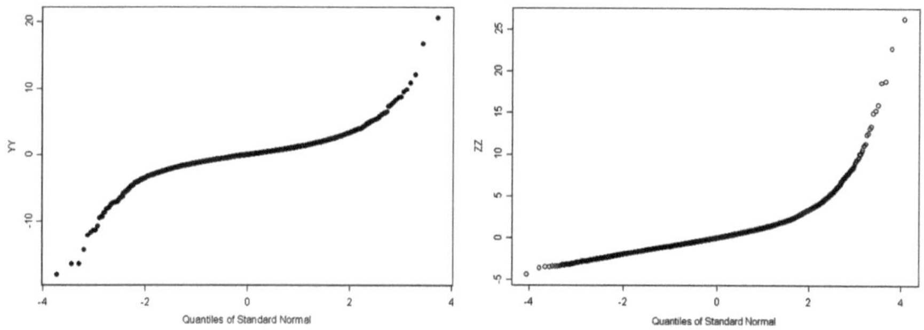

Fig. 1.30. Plots produced by the commands `qqnorm(YY)` (*left*) and `qqnorm(ZZ)` (*right*)

2. *In each of the following two scenarios, sketch the Q-Q plot of a bivariate sample of the form*

$$(x_1, y_1), (x_2, y_2), \cdots\cdots, (x_n, y_n)$$

when:

2.1. $x_1, x_2, \cdots x_n$ *is a typical sample from the Gaussian distribution* $N(0, 1)$ *and* y_1, $y_2, \cdots y_n$ *is a typical sample from the Cauchy distribution* $C(0, 1)$

2.2. $x_1, x_2, \cdots x_n$ *is a typical sample from the Gaussian distribution* $N(0, 1)$ *and* y_1, $y_2, \cdots y_n$ *is a typical sample from the Gaussian distribution* $N(1, 4)$

Explain your answers, label the horizontal and vertical axes of your plots, and mark them with numerical values to specify the range of your plots.

Ⓔ **Problem 1.7** *1. As explained in the caption, the plots of Fig. 1.31 were produced with the* `R` *command* `qqnorm`*. Articulate properties of the distributions of* `XX` *and* `YY` *which you can infer from these plots.*

2. *We now assume that*

$$x_1, x_2, \cdots\cdots, x_m \quad \textit{and} \quad y_1, y_2, \cdots\cdots, y_n$$

are two univariate samples from possibly different distributions. In each of the following situations, sketch the Q-Q plot of `Y` *against* `X` *as given by the* `R` *command* `qqplot(X,Y)` *when:*

2.1. $x_1, x_2, \cdots x_m$ *is a typical sample from the exponential distribution* $E(1)$ *with rate* 1*, and* $y_1, y_2, \cdots y_n$ *is a typical sample from the Cauchy distribution* $C(0,1)$

2.2. $x_1, x_2, \cdots x_n$ *is a typical sample from the Cauchy distribution* $C(0,1)$ *and* $y_1, y_2, \cdots y_n$ *is a typical sample from the Cauchy distribution* $C(1,4)$.

Explain your answers, label the horizontal and vertical axes of your plots, and mark them with numerical values to specify the range of your plots.

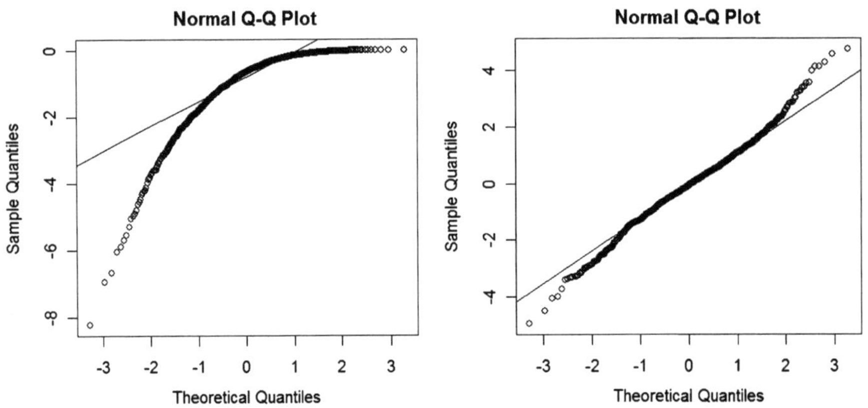

Fig. 1.31. Plots produced by the commands `qqnorm(XX)` (*left*) and `qqnorm(YY)` (*right*)

ⒺⓈ **Problem 1.8** *1. Create a sample of size* $N = 128$ *from the standard normal distribution and use Q-Q plots to assess the normality of the data, in other words, explain which features oft the Q-Q plot you produce suggest that the data are from a Gaussian distribution and which don't.*

2. *Create a sample of size* $N = 128$ *from the exponential distribution with parameter* 1*, and use Q-Q plots to assess as in question 1, the normality of the data. Describe and explain the differences with the results of question 1.*

Ⓢ **Problem 1.9** *The goal of this problem is to design and use a home-grown random number generator for the exponential distribution. For the first question of this problem, you are not allowed to use any of the functions* `dexp`, `pexp`, `qexp` *or obviously* `rexp`.

1. *Recall the formulas derived in the text for the cdf* $F_\lambda(x) = \mathbb{P}\{X \le x\}$ *of a random variable* X *with an exponential distribution with scale parameter* $\lambda > 0$. *and its inverse* F_λ^{-1} *and write an* R *function* `myrexp` *which takes the parameters* N, *and* LAMBDA, *and which returns a numeric vector of length* N *containing* N *samples of random variates from the exponential distribution with scale parameter* LAMBDA.
2. *Use your function* `myrexp` *to generate a sample of size* $N = 1{,}024$ *from the exponential distribution with mean* 1.5, *use the* R *function* `rexp` *to generate a sample of the size* $2N$ *from the same distribution, and produce a Q-Q plot of the two samples. Are you satisfied with the performance of your simulation function* `myrexp`*? Explain why.*

Ⓣ **Problem 1.10** *Let us assume that the c.d.f.* $F_X(x)$ *of a random variable* X *is a continuous function of* x.

1. *What is the distribution of the random variable* $F_X(X)$*? Give its c.d.f.*
2. *What can you say about the distribution of the random variable* $F_X(-X)$ *without having to make any extra assumptions on the distribution of* X*?*
3. *What can you say about the distribution of the random variable* $F_X(-X)$ *if* X *is positive?*

Ⓣ **Problem 1.11** *Let* X *be a random variable with probability density function given by*

$$f_X(x) = \begin{cases} \alpha\beta^{-\alpha}x^{\alpha-1}e^{-(x/\beta)^\alpha} & \text{if } x > 0 \\ 0 & \text{otherwise} \end{cases},$$

where α *and* β *are strictly positive constants. Such a random variable is said to have a* Weibull distribution *with* shape parameter $\alpha > 0$ *and* scale parameter $\beta > 0$, *and we denote this fact by* $X \sim W(\alpha, \beta)$.

1. *Compute the cdf.* F_X *of* X.
2. *The logarithm of a Weibull random variable has a distribution known as the* Gumbel distribution *(with the same parameters). Describe (in words) an algorithm to generate a random variable having such a Gumbel distribution.*

NOTES & COMPLEMENTS

The emphasis of this book is on graphical, computational and simulation methods for data analysis, with a view toward financial applications. Most introductory statistical textbooks spend a fair amount of time discussing the exploratory data analysis tools introduced in this chapter. An excellent reference in this spirit is the book of Bill Cleveland [23]. For books with applications using R, we refer the interested reader to the book of Venables and Ripley [94], for a thorough discussion of the properties of histograms and their implementations in R. A detailed discussion of the ASH variation can be found there. The introductory book [25] by Dalgaard gives a low key initiation to R.

Following its introduction in the mid 1990s, several mathematical models for the dynamics of the PCS index have been proposed in the literature. Most of these models try to capture the catastrophic events' arrivals, the initial "shocks", and the possible revisions to the damage estimates that arise in the ensuing weeks after a catastrophe occurs. Standard probabilistic arguments suggest to use a model of the form:

$$S(t) = \sum_{i=0}^{N(t)} X_i(t),$$

where $S(t)$ is the PCS (total) index, $X_i(t)$ is the total damage caused by the i-th catastrophe, and $N(t)$ is a Poisson process modeling catastrophe arrival. The catastrophe damage is further modeled as:

$$X_i(t) = \zeta + \theta(t - (T_i - T_\theta)),$$

where ζ is the time of the initial shock (it is usually assumed to be exponentially distributed), $\theta(t)$ is the revision function which is zero for $t < 0$ and equal to κ for $t \geq 0$, T_i is the arrival time of the i-th catastrophe, T_θ is the length of time until the catastrophe damage amount is revised (it is usually assumed to be a random variable with a Poisson distribution), and finally κ is a random variable used for the amount of revision of the index. In order to fit such a model, four parameters need to be estimated:

- The catastrophe arrival time: it is usually modeled as a Poisson process;
- The initial damage of the catastrophe: it is often modeled as an exponentially distributed random variable, but it may also be modeled with a heavy tail distribution;
- The delay time for the revision: it is usually modeled as an exponential distribution;
- The revision amount.

We shall not discuss any further this ambitious, though realistic, model.

The reader interested in applications of Monte Carlo techniques to financial engineering problems is referred to Glasserman's book [40] which is still the state of the art in the field. The most encyclopedic compilation of algorithms for the generation of random number with given distributions is the classic book by De Vroye [96]. Large scale computations on modern computers are often based on low discrepancy sequences and such computations are often called quasi Monte Carlo.

2

HEAVY TAIL DISTRIBUTIONS

Motivated by the instances of extreme events and heavy tail distributions encountered in the first chapter, we present the most important theoretical results underpinning the estimation of the probabilities of these extreme and rare events. The basics of extreme value theory are presented as they pertain to estimation and risk management of extremes observed in financial applications. Our goal is to explain the connection between the generalized extreme value distributions and the generalized Pareto distributions, and illustrate the implementation of the theory into a set of practical tools for the detection and estimation of heavy tail distributions. In preparation for some of the applications considered later in the book, the chapter concludes with a discussion of measures of risk, both from a theoretical and a practical point of view.

2.1 A Primer on Extreme Value Theory

We present the parametric families of Pareto and extreme value distributions, very much in the spirit of the parametric families discussed in Chap. 1, and we show how the properties of the latter can be used to detect and identify the characteristics of the former.

2.1.1 Empirical Evidence of Extreme Events

We already argued that histograms and kernel density estimators could not give a good account of the tail properties of distributions, and we insisted that Q-Q plots offered the best graphical way to get a reasonable feeling for these properties. We emphasize one more time the non-normality of the distribution of daily financial

R. Carmona, *Statistical Analysis of Financial Data in R*, Springer Texts in Statistics,
DOI 10.1007/978-1-4614-8788-3_2, © Springer Science+Business Media New York 2014

returns by considering their extreme values. Since we do not plan to give a precise mathematical definition of an extreme value, we shall simply say that a value is extreme if its distance to the mean location of the data (as given by the mean for example) is large when measured in standard deviation units, say greater that three or four standard deviations. For the purpose of illustration, we consider the daily log-returns on the S&P 500 index. Their values are encapsulated in the numeric vector `DSPLRet` included in the library `Rsafd`. Using the functions `mean` and `sd`, we compute the mean and the standard deviation of the daily log-returns.

```
> mean(DSPLRet)
[1] -0.0002729406
> sd(DSPLRet)
[1] 0.009727974
```

Looking at the sequential plot of the daily log-return (as reproduced in the right pane of Fig. 1.20) we notice a few very large negative values. Looking more closely at the largest of these down-moves we see that:

```
> min(DSPLRet)
[1] -0.2289972
> (min(DSPLRet)-mean(DSPLRet))/sd(DSPLRet)
[1]  -23.56813
```

which shows that this down move was over

23 standard deviations away from the mean

daily move! So much for the normal distribution as a model for the daily moves of this index. The log-return on this single day of October 1987, as well as many others since then (though less dramatic in sizes) cannot be accounted for if the Gaussian distribution is used as a model for the daily log-returns. The tails of the normal distribution are too thin to produce such extreme values. However, other families of distributions could be used instead, and stable or Pareto distributions have been proposed with reasonable success. Pareto distributions are studied in detail in this chapter. For the time being, it suffices to say that, like Pareto distributions, stable distributions have polynomial tails, and moreover, they have useful scaling properties. However, their usefulness as statistical models for heavy tail distribution is limited by the fact that the rates of polynomial decay of their densities are restricted to an interval. Moreover, their scaling properties are of very little use since at least in the first part of the book, we are mostly interested in marginal distributions of financial returns, and hence we rarely use dynamical models involving time evolution of prices. Finally, the main shortcoming of the stable distributions is the lack of a closed form formula for the density and/or the cdf. The Cauchy distribution, is the only exception. Recall formula (1.13) for the definition of the Cauchy distribution which is sometime used as an alternative to the Gaussian distribution in the presence of extreme values. Indeed, like the Gaussian density, it is bell-shaped, but unlike the Gaussian density, its tails are so *thick* that the moments of the distribution such as the mathematical expectation (or mean) and the variance do not even exist.

The theory of probability distributions giving rise to unusually large numbers of extremes in each sample is called the theory of extreme-value distributions, or extreme-value theory. It is a well developed mathematical theory. The remainder of this chapter is devoted to an informal presentation of the most fundamental facts of this theory. For the purpose of illustration we demonstrate the practical implementations in the library `Rsafd`, providing versatile tools to fit heavy tail distributions to data, and generate random Monte Carlo samples from the fitted distributions.

2.1.2 **Pareto Distributions**

We first introduce a class of distributions which will play a fundamental role in our modeling of heavy tails. The present subsection could have been included in the previous chapter and provide one more example of a family of distributions. However, because of its pivotal role in the theory presented in this chapter, we chose to introduce it here.

2.1.2.1 Ordinary Pareto Distributions

The classical Pareto distribution is a distribution on the positive axis $[0, \infty)$ (i.e. the distribution of a positive random variable) with density given by the formula

$$f_\alpha(x) = \begin{cases} (1 + \frac{x}{\alpha})^{-(1+\alpha)} = \frac{1}{(1+x/\alpha)^{1+\alpha}} & \text{if } x > 0, \\ 0 & \text{otherwise.} \end{cases} \tag{2.1}$$

for some positive real number $\alpha > 0$. Like the exponential and lognormal distributions, this distribution has only one tail extending to $+\infty$. For this reason, it is often called a *one-sided* Pareto distribution. The above definition of the one-sided Pareto distribution can be found in many probability textbooks. For geosciences applications, especially in hydrology where heavy tail distributions were introduced first in order to estimate the frequencies of floods, the Pareto distributions are parameterized by α. However for some strange reason, in financial applications, these distributions are parameterized by $\xi = 1/\alpha$ which is called the shape parameter of the distribution. Both parametrizations are implemented in the library `Rsafd`, but as we concentrate on the analysis of financial data, we shall use the ξ – parameterization in this book. This choice is *passed* to the routines of the library `Rsafd` by setting the parameter `SHAPE.XI` to `TRUE`. For the sake of convenience, we restate the definition of the classical Pareto distribution using the shape parameter ξ.

$$f_\xi(x) = \begin{cases} (1 + \xi x)^{-(1+1/\xi)} = \frac{1}{(1+\xi x)^{1+1/\xi}} & \text{if } x > 0, \\ 0 & \text{otherwise.} \end{cases} \tag{2.2}$$

We shall discuss later the role of the shape parameter ξ, and when we do, we shall emphasize that even though ξ will have to be a non-negative number in most of the applications we are interested in, from a mathematical point of view, we can extend the definition of the Pareto distribution to include negative values of the shape parameter ξ. See below for details.

In any case, it is easy to compute the cdf of an ordinary Pareto distribution in closed form. Indeed straightforward integration gives:

$$F_\xi(x) = \int_{-\infty}^{x} f_\xi(x')dx' = \begin{cases} 1-(1+\xi x)^{-1/\xi} = 1 - \frac{1}{(1+\xi x)^{1/\xi}} & \text{if } x > 0, \\ 0 & \text{otherwise.} \end{cases}$$

Changing location and scale, (remember our discussion of affine transformations), we can define and study ordinary (one-sided) Pareto distributions with location parameter $m \in \mathbb{R}$ and scale parameter $\lambda > 0$. Such a distribution is supported by the half line $[m, \infty)$ (i.e. it is the distribution of a random variable which is always greater than or equal to m). Its density will be denoted by $f_{m,\lambda,\xi}$. It is given by the formula:

$$f_{m,\lambda,\xi}(x) = \begin{cases} \frac{1}{\lambda}\left(1+\frac{\xi}{\lambda}(x-m)\right)^{-(1+1/\xi)} & \text{if } x \geq m, \\ 0 & \text{otherwise.} \end{cases}$$

As before, we can compute the corresponding cdf. It is given by:

$$F_{m,\lambda,\xi}(x) = \begin{cases} 1-(1+\xi\frac{x-m}{\lambda})^{-1/\xi} = 1 - \frac{1}{(1+\xi\frac{x-m}{\lambda})^{1/\xi}} & \text{if } x > m, \\ 0 & \text{otherwise.} \end{cases}$$

2.1.2.2 *More General Shape Parameters*

We now consider a first generalization of the parametric family of one-sided Pareto distributions which we shall call Generalized One-Sided Pareto distributions, GOSPD for short. It still relies on three parameters: a location parameter m, a scale parameter λ and a shape parameter ξ. The cumulative distribution function of a GOSPD is given by

$$F_{m,\lambda,\xi}(x) = \begin{cases} 1-\left(1+\xi\frac{x-m}{\lambda}\right)^{-1/\xi} & \text{for } \xi \neq 0, \\ 1-\exp\left\{-\frac{x-m}{\lambda}\right\} & \text{for } \xi = 0. \end{cases} \tag{2.3}$$

the above formulae defining the GOSPD on the domains:

$$\begin{array}{ll} m < x \leq m - \lambda/\xi & \text{for } \xi < 0, \\ m < x \leq \infty & \text{for } \xi \geq 0. \end{array}$$

In other words, we extended the family of one-sided Pareto distributions to include distributions with a negative shape parameter ξ. This is done for the sake of generality. It will not be used in the financial applications we consider in this book.

Notice that if $\xi > 0$, the generalized Pareto distribution with cdf $F_{m,\lambda,\xi}$ is nothing but the distribution of a random variable $m + \lambda X$ where X has the ordinary Pareto distribution with parameter α provided we set $\xi = 1/\alpha$ as shape parameter.

Notice also that the case $\xi = 0$, $m = 0$ corresponds to the exponential distribution with scale parameter λ. In general, the case $\xi = 0$ corresponds to an exponential distribution with scale λ *shifted* by the amount m, i.e. to the distribution of a random variable $X + m$ where $X \sim E(1/\lambda)$.

2.1.2.3 ***Existence of Moments (or Lack Thereof)***

Another important fact concerning the size of the tail of a one sided Pareto distribution (generalized or not) is given by the existence (or lack thereof) of moments. Indeed, the above definition implies that, if $X \sim F_{m,\lambda,\xi}$ with $\xi \geq 0$, then

$$\mathbb{E}\{|X|^p\} < \infty \Leftrightarrow p < \frac{1}{\xi}. \tag{2.4}$$

Here are a few consequences for a non-negative random variable X with a one-sided GOSPD.

- If $\xi = 0$, X has moments of all orders, i.e. $\mathbb{E}\{X^p\} < \infty$ for all $p > 0$;
- The mean of X exists (i.e. $\mathbb{E}\{X\} < \infty$) if and only if $\xi < 1$;
- The variance of X exists if and only if $\xi < 0.5$.

Figure 2.1 shows the graphs of the densities of three one-sided Pareto distributions with default values $m = 0$ and $\lambda = 1$ for the location and scale parameters, and values $\xi = 0.2$, $\xi = 1.6$ and $\xi = 2.5$ for the shape parameter. The plots were produced with the following commands.

```
> X <- seq(from=-.2,to=15,length=5000)
> plot(X,dpareto(X,xi=.2),type="l",ylab="GPD densities",
                                          ylim=c(-.05,1.1))
> points(X,dpareto(X,xi=1.6),type="l",lty=3)
> points(X,dpareto(X,xi=2.5),type="l",lty=5)
> abline(h=0); abline(v=0)
```

Remark. The number of finite moments of a distribution is a good indication of the *thickness* of its tail. This number has been estimated for the marginal distribution of financial returns over different periods ranging from minutes, to days, weeks, months, ... and there is a heated debate concerning the values of these estimates in the so-called eocnophysics community. Indeed, it is claimed by some that this number of finite moments is universal across financial indices and asset classes. Others use self-similarity arguments to claim that this exponent should not change with time horizon, and that it should remain the same when computed with returns over 1 day, 1 week, 1 month, ... The rationale behind the universality of this exponent is beyond the scope of this book. However, we shall give examples (both in the text and in the problem sets) indicating that this universality conjecture does not stand some of the empirical analyzes made possible by the tools presented in this book.

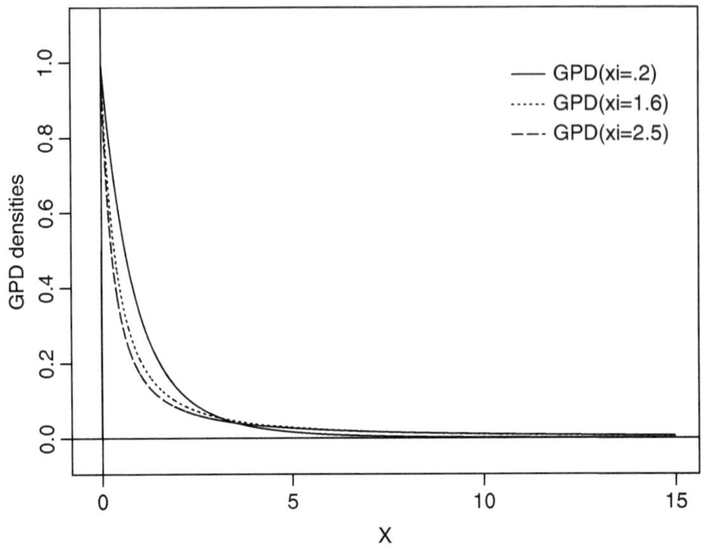

Fig. 2.1. One-sided Pareto densities with $m = 0$ and $\lambda = 1$ for the location and scale parameters, and values $\xi = 0.2$, $\xi = 1.6$ and $\xi = 2.5$ for the shape parameter

2.1.2.4 *Implementation*

As in the case of the classical distributions considered earlier, recall Table 1.2, on can compute values of the density, quantile, and cumulative distribution functions, as well as generating random samples with a set of functions adhering to the naming convention used in `R`. They are listed in Table 2.1

Distribution	Random samples	Density	cdf	Quantiles
One-sided Pareto	`rpareto`	`dpareto`	`ppareto`	`qpareto`

Table 2.1. `Rsafd` commands for the manipulation of one-sided Pareto distributions

2.1.2.5 *(Two-Sided) Generalized Pareto Distributions (GPD)*

We now define the class of (heavy tail) distributions which we fit to sample data exhibiting thick tails as detected by empirical Q-Q plots. At an intuitive level, our fitting procedures will search for heavy tails (typically densities with inverse polynomial decays) at $+\infty$, and $-\infty$ in the case of a left tail. Roughly speaking, these distributions should behave like

- The distribution of a **one-sided Pareto** random variable to the right of a specific threshold;

- The distribution of the negative of a **one-sided Pareto** random variable to the left of a specific threshold;
- **Nothing special in between.**

To be more specific, these distributions will be characterized by

- A location parameter m_+, a scale parameter λ_+ and a shape parameter ξ_+ specifying the one-sided Pareto distribution which applies to the right of the threshold m_+;
- A location parameter m_-, a scale parameter λ_- and a shape parameter ξ_- specifying the one-sided Pareto distribution which applies to the left of the threshold m_- whenever the distribution has a left tail;
- Any distribution in the interval $[m_-, m_+]$.

Clearly, estimating such a distribution amounts to the estimation of the three parameters (possibly six when the distribution has tails extending to both $+\infty$ and $-\infty$) of the one sided Pareto distribution(s), and to the estimation of the density in between the thresholds. The latter will be done by a plain histogram.

The class of (possibly two-sided) Generalized Pareto Distribution (GPD for short) defined above is used in all the applications of extreme value theory considered in this book. The mathematical results we state and use for GPDs hold for a slightly more general class of distributions, namely those distributions with densities at $+\infty$ and/or $-\infty$ which, up to a slowly varying function (concept which we define later), behave like inverse powers.

2.1.3 Tidbits of Extreme Value Theory

There are several ways to investigate the statistical properties of extremes. The classical approach is based on the analysis of the statistics of the maxima over large blocks of data. It is most elegant mathematically, and we briefly review it below. However, because it requires large data samples, and involves much too often inefficient computations, the *block maxima approach* fell out of grace with practitioners who prefer relying on *threshold exceedance models* which lead to a more efficient use of limited data. We present the former first.

2.1.3.1 The Fisher-Tippett Theorem

As usual we start from a sample of values

$$x_1, x_2, \cdots\cdots, x_n, \cdots$$

which we envision as realizations of independent identically distributed random variables $X_1, X_2, \cdots\cdots, X_n, \cdots$ with a common distribution which we try to estimate.

Remark 1. Most of the results reviewed in this chapter remain valid without this independence assumption. Indeed, under various forms of dependence between the X_j's, similar conclusions can be reached. These extensions have great practical relevance, as real life data, and especially financial returns, are not strictly independent from one period to the next. However, we refrain from considering these generalizations by fear that their technical nature may obscure the ideas underpinning the theory.

As earlier, we use numerical statistics computed from the data in order to infer properties of the common distribution of these random variables. The limit theorems discussed in the first chapter are involved with the limiting behavior of the partial sums $S_n = X_1 + \cdots + X_n$ for large values of n. In particular, the Central Limit Theorem (CLT) states that

$$\lim_{n\to\infty} \mathbb{P}\left\{\frac{S_n - m_n}{\lambda_n} \le x\right\} = \Phi(x), \qquad x \in \mathbb{R} \tag{2.5}$$

provided we define the normalizing (centering and scaling) constants m_n and $\lambda_n > 0$ as

$$m_n = n\mu \qquad \text{and} \qquad \lambda_n = \sigma\sqrt{n}$$

where μ and σ denote the mean and the standard deviation of the common distribution of the X_j's. Extreme Value Theory (EVT for short) is concerned with the search of centering and scaling constants m_n and $\lambda_n > 0$ for which limiting results of the form (2.5) hold for some limiting distribution functions $\Psi(x)$ when one replaces the partial sums S_n by partial maxima

$$M_n = \max\{X_1, \cdots, X_n\}. \tag{2.6}$$

Obviously, switching from partial sums to maxima shifts the emphasis from aggregation to extremes.

Remark 2. The theory presented below is geared toward the analysis of upper tails of statistical distributions as it is formulated in terms of maxima of random samples. Obviously, similar results hold true for minima, and the same theory can be used for the analysis of lower tails of statistical distributions. For the sake of simplicity, we focus our discussion on results on maxima of sequences of random variables, even though we shall eventually turn the results of this theory into computing tools for the analysis of both upper and lower tails of statistical distributions.

The cornerstone of the block maxima approach is the following theoretical result known as the Gnedenko or Fisher-Tippett theorem.

Theorem 1. *If the cdf*

$$x \hookrightarrow \mathbb{P}\left\{\frac{M_n - m_n}{\lambda_n} \le x\right\}$$

converges as $n \to \infty$ toward a (non-degenerate) cdf for some normalizing sequences $\{m_n\}_n$ and $\{\lambda_n\}_n$ of centering and positive scaling constants, then the limiting distribution necessarily belongs to the family of Generalized Extreme Values (GEV for short) distributions defined below in formula (2.7).

Before we give such a definition, we present a couple of enlightening examples, for which the result of this theorem can be checked with elementary calculus.

The Case of the Exponential Distribution. If the X_j are independent random variables with the exponential distribution with rate $r > 0$, then $F_X(x) = 1 - e^{-rx}$ for $x \geq 0$, and the cdf of normalized M_n is equal to

$$F_{(M_n - m_n)/\lambda_n}(x) = F_X(m_n + \lambda_n x)^n = (1 - \exp[-(m_n + \lambda_n x)])^n$$

for $x > -m_n/\lambda_n$, so that with the choices $m_n = (\log n)/r$ and $\lambda_n = 1/r$ for the normalizing constants we get

$$\lim_{n\to\infty} F_{(M_n - m_n)/\lambda_n}(x) = \lim_{n\to\infty} (1 - \frac{1}{n}\exp[-x])^n = 1 - \exp(-e^{-x})$$

for all $x \in \mathbb{R}$ since $-\log n = -m_n/\lambda_n$. This limiting distribution is known as the Gumbel distribution.

The Case of the Ordinary Pareto Distribution. Using the ordinary Pareto distribution with shape parameter $\xi > 0$ instead of the exponential distribution, we can still illustrate the result of the Fisher-Tippett-Gnedenko theorem with explicit computations. Indeed, if we choose the centering and the scaling constants m_n and λ_n as $m_n = n^\xi - 1$ and $\lambda_n = \xi n^\xi$, then:

$$F_{(M_n - m_n)/\lambda_n}(x) = F_X(m_n + \lambda_n x)^n = \left(1 - \frac{1}{n}(1 + \frac{x}{\alpha})^{-\alpha}\right)^n$$

for $\alpha = 1/\xi$ and $x > \alpha(-1 + n^{-1/\alpha})$, and consequently

$$\lim_{n\to\infty} F_{(M_n - m_n)/\lambda_n}(x) = \lim_{n\to\infty} \left(1 - \frac{1}{n}(1 + \frac{x}{\alpha})^{-\alpha}\right)^n = \exp\left[-(1 + \frac{x}{\alpha})^{-\alpha}\right]$$

for $x > -\alpha$. This distribution is known as the Fréchet distribution.

2.1.3.2 *Generalized Extreme Value Distributions (EVD)*

The families of extreme value distributions (EVD for short) which have been studied in the classical statistical literature comprise the Gumbel distribution (also known as EVI distribution), the Fréchet distribution (also known as EVII distribution), and the Weibull distribution (also known as EVIII distribution). These three distribution families can be combined into a single parametric family which is usually called the Generalized Extreme Value (GEV) distribution family. Its cumulative distribution function is given by the following formula:

$$G_{m,\lambda,\xi}(x) = \begin{cases} \exp\left[-\left(1 + \frac{\xi(x-m)}{\lambda}\right)^{-1/\xi}\right] & \text{for } \xi \neq 0, \\ \exp\left[-\left(e^{-\frac{(x-m)}{\lambda}}\right)\right] & \text{for } \xi = 0. \end{cases} \tag{2.7}$$

A GEV distribution is characterized by three parameters: a location parameter m, a scale parameter $\lambda > 0$, and a shape parameter ξ. In the above formula it is assumed that:

$$\begin{array}{ll} -\infty < x \le m - \lambda/\xi & \text{for } \xi < 0, \\ -\infty < x < \infty & \text{for } \xi = 0, \\ m - \lambda/\xi \le x < \infty & \text{for } \xi > 0. \end{array}$$

The Gumbel distribution corresponds to the case $\xi = 0$, the Fréchet distribution corresponds to the case $\xi > 0$, while the Weibull distribution corresponds to the case $\xi < 0$.

Figure 2.2 shows the graphs of the densities of three GEV distributions with default values $m = 0$ and $\lambda = 1$ for the location and scale parameters, and values $\xi = 0.2$, $\xi = 0.6$ and $\xi = 1.5$ for the shape parameter. Note that the left hand point of the distribution changes with the parameters. The plots were produced with the following commands.

```
> X <- seq(from=-2.2,to=8,length=5000)
> plot(X,dgev(X,xi=.2),type="l",ylab="GEV densities",
                                        ylim=c(-.05,.9))
> points(X,dgev(X,xi=.6),type="l",lty=3)
> points(X,dgev(X,xi=1.5),type="l",lty=5)
> abline(h=0); abline(v=0)
```

We use the fact that, like in the case of GPDs, the library `Rsafd` provides functions to generate random samples and compute densities, cdfs and quantiles of the GEV distributions. Table 2.2 gives these commands, and as we can see, they follow the

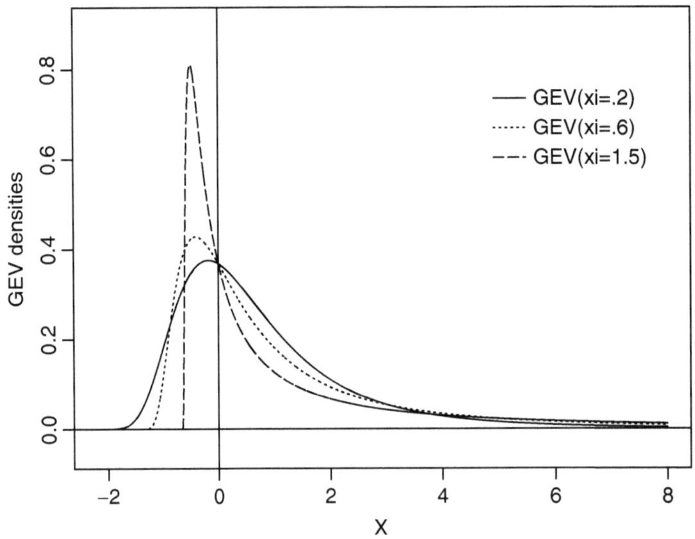

Fig. 2.2. Densities of GEV distributions with $m = 0$ and $\lambda = 1$ for the location and scale parameters, and values $\xi = 0.2$, $\xi = 0.6$ and $\xi = 1.5$ for the shape parameter

Distribution	Random samples	Density	cdf	Quantiles
GEV distribution	`rgev`	`dgev`	`pgev`	`qgev`

Table 2.2. Commands for the manipulation of the generalized extreme value distributions

standard `R` naming convention. We do not give plots of densities corresponding to negative values of ξ for they are typically not used in the analysis of financial data.

Formula (2.7) is explicit and simple enough to be inverted explicitly. Doing so, we obtain the following formula for the quantile function of a GEV distribution.

$$Q_{m,\lambda,\xi}(p) = \begin{cases} m - \frac{\lambda}{\xi}\left[1 - (-\log p)^{-\xi}\right] & \text{for } \xi \neq 0, \\ m - \lambda \log(-\log p) & \text{for } \xi = 0. \end{cases} \tag{2.8}$$

This closed form formula makes the generation of random samples from GEV distributions quite easy and efficient. It also shows that estimates of the quantiles of a GEV distribution can be obtained from estimates of the parameters m, λ and ξ of the distribution by substitution of these estimates in (2.8). We address the estimation of the parameters of a GEV distribution later in this chapter.

Remark 3. Roughly speaking, when the Gnedenko, Fisher-Tippett theorem holds, it says that if n is large enough

$$\mathbb{P}\{M_n \leq x\} \approx G_{m,\lambda,\xi}(x)$$

for some set $\{m, \lambda, \xi\}$ of parameters. But since the X_j's are assumed to be independent, we have

$$\mathbb{P}\{M_n \leq x\} = F_X(x)^n$$

and consequently

$$\mathbb{P}\{M_n \leq \pi_p\} = F_X(\pi_p)^n = p^n$$

(if we recall the notation π_p for the p-quantile of the common distribution of the X_j's) which in turn implies that

$$\pi_p \approx m + \frac{\lambda}{\xi}\left[\left(n \log \frac{1}{p}\right)^{-\xi} - 1\right].$$

This approximation for the quantiles of the distribution of the X_j's should be compared to the formula

$$\pi_p = \mu + \sigma\Phi^{-1}(p)$$

which holds in the Gaussian case. This remark should shed some light on the consequences of the Gnedenko, Fisher-Tippett theorem on tail sizes and properties of the quantiles of the common distribution of the individual random variables X_j. We shall revisit the significance of this remark when we discuss the estimation of Value at Risk in the presence of heavy tails.

The Gnedenko, Fisher-Tippett theorem is a very nice theoretical result, but in order to be useful in practical situations, we need to know for which distribution functions F_X it does hold, and even better, we need a hash table pointing out which distributions F_X lead to which GEV distributions. In mathematical terms, this last request is screaming for the identification of the so-called domain of attraction of a given GEV distribution. In other words, given a GEV distribution $G_{m,\lambda,\xi}$, can we characterize the distribution functions F_X for which the distributions of maxima M_n converge (after proper normalization), toward the GEV distribution in question as stated in Theorem 1. This *wishful thinking* is at the origin of several results of great practical usefulness. They go under the names of Gnedenko, Pickands, Balkema and de Haan. We state them in an informal way to avoid being distracted by the technical nature of some of the mathematical assumptions under which these results hold. The interested reader is directed toward the Notes and Complements at the end of the chapter for references of textbooks where these theories are presented in detail.

2.1.3.3 **Illustration**

The following commands illustrate the convergence of the distribution of block maxima of ordinary Pareto variates toward the Fréchet distribution, fact which we proved rigorously earlier.

```
> XX <- rpareto(1000000,xi=.4)
> dim(XX) <- c(1000,1000)
> MAX <- apply(XX,2,max)
> qqplot(MAX,rgev(1000,xi=.4))
> title("Q-Q Plot Evidence of Block Maxima Convergence,
                                                    xi=.4")
```

The first command creates a sample of size 10^6 of independent random samples from the ordinary Pareto distribution with location 0, scale 1 and shape parameter $\xi = 0.4$. The second command splits this sample into 1,000 blocks of lengths 1,000 each by organizing them in a 1,000 $\times$ 1,000 data matrix. The next command computes the maximum of each of these blocks, creating in this way a sample of size 1,000 of maxima M_n with $n = 1{,}000$. The `qqplot` command produces a Q-Q plot of this sample of maxima against a random sample from the GEV distribution with the same shape parameter $\xi = 0.4$. This plot is reproduced in the left pane of Fig. 2.3. The fact that the points line up on a straight line is an indication that we are in the limiting regime of the theorem of Gnedenko, Pickands, Balkema and de Haan. This fact is a particular case of a more general result which we state as a theorem for later reference.

Theorem 2. *The distribution of the maxima M_n converge after appropriate centering and scaling, toward a GEV distribution with shape parameter $\xi > 0$ if and only if the common cdf $F_X(x)$ of the X_j's converges toward 1 as $x \to \infty$ at the rate $x^{-1/\xi}$.*

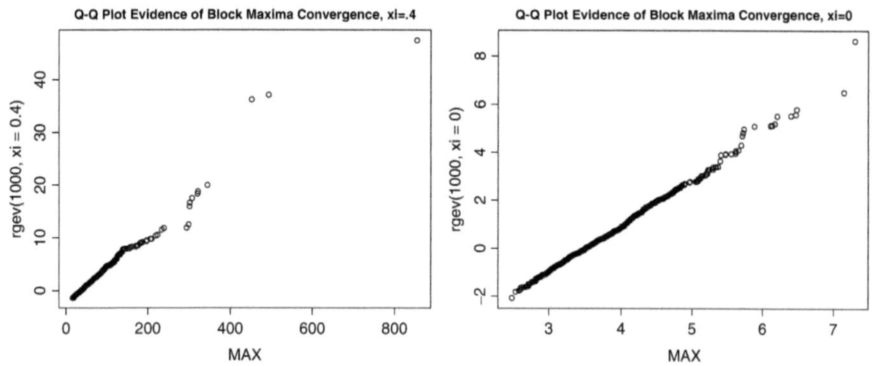

Fig. 2.3. Q-Q plots of a sample of 1,000 maxima over 1,000 disjoint blocks in a sample from a GPD, against a sample from the GEV distribution with the same shape parameter. *Left*: case of the Fréchet distribution with $\xi = 0.4$. *Right*: case of the Gumbel distribution (i.e. $\xi = 0$) from an exponential sample with rate $r = 2.0$

The precise mathematical statement is that the function $L(x) = x^{1/\xi}(1 - F_X(x))$ is slowly varying at $+\infty$ in the sense that

$$\lim_{x \to \infty} \frac{L(\lambda x)}{L(x)} = 1, \qquad \text{for all } \lambda > 0.$$

The case of the Gumbel distribution is unfortunately not as clearly delineated by a theoretical result such as Theorem 2 above. We proved in Sect. 2.1.3.1 that the Gumbel distribution was the limit of the distributions of block maxima of increasing sizes of independent exponential variates. As before, we can illustrate this theoretical fact with the help of random simulations.

```
> XX <- rexp(1000000,r=2)
> dim(XX) <- c(1000,1000)
> MAX <- apply(XX,2,max)
> qqplot(MAX,rgev(1000,xi=0.0))
> title("Q-Q Plot Evidence of Block Maxima Convergence,
                                                   xi=0")
```

The resulting plot is reproduced in the right pane of Fig. 2.3, and as before, the fact that the points line up on a straight line is an indication that we are in the limiting regime of the theorem of Gnedenko, Pickands, Balkema and de Haan. The exponential distribution is not the only distribution F_X for which the distributions of the block maxima converge toward the Gumbel distribution. These distributions F_X are not easily characterized. However, it can be proved that they all have finite moments of all orders in the sense that $\mathbb{E}\{X_j^p\} < \infty$ for all $p > 0$. So if the X_j's have a common density $f_X(x)$, then this density goes to zero faster than any inverse polynomial. Exponentials do, but Gaussian and log-normal densities do as well. So in the case $\xi = 0$, the information content of the fact that the limit distribution of the normalized block maxima is the Gumbel distribution is not as precise: we know that the tail decays faster than any inverse polynomial, but we cannot pin-point the exact rate of decay!

Remark. Notice that, because we are interested in extremes, and especially in rare and unexpected large values of financial returns or losses, we shall not consider the Weibull case $\xi < 0$ which forces the distribution to be limited, and prevents the tail from extending to infinity.

2.1.3.4 ***Block Maxima Approach to Extreme Values Estimation***

We now formulate in an algorithmic fashion, the tail size estimation procedure based on the Gnedenko, Pickands, Balkema and de Haan theory which we reviewed earlier and illustrated by examples. This will provide us with a natural transition to the topics presented later on.

- In order to infer properties of the upper tail of the common distribution of the entries of a data sample $x_1, x_2, \cdots\cdots, x_m$, we partition the sample into blocks B_1, B_2, $\cdots$, B_M, and we compute the maxima $M_n = \max_{j \in B_n} x_j$ in each of these blocks.
- Assuming that each block size is large enough, we treat the set $\{M_n\}_n$ of maxima as a sample from a GEV distribution, and assuming that the number of blocks is large enough, we estimate the parameters of this hypothetical GEV distribution from the sample $\{M_n\}_n$
- We infer the size of the tail of the common distribution of the x_j's (in particular the shape parameter ξ) from the values of the estimated parameters and the results of the Gnedenko, Pickands, Balkema and de Haan theory.

It is obvious from the second bullet point above that the inference procedure is justified if the block size is large since we rely on an asymptotic result holding in the limit of the block size going to ∞. Moreover, the estimation of the parameters of the limiting distribution also requires the blocks to be in large numbers. Having both large blocks, and a large number of maxima, requires a very large data set to start with. This sample size requirement is the major shortcoming of this block maxima method. Band-aids have been suggested, the most natural one being to use overlapping blocks. However, the gain in sample size is compensated by a loss in accuracy since the block maxima are not independent any longer, and as a consequence, the parameter estimation procedure looses efficiency. Quantifying the effects of dependencies due to block overlap as well as in the original data has been a concern of many researchers in the field, and the interested reader is referred to the books mentioned in the Notes and Complements at the end of the chapter.

For the time being, we note that the important second bullet point above stresses the need for procedures capable of estimating the parameters of a GEV distribution. This is the task we tackle next. Then, and only then, will we be able to implement the block maxima method and conclude on specific tail size alternatives.

2.2 GEV & GPD Parameter Estimation

The previous section has singled out two distribution families playing an important role in the analysis of extremes and heavy tails: the generalized extreme value and Pareto distributions. It also showed the need for estimating the parameters of these distributions. Maximum likelihood and method of moments are classical statistical procedures frequently used in estimating parameters. This section explains how these methods can be extended to fit these two important parametric distribution families.

2.2.1 The Method of L-Moments

Because many heavy tail distributions do not have enough finite moments (after all, the Cauchy distribution does not even have a first moment!) the classical method of moments cannot be used to estimate the parameters of GPD and GEV distributions. Keeping with the spirit of this time honored estimation procedure, researchers have devised work-arounds by *renormalizing* the traditional statistical moments in order to get analogs which could be used for data with extreme values. With this simplistic strategy in mind, we introduce the notion of theoretical L-moment.

2.2.1.1 Theoretical Definitions

L-moments are defined in terms of the so-called probability weighted moments. These generalized moments are defined for non-negative random variables X with finite expectations and continuous cdf $F(x)$ in the following way. For each integer $r \geq 0$, the r-th probability weighted moment α_r is defined as the number

$$\alpha_r = \mathbb{E}\{X F(X)^r\} = \int_0^\infty x\, F(x)^r\, dF(x), \qquad r = 0, 1, 2, \cdots. \tag{2.9}$$

In other words, in computing the r-th probability weighted moment, we sum the possible values x of the random variable X, but instead of weighting them by their probability of occurrence, we weight them by this probability times the cdf $F(x)$ raised to the power r. Recall that assuming that X is has finite expectation means that

$$\mathbb{E}\{X\} = \int_0^\infty x dF(x) < \infty,$$

which guarantees that all the probability weighted moments α_r make sense as finite numbers since $0 \leq F(x)^r \leq 1$.

As usual we denote the corresponding quantile function by $F^{-1}(x)$, and a simple substitution in the integral appearing in (2.9) gives:

$$\alpha_r = \int_0^1 F^{-1}(y) y^r dy.$$

The L-moments are defined as specific linear combinations of the probability weighted moments with the intent to capture the descriptive features of the distribution in question, namely location, dispersion and other shape parameters. The first few L-moments are defined by the following equations

$$\begin{aligned}
\lambda_1 &= \alpha_0 = \int_0^1 F^{-1}(p)\,dp \\
\lambda_2 &= 2\alpha_1 - \alpha_0 = \int_0^1 F^{-1}(p)(2p-1)\,dp \\
\lambda_3 &= 6\alpha_2 - 6\alpha_1 + \alpha_0 = \int_0^1 F^{-1}(p)(6p^2-6p+1)\,dp \\
\lambda_4 &= 20\alpha_3 - 30\alpha_2 + 12\alpha_1 - \alpha_0
\end{aligned}$$

The coefficients of these linear combinations are nothing but the coefficients of the "shifted Legendre polynomials"

$$P^*_{r-1}(y) = \sum_{k=0}^{r}(-1)^{r-k}\binom{r}{k}\binom{r+k}{k}y^k, \qquad r = 1, 2, \cdots$$

For the sake of definiteness we give the values of the first four Legendre polynomials $P^*_j(y)$:

$$\begin{aligned}
P^*_0(y) &= 1 \\
P^*_1(y) &= 2y - 1 \\
P^*_2(y) &= 6y^2 - 6y + 1 \\
P^*_3(y) &= 20y^3 - 30y^2 + 12y - 1
\end{aligned}$$

An alternative definition of L-moments can be given in terms of order statistics. Such form of the definition will be useful for empirical estimation from data samples. For any given integer $r \geq 1$ and sample $X_1, \cdots, X_r$ of i.i.d. random variables with the same distribution F, we use momentarily the notation

$$X_{(1:r)} \leq X_{(2:r)} \leq \cdots\cdots \leq X_{(r:r)}$$

for the order statistics which we usually denote by $X_{(1)} \leq X_{(2)} \leq \cdots \leq X_{(r)}$. We use this notation to emphasize the dependence of these order statistics on the sample size. Given these preliminaries, the r-th L-moment can be alternatively defined as

$$\lambda_r = \frac{1}{r}\sum_{k=0}^{r-1}(-1)^k\binom{r-1}{k}\mathbb{E}\{X_{((r-k):r)}\}, \qquad r = 1, 2, \cdots \tag{2.10}$$

and using this definition we get the formulae

$$\begin{aligned}
\lambda_1 &= \mathbb{E}\{X\} \\
\lambda_2 &= \frac{1}{2}\left(\mathbb{E}\{X_{(1:2)}\} - \mathbb{E}\{X_{(2:2)}\}\right) \\
\lambda_3 &= \frac{1}{3}\left(\mathbb{E}\{X_{(1:3)}\} - 2\mathbb{E}\{X_{(2:3)}\} + \mathbb{E}\{X_{(3:3)}\}\right),
\end{aligned}$$

Descriptive statistics such as skewness and kurtosis play an important role in the analysis of statistical distributions. Since they are defined in terms of moments and their ratios, they have natural analogs in the present framework. An L-moment ratio is a dimensionless quantity defined as the ratio of an L-moment to the second L-moment. L-skewness, τ_3, is the third L-moment ratio,

$$\tau_3 = \frac{\lambda_3}{\lambda_2},$$

and L-kurtosis, τ_4, is the fourth L-moment ratio,

$$\tau_4 = \frac{\lambda_4}{\lambda_2}.$$

Examples.

- For the uniform distribution $U(0,1)$ we have

$$\lambda_1 = 1/2, \quad \lambda_2 = 1/6, \quad \tau_3 = 0, \quad \tau_4 = 0.$$

- In the case of the standard normal distribution $N(0,1)$ we have

$$\lambda_1 = 0, \quad \lambda_2 = 1/\sqrt{\pi}, \quad \tau_3 = 0, \quad \tau_4 \approx 0.123.$$

- In the case of the exponential distribution with unit rate we have

$$\lambda_1 = 1, \quad \lambda_2 = 1/2, \quad \tau_3 = 1/3, \quad \tau_4 = 1/6.$$

2.2.1.2 *First L-Moments Empirical Estimation*

Given the ordered statistics

$$x_{(1)} \leq x_{(2)} \leq \ldots \leq x_{(n)}$$

of a sample $x_1, x_2, \ldots, x_n$ of size n, the estimate l_r defined by

$$l_r = \frac{1}{\binom{n}{r}} \sum_{0 \leq i_1 < i_2 < \ldots < i_r \leq n} r^{-1} \sum_{k=0}^{r-1} (-1)^k \binom{r-1}{k} x_{(i_{r-k})}$$

is an unbiased estimator of the theoretical r-th L-moment λ_r. Moreover, it has been shown that l_r can be computed, from the order statistics as

$$l_r = (-1)^r \sum_{k=0}^{r-1} (-1)^{r-k} \binom{r}{k} \binom{r+k}{k} a_k, \qquad r = 0, 1, 2, \ldots \tag{2.11}$$

where the numbers a_k are the so-called probability weighted moments defined by

$$a_0 = \overline{x} = \frac{1}{n}\sum_{i=1}^{n} x_i, \qquad a_k = \frac{1}{n}\sum_{j=k+1}^{n} \frac{(j-1)(j-2)\cdots(j-k)}{(n-1)(n-2)\ldots(n-k)} x_{(i)}, \quad k \geq 1. \tag{2.12}$$

Notice that, consistent with our earlier discussion, the coefficients appearing in expression (2.11) are the coefficients of the shifted Legendre polynomials introduced above.

The function `sample.LMOM` gives and implementation of formula (2.12). For the sake of illustration, we compute the L-moments of a Monte Carlo sample of size 1,000 from a GPD.

```
> X <- rpareto(1000)
> sample.LMOM(X)
Mean (l_1) L-mom 2 (l_2)    L-skewness    L-kurtosis
 1.0144528      0.5045187    0.3278689     0.1626310
```

For the sake of comparison we check with the theoretical L-moments of such a GPD. Indeed, since the GPD with location parameter $m = 0$, scale parameter $\lambda = 1$, and shape parameter $\xi = 0$ is nothing but the standard exponential distribution with rate one, we already gave its L-moments L-skewness and L-kurtosis. They are

$$\lambda_1 = 1, \qquad \lambda_2 = \frac{1}{2}, \qquad \tau_3 = \frac{1}{3}, \qquad \tau_4 = \frac{1}{6},$$

which shows that, at least in this case, the estimation procedure gets reasonable values for the parameters. Quite expectedly, the estimates t_3 of L-skewness and t_4 of L-kurtosis computed by the function `sample.LMOM` are obtained as the ratios l_3/l_2 and l_4/l_2, respectively.

Important Remark. Even though a sample mean can be computed from any sample irrespective of the distribution which governs the generation of the values appearing in the sample, it is used as an estimator, only when the theoretical distribution is at least of order one, namely when the mean actually exists. We recalled these facts in our discussion of the law of large numbers in Chap. 1. In particular, the empirical mean can always be computed for a sample from the Cauchy distribution, however, it cannot have the interpretation of an estimate of the mean in that case. A similar state of affairs holds in the case of L-moments. The empirical estimates introduced in this section can always be computed. However, as we said in their introduction, L-moments make sense only for distributions with a first moment. In particular, when we talk about L-moments of GPDs and GEV distributions, we shall always implicitly assume that $\xi < 1$ so the theoretical moment of order one does exist.

2.2.1.3 *Small Sample Alternative*

Because of the very definition of L-moments, estimation involves the approximation of an integral whose integrand depends upon the entire cdf. It is intuitively clear that the error produced by approximating the integral using a numerical quadrature method is much smaller than the error due to the approximation of the cdf from sample data when the sample size is small. For that reason, practitioners have searched for alternative estimates which could perform better with small samples.

The following procedure was proven to give good estimates for the L-moments of GPD in the case of small samples when the parameters γ and δ are chosen appropriately. It is based on the notion of *plotting position*. For each integer n (which will be chosen as the size of the sample under study) the plotting positions are defined as the numbers

$$p_i = \frac{i+\gamma}{n+\delta}$$

and the corresponding estimates of the r-th L-moments are given by

$$l_r = \frac{1}{n}\sum_{i=1}^{n}\sum_{k=0}^{r-1}(-1)^{r-1-k}\binom{r-1}{k}\binom{r-1+k}{k}p_i^k x_i.$$

It has been shown that the plotting position estimators with $\gamma = 0.35$ and $\delta = 0$ produce good approximations of the L-moments for small GPD samples. This method is implemented in the library `Rsafd` by the function `plotting.positions` whose use is illustrated by the following display.

```
> X <- rpareto(50,xi = 0.4)
> PPLM <- plotting.positions(X)
> PPLM
ell_1     ell_2     tau_3     tau_4
2.0628630 1.7630784 0.7029661 0.5526929
> SLM <- sample.LMOM(X)
> SLM
ell_1     ell_2     tau_3     tau_4
2.3626477 1.7845944 0.7196471 0.5867451
```

2.2.1.4 *Distribution Estimation by the Method of L-Moments*

We now explain how estimates of the L-moments can be used to estimate the parameters of generalized extreme value and Pareto distributions.

2.2.1.5 *Estimating the Parameters of a GEV Distribution*

We now concentrate on the case of GEV distributions. Recall that, since the existence of L-moments requires that the common distribution of the observations has at least a first moment, we need to restrict ourselves to the case $\xi < 1$. Under this condition, the L-moments of a GEV distribution can be computed in closed form, leading to the following expressions:

$$\lambda_1 = m - \frac{\lambda(1 - \Gamma(1-\xi))}{\xi}, \tag{2.13}$$

$$\lambda_2 = -\frac{\lambda(1 - 2^\xi)\Gamma(1-\xi)}{\xi}, \tag{2.14}$$

$$\lambda_3 = \frac{\lambda}{\xi}(1 - 3 \cdot 2^\xi + 2 \cdot 3^\xi)\Gamma(1-\xi), \tag{2.15}$$

where $\Gamma(\alpha)$ is the Gamma function whose definition was recalled in (1.12). Taking the ratio of (2.15) to (2.14) we get:

$$\tau_3 = \frac{2(1-3^\xi)}{(1-2^\xi)} - 3.$$

Assuming that the first three L-moments λ_1, λ_2 and λ_3 were estimated as $\hat{\lambda}_1$, $\hat{\lambda}_2$ and $\hat{\lambda}_3$ from an empirical sample, we set $\hat{\tau}_3 = \hat{\lambda}_3/\hat{\lambda}_2$ and we plug the latter in the above equation in lieu of τ_3. Since the equation so obtained involves only the unknown parameter ξ, we can use it to extract a value, say $\hat{\xi}$, for the shape parameter ξ. Obviously, this equation cannot be solved in a closed form, so we use a numerical method to do so. Once this is done, the computation of the remaining estimates is straightforward. The estimate of $\hat{\lambda}$ is easily derived from Eq. (2.14),

$$\hat{\lambda} = -\frac{\hat{\lambda}_2\hat{\xi}}{(1-2^{\hat{\xi}})\Gamma(1-\hat{\xi})}, \tag{2.16}$$

and after that, $\hat{m}$ is obtained from Eq. (2.13) by

$$\hat{m} = \hat{\lambda}_1 + \frac{\hat{\lambda}}{\hat{\xi}}\left(1 - \Gamma(1-\hat{\xi})\right). \tag{2.17}$$

Remark. Since we aim at computing the values of three parameters, we should only need three equations. Not surprisingly, the above methods requires only the knowledge of the first two L-moments l_1 and l_2 and the L-skewness τ_3.

The above method of L-moment estimation of a GEV distribution is implemented in the function `gev.lmom`. Starting with a set of L-moments (as produced for example by the functions `sample.LMOM` or even the function `plotting.positions` discussed above) this function computes estimates of the three parameters of the GEV distribution suspected to have produced these L-moments. We demonstrate its use with the following simulation example where we first estimate the L-moments from a random sample from a GEV distribution which we choose.

```
> X <- rgev(500, lambda = 3.5, xi = 0.4)
> LMOMX <- sample.LMOM(X)
> LMOMX
ell_1          ell_2          tau_3          tau_4
4.2065658 3.9795450 0.4372888 0.3242249
```

```
> gev.lmom(LMOMX)
 $param.est
m          lambda    xi
0.1417766 3.4847654 0.3781168
```

2.2.1.6 *Estimating the Parameters of a GPD*

In the case of GPDs, different methods are used depending upon whether or not the location parameter m is known. The reason for considering these two alternatives will become clear in the next section. When using the POT method to estimate the size of a tail, the estimation procedure consists in fitting a GPD to the exceedances over an appropriately chosen threshold. By construction, the location parameter of a sample of exceedances is automatically zero. If m is known, the GPD L-moment estimators are:

$$\hat{\xi} = 2 - \frac{l_1}{l_2}, \qquad \text{and} \qquad \hat{\lambda} = \left(\frac{l_1}{l_2} - 1\right) l_1 .$$

Notice that, since we assume that m is known, we need to compute values for two parameters only, and hence, two equations are sufficient. In this case, we need only the knowledge of the first two L-moments l_1 and l_2 to estimate the entire GPD.

If m is unknown, we need to compute three parameters. We expect to need three equations. However, instead of using the L-skewness as in the case of GEV distributions, the GPD L-moment estimation procedure which we implemented in `Rsafd` uses the first two L-moments l_1 and l_2 and the first order statistics $x_{(1)}$. The resulting estimates are given by the formulae:

$$\hat{\xi} = -\frac{2(n-1)l_2 - n(l_1 - x_{(1)})}{(n-1)l_2 - (l_1 - x_{(1)})}, \quad \hat{\lambda} = (1-\hat{\xi})(2-\hat{\xi})l_2, \text{ and } \hat{m} = x_{(1)} - \frac{\hat{\lambda}}{n-\hat{\xi}},$$

where $x_{(1)}$ is the smallest value of the sample.

The above method of L-moment estimation of a GPD is implemented in the function `gpd.lmom`. Starting with a set of L-moments and a value for the location parameter m or a sample data set (from which the first order statistic will be computed) this function computes estimates of the three parameters of the GPD suspected to have produced these L-moments. As before, we demonstrate its use with a simulation example where we first estimate the L-moments from a random sample from a GPD which we choose. We give two examples, showing the results both when the location argument is provided and when the sample is provided instead.

```
> X<- rpareto(500,xi = 0.4)
> SLM <- sample.LMOM(X)
> gpd.lmom(SLM,location=0)
$param.est
```

```
          m          lambda            xi
0.0000000 0.9178838 0.4531872
 > gpd.lmom(SLM,sample=X)
 $param.est
          m          lambda          xi
0.002611717 0.912422137 0.455593845
```

2.2.2 Maximum Likelihood Estimation

We now present the most widely used method of parameter estimation. In the situations of interest, the parameter θ is multivariate since it comprises the location parameter m, the scale parameter λ and the shape parameter ξ, so $\theta = (m, \lambda, \xi)$. Since explicit formulae for the density functions of GEV distributions and GPDs can be derived in a straightforward manner from the definition expressions we gave in (2.7) and (2.3), the strategy of the classical maximum likelihood estimation seems appropriate. The only slight difference with the classical cases handled by this method is the fact that the domain of definition of the density function f_θ changes with the parameter. This is a minor hinderance which can be overcome in practice.

2.2.2.1 Likelihood and Log-Likelihood Functions

We first consider the case of the GEV distributions. For the sake of notation, we give separate formulae for the cases $\xi = 0$ and $\xi \neq 0$. When $\xi = 0$, taking derivatives of both sides of (2.7) gives:

$$g_{m,\lambda,0}(x) = \frac{1}{\lambda} e^{-(x-m)/\lambda} \exp[-e^{-(x-m)/\lambda}] \tag{2.18}$$

which implies that the likelihood of a sample $x_1, \cdots, x_n$ is given by

$$L(m, \lambda | x_1, \cdots, x_n) = \frac{1}{\lambda^n} \exp[-\frac{1}{\lambda} \sum_{i=1}^{n} (x_1 - m)] \exp[-\sum_{i=1}^{n} e^{-(x_i - m)/\lambda}] \tag{2.19}$$

and the corresponding log-likelihood by:

$$\mathcal{L}(m, \lambda | x_1, \cdots, x_n) = -n \log \lambda + nm - \frac{1}{\lambda} \sum_{i=1}^{n} x_1 - \sum_{i=1}^{n} e^{-(x_i - m)/\lambda}. \tag{2.20}$$

The case $\xi \neq 0$ leads to similar computations. The density of the GEV distribution is given by:

$$g_{m,\lambda,\xi}(x) = \frac{1}{\lambda} \left(1 + \frac{\xi}{\lambda}(x - m)\right)^{-(1+1/\xi)} \exp\left[-\left(1 + \frac{\xi}{\lambda}(x - m)\right)^{-1/\xi}\right] \tag{2.21}$$

if $x \leq m - \lambda/\xi$ for $\xi < 0$ or $x \geq m - \lambda/\xi$ for $\xi > 0$, and 0 otherwise. This in turn implies that the likelihood of a sample $x_1, \cdots, x_n$ is given by

$$L(m, \lambda, \xi | x_1, \cdots, x_n) = \frac{1}{\lambda^n} \prod_{i=1}^{n} \left(1 + \frac{\xi}{\lambda}(x_i - m)\right)^{-(1+1/\xi)} \exp\left[-\sum_{i=1}^{n} \left(1 + \frac{\xi}{\lambda}(x_i - m)\right)^{-1/\xi}\right] \tag{2.22}$$

if $\max\{x_1, \cdots, x_n\} \leq m - \lambda/\xi$ for $\xi < 0$ or $\min\{x_1, \cdots, x_n\} \geq m - \lambda/\xi$ for $\xi > 0$, and 0 otherwise. Finally, the corresponding log-likelihood is given by:

$$\mathcal{L}(m, \lambda, \xi | x_1, \cdots, x_n) = -n \log \lambda - \left(1 + \frac{1}{\xi}\right) \sum_{i=1}^{n} \log\left(1 + \frac{\xi}{\lambda}(x_i - m)\right) - \sum_{i=1}^{n} \left(1 + \frac{\xi}{\lambda}(x_i - m)\right)^{-1/\xi} \tag{2.23}$$

with the same domain restrictions as before. Consequently, maximum likelihood estimates of the parameters of a GEV distribution are obtained by solving the optimization problem

$$(\hat{m}, \hat{\lambda}, \hat{\xi}) = \arg \sup_{\lambda > 0.\ \lambda + \xi(x_i - m) \geq 0,\ i=1,\cdots,n} L(m, \lambda, \xi | x_1, \cdots, x_n) \tag{2.24}$$

The above constraints guarantee that the density is non-negative at the observations $x_1, \cdots, x_n$. Such an optimization problem could have presented difficulties years ago, but with the advent of modern computers and the development of efficient solvers, it can be solved in a very reliable manner on most every platforms. `S` and `R` come with solvers for nonlinear optimization based on quasi-Newton methods. The library `Rsafd` uses these solvers to produce maximum likelihood estimates of the parameters.

Next, we consider the case of the GPDs. As before, we give separate formulae for the cases $\xi = 0$ and $\xi \neq 0$. The case $\xi = 0$ is well known since it reduces to the classical analysis of exponential samples. Indeed, the density function is given by:

$$f_{m,\lambda,0}(x) = \frac{1}{\lambda} e^{-(x-m)/\lambda} \tag{2.25}$$

if $x \geq m$ and 0 otherwise. This implies that the likelihood of a sample $x_1, \cdots, x_n$ is given by

$$L(m, \lambda | x_1, \cdots, x_n) = \frac{1}{\lambda^n} \exp[-\frac{1}{\lambda} \sum_{i=1}^{n} (x_1 - m)] \tag{2.26}$$

if $\min\{x_1, \cdots, x_n\} \geq m$ and 0 otherwise. Hence, the corresponding log-likelihood is given by:

$$\mathcal{L}(m, \lambda | x_1, \cdots, x_n) = -n \log \lambda + \frac{nm}{\lambda} - \frac{1}{\lambda} \sum_{i=1}^{n} x_i \tag{2.27}$$

which leads to the classical maximum likelihood estimates of the location and scale of an exponential sample.

Computations are simpler in the case $\xi \neq 0$. Indeed, taking derivatives on both sides of (2.3) gives a density of the form:

$$f_{m,\lambda,\xi}(x) = \frac{1}{\lambda} \left(1 + \frac{\xi}{\lambda}(x - m)\right)^{-(1+1/\xi)} \tag{2.28}$$

if $x \leq m - \lambda/\xi$ for $\xi < 0$ or $x \geq m - \lambda/\xi$ for $\xi > 0$, and 0 otherwise. This in turn implies that the likelihood of a sample $x_1, \cdots, x_n$ is given by

$$L(m, \lambda, \xi | x_1, \cdots, x_n) = \frac{1}{\lambda^n} \prod_{i=1}^{n} \left(1 + \frac{\xi}{\lambda}(x_i - m)\right)^{-(1+1/\xi)} \tag{2.29}$$

if $\max\{x1, \cdots, x_n\} \leq m - \lambda/\xi$ for $\xi < 0$ or $\min\{x_1, \cdots, x_n\} \geq m - \lambda/\xi$ for $\xi > 0$, and 0 otherwise. Finally, the corresponding log-likelihood is given by:

$$\mathcal{L}(m, \lambda, \xi | x_1, \cdots, x_n) = -n \log \lambda - \left(1 + \frac{1}{\xi}\right) \sum_{i=1}^{n} \log \left(1 + \frac{\xi}{\lambda}(x_i - m)\right) \tag{2.30}$$

with the same domain restrictions.

2.2.2.2 *MLE of the Parameters of a GPD and GEV Distributions*

Maximum Likelihood Estimates (MLE for short) of the parameters of a GEV distribution and a GPD are provided by the functions `gev.ml` and `gpd.ml`. Since by definition of a maximum likelihood estimate, the result is obtained by solving an optimization problem, one needs to initialize the procedure with a first guess for the set of arguments (i.e. the three parameters of the distribution family). In the `gev.ml` and `gpd.ml` implementations, if no initial guess is provided, a vector of parameter estimates obtained by a different method is used by the function as starting point for the optimization routine attempting to maximize the likelihood. Indeed, if no such argument is specified, L-moment estimates are computed by the functions `gev.ml` and `gpd.ml` and used for initialization purposes. As in the case of L-moment estimation, if the location parameter `m` of a GPD is known, it may be specified, in which case, only the remaining two parameters will be estimated by maximum likelihood.

As before, we demonstrate the use of the functions of the package `Rsafd` with a simulation example where we choose the GEV distribution.

```
> X <- rgev(500, lambda = 3.5, xi = 0.4)
> gev.ml(X)
```

```
$param.est
         m       lambda          xi
-0.1714924  3.4420736  0.4243908

$converged
[1] TRUE
```

Similarly, in the case of a GPD:

```
> X <- rpareto(500, lambda = 3.5, xi = 0.4)
> gpd.ml(X)$param.est
$param.est
          m      lambda          xi
0.001238288 3.171523526 0.467466969
```

2.2.3 An Example Chosen for Pedagogical Reasons

It is possible to propose mathematical models for the time evolution of the PCS index. We described one of them in the Notes & Complements at the end of Chap. 1. These models are most often quite sophisticated, and they are difficult to fit and use in practice. Instead of aiming at a theory of the dynamics of the index, a less ambitious program is to consider the value of the index on any given day, and to perform a static analysis of its marginal distribution. This gives us a chance to illustrate how one uses the tools introduced above to fit a Pareto distribution to the data. The purpose of this exercise is to emphasize the limitations of a blind application of the general theory, and to motivate the modifications introduced and implemented in the following section on semi-parametric estimation.

The Q-Q plots produced in Chap. 1 clearly showed that the upper tail of the PCS index data was heavier than the tail of the exponential distribution. We use the function `gpd.lmom` to fit a GPD to the `PCS.index`, and we print the estimated location, scale and shape parameters with the following commands:

```
> PCS.lmom <- gpd.lmom(PCS.index)$param.est
> PCS.lmom
         m     lambda          xi
0.06824616 0.66521009 0.71314021
```

To visualize the properties of the fit we choose to plot the histogram of the original data set `PCS.index` together with the density of the estimated GPD.

```
> hist(PCS.index,breaks=25,density=20,freq=F)
> X <- seq(from=-1,to=160,length=1000)
> points(X,dpareto(X,m=PCS.lmom[1],lambda=PCS.lmom[2],
                   xi=PCS.lmom[3]),type="l")
```

The plot is given in Fig. 2.4. The fit does not look very good, especially in the left part of the plot where the histogram shows significant positive values.

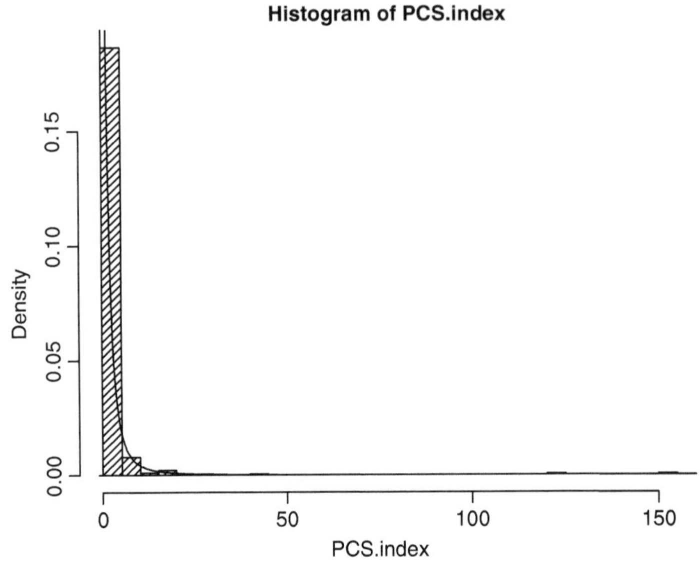

Fig. 2.4. Histogram of the PCS index, together with the density of the Pareto distribution estimated by the method of L-moments

As we explained in the first chapter, histograms and density plots do not give a clear picture of what is happening *in the tail.* So in order to check the goodness of the fit in the tail, we generate a large random sample from the distribution fitted to the data, and we produce a Q-Q plot of the Monte Carlo sample against the original data set PCS.index.

```
> PCS.rlmom <- rpareto(n=10000,m=PCS.lmom[1],
                       lambda=PCS.lmom[2],xi=PCS.lmom[3])
> qqplot(PCS.index,PCS.rlmom)
```

The result is reproduced in the left pane of Fig. 2.5.

As with Fig. 2.4, the result is disappointing. However, the plot in Fig. 2.5 points to a possible reason for the poor fit. Up until the large values, the quantiles of the simulated sample seem to align reasonably well with the quantiles of PCS.index. However the last quantile – quantile point being out of line seems to indicate that the thickness of the tail was not captured properly by the estimated distribution. It happens often that moment estimates are not as good as maximum likelihood estimates, so knowing that, we compute the GPD estimate produced by the function gpd.ml.

```
> PCS.ml <- gpd.ml(PCS.index)
> PCS.ml <-  PCS.ml$param.est
> PCS.ml
        m    lambda        xi
0.0700000 0.7095752 0.6359470
```

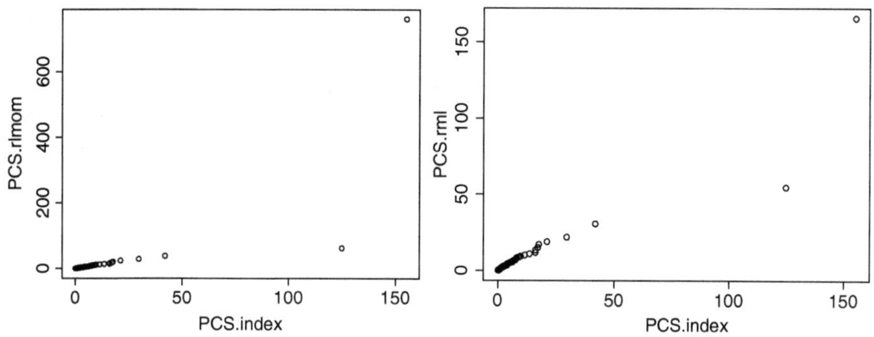

Fig. 2.5. Q-Q-plot of the sample of the one-sided Pareto distribution generated from the parameters estimated with the method of L-moments (*left*) and by maximum likelihood (*right*) against the original PCS index data

As before, we can try the same random generation experiment as before, using the maximum likelihood estimates of the location, scale and shape parameter instead.

```
> PCS.rml <- rpareto(n=10000,m=PCS.ml[1],lambda=PCS.ml[2],
                                                xi=PCS.ml[3])
> qqplot(PCS.index,PCS.rml)
```

The result shown in the right pane of Fig. 2.5 are much better, strikingly good in fact. But a warning is in order as these results are very much dependent upon the actual random sample generated, and as such, they vary from one Monte Carlo experiment to another.

The following final remark uses the example of the PCS index given above to explain some of the reasons why one should not be surprised by the poor performance of these statistical estimation procedures.

Final Remark. Fitting a parametric distribution family as specific as the Pareto family cannot accommodate at the same time the features of the bulk of the data (i.e. the small values of the index in the example treated above), and of the tail (i.e. the extremely large values of the index). It is quite conceivable that the tail of the distribution has a polynomial decay while the left part of the distribution behaves in a non-polynomial way. The estimation procedure tries to find one single set of parameters to fit all the different parts of the distribution, and the resulting compromise often penalizes the tail because by definition, the latter is represented by a small number of data values. This conundrum is at the root of the semi-parametric approach presented in the next section.

2.2.4 Implementation of the Block-Maxima Method

We closed the previous section with a discussion of the block-maxima method, and we explained why its implementation required the estimation of the parameters of a

GEV distribution. The maximum likelihood method and the method of L-moments can now be brought to bear to solve a problem which we could not resolve then. We use the function `block.max` of the package `Rsafd` to illustrate the performance of the method on simulated data sets. Knowing the true shape parameter, and being able to afford as large a data set as needed make it possible to illustrate the shortcomings of the block-maxima method.

We first generate a sample of size $n = 5{,}000$ from the Pareto distribution with shape parameter $\xi = 0.3$. In the context of daily financial data, such a sample size would correspond approximately to 20 years worth of daily data.

Besides the data vector, the function main parameters of `block.max` are the variable `overlap` which is 0 by default and which should be an integer between 0 and 50, and the common length of all the block passed to the function as parameter `block.size` which needs to be an integer greater than or equal to 100. We study the influence of these parameters separately.

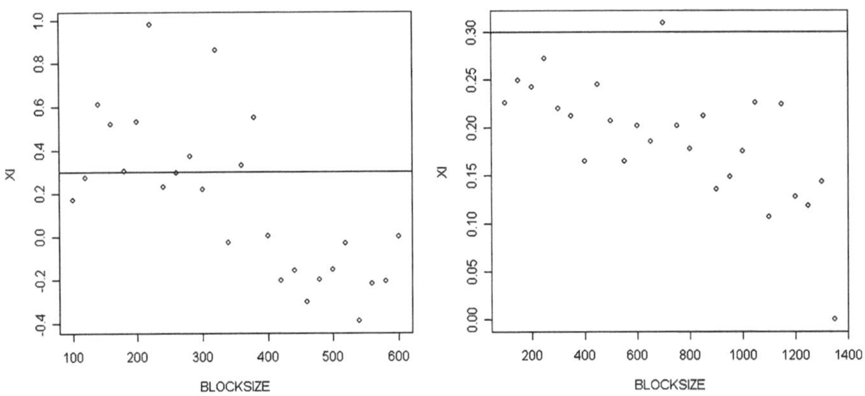

Fig. 2.6. Block-maxima shape parameter estimate as a function of the block size, for a sample of size 5,000 (*left*) and 50,000 (*right*). In both cases the true parameter was $\xi = 0.3$

The left pane of Fig. 2.6 shows the estimates $\hat{\xi}$ given by the block-maxima method when non-overlapping blocks are used. We vary the common length of the blocks from 100 to 600 by increments of 20, and for each fixed block size, we compute and plot the estimate of the shape parameter. The resulting points are scattered, indicating that the method fails in most cases: either the block size is not large enough, or when it is large enough, we do not have enough blocks to get a good estimate of the GEV shape parameter. The right pane gives the plot of the shape parameter estimates for the same block sizes when the data sample is 50,000. The results are obviously much better (notice the differences of the ticks on the vertical axes). However, if the data were arising from daily measurements, one would have to collect 200 years worth of data to have such a sample size. Needless to say, this does not happen often in financial applications.

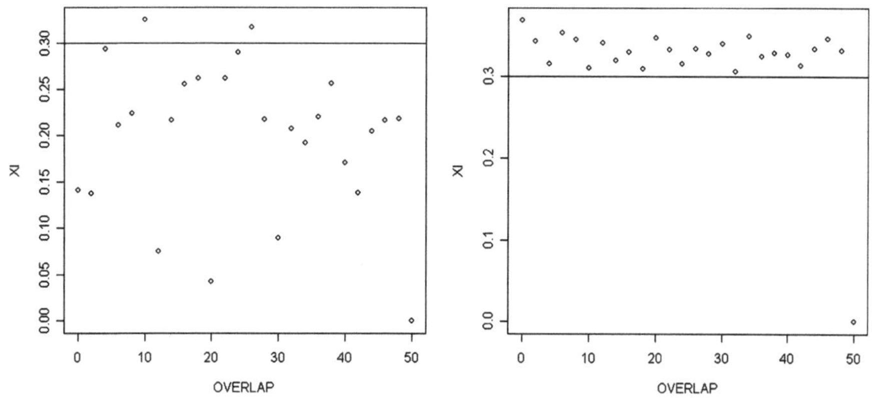

Fig. 2.7. Block-maxima shape parameter estimate as a function of the block overlap, for a sample of size 5,000 (*left*) and 50,000 (*right*). In both cases we plotted a horizontal line at the true value $\xi = 0.3$ of the parameter

The left pane of Fig. 2.7 shows the estimates $\hat{\xi}$ given by the block-maxima method when overlapping blocks are used. We vary the overlap of the blocks from 0 to 50 by increments of 2, and for each overlap, we compute and plot the estimate of the shape parameter ξ. The results are not very good, for essentially the same reasons as before. They improve dramatically when we increase the sample size to 50,000 as shown in the right pane.

2.3 Semi Parametric Estimation

This section is the culmination of the density estimation procedures introduced in this chapter. It combines the benefits of the non-parametric estimation when data are plentiful, and of the parametric methods to estimate generalized Pareto distributions in the tails when the latter are heavier than normal.

2.3.1 Threshold Exceedances

As before, we consider a sample $x_1, \cdots, x_n$ from the distribution of a random variable X with cdf F which we try to *estimate.* In most insurance and financial applications, F is the loss distribution of a portfolio of contracts.

For any given level ℓ, we define the excess distribution over the threshold ℓ as the conditional distribution of $X - \ell$ given $X > \ell$. The corresponding cdf is given by

$$F_\ell(x) = \mathbb{P}\{X - \ell | \leq xX > \ell\} = \frac{F(x+\ell) - F(\ell)}{1 - F(\ell)}, \qquad x \geq 0.$$

The mean of F_ℓ is called the mean excess over the level ℓ, and viewed as a function of the level ℓ, it is called the mean excess function.

$$\ell \hookrightarrow e(\ell) = \mathbb{E}\{X - \ell | X > \ell\}.$$

In the next section, we study risk measures computed from loss distributions. When X represents a loss, the mean excess function gives the expected loss above a given level ℓ.

Examples. When F is an exponential distribution, the memoryless property implies that the excess distribution F_ℓ does not depend upon the level ℓ since $F_\ell(x) \equiv F(x)$. The excess distribution can also be computed explicitly in the case of GPD's. Indeed, for any ℓ we have:

$$\begin{aligned}
F_\ell(x) &= \frac{F_{m,\lambda,\xi}(x+\ell) - F_{m,\lambda,\xi}(\ell)}{1 - F_{m,\lambda,\xi}(\ell)} \\
&= \frac{(1+\xi(x+\ell-m)/\lambda)^{-1/\xi} - (1+\xi(\ell-m)/\lambda)^{-1/\xi}}{(1+\xi(x+\ell-m)/\lambda)^{-1/\xi}} \\
&= 1 - \left[\frac{1+\xi(x+\ell-m)/\lambda}{1+\xi(\ell-m)/\lambda}\right]^{-1/\xi} \\
&= F_{m',\lambda',\xi'}(x)
\end{aligned}$$

with $m' = 0$, $\lambda' = \lambda + \xi(\ell - m)$ and $\xi' = \xi$. So for a GPD, the excess distribution is another GPD located at 0 and with the same shape parameter ξ. This stability property is a remarkable property of the GPD's. Notice also that the mean excess function can only be defined when $\xi < 1$. It can be shown that in this case, it is linear in ℓ since

$$e(\ell) = \frac{\xi}{1-\xi}\ell + \text{cst} \tag{2.31}$$

as can be seen by a direct integration from the explicit form of $F_\ell(x)$ given above.

Empirical Estimation. Given a sample $x_1, \cdots, x_n$ and a level ℓ, we denote by n_ℓ the number of x_j's which are greater than ℓ, i.e. the number of exceedances above the level ℓ, and we denote by $x_1^{(e,\ell)}, \cdots, x_{n_\ell}^{(e,\ell)}$ the actual overshoots over the level ℓ obtained by subtracting ℓ from the n_ℓ values x_j's which are greater than ℓ. In this way, we can think of $x_1^{(e,\ell)}, \cdots, x_{n_\ell}^{(e,\ell)}$ as a sample from the excess distribution above ℓ and the excess function $e(\ell)$ can be estimated by the empirical mean

$$\widehat{e_n}(\ell) = \frac{1}{n_\ell}\sum_{j=1}^{n_\ell} x_j^{(e,\ell)} \tag{2.32}$$

Formula (2.31) shows that, when the sample $x_1^{(e,\ell)}, \cdots, x_{n_\ell}^{(e,\ell)}$ comes from a GPD, then the empirical estimate of the mean excess function given by formula (2.32) should be approximately **linear in the level** ℓ.

We now state in a rather informal way, the main theoretical result of this section. It is known as the Balkema-de Hann-Pickands theorem. It is in the same vein as the main result of the block maxima approach presented in the previous section. However, the practical estimation procedure which it leads to makes a more parsimonious use of the data, hence the reason of its success with practitioners.

Theorem 3. *The distribution of the block maxima M_n converge toward a GEV with shape parameter ξ if and only if the excess distribution $F_\ell(x)$ over a level ℓ converges uniformly in x as ℓ increases, toward a GPD with shape parameter ξ and a scale parameter possibly varying with ℓ.*

The above result is at the root of the Peaks Over Threshold (POT for short) method described below. In particular, it has the following consequence. If the excess distribution $F_\ell(x)$ is essentially a GPD with shape parameter ξ, then the mean excess function over the levels higher than ℓ should be approximately linear. This justifies the use of the *mean excess plot* as a diagnostic for the POT approach. This plot is obtained by graphing the couples

$$(x_j, \widehat{e_n}(x_j))_{j=1,\cdots,n} \tag{2.33}$$

of the empirical estimate of the mean excess function computed at the sample values. Except for the expected fact that the right most points may be randomly varying because of the smaller number of exceedances used to compute the mean excess estimate, this plot should show a linear trend in case the Balkema-de Hann-Pickands theorem holds. We shall use this graphic diagnostic extensively in what follows.

2.3.1.1 ***Peaks Over Threshold Modelling***

We now explain how the theoretical facts reviewed above can be used to estimate the tail of a distribution function which behaves like a GPD beyond a certain threshold. So, if we remember the disappointing results obtained in Sect. 2.2.3 when we tried to fit a GPD to the whole PCS data, the main difference is that instead of forcing a GPD on the entire range of the random samples, we only fit a GPD to the large values in the sample. This seemingly innocent difference will turn out to have drastic effects on the usefulness of the estimates.

As usual we describe the statistical procedure starting from a sample $x_1, \cdots, x_n$ of realizations of random variables $X_1, \cdots, X_n$ which we assume to be independent and with the same cdf F. Our main assumption will be that the Balkema-de Haan-Pickands result stated above as Theorem 3 applies to this distribution. In other words, this common distribution gives rise to a distribution of block maxima converging toward a GEV with shape parameter ξ. The theory presented in the previous section says that this shape parameter ξ determines the size of the upper tail of the distribution, and controls the size and the frequency of the extreme values occurrences.

- The first step is a graphical check that the method is appropriate for the data at hand. Based on the rationale identified in the previous subsection, we check that we are dealing with a generalized Pareto distribution by checking that the mean excess plot is mostly linear (except may be for the few right most points).

- Now, according to Balkema-de Haan-Pickands theorem, for each threshold ℓ high enough, the sample $x_1^{(e,\ell)}, \cdots, x_{n_\ell}^{(e,\ell)}$ of exceedances over the level ℓ form a sample from a distribution which is uniformly close to a GPD with shape parameter ξ and a scale parameter $\lambda = \lambda(\ell)$ which may depend upon ℓ.
- Using this sample of exceedances, we estimate the shape and scale parameters ξ and λ by the method of L-moments, or by maximum likelihood. Let us denote by $\hat{\xi}$ the estimate of the shape parameter and by $\hat{\lambda}$ the estimate of the scale parameter. Note that the location estimate is irrelevant since we consider only exceedances, so the location parameter is necessarily 0.
- The final estimate of the unknown cdf above the level ℓ is then given by the formula

$$\hat{F}(x) = 1 - \frac{n_\ell}{n}\left(1 + \hat{\xi}\frac{x - \ell}{\hat{\lambda}}\right)^{1/\hat{\xi}}, \qquad x \geq \ell \tag{2.34}$$

  The rationale for this estimate is the following. If $x \geq \ell$ we have

$$\begin{aligned} 1 - F(x) &= \mathbb{P}\{X > x | X \geq \ell\}\mathbb{P}\{X \geq \ell\} \\ &= \mathbb{P}\{X - \ell > x - \ell | X \geq \ell\}(1 - F(\ell)) \\ &= (1 - F_\ell(x - \ell))(1 - F(\ell)) \end{aligned} \tag{2.35}$$

  from which the choice of formula (2.34) is now clear. The factor $1 - F(\ell)$ appearing in the right hand side of (2.35) is estimated empirically by the ratio n_ℓ/n giving the empirical frequency of the exceedances. This is usually a reasonable estimate since by definition of the tail of a distribution, most of the data values in the sample are below the level ℓ. The estimate of the first factor of (2.35) is taken from the fact that the sample of exceedances above ℓ is a sample from a GPD whose shape and scale parameters have been estimated.

This estimation strategy is extended in the next subsection to handle the estimation of entire distributions.

2.3.2 Semi Parametric Estimation

After reviewing the classical parametric and non-parametric methods of density estimation, we introduce our method of choice to estimate heavy tail distributions.

By definition of the tails of a distribution, most of the sample values do bundle up in the *center* or *bulk* of the distribution. On this part of the domain, the density and the cumulative distribution functions can efficiently be estimated by non-parametric methods. Appealing again to the definition of the tails of a distribution, one knows that observations in the tails, even if they end up being extreme, may not be plentiful, and as a consequence, parametric estimation methods will make a better use of the scarce data. This is exactly the philosophy promoted by the POT approach: identify a threshold to the left of which the distribution can be estimated non-parametrically, and beyond which it is estimated parametrically.

Remark: Going beyond the Data. Another advantage of the parametric estimation of the tails is the possibility to *go beyond the data.* Indeed, non-parametric methods are limited by the scope of the data. Except for minor leakage produced by the smoothing of kernel-like methods, (especially when the bandwidth is too large) a non-parametric estimate of a distribution will not assign probability to values which are not part of the sample (i.e. *have not been observed in the past*). So in terms of extreme events, nothing more extreme than what has already been observed will carry any probability: so non-parametric methods cannot foresee events more extreme than those that have already been observed. Parametric methods can. Indeed, having used the data at hand to estimate the shape parameter ξ, the density estimate will extend beyond the most extreme observed data values, and extreme events will be given a positive probability (depending on the estimate of ξ) even if they never occurred in the past.

Identifying the Tails. Before getting into the gory details of the estimation procedures, the first question to address is:

where does the center of the distribution end, and where do the tails start?

As in most cases, common sense will be required to make sure that poor choices do not bias the estimates of the shape parameters in a significant way. The POT implementation of `fit.gpd` can be used without having to make this delicate choice. If values of the thresholds are not provided, the program uses values which guarantee that the tail contains 15 % of the points when the data set is small, and 150 observations when the original data set is large. But we should be clear on the fact that there is no panacea, and that any automatic threshold choice will fail from time to time. The solution we recommend is to use the plots provided by the function `shape.plot` to choose the thresholds.

As we shall see throughout the remainder of this chapter, the results of many analyzes depend upon the choices of these thresholds. So we encourage the reader to get a sense of the sensitivities of his or her results with respect to the choices of the thresholds. To this effect we propose an enlightening simulation example in Problem 2.8 below. It was designed for pedagogical reasons to illustrate the possible biases in the estimates of the tail shape parameters with poor choices of the thresholds. We show that the POT method can fail in two ways: either by not including enough observations in the tail (this is typically the case when the absolute value of the threshold is too large), or by including too many observations from the center of the distribution in the tails when the absolute value of the threshold is not large enough. This simulation example also shows that the graphical diagnostics offered by the function `shape.plot` are our best weapon against the dangers of poor threshold choices.

2.3.2.1 ***The POT Strategy***

We first recall the main steps in this strategy to estimate the tail(s) of a distribution. We concentrate on the upper tail for the sake of definiteness. Let X be a random variable with distribution F_X, let $x_1, x_2, \ldots, x_n$, be a random sample of observations of X, and let $x_{(1)} \leq x_{(2)} \leq \ldots \leq x_{(n)}$ be its order statistics. We assume that we already gathered evidence (usually from descriptive statistics and plots such as Q-Q plots) that the tail of the distribution is of a *generalized Pareto* type. Not only does that implies that F_X is in the domain of attraction of a GEV distribution in the sense that the distributions of properly normalized block maxima converge toward a GEV distribution, but the POT theory does also apply. In other words, we can use the fact that, provided the level ℓ is appropriately chosen, the conditional distribution of excesses over ℓ can be closely approximated by a GPD:

$$F_\ell(x) = \mathbb{P}\{X \leq x + \ell | X > \ell\} = \frac{F_X(\ell + x) - F_X(\ell)}{1 - F_X(\ell)} \sim F_{m=0,\lambda(\ell),\xi}(x).$$

As explained in formula (2.34), given a threshold level ℓ, $F_X(x)$ is estimated by a non-parametric empirical cdf for $x \leq \ell$, and by a GPD for $x > \ell$. To be specific, we choose the estimate

$$\hat{F}(x) = \begin{cases} \frac{i-0.5}{n} & \text{if } x_{(i)} \leq x < x_{(i+1)} \text{ and } x \leq \ell, \\ 1 - \frac{n_\ell}{n}\left(1 - \frac{\hat{\xi}(x-\ell)}{\hat{a}}\right)^{1/\hat{k}} & \text{if } x > \ell, \end{cases}$$

where n_ℓ is the number of points greater than ℓ in the sample. This estimate is implemented in the function `fit.gpd` of the package `Rsafd`. Strictly speaking, the above non-parametric part is implemented in the way described above when the optional parameter `linear` is set to `FALSE`. If `linear = TRUE`, then $\hat{F}(x)$ is linearly interpolated for $x \leq \ell$.

Moreover, as we can see from a quick look at the explanations in the help file, this function can handle distributions with two tails. In that case, instead of one single level ℓ, we need to identify two thresholds which we call `upper` and `lower`. The non-parametric estimation of the cdf is now restricted to the interval limited by the thresholds `lower` and `upper`. Furthermore, the exceedances above the threshold `upper` are used as described above to estimate the shape parameter of the upper tail, while similarly, the excursions below the threshold `lower` are treated in the same way to estimate the shape parameter of the lower tail. Obviously the two shape parameter estimates can be different, this is a result of the flexibility of the method of estimation.

2.3.3 The Example of the PCS Index Revisited

We now revisit the estimation of the distribution of `PCS.index` already considered in Sect. 2.2.3 where we attempted to fit a one-sided ordinary Pareto distribution. Here, we try to fit a GPD with the tools of the library `Rsafd`. As noticed at the start of Sect. 2.3.2, the first order of business is to choose a cut-off value to separate the tail from the bulk of the distribution. This choice should be driven by the following two seemingly contradictory requirements. The cut-off point should be large enough so that the behavior of the tail is homogeneous beyond this threshold. But at the same time, it should not be too large, as we need enough data points beyond the threshold to guarantee a reasonable estimation of ξ by the POT method. For the sake of the discussion, we make a specific choice without justification, leaving the discussion of a reasonable procedure to choose the threshold to our explanations about the function `shape.plot` later in this subsection.

```
> PCS.est <- fit.gpd(PCS.index, tail="upper",upper=4)
```

This command creates an object `PCS.est` of class gpd which contains all we need to know about the estimation results. As a side effect, it also generates a plot. We reproduce the latter in Fig. 2.8. We shall also give examples of ways to extract information from the objects thus created. We used the parameter `tail="upper"` because the distribution does not have a lower/left tail (remember that all the values of the index are positive). According to our earlier discussion of the mean excess plots, the fact that the points appearing in the left part of the plot in Fig. 2.8 are essentially in a straight line is an indication that a generalized Pareto distribution may be appropriate.

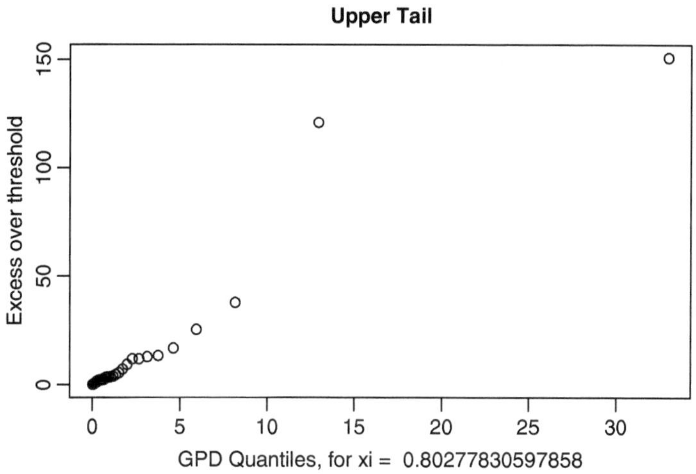

Fig. 2.8. Mean excess plot from the use of the function `fit.gpd` on the PCS index data

Plotting an object of class gpd with the command `plot(PCS.est)` would produce four plots: a plot of the excesses, a plot of the tail of the underlying distribution, and also a scatterplot and a Q-Q plot of the residuals. Since we are mostly interested in the second of these plots, we use instead the command `tailplot` to visualize the quality of the fit. For the sake of illustration we run the commands:

```
> tailplot(PCS.est)
```

and reproduce the result in the bottom pane in Fig. 2.9. Notice that the vertical axis is for the *survival function* $1 - F(x)$, instead of the cdf $F(x)$. The use of the option `optlog` forces `R` to use the natural scale instead of the logarithmic scale which is used by default. This is done for the first plot reproduced on the top of Fig. 2.9. Unfortunately, the curve sticks very early to the horizontal axis and it is extremely difficult to properly quantify the quality of the fit. In other words, this plot is not very instructive. It was given for illustration purposes only. Plotting both the values of the index, and the values of the survival function on a logarithmic scale makes it easier to see how well (or possibly how poorly) the fitted distribution gives an account of the data. The second command (using the default value of the parameter `optlog`) gives the plot of the survival function in logarithmic scales. Both plots show that the fit is very good. Our next inquiry concerns the value of the shape parameter ξ. Remember that this number is what controls the *power decay* of the density in the tail of the distribution at ∞. The choice of a threshold indicating the beginning of the tail, forces an estimate of ξ. The value of this estimate is printed on the plot produced by the function `fit.gpd` and it can be read off Fig. 2.8. Since the location parameter is passed to the function as the (upper) threshold determining the beginning of the tail, only two parameters are fitted. The estimated values for the parameters are included in the object `PCS.est` and can be extracted in the following way:

```
> PCS.est@upper.par.ests
   lambda          xi
4.5014927 0.8027783,
```

the command `$upper.par.ests[2]` giving the single shape parameter `xi`.

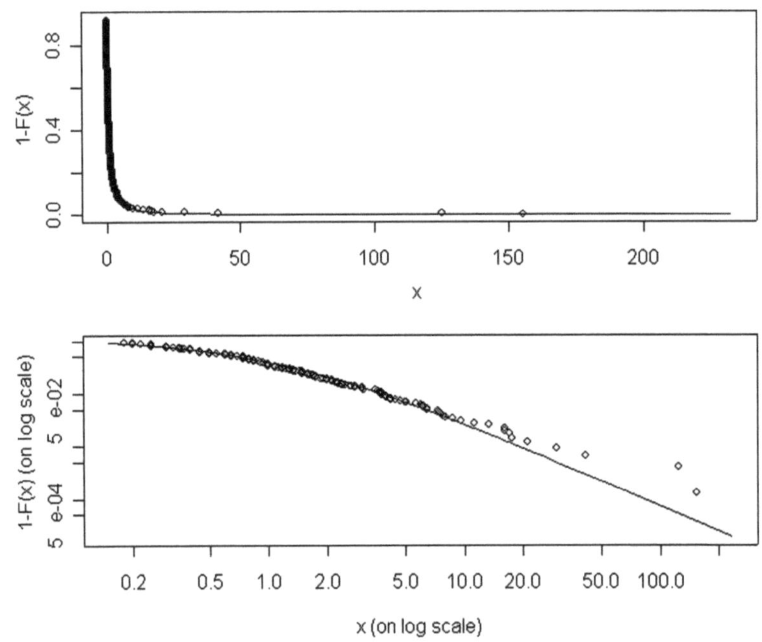

Fig. 2.9. Plot of the tail of the GPD fitted to the PCS data together with the empirical tail given by the actual data points, in the natural scale (*top*) and in logarithmic scale (*bottom*)

Changing the value of the threshold `upper` in the call of the function `fit.gpd` changes the value of the estimate of ξ, so we should be concerned with the stability of the result: we would not want to rely on a procedure that is too sensitive to small changes in the choice of the threshold. Indeed, since there is no obvious way to choose this threshold, the result of the estimation of the shape parameter should remain robust to reasonable errors/variations in the choice of this threshold. The best way to check that this is indeed the case is graphical. It relies on the use of the function `shape.plot` which gives a plot of the estimates of the shape parameter ξ as they change with the values of the threshold used to produce these estimates. The command:

```
> shape.plot(PCS.index)
```

produces a plot of all the different estimates of ξ which can be obtained by varying the threshold parameter `upper`. This plot is reproduced in Fig. 2.10. The leftmost part of the plot should be ignored because, if the threshold is too small, too much of the bulk of the data (which should be included in the center of the distribution) contributes to the estimate of the tail, biasing the result. The rightmost part of the plot should be ignored as well because, if the threshold is too large, not enough points contribute to the estimate. A horizontal axis was added to the upper part of the plot to give the percentage of points included in the estimate. This information is extremely

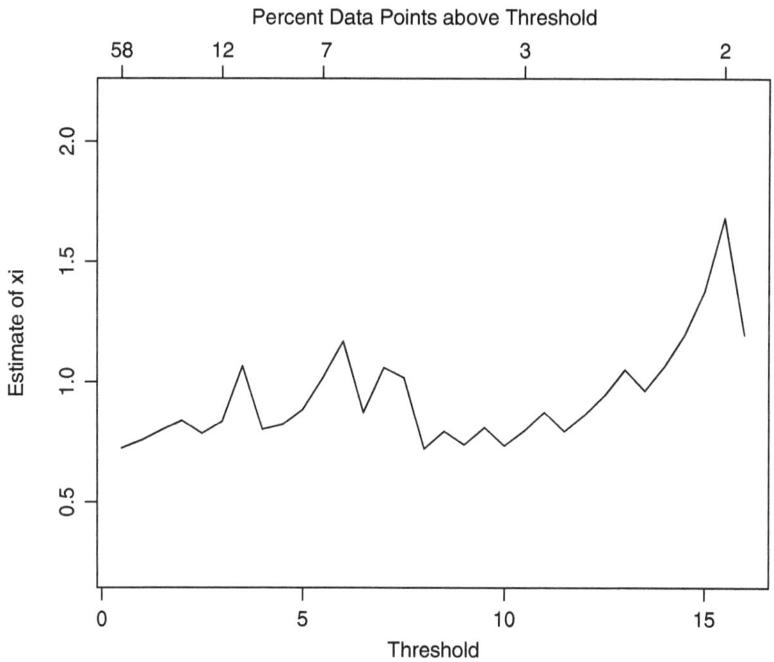

Fig. 2.10. PCS data shape parameter ξ (*vertical axis*) as function of the upper threshold (*lower horizontal axis*) and the corresponding percentage of point in the subsequently defined tail (*upper horizontal axis*)

useful when it comes to deciding whether one should take seriously some of the estimates of ξ which appear on the left and right ends of the plot. The central part of the graph should be essentially horizontal (though not always a straight line) when the empirical distribution of the data can be reasonably well explained by a GPD. This is indeed the case in the present situation, and a value of $\xi = 0.8$ seems to be a reasonable estimate for the intercept of a horizontal line fitting the central part of the graph. Also from this plot we see that the particular choice `upper=4` we made for the location threshold gives a tail containing approximately 10 % of the sample points, which gives a sample of size 38 (since the size of the vector `PCS.index` is 381) for the estimation of the scale and shape parameters λ and ξ which is reasonable.

Our last test of the efficiency of our extreme value toolbox is crucial for risk analysis and stress testing of stochastic systems suspected to carry extreme rare events. It addresses the following important question: can we generate random samples from a generalized Pareto distribution fitted to a data set? The function `qgpd` was included in the library `Rsafd` for the sole purpose of answering this question. If `X` is a vector of numerical values, and `gpd.object` is a gpd.object, then `qgpd(gpd.object,X)` gives the vector of the values computed at the entries of `X`, of the quantile function (i.e. the inverse of the cdf) of the GPD whose characteristics are given by `gpd.object`. If we recall our discussion in Chap. 1 of the way Monte Carlo samples from a given distribution can be generated if one can evaluate the quantile function, we see that, replacing the numerical vector `X` by a sample from the uniform distribution will give a sample from the desired distribution. We now show how this is done in the case of the PCS index. The command

```
> PCSsim <- qgpd(PCS.est,runif(length(PCS.index)))
```

produces a random sample of the same size as the original PCS data from the GPD fitted to the data. The plots produced by the following commands are reproduced in Fig. 2.11.

```
> par(mfrow=c(1,2))
> plot(PCS[,1],PCS.index)
> plot(PCS[,1],PCSsim)
> par(mfrow=c(1,1))
```

When the `R` function `plot` is called with a couple of numerical vectors with the same numbers of rows say n, as arguments, it produces a plot of n points whose coordinates are the entries found in the rows of the two vectors. Putting next to each other the sequential plots of the original data, and of this simulation, shows that our simulated sample seems to have the same statistical features as the original data. This claim is not the result of a rigorous test, but at this stage, we shall consider ourselves as satisfied! See nevertheless Problem 2.4 for an attempt at quantifying the goodness of fit.

2.3.4 The Example of the Weekly S&P Returns

The following analysis is very similar to the previous one, the main difference being the presence of two tails instead of one. We include it in the text to show the details

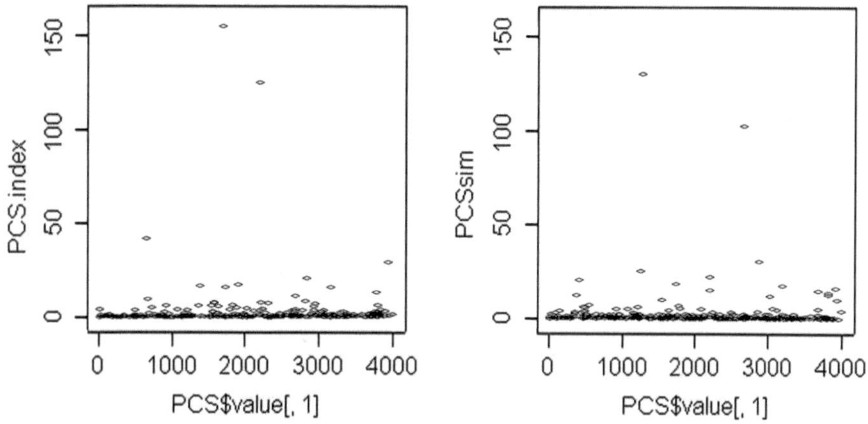

Fig. 2.11. PCS original data (*left*) and simulated sample (*right*)

of all the steps necessary to perform a complete analysis in this case, i.e. when the distribution is unbounded both from above *and* below. We choose the thresholds designating the end points of the tails from the output of the function `shape.plot`. From the results of the command:

```
> shape.plot(WSPLRet, tail="two")
```

reproduced in Fig. 2.12 we see that 0.02 and −0.02 are reasonable choices for the upper and lower thresholds to be fed to the function `fit.gpd`. So the fundamental object of the fitting procedure is obtained using the command:

```
> WSPLRet.est <- fit.gpd(WSPLRet,lower=-0.02,upper=0.02)
```

Notice also that the shape plots in Fig. 2.12 confirm the differences in the sizes of the left and right tails: the frequency and the size of the negative weekly log-returns are not the same as the positive ones. The threshold parameters `lower` and `upper` do not have to be given "opposite" values, i.e. they do not need to have the same absolute values. This is likely to be the case for symmetric distributions, but it does not have to be the case in general. Finally, notice that we did not have to set the parameter `one.tail` by including `one.tail=FALSE` in the command because this is done by default. The above command produced the two plots given in Fig. 2.13.

Both sets of points appear to be essentially in a straight line, so a generalized Pareto distribution is a reasonable guess. Notice that the two estimates of the shape parameter ξ are not the same. The estimates obtained from the particular choices of the threshold parameters `lower` and `upper` are $\xi_{left} = 0.24$ and $\xi_{right} = -0.01$. If the distribution is not symmetric, there is no special reason for the two values of ξ to be the same, in other words, there is no particular reason why in general the polynomial decays of the right and left tails should be identical! As before, we can check visually the quality of the fit by superimposing the empirical distribution of the points in the tails onto the theoretical graphs of the tails of the fitted distributions. This is done with the command:

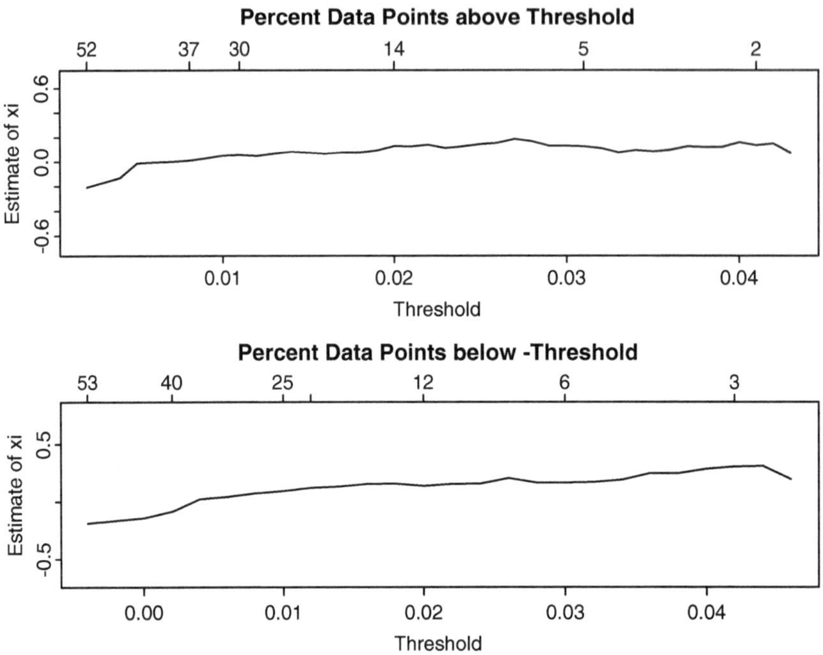

Fig. 2.12. Values of the shape parameter ξ for the right tail (*top*) and left tail (*bottom*) of the distribution of the weekly log-returns of the S&P 500 index, as functions of the values of the thresholds marking the ends of the tails

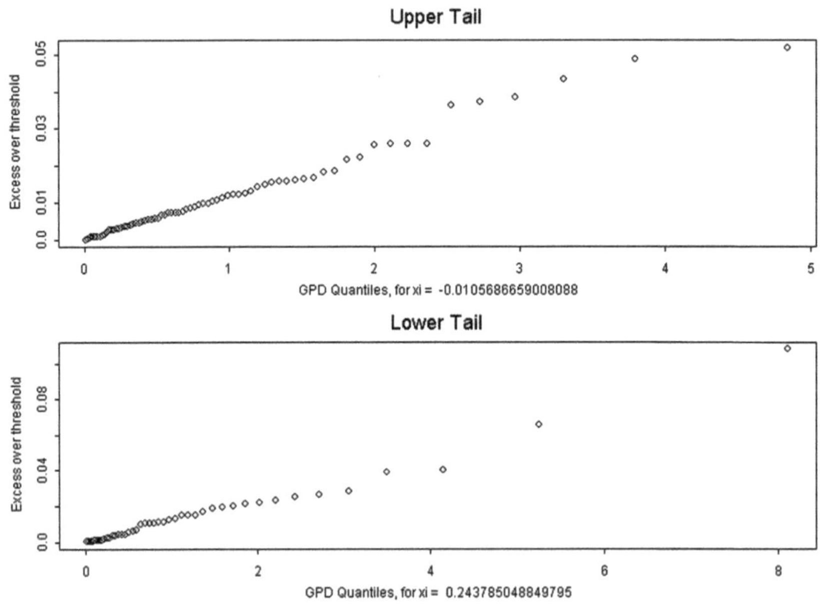

Fig. 2.13. Mean excess plots for the right/upper tail (*top*) and left/lower tail (*bottom*) resulting from the fit of a GPD distribution to the weekly S&P log-return data

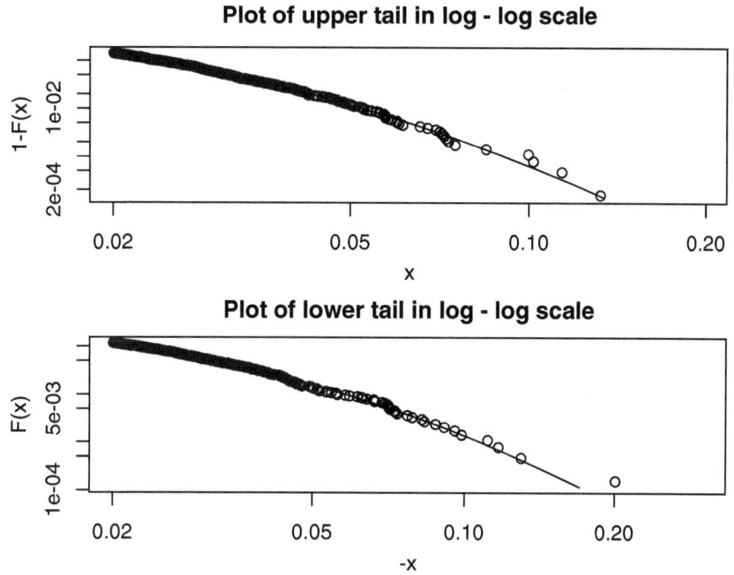

Fig. 2.14. Plot of the tails of the GPD fitted to the `WSPLRet` data together with the empirical tails given by the actual data points, for the upper tail (*top*) and the lower tail (*bottom*)

```
> tailplot(WSPLRet.est,tail="two")
```

which produces the plots given in Fig. 2.14, showing the results (in logarithmic scale) for both tails. Using the quantile function `qgpd(WSPLRet.est, . )` as before, we can generate a sample of size `N` from the fitted distribution with the command:

```
> WSPLRetsim <- qgpd(WSPLRet.est,runif(N))
```

Appendix: Risk Measures: Why and What For?

The goal of this appendix is to give a more mathematical account of the notion of measure of risk as it emerged in the development of mathematical models for applications in the financial and insurance industries.

Historically, and especially in the financial and insurance industries, risk has been equated to the size of the fluctuations of random outcomes as quantified by the standard deviations of these outcomes. Markowitz' mean-variance portfolio theory is the epitome of such a risk-reward modelling. In line with our introduction of the value at risk, the modern approach to risk measure is based on efforts to quantify capital requirements of financial institutions, and risk measures are now used as yardsticks for insurance underwriting, to allocate capital, to identify prudent investment strategies and acceptable future net worths. It is now a commonly accepted view that one should think of a risk measure as a way to estimate the

minimum extra capital which makes the future position acceptable.

Axiomatic Set-Up

The basic objects of the theory are random quantities intended to describe all the states of nature at a future date (for example the future values of a portfolio). The possible outcomes of these random variables represent the scenarios which could occur depending upon market changes and other random events. The set of *acceptable* positions and portfolios is decided by the regulator (states of the world requiring government resources – *guarantor of last resort*), an exchange or clearing firm, the investment manager in charge of the portfolio, the board of Directors, etc. For the purpose of illustration of the importance of heavy tail distributions in the quantification of risk, we use a very simple model in order to capture some of the most important stylized facts needed to make our point. We only consider one period static models and we denote by Ω the set of all possible outcomes/scenarios. A *risk* X is a function on Ω giving the possible values $X(\omega)$ of a position at the end of the period. The set $\mathcal{A}$ of *acceptable risks* is a subset of the set of risks satisfying a set of *axioms* which will be articulated later, and a *risk measure* ρ is a function associating a real number $\rho(X)$ to each risk X. The interpretation of $\rho(X)$ is captured in the following bullet points:

- If $\rho(X) > 0$, $\rho(X)$ is the minimum extra cash one has to add to the position (and invest prudently in the instrument) in order to make the position acceptable;
- If $\rho(X) < 0$, as much as $-\rho(X)$ can be withdrawn from the position without making it unacceptable.

With this interpretation in mind, the following theoretical properties become natural:

- **Shift Invariance** for all real number m and X
$$\rho(X + m) = \rho(X) - m$$
Intuitively, this axiom means that adding cash to a position reduces the risk by the same amount;
- **Monotonicity** for all X_1 and X_2
$$\text{if} \quad X_1 \leq X_2 \quad \text{then} \quad \rho(X_2) \leq \rho(X_1)$$
Intuitively, this condition means that the higher the value of the asset or portfolio, the smaller the risk;
- **Convexity** for all X_1 and X_2 and real numbers $\lambda_1 \geq 0$ and $\lambda_2 \geq 0$ such that $\lambda_1 + \lambda_2 = 1$,
$$\rho(\lambda_1 X_1 + \lambda_2 X_2) \leq \lambda_1 \rho(X_1) + \lambda_2 \rho(X_2)$$

This last axiom has a clear geometric interpretation. Financially speaking, this axiom has a very important consequence: it implies that a measure of risk satisfying this axiom should **encourage diversification** as the risk of an aggregate portfolio (in the left hand side of the above inequality) is lower than the aggregation of the risks of the individual components (the above right hand side).

In modern textbooks on quantitative risk management, a risk measure satisfying these four axioms is called a *convex risk measure*.

First Example

After choosing a benchmark instrument whose returns over the given horizon we denote R and a set $\mathcal{P}$ of probabilities (agent beliefs) on the set Ω of outcomes, for each (random variable) risk X we set:

$$\rho_{\mathcal{P}}(X) \;=\; \sup_{\mathbb{P}\in\mathcal{P}} \mathbb{E}_{\mathbb{P}}\{-X/R\}$$

The interpretation of $\rho_{\mathcal{P}}(X)$ is the following: for a given risk X, $\rho_{\mathcal{P}}(X)$ represents the worst expected discounted loss computed from an a priori set of beliefs. So-defined, $\rho_{\mathcal{P}}(X)$ is a coherent measure of risk.

Second Example: Value at Risk (VaR)

We jump in directly to the formal definition of a mathematical notion of value at risk, referring to the discussion of Sect. 1.1.3 in Chap. 1 for a discussion of the practical applications leading to the abstract definition in terms of quantile of a distribution. Given a probability level $p \in (0,1)$, the value at risk VaR_p (at level p and for the given horizon) of the final net worth X is the negative of the $100p$ percentile of X

$$VaR_p(X) = -\inf\{x;\ \mathbb{P}\{X \le x\} > p\}$$

Clearly this measure of risk is given by the amount of capital needed to make the position X acceptable with probability $1-p$ if acceptability is understood as being positive. Value at Risk is widely used as internal risk control (in accordance with Basel I), however in practice, it is **not clear** which probability to use in order to compute the percentile quantifying the risk: should one use a quantile estimated from historical data, or should one use a probability model calibrated to be risk neutral? This is only one of the very many practical problems associated with the use of VaR as a measure of financial risk.

In order to emphasize the dramatic effect that the choice of a particular distribution can have, we recall the comparison of the percentiles of two distributions presented given in Chap. 1. Choosing the probability level $p = 2.5\,\%$ for the sake of definiteness, if a portfolio manager assumes that the P&L distribution is Gaussian, then she will report that the value at risk is 1.96 (never mind the units, just bear with me), but if she assumes that the P&L distribution is Cauchy, the reported value at risk will jump to 12.71. Quite a difference!!! This shocking example illustrates the crucial importance of the choice of a model for the P&L distribution. As we just proved, this choice is not innocent, and as a consequence, open to abuse.

VaR Computation from Empirical Data

For the sake of illustration, we give a simple example based on data already analyzed in this chapter. We shall discuss less stylized and less contrived examples in Chap. 3 next. Even in this simple situation, several avenues are possible:

- One can use empirical VaR given by the empirical estimate of the percentile;
- One can also assume that the portfolio returns are reasonably explained by a Gaussian model in which case we
  - Estimate the mean and the variance of the sample returns;
  - Compute the quantile of the corresponding Gaussian cdf;
- Finally, one can also use the tools developed earlier in the chapter to fit heavy tail distributions, in which case we
  - Fit a GPD to the returns;
  - Compute the quantile of the estimated distribution.

To illustrate the differences between these three procedures on a specific example, we choose the weekly S&P 500 log return data already studied in this chapter. The numerical results reported in Table 2.3 were obtained by running the commands:

```
> -quantile(WSPLRet,0.01)
> -qnorm(0.01,mean=mean(WSPLRet),sd=sd(WSPLRet))
> -qgpdt(WSPLRet.est,0.01)
```

	Empirical quantile	**Gaussian** model	**GPD** model
$VaR_{0.01}$	0.05582396	0.0471736	0.0582578

Table 2.3. One week 1 % Values at Risk from the S&P 500 index data

Clearly VaR computed under the Gaussian hypothesis is the smallest of the three, offering the most optimistic vision of the risk over a period of one week. The most conservative vision is offered by the fit of a GPD to the weekly returns, while the empirical VaR is reasonable because of the large size of the data set and the presence of a few crashes. The reader is encouraged to rerun this analysis with data prior to October 1987 to understand the role of the semi-parametric fitting procedure in the estimation of the tail of the distribution.

The above example is still rather academic as most practical situations involve multiple underlying instruments (baskets including stocks, bonds, and derivatives)

or even aggregations at the fund or company level of the risk profiles of many desks or business units. The estimation is more difficult in this case. Indeed, as we shall see in the next chapter, estimating separately the risks of the individual desks or units does not help much in deciding how to aggregate these risks while integrating their interdependencies. This is a very *touchy business* and except for the Gaussian case for which one can perform analytic computations, not many tools are available and we will rely on the theory of copulas developed in Chap. 3 and on Monte Carlo computations. In any case, we want to issue the following warning: Using a Gaussian computation when heavy tails are present GROSSLY UNDERESTIMATES the value at risk!

Troubling Example

We illustrate the main shortcomings of VaR with an example which, despite its rather artificial nature, captures well the features of VaR which we want to emphasize. Let us assume that the short interest rate is zero, that the spread on ALL corporate bonds is $2\,\%$, and that corporate bonds default with probability $1\,\%$, independently of each other. **First scenario** We assume that 1,000,000 is borrowed at the base rate and invested in the bond of a single company. In this case:

$$VaR_{0.05} = -20{,}000$$

in other words, there is no risk. **Second scenario** Now let us assume that searching for risk diversification, the same type of investment is set up so that 1,000,000 are borrowed at the base rate and invested in equal parts in the bonds of 100 different companies. Then

$$\mathbb{P}\{\text{at least two companies default}\} > 0.18$$

So $\mathbb{P}\{X < 0\} > 0.05$ and consequently $VaR_{0.05}(X) > 0$ and the portfolio now appears to be risky. In conclusion we see that

- VaR did not detect **over-concentration** of risk
- VaR did not encourage (in fact, it sometimes discourages) **diversification**

Summarizing the shortcomings of VaR we see that

- At the intuitive level VaR only captures the minimal size of a "one in a hundred" event, and highlights merely the best of the rare extreme events to be feared;
- At the mathematical level, VaR is not sub-additive, so VaR is not a convex measure of risk as it does not encourage diversification.

Conditional Value at Risk (CVaR) and Expected Shortfall (ES)

Despite its popularity, VaR's not encouraging diversification pushed academics and some practitioners to design and adopt risk measures free of this shortcoming. The natural candidate for taking over VaR is the Expected Short Fall which, while keeping the spirit of VaR in considering only rare losses, takes into account the actual sizes of these losses, something that VaR does not do. This measure of risk is also called *TailVaR* or *Tail Conditional Expectation*

For a given probability level p, the shortfall distribution is the cdf Θ_p defined by:

$$\Theta_q(x) = \mathbb{P}\{X \leq x | X > VaR_p\}. \qquad (2.36)$$

This distribution is just the conditional loss distribution given that the loss exceeds the Value at Risk at that level. The mean or expected value of this distribution is called the expected shortfall, and is denoted by ES_p. Mathematically, ES_p is given by:

$$ES_p = \mathbb{E}\{X | X > VaR_p\} = \int x \, d\Theta_p(x) = \frac{1}{q} \int_{x > VaR_p} x \, dF(x). \qquad (2.37)$$

It gives the expected loss size given that the loss is more extreme than VaR at the same level. Defined this way, it fixes most of the problems of VaR

- At the intuitive level, the sizes of the losses are taken into account,
- At the *theoretical* level it can be proven that it is *essentially* a coherent measure of risk.

A good part of risk analysis concentrates on the estimation of the value at risk VaR_p and the expected shortfall ES_p of various portfolio exposures. The main difficulty comes from the fact that the theoretical cdf F is unknown, and its estimation is extremely delicate since it involves the control of rare events. Indeed, by the definition of a tail event, very few observations are available for that purpose. More worrisome is the fact that the computation of the expected shortfall involves integrals which, in most cases, need to be evaluated numerically.

PROBLEMS

Ⓣ **Problem 2.1** *Explain (in two short sentences) the conflicting conditions which you try to satisfy when choosing the threshold in fitting a GPD to the tail of a distribution using the POT (Peak over Threshold) method.*

Ⓣ **Problem 2.2**

1. *For this first question we assume that X is a random variable with standard Pareto distribution with shape parameter ξ (location parameter $m = 0$, scale parameter $\lambda = 1$).*
   1.1. *Give a formula for the c.d.f. of X. Explain.*
   1.2. *Derive a formula for the quantile function of X.*
   1.3. *How would you generate Monte Carlo samples from the distribution of X if you only had a random generator for the uniform distribution on $[0, 1]$ at your disposal?*

2. *Give a formula for the density $f_Y(y)$ of a random variable Y which is equal to an exponential random variable with mean 2 with probability $1/3$ and to the negative of a classical Pareto random variable with shape parameter $\xi = 1/2$ (location $m = 0$ and scale $\lambda = 1$) with probability $2/3$. Explain.*
3. *How would you generate Monte Carlo samples from the distribution of Y?*

Ⓣ **Problem 2.3** *In this problem, we study the loss distribution of a portfolio over a fixed period whose length does not play any role in the analysis. Loss is understood as the negative part of the return defined as $L = \max(-R, 0)$. We assume that a fixed level $\alpha \in (0, 1)$ is given, and we denote by VaR_α the Value at Risk (VaR) at the level α of the portfolio over the period in question. In the present context, this VaR is the $100(1-\alpha)$-percentile of the loss distribution. This is consistent with the definition used in the text. The purpose of the problem is to derive a formula for the expected loss given that the loss is assumed to be larger than the value at risk.*

1. *For this question, we assume that the loss distribution is exponential with rate r.*
   1.1. *Give a formula for the c.d.f. of L. Explain.*
   1.2. *Derive a formula for VaR_α.*
   1.3. *Give a formula for the expected loss given that the loss is larger than VaR_α. Recall that, if a random variable X has density f, its expected value given the fact that X is greater than or equal to a level x_0 is given by the formula*

$$\frac{1}{\mathbb{P}\{X \geq x_0\}} \int_{x_0}^{\infty} x f(x) dx, \qquad \text{or equivalently} \qquad \frac{\int_{x_0}^{\infty} x f(x) dx}{\int_{x_0}^{\infty} f(x) dx}.$$

2. *For this question, we assume that the loss distribution is the standard Pareto distribution with shape parameter ξ, location parameter $m = 0$ and scale parameter $\lambda = 1$.*
   2.1. *Give a formula for the c.d.f. of L. Explain.*
   2.2. *Derive a formula for VaR_α.*
   2.3. *Give a formula for the expected loss given that the loss is larger than VaR_α.*
3. *The expected short fall (also known as the conditional VaR) at the level α is the expected loss conditioned by the fact that the loss is greater than or equal to VaR_α. The goal of this question is to quantify the differences obtained when using it as a measure of risk in the two loss models considered in questions 1 and 2.*
   3.1. *For each $\alpha \in (0, 1)$, derive an equation that the rate parameter r and the shape parameter ξ must satisfy in order for the values of VaR_α computed in questions 1.2 and 2.2 to be the same.*
   3.2. *Assuming that the parameters r and ξ satisfy the relationship derived in question 3.1 above, compare the corresponding values of the expected short fall in the models of questions 1 and 2 and comment on the differences.*

Ⓔ **Problem 2.4** *This problem attempts to quantify the goodness of fit resulting from our GPD analysis of samples with heavy tails.*

1. *Use the method described in the text to fit a GPD to the PCS index, and generate a Monte Carlo random sample from the fitted distribution five times the size of the original data sample.*
2. *Produce a Q-Q plot to compare graphically the two samples and comment.*

3. *Use a two-sample Kolmogorov-Smirnov goodness-of-fit test to quantify the qualitative results of the previous question.* **NB**: *Such a test is performed in* `R` *with the command* `ks.test`. *Check the help files for details on its use and the returned values.*
4. *Same questions as above for the weekly log-returns on the S&P data.*

(E) **Problem 2.5** *This problem uses the data set* `PSPOT` *included in the library* `Rsafd`. *The entries of this vector represent the daily Palo Verde (firm on peak) spot prices of electricity between January 4, 1999 and August 19, 2002. Use exploratory data analysis tools to argue that the tails of the distribution are heavy, fit a GPD to the data, and provide estimates of the shape parameters.*
NB: *We usually refrain from talking about the distribution of a financial time series, reserving fitting a distribution to the returns instead of the entries of the original series. You are asked to do just that in this problem. Even though an analysis of the returns in the spirit of what is done in the text would make perfectly good sense, a look at a time series plot of* `PSPOT` *shows a form of stationarity of the data (to be explained later in the book) justifying the analysis asked of you in this problem.*

(E) **Problem 2.6** *This problem requires the data set* `DSP`. *The entries of this numeric vector represent the daily closing values of the S&P 500 index between the beginning of January 1960 and September 18, 2001.*

1. *Compute the vector of log-returns and call it* `DSPLRet`.
2. *We now use the data set* `MSP`. *The entries of this numeric vector represent minute by minute quotes of the* `S&P 500` *on September 10, 1998. Compute the corresponding log-return vector and call it* `MSPLRet`.
3. *Produce a Q-Q plot of the empirical distributions of the two log-return vectors, and comment. In particular, say if what you see is consistent with the claim that the properties of the daily series are shared by the minute by minute series. Such an invariance property is called* self-similarity. *It is often encountered when dealing with fractal objects.*
4. *Compute the empirical means and variances of the* `DSPLRet` *and* `MSPLRet` *data. Assuming that these data sets are Gaussian, would you say that the two distributions are the same in view of these two statistics?*
5. *Fit GPDs to the* `DSPLRet` *and* `MSPLRet` *data, and compare the distributions one more time by comparing the shape parameters.*

(E) **Problem 2.7** *This problem deals with the analysis of the daily S&P 500 index closing values.*

1. *Create a vector* `DSPRET` *containing the daily raw returns. Recall that the raw return on a given day is the difference between the value on that day and the day before divided by the value on the previous day. Compute the mean and the variance of this daily raw return vector.*
2. *Fit a GPD to the daily raw returns, give detailed plots of the fit in the two tails, and discuss your results.*
3. *Generate a sample of size* 10,000 *from the GPD fitted above. Call this sample* `SDSPRET`, *produce a Q-Q plot of* `DSPRET` *against* `SDSPRET`, *and comment.*
4. *Compute the VaR (expressed in units of the current price) for a horizon of 1 day, at the level* $\alpha = 0.005$ *in each of the following cases:*
   4.1 *Assuming that the daily raw return is normally distributed;*
   4.2 *Using the object of class* `gpd` *which you created in question 2 to fit a GPD distribution to the data;*

4.3 Using the Monte Carlo sample SDSPRET *you generated.*
Explain the differences and the similarities between the three estimates of the VaR so obtained.

5. *Redo the questions above after replacing the vector* DSP *with the vector* SDSP *containing only the first* 6,000 *entries of* DSP. *Compare the results, and especially the VaR's. Explain the differences.*

Ⓔ Ⓣ **Problem 2.8** *The goal of this problem is to highlight some of the properties of the estimates obtained with the command* fit.gpd *when fitting a GPD to a data sample* $x_1, \cdots, x_n$. *We assume that the distribution of the data has two tails (one extending to* $-\infty$ *and the other one to* $+\infty$*), and we are interested in understanding the effect of the choice of the thresholds* lower *and* upper.
Remember that a distribution with an upper tail is a GPD if its density $f(x)$ *is well approximated in the tail by a function of the form*

$$f_{\xi_+, m_+, \lambda_+}(x) = \frac{1}{\lambda_+ \xi_+}(1 + \frac{x - m}{\lambda_+})^{-(1+\frac{1}{\xi_+})} \tag{2.38}$$

at least when $x > m_+$ *for some large enough threshold* m_+, *where* λ_+ *is interpreted as a scale parameter, and where* $\xi_+ > 0$ *is called the shape parameter governing the size of the upper tail. If the distribution has a lower tail, one requires a similar behavior for* $x < m_-$ *for possibly different parameters* m_-, λ_- *and* ξ_-.

For the purpose of the problem, we assume that the true density of the sample $x_1, \cdots, x_n$ *is given in Fig. 2.15. It is exactly equal to the function* $f_{\xi_+, m_+, \lambda_+}(x)$ *for* $x > 2$ *with* $m_+ = 2$ *and some value* $\xi_+ > 0$ *(to be estimated), and equal to the corresponding function* $f_{\xi_-, m_-, \lambda_-}(x)$ *for* $x < -2$ *with* $m_- = -2$ *and some* $\xi_- > 0$ *(to be estimated as well).*

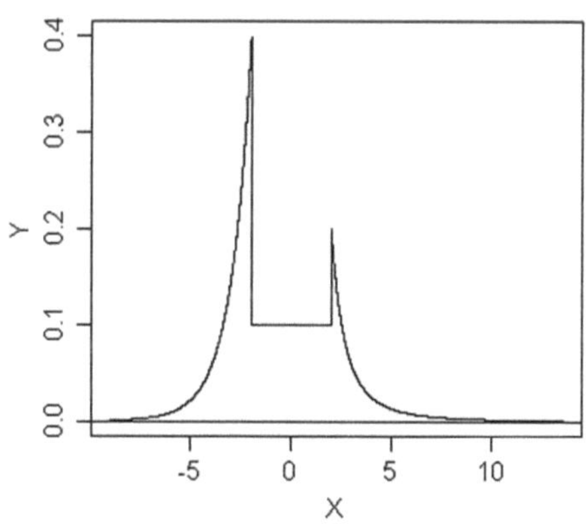

Fig. 2.15. Density of the GPD from which the sample $x_1, \cdots, x_n$ is generated

1. *What should you expect from the estimate* $\hat{\xi}_+$ *given by the function* fit.gpd *if you use a threshold* upper
   1.1. Exactly equal to 2.
   1.2. Greater than 5.
   1.3. Between 0 *and* 1.

2. *What should you expect from the estimate* $\hat{\xi}_-$ *given by the function* `fit.gpd` *if you use a threshold* `lower`
   2.1. *Exactly equal to* -2.
   2.2. *Smaller than* -8.
   2.3. *Between* 0 *and* -1

   and in each case, say how the estimate $\hat{\pi}_{0.01}$ *of the 1 percentile compares to the true value* $\pi_{0.01}$.

NOTES & COMPLEMENTS

What distinguishes our presentation of exploratory data analysis from the treatments of similar material found in most introductory statistics books, is our special emphasis on heavy tail distributions. Mandelbrot was presumably the first academic to stress the importance of the lack of normality of the financial returns. See [64, 65], and also his book [66]. He proposed the Pareto distribution as an alternative to the normal distribution. The theory of extreme value distributions is an integral part of classical probability calculus, and there are many books on the subject. We refer the interested reader to [29] because of its special emphasis on insurance applications. In particular, the discussion given in the Notes & Complements section of Chap. 1 of a possible mathematical model for the PCS index dynamics fits well in the spirit of [29].

The Fisher-Tippett theory reviewed in this chapter was enhanced and brought to the level of a complete mathematical theory in the fundamental works of Gnedenko. Many textbooks give a complete account of this theory. We refer the reader to the books of Leadbetter, Lindgren and Rootzen [63], Resnick [79] and Embrechts, Klüppelberg and Mikosch [29]. This last reference emphasizes the notion of maximum domain of attraction to delineate which distributions give rise to block maxima convergence, after proper normalization, toward a specific GEV distribution. This more modern point of view is also chosen in the more recent account of McNeil, Frey and Embrechts [72]. There are other reasons why a reader interested in the applications of the block maxima method should consult this text. Indeed, he or she will find there a detailed discussion of the effects of dependencies upon the estimates of the shape of the tail. These dependencies occur in two different and non-exclusive ways: as temporal correlation already contained in the data, or as artifacts of the overlap of blocks. Both these issues are addressed and further references to the relevant literature are given.

The block maxima approach to the estimation of extremes has its origin in hydrology where extreme value theory was used to study and predict flood occurrences. There the shape parameter ξ is replaced by its negative $k = -\xi$. The package `Rsafd` gives the user the option to choose which parametrization of the GEV distributions and GPD's he would rather work with by setting a global variable `SHAPE.XI` to `TRUE` or `FALSE`. We did not mention the k-parametrization in the text because of our overwhelming interest in financial applications. Early examples of the use of the block maxima approach in the analysis of financial data were introduced by Longin in [61]. See the book by Embrechts, Frey and McNeil [28] for more examples of in the same spirit.

Details on the maximum likelihood fitting of GEV distributions can be found in Hosking [46] and Hosking, Wallis and Wood [48]. Asymptotic normality was proved by Smith in [89] in the case $\xi > -0.5$.

The method of L-moments seems to have its origin in hydrology. It was introduced by Hosking, Wallis and Wood in [48], and further developed in [47] and [46]. The probability weighted moments, introduced by Greenwood and collaborators in [33] were the precursors of the L-moments. This method of estimation of the parameters of GPD's and GEV distributions does not seem to have permeated the insurance and finance literature, and we purposely chose to include it in our analysis in order to add diversity to our estimation toolbox. Moreover, even if we decide to rely exclusively on maximum likelihood estimators, initializing the maximization search algorithm with the values of the empirical L-moments has proven to be an efficient method of increasing the chances of convergence, speeding up this convergence, and even converging toward a more reasonable local maximum. The derivation of formulae (2.13)–(2.15) giving the L-moments of a GEV distribution in terms of its natural parameters can be found in Hosking [46].

Except possibly for the maximum likelihood estimation of GPD's and GEV distributions, the material presented in this chapter is not systematically covered in the literature devoted to insurance and financial applications. This is especially true with the use of L-moments. See nevertheless the recent work of Seco et al. [70] which may indicate a renewal of interest for these methods for financial applications. We were made aware of the importance of L-moments via numerous enlightening discussions with Julia Morrison.

The fundamental result of this chapter is due to Pickands [75] and Balkema and de Haan. The estimation procedures presented in this chapter rely on the assumption that the data points $x_1, \cdots, x_n$ are realizations of independent and identically distributed random variables. The independence assumption is rarely satisfied in real life applications, and especially with series of financial returns which are of interest to us. However, in many instances, this assumption is not as restrictive as it may seem. Indeed, for many stationary time series, the exceedances over increasing levels can be shown to have a limiting Poisson distribution. So it seems that the independence assumption is restored in the limit of exceedances over high levels. However, most financial return data exhibit clustering properties captured by ARCH and GARCH models, and incompatible with the independence assumption. The reader concerned by these issues is referred to the book of Mc Neil, Frey and Embrechts [72], where further references to the literature can be found.

A time honored method to estimate the size of *power tails* is to compute the Hill's estimator of the exponent α (essentially the inverse of the shape parameter ξ). We purposely chose to ignore this method of estimation, because of horror stories about the misleading conclusions one can reach with this estimation method. The interested reader is referred to the textbook [29] for a discussion of the Hill estimator, and for a series of examples showing clearly its limitations.

Financial institutions started worrying about risk exposures long before regulators got into the act. The most significant initiative was RiskMetrics span off by J.P. Morgan in 1994. Even though the original methodology was mostly concerned with market risk, and limited to Gaussian models, the importance of Value at Risk (VaR) calculations was clearly presented in a set of technical documents made available on the web at the URL `www.riskmetrics.com`. A more academic discussion of the properties of VaR can be found in the book by C. Gourieroux and J. Jasiak [42], and the less technical book by Jorion [53]. VaR is one among many possible ways to quantify a risky exposure to possible adverse moves. The seminal paper of Artzner, Delbaen, Eber and Heath [4] was the first instance of an attempt to formalize mathematically the notion of financial risk measure. Their original set of axioms included positive homogeneity stating that for all $\lambda \geq 0$ and X, one should have $\rho(\lambda X) = \lambda\rho(X)$, and a sub-additivity condition slightly weaker than convexity. Risk measures satisfying their four axioms were called coherent risk measures. Because positive homogeneity is not obviously a natural requirement for a measure of risk, it was gradually abandoned in favor of the smaller set of axioms given in the text, and which was systematically advocated by Föllmer and Schied. For a set of axioms to capture properly the desirable properties of a rigorous risk quantification, some form of convexity or sub-additivity should be included in order for the risk of a diversified portfolio to be less than the sum of the individual risks. Unfortunately, VaR does not encourage diversification in this sense. However, Conditional VaR and Expected Shortfall (at least when the distribution is continuous) do.

A clear exposé of risk measures and their mathematical theory can be found in Foellmer and Schied's book [36]. The risk measures discussed in the text are static in the sense that they are based on models of the sources of risk over a fixed period limited by a fixed horizon, and that no provision is made to update the quantification of the risk as time goes by. In this sense they can be viewed as a first generation of risk measures. Very active research is now dealing with a new generation of risk measures which can capture the time evolution of risk. Such multi-period models have to deal with very technical consistency issues, and easy implementations of the first theoretical results which appeared in this area are still a long way.

The `R` methods used in this chapter to estimate heavy tail distributions, simulate random samples from these distributions, and compute risk measures are taken from the `R` package `Rsafd` based in part on the library `EVANESCE` originally developed in `S` by J. Morrison and the author.

3

DEPENDENCE & MULTIVARIATE DATA EXPLORATION

This chapter extends some of the exploratory data analysis techniques introduced in the case of univariate samples to several variables. In particular, we discuss multivariate versions of kernel density estimators. Then we review the properties of the most important multivariate distribution of all, the normal or Gaussian distribution. For jointly Gaussian random variables, dependence can be completely captured by the classical Pearson correlation coefficient. In general however, the situation can be quite different. We review the classical measures of dependence, and emphasize how inappropriate some of them can become in cases of significant departure from the Gaussian hypothesis. In such situations, quantifying dependence requires new ideas, and we introduce the concept of copula as a solution to this problem. For graphical and tractability reasons, most of the discussion is focused on the bivariate case. We show how copulas can be estimated, and how one can use them for Monte Carlo computations and random scenarios generation. Later in the chapter, we consider higher dimensional cases and discuss application to large portfolio risk management and the valuation of baskets of credit derivatives. Finally, the last section deals with principal component analysis, a classical technique from multivariate data analysis, which is best known for its use in dimension reduction. We demonstrate its usefulness on data from the fixed income markets.

3.1 Multivariate Data and First Measure of Dependence

We begin the chapter with an excursion into the world of multivariate data, where dependencies between variables are important, and where analyzing variables separately would cause significant features of the data to be missed. We illustrate this point with several numerical examples, but we shall focus most of our attention to the specific example of the daily closing prices of futures contracts on Brazilian and Colombian coffee which we describe in full detail in Sect. 3.2.5 below. We reproduce the first seven rows of the data set to show how the data look like after computing the daily log-returns.

R. Carmona, *Statistical Analysis of Financial Data in R*, Springer Texts in Statistics, DOI 10.1007/978-1-4614-8788-3_3, © Springer Science+Business Media New York 2014

```
           [,1]    [,2]
[1,] -0.0232 -0.0146
[2,] -0.0118 -0.0074
[3,] -0.0079 -0.0074
[4,]  0.0275  0.0258
[5,] -0.0355 -0.0370
[6,]  0.0000  0.0000
[7,]  0.0000 -0.0038
```

Each row corresponds to a given day, the daily log return on the Brazilian contract being given in the first column of that row, the log return of the Colombian contract being given in the second one. As we shall see in Sect. 3.2.5, the original data came with time stamps, but as we already explained, the latter are irrelevant for the type of analysis conducted in the first part of this book. Indeed, for the time being, the dependence of the log-returns upon time does not play any role, and we could shuffle the rows of the data set without affecting the results of the analysis.

The data set described above is an example of *bivariate* data. We consider examples of multivariate data sets in higher dimensions later in the chapter, but in the present situation, the data can be abstracted in the form of a bivariate sample:

$$(x_1, y_1),\ (x_2, y_2), \ldots\ldots, (x_n, y_n),$$

which is to be understood as a set of realizations of n independent couples

$$(X_1, Y_1),\ (X_2, Y_2),\ \ldots\ldots,\ (X_n, Y_n)$$

of random variables with the same joint probability distribution. The goal of this chapter is the analysis of the statistical properties of this joint distribution, and in particular of the dependencies between the components X and Y of these couples. If X and Y are real valued random variables, then their joint distribution is characterized by their joint cdf which is defined by:

$$(x, y) \hookrightarrow F_{(X,Y)}(x, y) = \mathbb{P}\{X \le x,\ Y \le y\}. \tag{3.1}$$

This joint distribution has a density $f_{(X,Y)}(x, y)$ if the joint cdf can be written as an indefinite (double) integral:

$$F_{(X,Y)}(x, y) = \int_{-\infty}^{x} \int_{-\infty}^{y} f_{(X,Y)}(x', y')\, dx'dy',$$

in which case the density is given by the (second partial) derivative:

$$f_{(X,Y)}(x, y) = \frac{\partial^2 F_{(X,Y)}(x, y)}{\partial x \partial y}.$$

Setting $y = +\infty$ in (3.1) leads to a simple expression for the marginal density $f_X(x)$ of X. It reads:

$$f_X(x) = \int_{-\infty}^{+\infty} f_{(X,Y)}(x, y')dy' \tag{3.2}$$

and similarly

$$f_Y(y) = \int_{-\infty}^{+\infty} f_{(X,Y)}(x', y)dx'. \tag{3.3}$$

More generally, in the multivariate case (in dimension k to be specific), the cdf of the joint distribution of k random variables $X_1, \cdots, X_k$ is defined as:

$$F_{(X_1,\cdots,X_k)}(x_1, \cdots, x_k) = \mathbb{P}\{X_i \le x_1, \cdots, X_k \le x_k\},$$

and whenever it exists, the density is given by the k-th order partial derivative:

$$f_{(X_1,\cdots,X_k)}(x_1, \cdots, x_k) = \frac{\partial^k F_{(X_1,\cdots,X_k)}(x_1, \cdots, x_k)}{\partial x_1 \cdots \partial x_k},$$

in which case

$$F_{(X_1,\cdots,X_k)}(x_1, \cdots, x_k) = \int_{-\infty}^{x_1} \cdots \int_{-\infty}^{x_1} f_{(X_1,\cdots,X_k)}(x'_1, \cdots, x'_k)\, dx'_1 \cdots dx'_k.$$

3.1.1 Density Estimation

The notions of histogram and empirical cdf used earlier can be generalized to the multivariate setting. Let us discuss the bivariate case for the sake of definiteness. Indeed, one can divide the domain of the couples (x_i, y_i) into plaquettes or rectangular bins, and create a surface plot by forming cylinders above these plaquettes, the height of each cylinder being proportional to the number of couples (x_i, y_i) falling into the base. If the lack of smoothness of the one-dimensional histograms was a shortcoming, this lack of smoothness is even worse in higher dimensions. The case of the empirical cdf is even worse: the higher the dimension, the more difficult it becomes to compute it, and use it in a reliable manner. The main drawback of both the histogram and the empirical cdf is the difficulty in adjusting to the larger and larger proportions of the space without data points. However, they can still be used effectively in regions with high concentrations of points. As we shall see later in this chapter, this is indeed the case in several of the R objects used to code multivariate distributions.

3.1.1.1 The Kernel Estimator

The clumsiness of the multivariate forms of the histogram is one of the main reasons for the extreme popularity of kernel density estimates in high dimension. Given a sample $(x_1, y_1), \ldots, (x_n, y_n)$ from a distribution with (unknown) density $f(x, y)$, the formal kernel density estimator of f is the function $\hat{f}_b$ defined by:

$$\hat{f}_b(x,y) = \frac{1}{nb_1b_2}\sum_{i=1}^{n} K\left(\frac{x-x_i}{b_1}, \frac{y-y_i}{b_2}\right) \tag{3.4}$$

where the function K is a given non-negative function of the couple (x, y) which integrates to one (i.e. a probability density function) which we call the kernel, and the numbers b_1 and b_2 are positive numbers which we call the bandwidths. The interpretation of formula (3.4) is exactly the same as in the univariate case. If (x, y) is in a region with many data points (x_i, y_i), then the sum in the right hand side of (3.4) will contain many terms significantly different from 0 and the resulting density estimate $\hat{f}_b(x, y)$ will be large. On the other hand, if (x, y) is in a region with few or no data points (x_i, y_i), then the sum in the right hand side of (3.4) will contain only very small numbers and the resulting density estimate $\hat{f}_b(x, y)$ will be very small. This intuitive explanation of the behavior of the kernel estimator is exactly what is expected from any density estimator. Notice that the size of the bandwidths $b = (b_1, b_2)$ regulates the extent to which this statement is true by changing how much the points (x_i, y_i) will contribute to the sum.

It is reasonable to take $b_1 = b_2$ when the two *variables* x and y are on the same scale. However, we will need different bandwidths if the variables are on different scales and if we do not want to standardize them. More on that later in this chapter.

3.1.1.2 ***Implementation***

There is no function for multivariate histogram or kernel density estimation in the standard distribution of R, so we added to our library the function `kdest` which takes a bivariate sample as argument, and produces an R object (a data frame to be specific) containing a column for the values of the density estimator, and two columns for the values of the coordinates of the points of the grid at which the estimate is computed. To be specific, if `X` and `Y` are numeric vectors with equal lengths, the command:

```
> DENS <- kdest(X,Y)
```

outputs the values of the two bandwidths used in the *smoothing* of the couples (x_i, y_i) into a surface, produces the plot of the surface proposed as an estimate of the density of the joint distribution of X and Y, and creates a list `DENS` with the components: `deltax` (resp. `deltay`) for the mesh of the subdivision of the x-axis (resp. y-axis) at which the density is actually computed and plotted, `gridx` (resp. `gridy`) for the vector of points of the subdivision on the x-axis (resp. y-axis) and `z` for the matrix of computed values. The value `z[i,j]` is the value of the density estimate at the point `(gridx[i],gridy[j])`. By default, the number of points in `gridx` and `gridy` is `n=256` but this number can be passed to the function `kdest` as a parameter. We illustrate the results of the (bivariate) kernel density estimation with a couple of examples.

- The first example concerns part of a data set which we will study thoroughly in the next chapter. The surface plot of Fig. 3.1 is the result of running the command `kdest` on two data vectors `X` and `Y` derived from the values of indexes computed from the share values and the capitalizations of ENRON and DUKE over the period ranging from January 4, 1993 to December 31, 1993. The well-separated bumps show clearly that the observations (x_i, y_i) can be divided into several subsets which can be discriminated from each other on the basis of the values of the two variables. This situation is very much sought after in pattern recognition applications where the goal is to subdivide the population into well-defined, and hopefully well separated, clusters which can be identified by their local means, for example.

- Our second example concerns, once more, the daily closing values of the S&P 500 index. The goal is to estimate the joint probability density of the log-return computed on a period of 5 days on a given day, and the log-return computed on a period of 15 days ending the same day. The scatterplot of these two variables is given in the left pane of Fig. 3.2. From a central blob of points two sparse clouds extend in the direction of the negative x-axis and the positive y-axis. The most interesting feature of this scatterplot, however, is the following: the large positive values of the 5 days log-returns follow large negative values of the 15 days log-returns. Anticipating the discussion of the correlation coefficient introduced in the next subsection, we suspect there being a negative correlation between the two returns: indeed computing the correlation between these two variables gives a value approximately equal to -0.57. The density estimate reproduced in the right pane shows the central blob of points appearing in the scatterplot fails to reproduce the trail of isolated points in the scatterplot. As we explained earlier, we believe that these points are responsible for the significant negative correlation, and it is worrisome to see them ignored by the kernel density estimator. The problem is very delicate. A smaller bandwidth restores the presence of these points, but the surface becomes so rough that the density estimate ends up being less instructive than the scatterplot itself. On the other hand a larger bandwidth

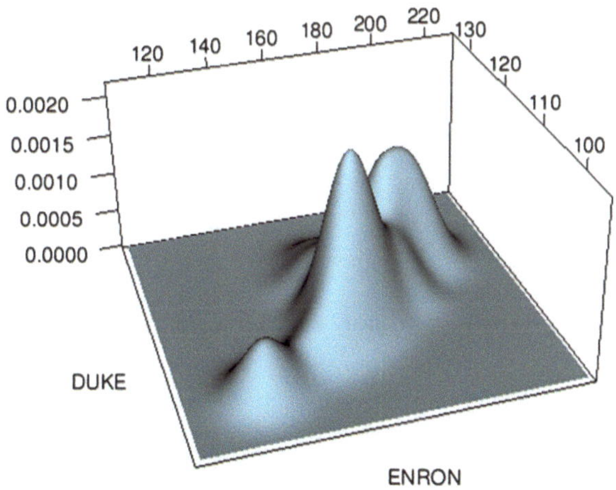

Fig. 3.1. Kernel density estimate for the utility data

gives a smoother surface, wiping out the signs of possible separate trails of points away from the center of the distribution. We chose the bandwidth to reach a compromise between these extremes, but as we already explained, we lost the trail of days responsible for the negative correlation. Unfortunately, the serious difficulties experienced in the analysis of this example are typical of many of the real-life applications in which one would like to use density estimation.

3.1.2 The Correlation Coefficient

Motivated by the previous discussion of the evidence of a possible linear dependence between variables, we introduce the correlation coefficient between two random variables. This theoretical concept and its empirical counterpart are designed to capture this type of linear dependence. It is the most widely-used measure of dependence between two random variables. It is called the Pearson correlation coefficient. For random variables X and Y it is defined as:

$$\rho_P\{X,Y\} = \frac{\mathrm{cov}\{X,Y\}}{\sigma_X \sigma_Y} \tag{3.5}$$

where the covariance $\mathrm{cov}\{X,Y\}$ is defined by:

$$\mathrm{cov}\{X,Y\} = \mathbb{E}\{(X-\mathbb{E}\{X\})(Y-\mathbb{E}\{Y\})\} = \mathbb{E}\{XY\} - \mathbb{E}\{X\}\mathbb{E}\{Y\} \tag{3.6}$$

and where σ_X and σ_Y denote as usual the standard deviations of X and Y i.e.

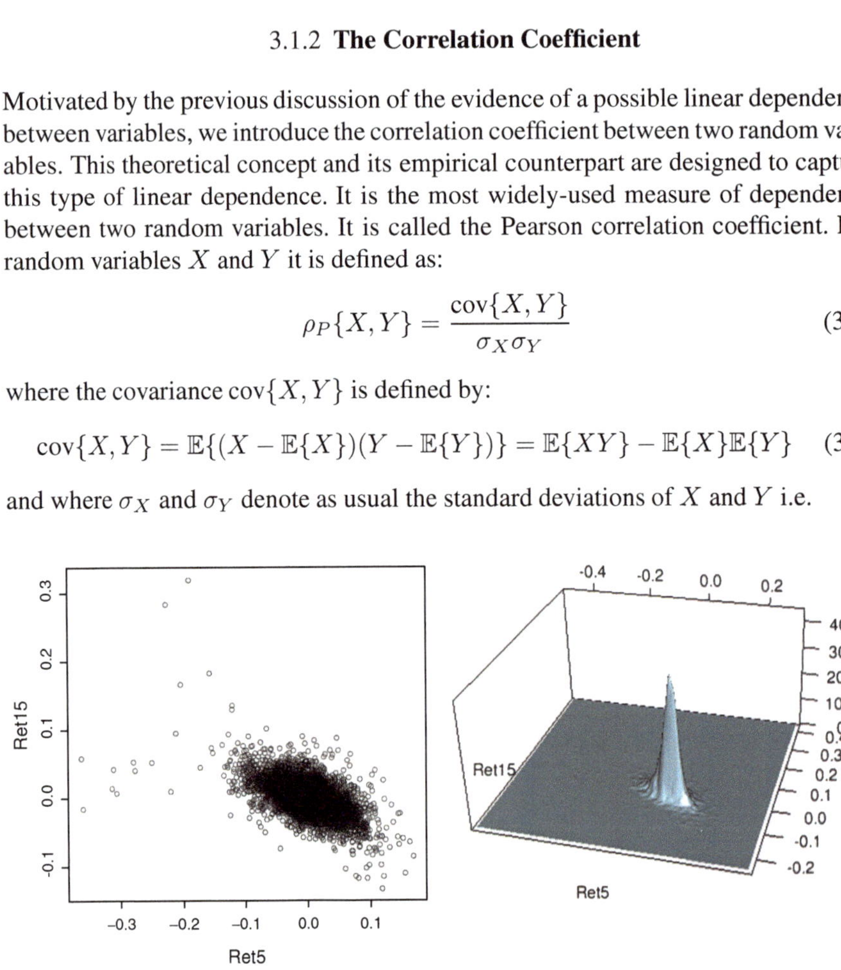

Fig. 3.2. Scatterplot (*left*) and kernel density estimate (*right*) for the 5 and 15 days S&P log-returns

$$\sigma_X = \sqrt{\mathbb{E}\{(X-\mathbb{E}\{X\})^2\}} = \sqrt{\mathbb{E}\{X^2\} - \mathbb{E}\{X\}^2} \tag{3.7}$$

and similarly for σ_Y. If X and Y have a joint density $f(x,y)$ then the definition of the covariance can be rewritten in terms of a double integral as:

$$\mathrm{cov}\{X,Y\} = \int\int xyf(x,y)\,dxdy - \left(\int xf_X(x)dx\right)\left(\int yf_Y(y)dy\right),$$

where f_X and f_Y are the marginal densities of X and Y as given by (3.2) and (3.3). Because of its frequent use, the subscript P is often dropped from the notation, and the Pearson correlation coefficient is commonly denoted by ρ. The empirical analog of this measure of dependence is defined for samples $x_1, \ldots, x_n$ and $y_1, \ldots, y_n$. By analogy with formula (3.5) it is defined as:

$$\hat{\rho}\{X,Y\} = \frac{\widehat{\mathrm{cov}\{X,Y\}}}{\hat{\sigma}_X \hat{\sigma}_Y} \tag{3.8}$$

and it is called the empirical correlation between the samples. Here, the empirical covariance $\widehat{\mathrm{cov}\{X,Y\}}$ is defined by:

$$\widehat{\mathrm{cov}\{X,Y\}} = \frac{1}{n}\sum_{i=1}^{n}(x_i - \overline{x})(y_i - \overline{y}) = \frac{1}{n}\sum_{i=1}^{n} x_i y_i - \overline{x}\,\overline{y} \tag{3.9}$$

where we used the notations $\overline{x}$ and $\overline{y}$ for the sample means of x and y defined by:

$$\overline{x} = \frac{1}{n}\sum_{i=1}^{n} x_i \quad \text{and} \quad \overline{y} = \frac{1}{n}\sum_{i=1}^{n} y_i, \tag{3.10}$$

and where the sample standard deviations $\hat{\sigma}_X$ and $\hat{\sigma}_Y$ are defined by:

$$\hat{\sigma}_X = \sqrt{\frac{1}{n}\sum_{i=1}^{n}(x_i - \overline{x})^2} = \sqrt{\frac{1}{n}\sum_{i=1}^{n} x_i^2 - \overline{x}^2} \tag{3.11}$$

and similarly for $\hat{\sigma}_Y$. Some of the properties of these correlation coefficients are well known. Others are less so. We review them in order to emphasize the usefulness of the correlation coefficient, and at the same time to stress its limitations.

3.1.2.1 *Properties of the Correlation Coefficient*

The most immediate properties of the correlation coefficient are:

- The real numbers ρ and $\hat{\rho}$ are always between -1 and $+1$
- $\rho = 0$ when the random variables X and Y are independent
- $\rho = 1$ when Y is a linear function of X.

These simple properties have lead to the following usage of the sample correlation coefficient $\hat{\rho}$. The samples are regarded as independent when $\hat{\rho}$ is small, while the samples are regarded as strongly dependent when $\hat{\rho}$ is close to 1 or -1. We shall see below that this practice is okay when the samples come from a multivariate Gaussian distribution, but it can be very misleading for other distributions.

The properties listed in the three bullets above are well known. Their intuitive content is the main reason for the enormous popularity of the correlation coefficient as a measure of dependence.

What is often overlooked is the fact that the Pearson correlation coefficient is only a measure of linear dependence between two random variables. In fact, ρ measures the relative reduction of the response variation by a linear regression. Indeed, anticipating our upcoming discussion on least squares linear regression, we can use the following general formula

$$\rho\{X, Y\} = \frac{\sigma^2\{Y\} - \min_{\beta_0,\beta_1} \mathbb{E}\{|Y - \beta_0 - \beta_1 X|^2\}}{\sigma^2\{Y\}}$$

to justify this claim. The numerator of the right hand side is the difference between the variation in the variable Y, and the smallest possible remaining variation after removing a linear function $\beta_0 + \beta_1 X$, of X. This formula is to be compared to the formula giving the slope of the least squares regression line of Y against X in terms of ρ_P.

Shocking Remark. We close this section with a very surprising property of the Pearson correlation coefficient. Strangely enough, this property is little known despite its important practical implications, especially in the world of finance. If the marginal distributions of X and Y are given, but no information is given on the nature of their dependence or lack thereof, the possible values of the correlation coefficient ρ are limited to an interval $[\rho_{min}, \rho_{max}]$. However, contrary to popular belief, this interval is not always the whole interval $[-1, +1]$. There are cases for which this interval is much smaller, even for frequently-used distributions. See for example Problems 3.10 and 3.18 at the end of this chapter, where the case of lognormal random variables is analyzed in detail.

3.2 The Multivariate Normal Distribution

We start our analysis of multivariate statistical distributions with the case of the well-known Gaussian family. All the reasons we gave for the popularity of the univariate Gaussian distribution still hold in the multivariate case. Moreover, the possible competition from other distribution families disappears. Indeed, the Gaussian family is essentially the only one for which explicit analytic computations are possible. We first give an abstract definition and concentrate on the interpretation of the consequences of such a definition. Even though most of the explicit computations done in the book will be limited to the bivariate case, we start with the general definition of the multivariate Gaussian distribution because of its widespread use in portfolio theory where realistic situations involve very large numbers of instruments. Because of this general setup, the discussion which follows is of rather abstract nature, and a quick look at the contents of Appendix 1 at the end of the chapter may help with some of the mathematics.

One says that k real valued random variables $Z_1, \ldots, Z_k$ are jointly Gaussian, or that their joint distribution is a multivariate Gaussian distribution, if any linear combination of $Z_1, \ldots, Z_k$ is a univariate Gaussian (i.e. normal) random variable, in other words, if for every choices $\ell_1, \ldots, \ell_k$ of real numbers, the random variable

$$\xi = \ell_1 Z_1 + \cdots + \ell_k Z_k \tag{3.12}$$

is Gaussian. Notice that this definition implies that each individual Z_i is itself Gaussian (just set $\ell_i = 1$ and all the other ℓ_js to 0 in which case the linear combination reduces to Z_i). However, it says much more, and it is important to understand how much more does this definition imply.

According to the standard practice of probability calculus with random vectors and matrices, which we recall in Appendix 1 at the end of the chapter, we denote by $\boldsymbol{\mu}$ the $k \times 1$ vector of means $\mu_i = \mathbb{E}\{Z_i\}$, and by $\boldsymbol{\Sigma}$ is the $k \times k$ variance/covariance matrix whose entries are $\boldsymbol{\Sigma}_{i,j} = \mathrm{cov}\{Z_i, Z_j\}$. Using the convention introduced in the appendix, this reads:

$$\mathbb{E}\{\boldsymbol{Z}\} = \boldsymbol{\mu} \qquad \text{and} \qquad \Sigma_{\boldsymbol{Z}} = \boldsymbol{\Sigma}.$$

If we use the notation $\boldsymbol{Z}$ for the k-dimensional vector whose components are the Z_i's, the above definition is usually encapsulated in the notation:

$$\boldsymbol{Z} \sim N_k(\boldsymbol{\mu}, \boldsymbol{\Sigma})$$

to signify that the random vector $\boldsymbol{Z}$ has the k-variate Gaussian distribution with mean vector $\boldsymbol{\mu}$ and variance/covariance matrix $\boldsymbol{\Sigma}$. Notice that if we denote by L the $k \times 1$ vector whose entries are the ℓ_is, then the linear combination (3.12) defining the random variable ξ is in fact equal to $L^t \boldsymbol{Z}$, and the computations of Appendix 1 give

$$\xi \sim N(\mu_\xi, \sigma_\xi^2) \quad \text{with} \quad \mu_\xi = L^t \boldsymbol{\mu}, \quad \text{and} \quad \sigma_\xi^2 = L^t \boldsymbol{\Sigma} L.$$

According to its definition, the entries of the covariance matrix $\Sigma_{\boldsymbol{Z}}$ are the covariances $\mathrm{cov}\{Z_i, Z_j\}$, and consequently, the knowledge of all the marginal (bivariate) distributions of the couples (Z_i, Z_j) is enough to determine the entire joint distribution. This particular property is specific to the multivariate Gaussian distribution. It does not hold for general distributions. Moreover, using again the matrix calculus developed for random vectors in Appendix 1, we see that:

$$\boldsymbol{Z} \sim N_k(\boldsymbol{\mu}, \boldsymbol{\Sigma}) \quad \text{when} \quad \boldsymbol{Z} = \boldsymbol{\mu} + \boldsymbol{\Sigma}^{1/2} \boldsymbol{X} \quad \text{and} \quad \boldsymbol{X} \sim N_k(0, \boldsymbol{I}_k) \tag{3.13}$$

where $\boldsymbol{I}_k$ denotes the $k \times k$ identity matrix, and $\boldsymbol{\Sigma}^{1/2}$ denotes the square root of the symmetric nonnegative-definite matrix $\boldsymbol{\Sigma}$. See Problem 3.27 at the end of the chapter for details on the definition and the first properties of this square root matrix. In other words, starting from a random vector $\boldsymbol{X}$ with independent $N(0,1)$ components (see the following remark) we can get to a vector $\boldsymbol{Z}$ with the most general multivariate Gaussian distribution just by linear operations: multiplying by a matrix and adding a vector. This simple fact is basic for the contents of the following subsection.

When the variance covariance matrix $\boldsymbol{\Sigma}$ is invertible, multivariate calculus shows that the random variables $Z_1, \ldots, Z_k$ have a joint density $f_{(Z_1,\ldots,Z_k)}$ given by:

$$f_{(Z_1,\ldots,Z_k)}(z_1,\ldots,z_k) = \frac{1}{\sqrt{(2\pi)^k \det(\boldsymbol{\Sigma})}} \exp\left(-\frac{1}{2}[\boldsymbol{z}-\boldsymbol{\mu}]^t \boldsymbol{\Sigma}^{-1}[\boldsymbol{z}-\boldsymbol{\mu}]\right). \tag{3.14}$$

3.2.1 Important Remark about Independence

If the random variables Z_i are independent, then obviously all the covariances $\text{cov}\{Z_i, Z_j\}$ are zero when $i \neq j$, and the variance/covariance matrix $\Sigma_{\boldsymbol{Z}}$ is diagonal. The converse is not true in general, *even when the random variable Z_i's are Gaussian!* See for example Problems 1.6 and 3.16 for counter-examples. But *the converse is true when the Z_i's are jointly Gaussian!* This striking fact highlights what a difference it makes to assume that the marginal distributions are Gaussian, versus assuming that the joint distribution is Gaussian. The proof of this fact goes as follow: if $\Sigma_{\boldsymbol{Z}}$ is diagonal, then:

$$\frac{1}{2}[\boldsymbol{z}-\boldsymbol{\mu}]^t \boldsymbol{\Sigma}^{-1}[\boldsymbol{z}-\boldsymbol{\mu}] = \frac{(z_1-\mu_1)^2}{2\sigma_1^2} + \frac{(z_2-\mu_2)^2}{2\sigma_2^2} + \cdots\cdots + \frac{(z_k-\mu_k)^2}{2\sigma_k^2},$$

if we denote by $\sigma_1^2, \sigma_2^2, \ldots, \sigma_k^2$ the elements which appear on the diagonal of $\Sigma_{\boldsymbol{Z}}$. So using the definition (3.14) of the multivariate Gaussian distribution and the fact that the exponential of a sum is the product of the exponentials, this implies that:

$$f_{(Z_1,Z_2,\ldots,Z_k)}(z_1,z_2,\ldots,z_k) = f_{Z_1}(z_1) f_{Z_2}(z_2) \cdots f_{Z_k}(z_k),$$

which in turn implies that the joint cdf is the product of the marginal cdfs, proving the desired independence property. So the conclusion is that:

> *For jointly Gaussian random variables, independence is equivalent to the variance/covariance matrix being diagonal!*

3.2.2 Simulation of Random Samples

We now show how one can use formula (3.13) to generate random samples from a multivariate Gaussian distribution. To that end, we assume that we are given a $k \times 1$ vector $\boldsymbol{\mu}$ of means, and a $k \times k$ variance/covariance matrix $\boldsymbol{\Sigma}$, and that we want to generate a sample of size N from the distribution $N_k(\boldsymbol{\mu}, \boldsymbol{\Sigma})$. We proceed as follows:

1. We create a $k \times N$-matrix whose columns are all identical, any column being a copy of the mean vector $\boldsymbol{\mu}$;
2. We generate a sample of size Nk from the standard univariate Gaussian distribution and reshape this $(N \times k) \times 1$ vector into a $k \times N$-matrix;

3. We compute a square root for the variance/covariance matrix, then we multiply each column of the random matrix constructed in Step 2, by the square root of the variance/covaiance matrix;
4. We add the matrix of means constructed in Step 1 to the random matrix constructed in Step 3 above. VOILA!

After Step 2, we have a sample of size N from the k-dimensional Gaussian distribution whose k components are independent $N(0,1)$. Step 3 (which is the most involved) is not needed when $\boldsymbol{\Sigma} = \boldsymbol{I}_k$. Its role is to build the dependence among the k components of the N vectors of dimension k forming the sample. The details of this random generation algorithm are given in Problem 3.27 at the end of the chapter. There, we show how to compute the square root of a covariance matrix and we develop the code for a *home grown* function capable of generating samples from a multivariate Gaussian distribution. Writing this code is just for the sake of illustration, since `R` provides the function `rmvnorm` whose use we illustrate in Sect. 3.2.4 below.

3.2.3 The Bivariate Case

In the bivariate case we have:

$$\boldsymbol{\mu} = \begin{bmatrix} \mu_1 \\ \mu_2 \end{bmatrix} \qquad \text{and} \qquad \boldsymbol{\Sigma} = \begin{bmatrix} \sigma_1^2 & \rho\sigma_1\sigma_2 \\ \rho\sigma_1\sigma_2 & \sigma_2^2 \end{bmatrix}.$$

Notice that $\det(\boldsymbol{\Sigma}) = \sigma_1^2\sigma_2^2(1-\rho^2)$, so that $\boldsymbol{\Sigma}$ is invertible if $\rho \neq \pm 1$, in which case the joint density can be written as:

$$f_{(Z_1,Z_2)}(z_1,z_2) = \frac{1}{2\pi\sigma_1\sigma_2\sqrt{1-\rho^2}}$$
$$\exp\left(-\frac{1}{1-\rho^2}\left[\frac{(z_1-\mu_1)^2}{2\sigma_1^2} - \rho\frac{(z_1-\mu_1)(z_2-\mu_2)}{\sigma_1\sigma_2} + \frac{(z_2-\mu_2)^2}{2\sigma_2^2}\right]\right).$$

This formula shows that, if we know the marginal distributions of Z_1 and Z_2, in other words, if we know μ_1, σ_1, μ_2 and σ_2, then the joint distribution is entirely determined by the correlation coefficient ρ. Also, we clearly see from this formula that when $\rho = 0$ we have:

$$\begin{aligned} f_{(Z_1,Z_2)}(z_1,z_2) &= \frac{1}{2\pi\sigma_1\sigma_2}\exp\left(-\frac{(z_1-\mu_1)^2}{2\sigma_1^2} - \frac{(z_2-\mu_2)^2}{2\sigma_2^2}\right) \\ &= \frac{1}{\sqrt{2\pi}\sigma_1}\exp\left(-\frac{(z_1-\mu_1)^2}{2\sigma_1^2}\right)\frac{1}{\sqrt{2\pi}\sigma_2}\exp\left(-\frac{(z_2-\mu_2)^2}{2\sigma_2^2}\right) \\ &= f_{Z_1}(z_1)f_{Z_2}(z_2) \end{aligned}$$

which shows the independence of Z_1 and Z_2. So we recover the fact that if Z_1 and Z_2 are jointly Gaussian, their independence is equivalent to their being uncorrelated. As we already pointed out, this fact is not true in general, not even when Z_1 and Z_2 are (separately) Gaussian. See Problems 1.6 and 3.16 for counterexamples.

3.2.4 A Simulation Example

For the sake of illustration, we consider the case of the distribution of a couple of (slightly correlated) Gaussian random variables X and Y, and we generate one sample of size $n = 128$ from the joint distribution of (X, Y). We use the R commands:

```
> TSAMPLE<-rmvnorm(n=128,mean=rep(0,2),sd=rep(1,2),rho=.18)
> TDENS <- kdest(TSAMPLE[,1],TSAMPLE[,2])
```

The function rmvnorm is the multivariate analog of rnom. It is designed to generate multivariate Gaussian samples. We chose the vector $[0, 0]$ for the mean by using the command rep(0,2) which creates a vector by repeating the number zero twice, and we used the command rep(1,2) to specify that both components have standard deviations equal to one. Finally, we decided on the correlation coefficient $\rho = 0.18$ by setting the parameter rho. We could have given the entire variance/covariance matrix by specifying the parameter cov instead of giving the vector of standard deviations and the correlation coefficient separately. See the help file for details. The second command computes a kernel density estimator (with the default kernel and bandwidth choices) and plots the resulting *surface*.

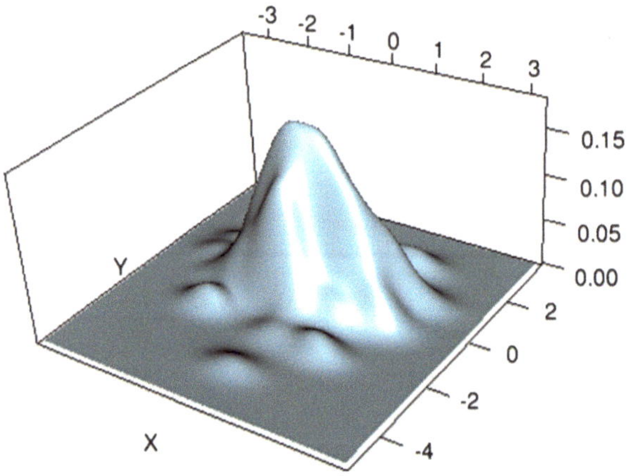

Fig. 3.3. Kernel density estimator for a (bivariate) Gaussian sample

The output is given in Fig. 3.3. We see that the unimodality of the density is violated by the estimate. Increasing the values of the bandwidths would resolve this problem, at the cost of a looser fit, by somewhat flattening the central bump. The poor quality of the estimation is to be blamed on the small size of the sample: in general, the higher the dimension, the larger the sample size needed to get reasonable density estimates.

3.2.5 Let's Have Some Coffee

We use Paul Erdös' famous quote (often attributed to Alfred Renyi):

A mathematician is a machine for turning caffeine into theorems

as a justification for our interest in the price of coffee. As explained in the abstract at the beginning of the chapter, we chose to illustrate the analysis of multivariate distributions with a simple example of two quantities which are obviously correlated. We use samples of log-returns of Brazilian and Colombian coffee spot prices. The original data are plotted in Fig. 3.4, which shows the daily prices of coffee in Brazil and Colombia between January 9, 1986 and January 1, 1999. This plot involves time

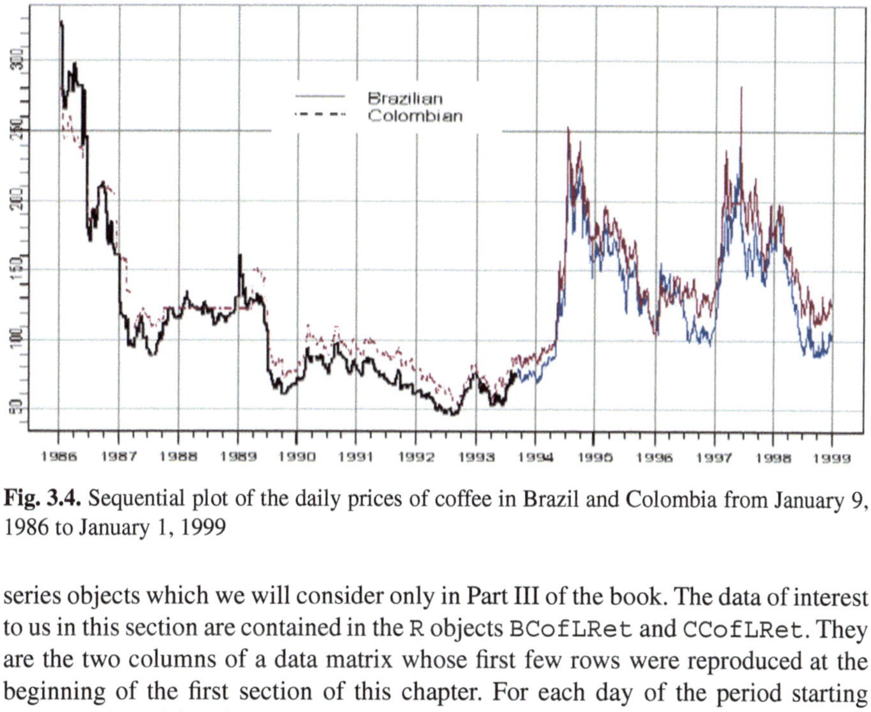

Fig. 3.4. Sequential plot of the daily prices of coffee in Brazil and Colombia from January 9, 1986 to January 1, 1999

series objects which we will consider only in Part III of the book. The data of interest to us in this section are contained in the R objects `BCofLRet` and `CCofLRet`. They are the two columns of a data matrix whose first few rows were reproduced at the beginning of the first section of this chapter. For each day of the period starting January 10, 1986 and ending January 1, 1999, we computed the logarithms of the daily returns from the nearest futures contract active on that day. The scatterplot of these two variables is given in Fig. 3.5. A close look at this scatterplot shows that many points are on the vertical axis and on the horizontal axis. This means that quite often, the price does not change from one day to the next, forcing the log returns to vanish on these days. The presence of so many of these zeroes indicates that the probability distributions are singular, in the sense that the cumulative distribution functions have jumps at 0. These jumps can be a hindrance to the analysis, so we choose to remove them by removing the zeroes from the data samples.

```
> NZ <- (BCofLRet != 0 & CCofLRet  != 0)
> BLRet <- BCofLRet[NZ]
> CLRet <- CCofLRet[NZ]
```

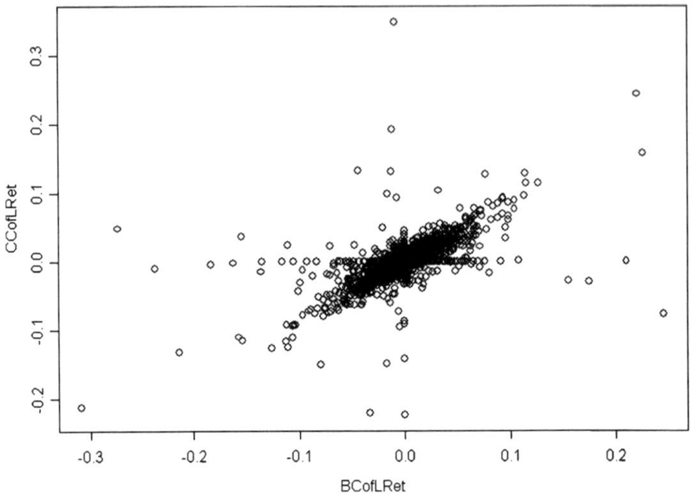

Fig. 3.5. Scatterplot of the daily log-returns of the coffee futures contracts in Brazil and Colombia from January 10, 1986 to January 1, 1999

The vector `NZ` created by the first command is a boolean vector with the same length as `BCofLRet` and `CCofLRet`. It is true (equal to `TRUE`) when both the daily Brazilian and Colombian log-returns are non-zero. The next two commands show the power of the sub-scripting capabilities of the `R` language. `BLRet` and `CLRet` are the vectors obtained by keeping the entries of `BCofLRet` and `CCofLRet` whose indices are those for which the value of `NZ` is `TRUE`. The scatterplot of `BLRet` and `CLRet` is reproduced in the left pane of Fig. 3.6.

We shall work with this new bivariate sample from now on, but we should keep in mind that, if we want to compute statistics of the actual log-returns, we need to put the zeroes back. The command

```
> PNZ <- mean(NZ)
> PNZ
[1] 0.4084465
```

gives the proportion of `TRUE`'s in the vector `NZ`, and it should be viewed as an estimate of the probability not to have a zero in the data. This probability could be used to add a random number of zeroes should we need to create random samples from the original distribution starting with samples from the modified distribution.

3.2.6 Is the Joint Distribution Normal?

"Is the joint distribution of `BLRet` and `CLRet` Gaussian" is the first question we address. Our first step is quite mundane: we compare graphically the joint (empirical) distribution of `BLRet` and `CLRet` to the distribution of a bivariate Gaussian sample which has the same five parameters (i.e. means, standard deviations and correlation coefficient). In order to do so, we first compute these parameters. We use the commands:

```
> XLIM <- c(-.4,.3)
> YLIM <- c(-.2,.4)
> Mu <- c(mean(BLRet), mean(CLRet))
> Sigma <- var(cbind(BLRet,CLRet))
> N <- length(BLRet)
```

We defined the vectors `XLIM` and `YLIM` as the limits of the ranges of `BLRet` and `CLRet`, respectively. We use these values each time we want to make sure that the scatterplot of `BLRet` and `CLRet` on one hand and the scatterplot of the simulated data on the other are on the same scale. As defined, `Mu` is the mean vector since it is defined as the vector of the means. Next we use the `R` function `cbind` to bind the columns `BLRet` and `CLRet` into one single matrix, then applying the function `var` to this data matrix produces the variance/covatiance matrix of the columns. So `Sigma` is the variance/covariance matrix, and `N` is the sample size. We use the `R` function `rmvnorm` to generate the desired bivariate sample $(x_1, y_1), \dots, (x_N, y_N)$ of size N from the bivariate Gaussian distribution with mean `Mu` and variance/covariance matrix `Sigma`.

```
> CNsim <- rmvnorm(N,mean=Mu,cov=Sigma)
> par(mfrow=c(2,1))
> plot(BLRet,CLRet, xlim=XLIM, ylim=YLIM)
> plot(CNsim,xlim=XLIM, ylim=YLIM)
> par(mfrow=c(1,1))
```

Notice that we could have used the home-grown function `vnorm` developed in Problem 3.27 at the end of the chapter. The results are given in Fig. 3.6. Both scatterplots comprise an ellipsoidal cloud of points around the origin. Clearly, this cloud seems to be thinner for the coffee data. However, even if we were to consider that the bulk of the distribution had been reproduced in a reasonable manner, the presence of isolated points in the empirical coffee data is a distinctive feature which has not been reproduced by the simulation. This is a clear indication that the joint distribution of `BLRet` and `CLRet` is not Gaussian. There are many reasons. In general, it is because at least one of the variables, `BLRet` or `CLRet`, is not Gaussian. But these variables could be Gaussian even when the joint distribution is not Gaussian. We now check that this is not the case by showing that the marginal distributions of `BLRet` and `CLRet` are not univariate Gaussian either.

3.3 Marginals and More Measures of Dependence

Trying to fit a multivariate distribution to a multivariate sample, as we did when trying to fit a bivariate Gaussian distribution to the sample of `BLRet` and `CLRet`, may be trying to tackle all the difficulties at once, and may be overwhelming. So instead, we opt for the "*divide and conquer*" approach: we first consider the estimation of the

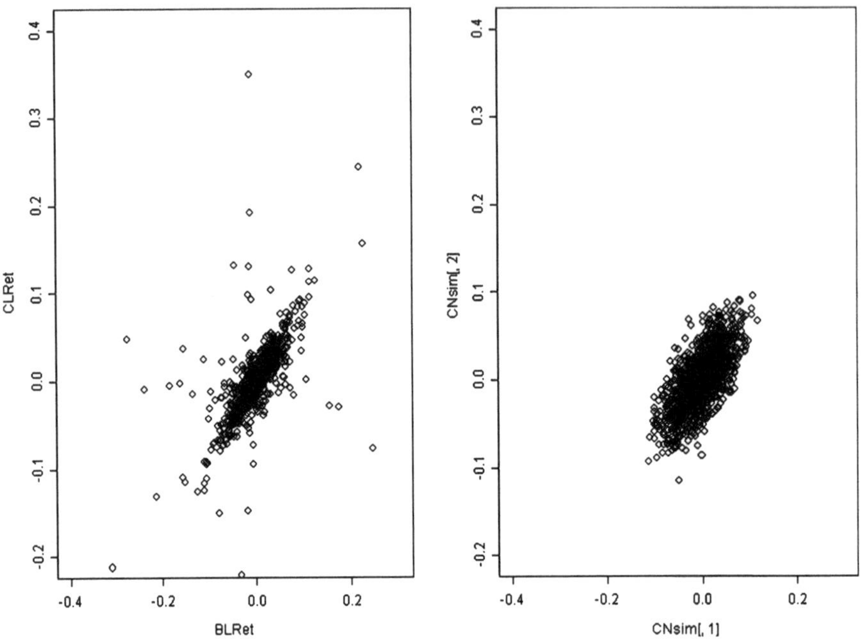

Fig. 3.6. Comparison of the empirical scatterplot of the coffee log-returns after the removal of the zeroes (*left*) and of the scatterplot of a sample of the same size simulated from a jointly Gaussian distribution with the same mean and covariance structure (*right*)

univariate marginal distributions separately, and only after this has been done, do we consider the issue of dependence. In the case of a bivariate sample from a Gaussian distribution, this would amount to first estimating the means and the variances of the two variables separately, and then estimating the correlation coefficient. Since we are interested in more general distributions, possibly with marginals having heavy tails, we may not be able to use the correlation coefficient as a way to quantify dependence. So once the identification of the marginals is over, we review the most commonly used statistics measuring the dependence of two samples, and we prepare for the concept of copula which will be introduced and analyzed in the next section.

As before, we try to sprinkle the presentation of the mathematical concepts with numerical examples, and we still use the example of the coffee data for that.

3.3.1 Estimation of the Coffee Log-Return Distributions

We use the graphical tools introduced in Chap. 1. Plotting histograms and kernel density estimators of `BLRet` and `CLRet` would vouch for unimodular distributions, possibly with extended tails on both sides. The presence of tails which are heavier than Gaussian is confirmed by the Q-Q plots reproduced in Fig. 3.7. The departures from the Q-Q lines are a clear indication that the tails are much heavier than the tails

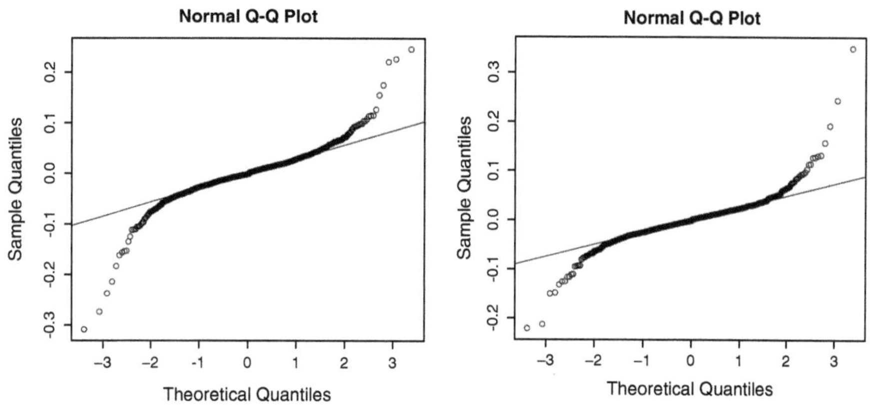

Fig. 3.7. Q-Q plots (against the Gaussian distribution) of the Brazialian (*left*) and Colombian (*right*) coffee daily log-returns for the period from January 9, 1986 to January 1, 1999 after removal of the zeroes

of the Gaussian distributions with the same means and variances, suggesting that fitting generalized Pareto distributions may be appropriate. Since the analysis of the marginal distribution of the daily log returns of the Colombian coffee is essentially the same, we only report the R commands used in the analysis for the Brazilian coffee.

Remark. Similar results would have been obtained with the original log return samples (prior to the removal of the zeroes) since the zeroes affect only the center of the distribution and not the tails.

As in the case of our analysis of the S&P 500 daily log-returns, we use the function `fit.gpd` to fit a generalized Pareto distribution to both `BLRet` and `CLRet`. This function requires information on the locations of the tails in the form of two parameters telling the program where to start fitting regular Pareto distributions. We use the function `shape.plot` in order to choose the locations of the thresholds. The command

```
> shape.plot(BLRet,tail=two)
```

produce the results given in Fig. 3.8. From these plots we decide that the estimation of the upper tail could be done to the right of the value 0.04, while the estimation of the lower tail could be done to the left of the value -0.045. Among other things, this will guarantee that each tail contains a bit more than 8 % of the data, namely more than 115 points, which is not unreasonable. With this choice we proceed to the actual estimation of the GPD with the command:

```
> B.est <- fit.gpd(BLRet, upper = 0.04, lower = -0.045)
```

The function `fit.gpd` creates Q-Q plots (reproduced in Fig. 3.9) of excesses over upper and lower thresholds against the quantiles of a GPD. As we mentioned several

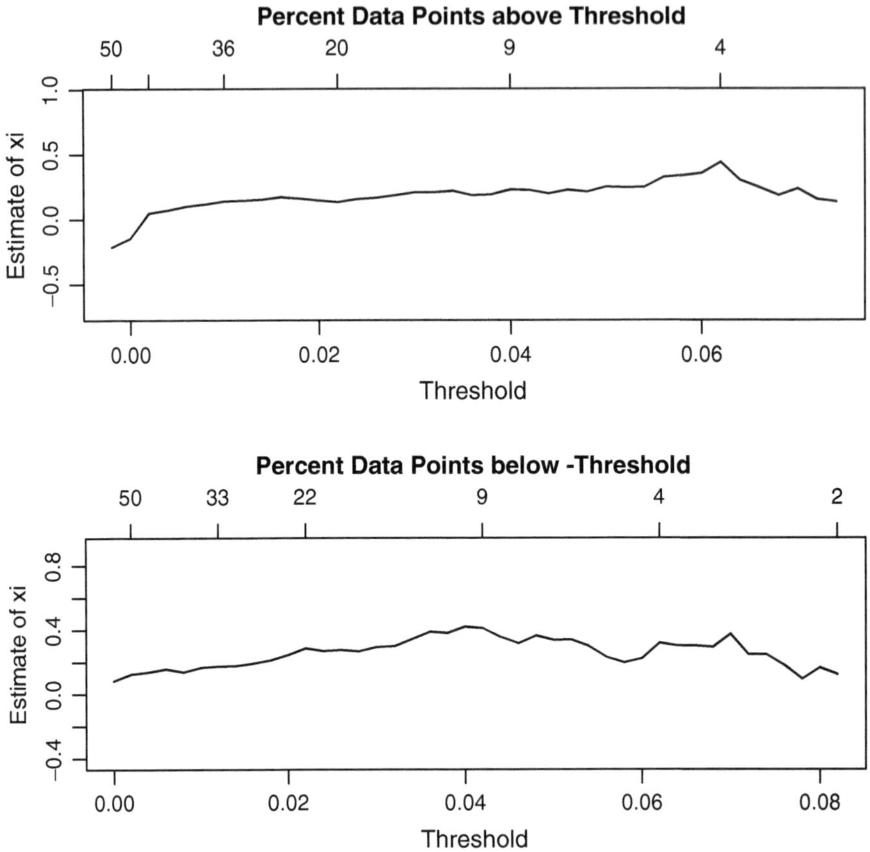

Fig. 3.8. Estimates of the shape parameters as functions of the thresholds for the right tail (*top*) and left tail (*bottom*) of the distribution of the daily log returns of the Brazilian coffee

times already, if the left parts of these plots are approximately linear, the estimation of the tails is expected to be good. An extensive discussion of the mathematical results underpinning these empirical procedures was given in Chap. 2. We check graphically the quality of the fit with the plots of the tails on a logarithmic scale.

```
> tailplot(B.est)
```

The results are given in Fig. 3.10. Given the point patterns, the fit looks very good. We perform the analysis of the heavy tail nature of the distribution of the daily log-returns of the Colombian coffee in exactly the same way.

3.3.1.1 ***First Monte Carlo Simulations***

Motivated by the desire to perform a simulation analysis of the risk associated with various coffee portfolios containing both Brazilian and Colombian futures contracts,

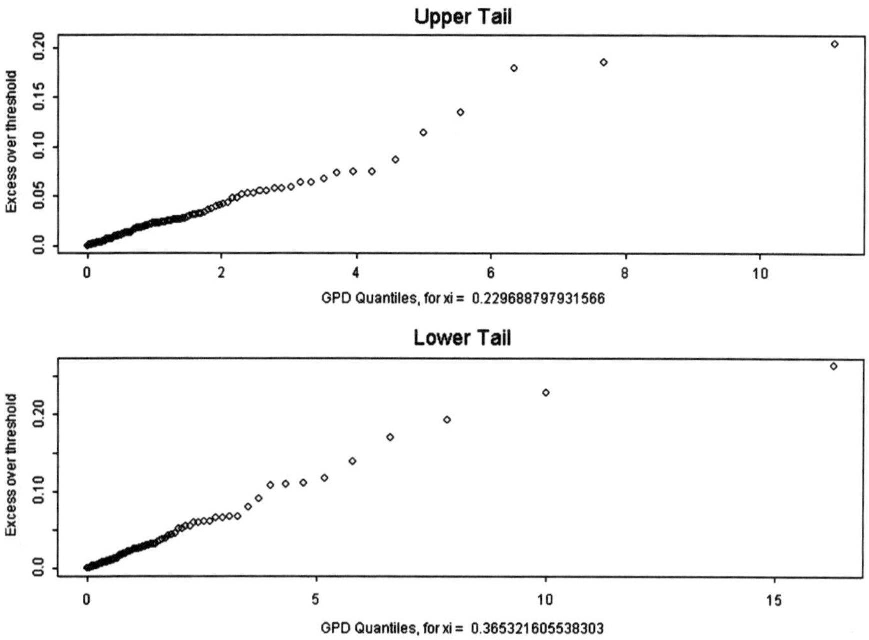

Fig. 3.9. Checks that the estimation of the distribution of the daily log returns of the Brazilian coffee by a generalized Pareto distribution is reasonable

we proceed to the Monte Carlo simulation of samples of log-returns using the tools developed in the previous chapter. The commands:

```
> BLRet.sim <- rgpd(B.est, length(BLRet))
> CLRet.sim <- rgpd(C.est, length(CLRet))
```

generate samples `BLRet.sim` and `CLRet.sim` of the same sizes as the original data, and from the GPD's fitted to the data. To make sure that these samples have the right distributions, we check that their Q-Q plots against the empirical data are concentrated along the main diagonal. This is clear from Fig. 3.11 which was produced with the `R` commands:

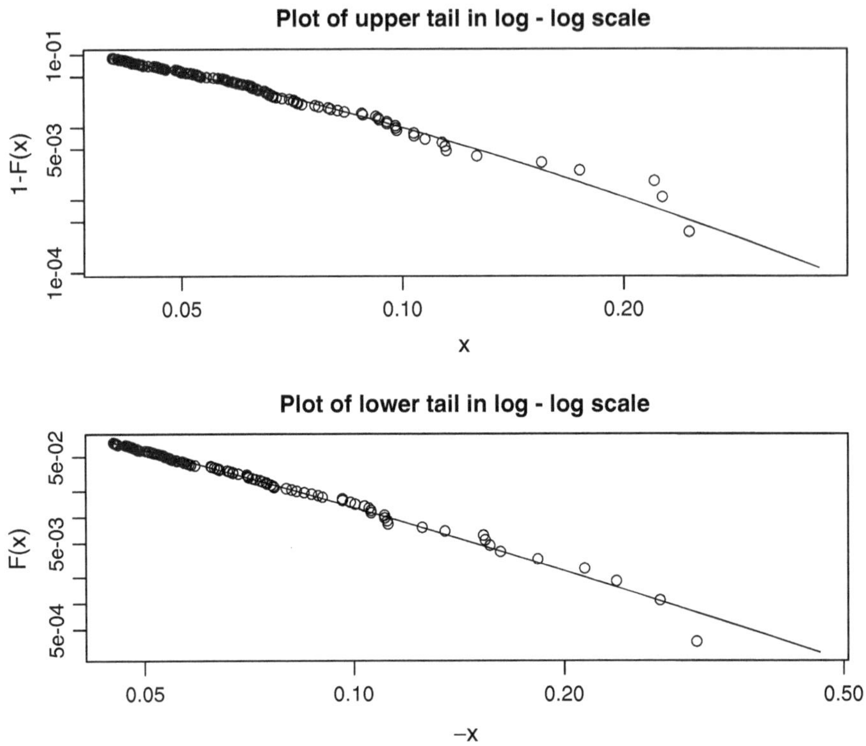

Fig. 3.10. Goodness of the fits for the right tail (*top*) and the left tail (*bottom*)

```
> qqplot(BLRet,BLRet.sim)
> abline(0,1)
> qqplot(CLRet,CLRet.sim)
> abline(0,1)
```

So it is clear that the distributions of the simulated samples are as close as we can hope for from the empirical distributions of the Brazilian and Colombian coffee log-returns. Since they capture the marginal distributions with great precision, these simulated samples can be used for the computations of statistics involving the log-returns separately. However, they cannot be used for the computations of joint statistics since they do not capture the dependencies between the two log-returns. Indeed, the simulated samples are statistically independent. This is clearly illustrated by plotting them together in a scatterplot as in Fig. 3.12. We need to work harder to understand better the dependencies between the two log-return variables, and to be able to include their effects in Monte Carlo simulations.

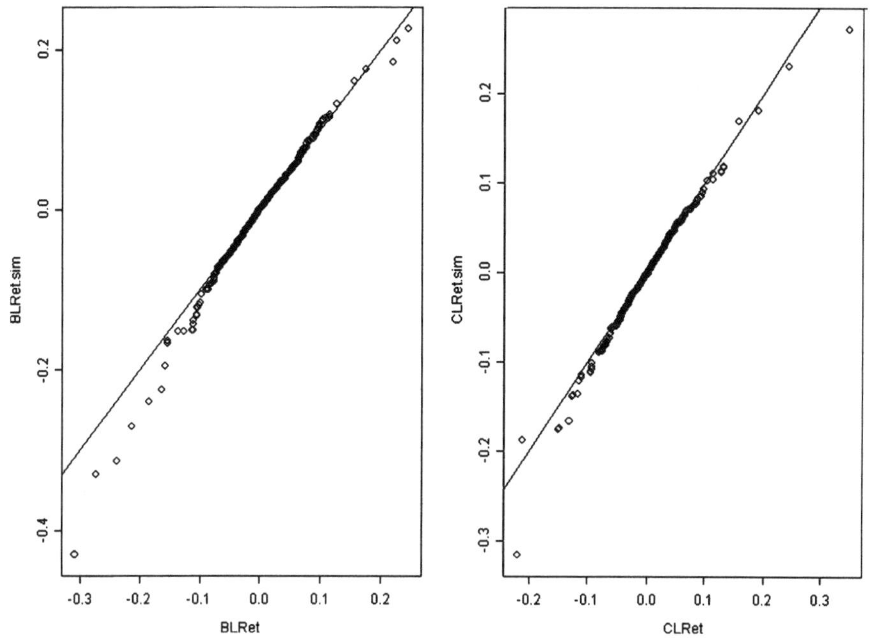

Fig. 3.11. Empirical Q-Q plot of the Monte Carlo sample against the empirical coffee log-return sample in the case of the Brazilian futures (*left pane*) and the Colombian futures prices (*right pane*)

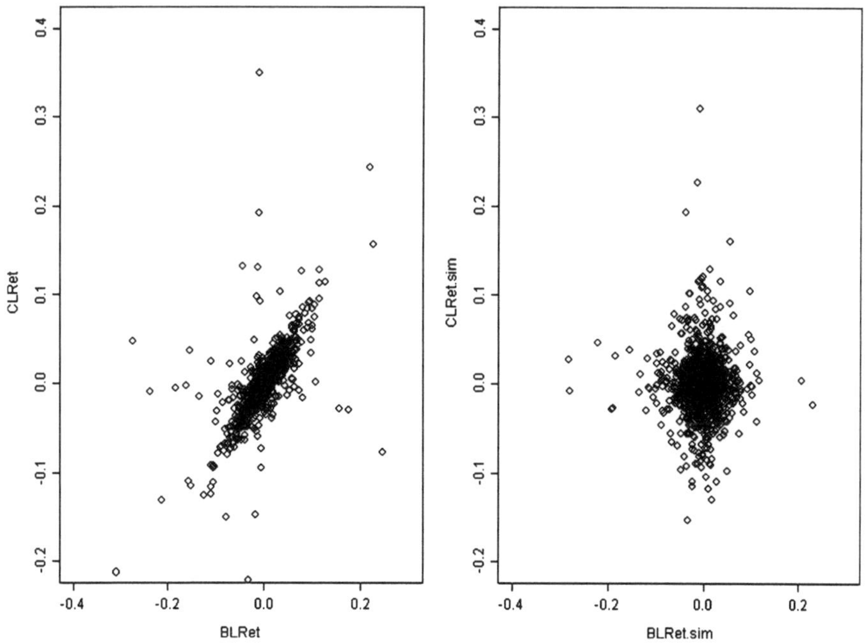

Fig. 3.12. Scatterplot of the Colombian coffee log-returns against the Brazilian ones (*left pane*), and scatterplot of the Monte Carlo samples (*right pane*)

3.3.2 More Measures of Dependence

Because of the limitations of the correlation coefficient ρ as a measure of the dependence between two random variables, other measures of dependence have been proposed and used throughout the years. They mostly rely on sample order statistics. For the sake of completeness, we shall quote two of the most commonly used: the Kendall's τ and the Spearman's ρ. Given two random variables X and Y, their Kendall's correlation coefficient $\rho_K(X,Y)$ is defined as:

$$\rho_K(X,Y) = \mathbb{P}\{(X_1 - X_2)(Y_1 - Y_2) > 0\} - \mathbb{P}\{(X_1 - X_2)(Y_1 - Y_2) < 0\} \quad (3.15)$$

provided (X_1, Y_1) and (X_2, Y_2) are independent random couples with the same joint distribution as (X, Y). Notice that if α and β are positive constants, then

$$\rho_K(\alpha X, \beta Y) = \rho_K(X,Y).$$

In fact a much stronger invariance result holds in the sense that whenever g and h are monotone increasing functions then

$$\rho_K(g(X), h(Y)) = \rho_K(X,Y),$$

as the Kendall correlation coefficient is a measure of the rank dependence between two random variables. Even though the notation ρ_K should be used for consistency, we shall often use the notation $\tau(X,Y)$ because this correlation coefficient is usually called Kendall's tau.

The dependence captured by Kendall's tau is better understood on sample data. Given samples $x_1, \ldots, x_n$ and $y_1, \ldots, y_n$, the empirical estimate of the Kendall correlation coefficient is given by:

$$\hat{\rho}_K(X,Y) = \frac{1}{\binom{n}{2}} \sum_{1 \leq i \leq j \leq n} \text{sign}\left((x_i - x_j)(y_1 - y_j)\right)$$

which shows clearly that what is measured here is merely the relative frequency with which a change in one of the variables is accompanied by a change in the same direction of the other variable. Indeed, the `sign` appearing in the right hand side is equal to one when $x_i - x_j$ has the same sign as $y_i - y_j$, whether this sign is plus or minus, independently of the actual sizes of these numbers. Computing this coefficient with `R` can be done in the following way:

```
> cor(BLRet,CLRet,method=k)
[1] 0.688215
```

The Spearman rho of X and Y is defined by:

$$\rho_S(X,Y) = \rho\{F_X(X), F_Y(Y)\}, \quad (3.16)$$

and its empirical estimate from sample data is defined as:

$$\hat{\rho}_S(X,Y) = \frac{12}{n(n^2-1)} \sum_{i=1}^{n} \left(\text{rank}(x_i) - \frac{n+1}{2}\right)\left(\text{rank}(y_i) - \frac{n+1}{2}\right).$$

The value of this correlation coefficient depends upon the relative rankings of the x_i and the y_j. However, the interpretation of the definition is better understood from the theoretical definition (3.16). Indeed, this definition says that the Spearman's correlation coefficient between X and Y is exactly the Pearson's correlation coefficient between the uniformly distributed random variables $F_X(X)$ and $F_Y(Y)$. This shows that Spearman's coefficient attempts to remove the relative sizes of the values of X among themselves, similarly for the relative values of Y, and then to capture what is left of the dependence between the transformed variables. We shall come back to this approach to dependence below. In any case, The Spearman correlation coefficient is also a measure of rank dependence, and as with the Kendall correlation coefficient, it is invariant under monotone transformations in the sense that whenever g and h are monotone increasing functions then

$$\rho_S(g(X),h(Y)) = \rho_K(X,Y),$$

and as a particular case we also have $\rho_S(\alpha X, \beta Y) = \rho_K(X,Y)$ whenever α and β are positive numbers. In R, Spearman's rho is computed with the command:

```
> cor(BLRet,CLRet,method="s")
[1] 0.8356541
```

Analytic computations of the Spearman and Kendall correlation coefficients are typically impossible, except in the jointly Gaussian case as usual. Indeed if X and Y are jointly Gaussian random variables with Pearson correlation coefficient ρ, then we have the formulae:

$$\rho_S(X,Y) = \frac{6}{\pi}\text{arcsin}\frac{\rho}{2}, \qquad \text{and} \qquad \rho_K(X,Y) = \frac{2}{\pi}\text{arcsin}\rho. \qquad (3.17)$$

Notice that we did not specify the means and the variances of X and Y since they do not affect the Spearman and Kendall correlation coefficients as we noticed earlier. So for the purpose of the above statement, we might as well assume that $X \sim N(0,1)$ and $Y \sim N(0,1)$ without any loss of generality. Figure 3.13 gives the plots of $\rho_S(X,Y)$ and $\rho_K(X,Y)$ as functions of ρ. In the left pane of the figure, the plot coincides almost perfectly with the diagonal (which is superimposed as a dashed line), suggesting that the value of $\rho_S(X,Y)$ is very close to ρ. In fact, from formula (3.17) one can derive the error bound $|\rho_S(X,Y) - \rho| \leq (1-(3/\pi))|\rho|$ which confirms the graphical evidence.

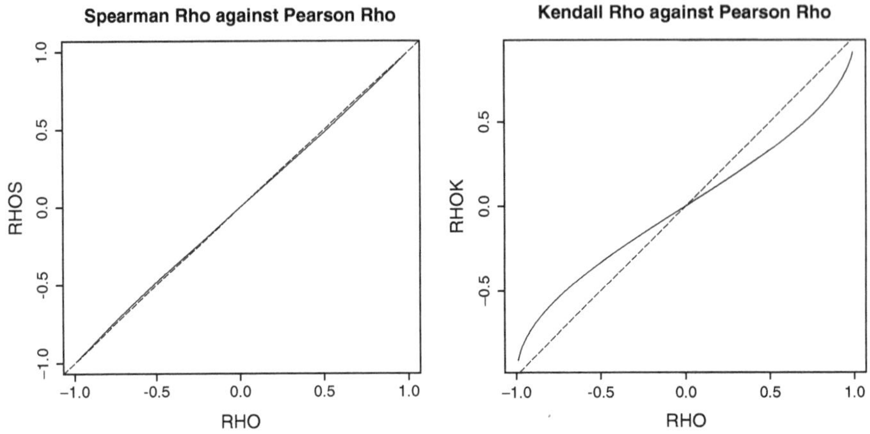

Fig. 3.13. Spearman (*left*) and Kendall (*right*) correlation coefficients against the Pearson correlation coefficient of two jointly Gaussian random variables

3.4 COPULAS

The first part of this section elaborates on the rationale behind the introduction of Spearman's correlation coefficient. As a warm up to the introduction of the abstract concept of copula, we consider first the practical implication of the first of the two fundamental facts of the theory of random generation as presented in Sect. 1.3. Because of Fact 1, which reads:

$$X \text{ r.v. with cdf } F_X(\cdot) \quad \Longrightarrow F_X(X) \text{ uniform on } [0,1],$$

we can transform the original coffee log-return data and create a bivariate sample in the unit square in such a way that both marginal point distributions are uniform. Indeed, the above theoretical result says that this can be done by evaluating each marginal cdf exactly at the sample points. In some sense, this wipes out the dependence of the point cloud pattern, seen for example in the left pane of Fig. 3.6, upon the marginal distributions, leaving only the intrinsic dependence between the variables. We use our estimates of the distribution functions of the coffee log-returns as proxies for the theoretical distributions.

```
> U <- pgpd(B.est, BLRet)
> V <- pgpd(C.est, CLRet)
> plot(U,V)
> EMPCOP <- empirical.copula(U,V)
```

The first two commands use the function `pgpd` from the library `Rsafd`, to compute the estimate of the cdf of the GPD, identified by the object of class `gpd`, at the points given in its first argument. Figure 3.14 shows the result of the above plot command. As expected all the data points are in the unit square. Moreover, the first coordinates of the points seem to be uniformly distributed on the unit interval of the horizontal axis, as they should according to Fact 1, which was recalled above. Similarly for the second coordinates. The fact that the marginal distributions are now uniform is a sign that the influences of the original marginal distributions have been removed from the data. The only remaining feature is the way the numbers u_i and v_i are paired, and we claim that the dependence between the two log-returns is captured by the way these couplings are done. The dense point concentration around the diagonal of the unit square is a consequence of this pairing, and it should be viewed as a graphical representation of the intrinsic dependence between the two random variates.

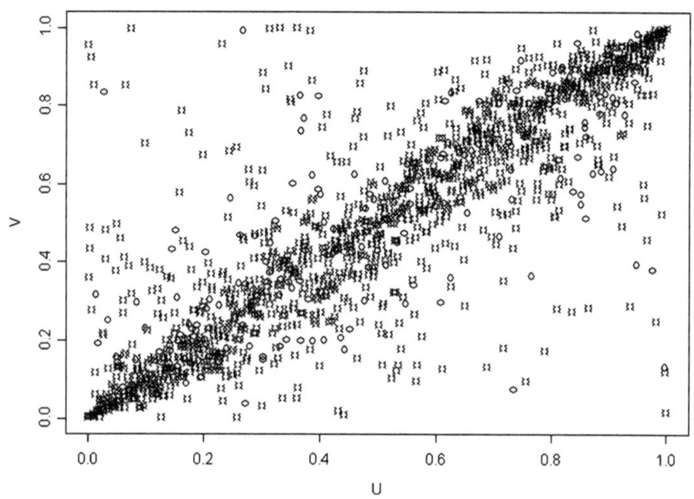

Fig. 3.14. Dependence between the coffee log-returns after removing the effects of the marginal distributions

3.4.1 Definitions and First Properties

We interrupt the analysis of the coffee data to introduce theoretical concepts intended to capture the dependence between several random variates. The analysis of the coffee data will be continued in Sect. 3.4.4. For the sake of simplicity, we limit ourselves to the bivariate case. The reader interested in the general multidimensional case can jump to Sect. 3.4.8 to avoid repetitions.

Definition 1. *A copula is the joint distribution of uniformly distributed random variables.*

So if U and V are $U(0,1)$, the function C defined on $[0,1] \times [0,1]$ by:

$$C(u,v) = \mathbb{P}\{U \leq u, V \leq v\}$$

is a copula. Moreover, if X and Y are r.v.'s with cdfs F_X and F_Y, then the joint distribution of the uniform random variables

$$U = F_X(X), \quad V = F_Y(Y)$$

is called the copula of (X, Y).

3.4.1.1 First Properties of Copulas

It is straightforward to check that:

- C does not change if one replaces X or Y by non-decreasing functions of X and Y;
- The joint cdf of (X, Y) can be recovered from the copula and the marginal cdfs via the formula:
$$F_{(X,Y)}(x, y) = C(F_X(x), F_Y(y));$$
- C is unique if $F_{(X,Y)}(x, y)$ is continuous (i.e. does not jump).

Moreover:

- $C(u, v)$ is non-decreasing in each variable;
- $C(u, 1) = u$ and $C(1, v) = v$ since the marginal distributions of a copula are uniformly distributed;
- if $u_1 \leq u_2$ and $v_1 \leq v_2$, then:
$$C(u_2, v_2) - C(u_2, v_1) - C(u_1, v_2) + C(u - 1, v_1) \geq 0.$$

This last inequality formalizes mathematically the fact that a copula is a multivariate cdf, and as such, it has to satisfy some positivity and monotonicity properties. It is best understood on a picture!

3.4.2 Examples of Copula Families

◇ ***Independent copula*** The first example corresponds to the case when X and Y are independent. So from the point of view of measure of dependence, it is degenerate.
$$C_{ind}(u, v) = uv,$$
is the copula of independent random variables.

◇ ***Gaussian copula*** For each $\rho \in (0, 1)$ the function defined by:
$$C_{Gauss,\rho}(u, v) = \frac{1}{2\pi\sqrt{1-\rho^2}} \int_{-\infty}^{\Phi^{-1}(u)} \int_{-\infty}^{\Phi^{-1}(v)} e^{-[s^2 - 2\rho st + t^2]/2(1-\rho^2)} ds\, dt$$
is a copula called the Gaussian copula with parameter ρ. This is the copula of random variables which, whether or not their marginal distributions are Gaussian, depend upon each other as jointly Gaussian random variables with Pearson correlation coefficient ρ do. The family of Gaussian copulas is parameterized by the parameter $\rho \in (0, 1)$. Figure 3.15 gives the surface plot of this copula when $\rho = 0.7$, i.e. the plot of the graph of the function $(u, v) \hookrightarrow C_{Gauss,0.7}(u, v)$, together with the plot of its density. The fact that the marginals of a copula

are uniform is clearly seen on this plot. Indeed, the facts $C(u,1) = u$ and $C(1,v) = v$ force edges of the surface to be linear (coinciding with the second diagonal) and to meet at height 1 above the point $(1,1)$.

Varying the parameter ρ is a way of controlling the degree of dependence between the two random variables. Notice that two Gaussian random variables X and Y may not be *jointly Gaussian* if their copula is not in the Gaussian family. They are jointly Gaussian when the copula is from the Gaussian family, in which case the parameter ρ has a simple interpretation since it is the correlation coefficient of X and Y. This interpretation is not valid in general. Indeed, if X and Y have Cauchy marginal distributions, their (Pearson) correlation coefficient does not exist since Cauchy random variables do not have means or variances ..., but nevertheless, it is quite possible for their copula to be in the Gaussian family, i.e. to be equal to $C_{Gauss,\rho}$ for some $\rho \in (0,1)$. However, in this case, the parameter ρ cannot have the interpretation of correlation coefficient.

If two random variables have the copula $C_{Gauss,\rho}$, monotone increasing functions of these random variables are jointly Gaussian with correlation coefficient ρ. Consequently, formulae (3.17) give the Spearman and Kendall correlation coefficients of these random variables. According to our previous discussion, the analytic formula (3.17) for the Spearman correlation coefficient of jointly Gaussian random variables, together with the plot in the left pane of Fig. 3.13 confirm the fact often noticed empirically that the Spearman correlation coefficient of two random variables with copula $C_{Gauss,\rho}$ is very close to ρ. Nevertheless, however similar, they are different, and *contrary to common belief, the Spearman correlation coefficient of two random variables with a Gaussian copula with parameter ρ,* ***is not*** *equal to ρ.*

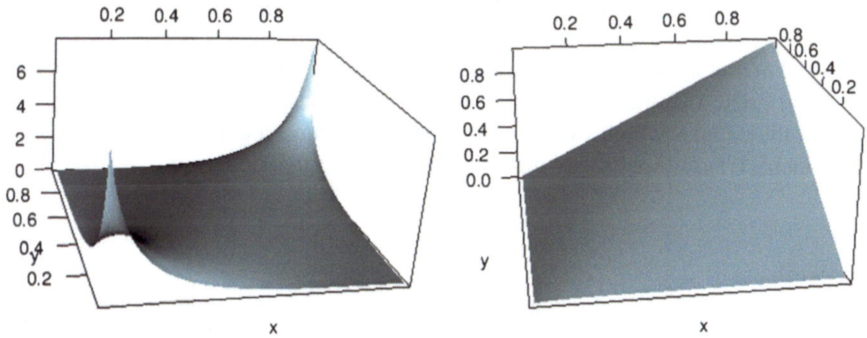

Fig. 3.15. Surface plot of the Gaussian copula with parameter $\rho = 0.7$ (*right*), and of its density (*left*)

◇ For each $\delta \geq 1$ the function

$$C_{Gumbel,\delta}(u,v) = e^{-[(-\log u)^{1/\delta} + (-\log v)^{\delta}]^{1/\delta}}$$

is a copula called the ***Gumbel*** (or ***logistic***) copula with parameter δ. The Gumbel copula family is parameterized by the parameter $\delta \geq 1$. However this parameter does not have as nice an interpretation as the parameter ρ of the Gaussian copula family. Figure 3.16 gives the surface plot of this copula when $\delta = 1.5$ together with the plot of its density. As before, varying the parameter is a way of varying the *strength* of the dependence between two random variables. This family appears naturally in the analysis of extreme events, as it is quite often the case that the copula of random variables with heavy tail distributions is of this family.

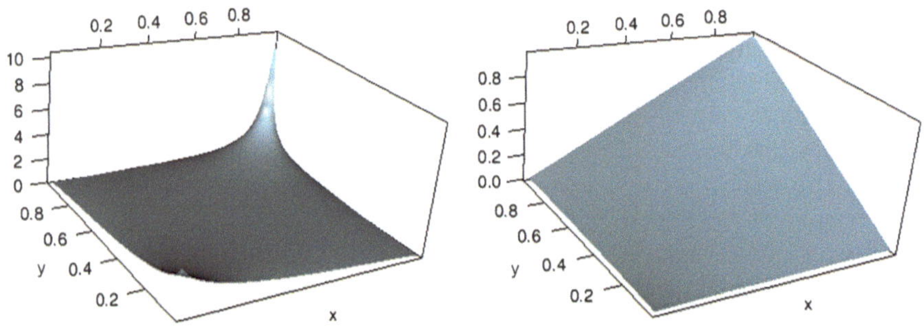

Fig. 3.16. Surface plot of the Gumbel copula with parameter $\delta = 1.5$ (*right*), and of its density (*left*)

A complete list of the parametric copula families supported by the library `Rsafd` is given in Appendix 2, where the reader will also find the defining formulae together with some of the most important properties of these families.

3.4.3 **Copulas and General Bivariate Distributions**

The goal of this subsection is to show how copulas and univariate cdfs come together to characterize ALL the bivariate statistical distributions.

All the copulas which we consider in this book have a density. In other words, the copulas $C(u, v)$ we use are differentiable, and the function:

$$c(u, v) = \frac{\partial^2}{\partial u \partial v} C(u, v)$$

is the density of the copula. Notice that since we are dealing with bivariate distributions, we need a second order derivative in order to get a density from its cdf. Instead of limiting ourselves to distributions with uniform marginals, we can apply this remark to a general bivariate distribution as well. This leads to some interesting formulae.

Let us denote by $F_{(X,Y)}$ the joint cdf of a couple (X, Y) of random variables, and let us denote by $C_{(X,Y)}$ their copula. For the sake of simplicity we momentarily drop the subscript (X, Y) from the notation. According to our definition, we have:

$$F(x, y) = C(F_X(x), F_Y(y)) \tag{3.18}$$

if we denote by F_X and F_Y the cdfs of X and Y respectively. We can compute the joint density $f_{(X,Y)}$ of X and Y by taking partial derivatives on both sides of (3.18). We get:

$$\begin{aligned} f(x,y) &= \frac{\partial^2}{\partial x \partial y} F(x,y) \\ &= \frac{\partial^2}{\partial u \partial v} C(F_X(x), F_Y(y)) \frac{\partial F_X(x)}{\partial x} \frac{\partial F_Y(y)}{\partial y} \end{aligned}$$

which gives the following formula for the joint density of X and Y:

$$f(x,y) = c(F_X(x), F_Y(y))\, f_X(x)\, f_Y(y) \tag{3.19}$$

in terms of the density of their copula, their marginal cdfs and their marginal densities. Obviously we used the formulae

$$f_X(x) = \frac{d}{dx} F_X(x) \qquad \text{and} \qquad f_Y(y) = \frac{d}{dy} F_Y(y)$$

giving the densities of X and Y in terms of their respective cdfs. Formula (3.19) has the following interesting consequence. Contrary to what can be done with the correlation coefficient (see Problems 3.10 and 3.18 for the striking example of the lognormal distributions), it is *always* possible to specify a bivariate distribution by specifying:

- The marginal distributions
- A copula

without having to worry about the existence of the distribution, Moreover, as formulae (3.18) and (3.19) show, formulae for the components can be used to get formulae for the cdf and the density of the bivariate distribution. Figure 3.17 shows an example where we computed the density of a joint distribution specified by the Gumbel copula with parameter 1.4, and with the Gaussian distribution $N(3, 4)$ and the Student t-distribution $t(3)$ as marginals. A bivariate distribution can be created with the

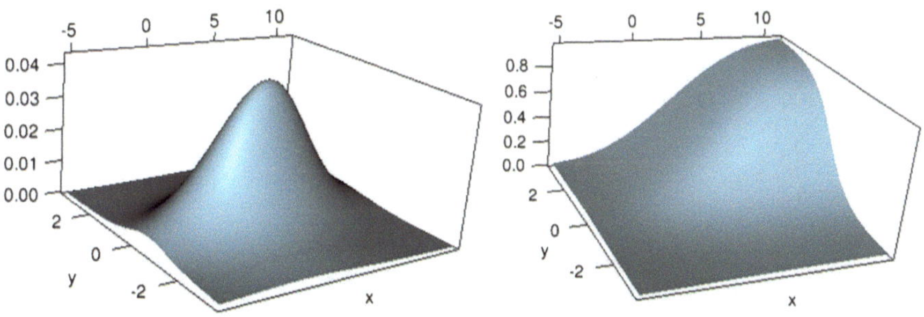

Fig. 3.17. Surface plot of the density (*left*) and cdf (*right*) of the bivariate distribution with Gumbel copula with parameter 1.4 and marginal distributions $N(3, 4)$ and $t(3)$

command bivd, and the function persp.dbivd can be used to produce a 3-d surface plot of the density of a bivariate distribution. The plot in the left pane of Fig. 3.17 was obtained with the R commands:

```
> BIV <- bivd(gumbel.copula(1.4), "norm", "t", c(3,4), 3)
> persp.dbivd(BIV)
```

The function persp.pbivd produces a surface plot of the cdf of the bivariate distribution. The plot in the right pane of Fig. 3.17 was obtained with the command persp.pbivd(BIV). Like in the univariate case, the plots of the two dimensional cumulative distribution functions are not very instructive, especially for copulas, as we can see from Figs. 3.15 and 3.16. Indeed, all these copula surface plots show a *tent* tied to the segments going from the origin $(0,0)$ to the points $(0,1)$ and $(1,0)$. It is also tied at the point $(1,1)$ where its value is always 1, and it is linear on the two coordinate planes. These properties are mere re-statements of the first properties of copulas given in Sect. 3.4.1. These constraints are common to all the copula surface plots, so it is extremely difficult to differentiate between copulas from the plots of their cdfs. For this reason, one very often uses contour plots to get a sense of the shape of the distribution and possibly to compare several copulas, or more general bivariate distributions. The commands:

```
> par(mfrow=c(1,2))
> contour.dbivd(BIV)
> contour.pbivd(BIV)
> par(mfrow=c(1,1))
```

were used to produce the plots of Fig. 3.18.

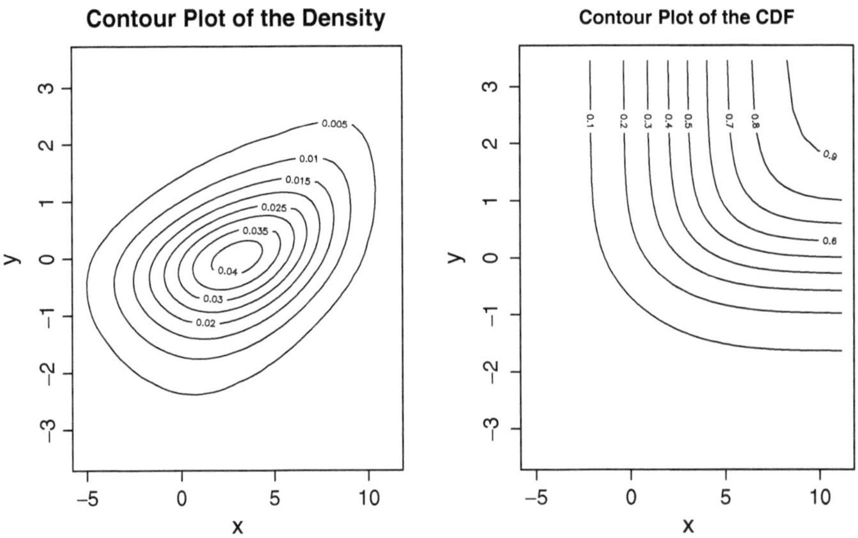

Fig. 3.18. Contour plot of the density (*left*) and the cdf (*right*) of the bivariate distribution with Gumbel copula with parameter 1.4 and marginal distributions $N(3,4)$ and $T(3)$

3.4.4 **Fitting Copulas**

Because of the serious difficulties resulting from the lack of data in the tails of the marginal distributions, copulas are best estimated by parametric methods. In order to do so, we choose a family of copulas (see Appendix 2 for a list of the parametric families supported by the library `Rsafd`) and we estimate the parameter(s) of the family in question by a maximum likelihood standard estimation procedure. The function `fit.copula` which we use below returns an object of class `copula`, and creates a contour plot of the level sets of the empirical copula and of the fitted copula. Since it is so difficult to compare copulas using only their graphs, in order to visualize the goodness of the fit, we chose to put a selected ensemble of level sets for the two surfaces on the same plot. Differences in these level curves can easily be interpreted as differences between the two surfaces. The fitting procedure can be implemented in the case of the coffee log-return data by the following commands.

```
> FAM <- "gumbel"
> ESTC <- fit.copula(EMPCOP,FAM)
```

Recall that `EMPCOP` was the `R` object constructed as the empirical copula of the Brazilian and Colombian coffee daily log-returns. The results are shown in Fig. 3.19. The level sets of the Gumbel copula fitted to the data are very close to the level sets of the empirical copula. This graphical check shows that the fit is very good.

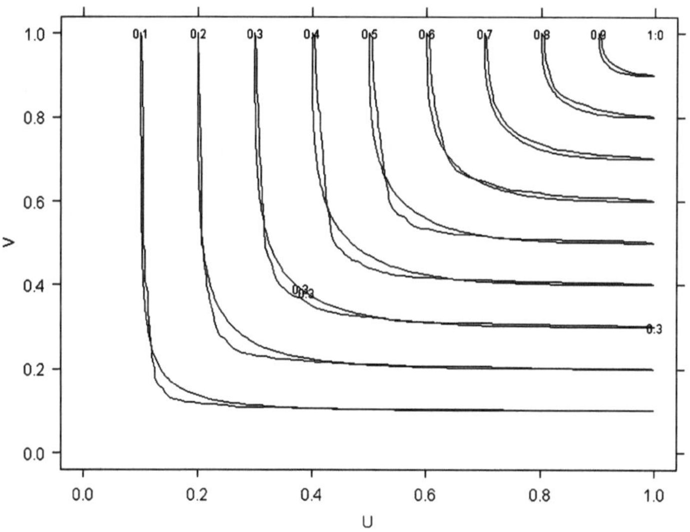

Fig. 3.19. Contour plot of the empirical copula with the contours of the fitted Gumbel copula superimposed

3.4.5 Monte Carlo Simulations with Copulas

We learned in Chap. 1 how to generate random samples from a univariate distribution. We introduced and tested a set of tools to do just that, even when the distribution in question had to be estimated from data, and even when the distribution was suspected to have heavy tails. But as we saw earlier (recall Fig. 3.12) having separate univariate random samples is not enough if we want to have a realistic rendering of how the variables in a bivariate sample relate to each other. In this subsection, we consider the problem of the generation of random samples from a bivariate distribution, which we assume to be given by its marginal distributions and a copula.

Let us imagine that we have a tool capable of producing bivariate samples from a copula. We shall not enter into the details of the construction of such a tool, we shall merely indicate that it can be built by aggregation of one dimensional random generators for the various conditional distributions. The gory details of such a construction are too technical for this book, so we shall leave them aside. *Armed with such a weapon*, it is very simple to generate samples from all the distributions having this specific copula as their own copula. Indeed, the first components of a random sample from a copula form a univariate sample uniformly distributed on $[0, 1]$. So transforming this sample by computing the quantile function of the first marginal distribution will turn this uniform sample into a sample from the first marginal distribution. This is an instance of our favorite method for generating random samples. Similarly, transforming the second components (which also form a uniform sample, by definition of a copula) by computing the quantile function of the second marginal distribution will give a random sample from the second marginal distribution. Now, by the very definition of the copula, these two univariate samples have not only the right marginals, but they also have the right copula! So put together, they form a bivariate sample from the desired bivariate distribution.

We implement this strategy on our example of the coffee log returns. The following set of commands produce a bivariate sample of the same size as the data, from our estimation of the joint distribution of the coffee log-returns. Remember that this estimate is comprised of the estimates of the marginal distributions of the two random quantities together with the parametric estimate of their copula.

```
> N <- length(BLRet)
> SD <- rcopula(ESTC,N)
> Xsim <- qgpd(B.est, SD$x)
> Ysim <- qgpd(C.est, SD$y)
```

We review the main steps of the simulation before plotting the results. The function `rcopula` produces bivariate samples from the copula whose information is encapsulated in the argument `ESTC`, which needs to be an object of class `copula`. We extract the two elements (columns) of `SD` by means of the dollar signs $ followed by the lower case `x` for the first column, and by the lower case `y` for the second column. By definition of a copula, `SD$x` and `SD$y` are random samples uniformly distributed over the unit interval. Consequently, the third and fourth commands result in samples `Xsim` and `Ysim` from the GPD's given by the `gpd` objects `BEST` and `CEST`. This is because we compute quantile functions on uniform samples. We already used this trick several times to generate random samples from a given distribution. But the situation is quite different from those earlier in that the uniform samples are paired by the copula used to generate them. So the copula of the resulting bivariate sample is the copula we started from. The loop is closed, and we have produced a bivariate sample from the desired distribution. In the same way we produced Fig. 3.12, we can place the scatterplot of the simulated samples `Xsim` and `Ysim` to the right of the scatterplot of the original samples `BLRet` and `CLRet`. The result is reproduced in Fig. 3.20. The differences with Fig. 3.12 are striking. This plot shows that our model and the ensuing simulations capture rather well the characteristics of the point distribution in the plane. As further evidence, the numerical measures of dependence which we introduced earlier confirm that the results are very satisfactory. This is clear from the comparison of the values of the Kendall's tau and Spearman's rho statistics computed for the empirical copula (directly from the data) and from the fitted copula. We reproduce the commands and the results:

```
> print(ESTC)
Gumbel copula family; Extreme value copula.
Parameters :
   delta  =  2.98875657681924
> Kendalls.tau(EMPCOP)
[1] 0.6881483
```

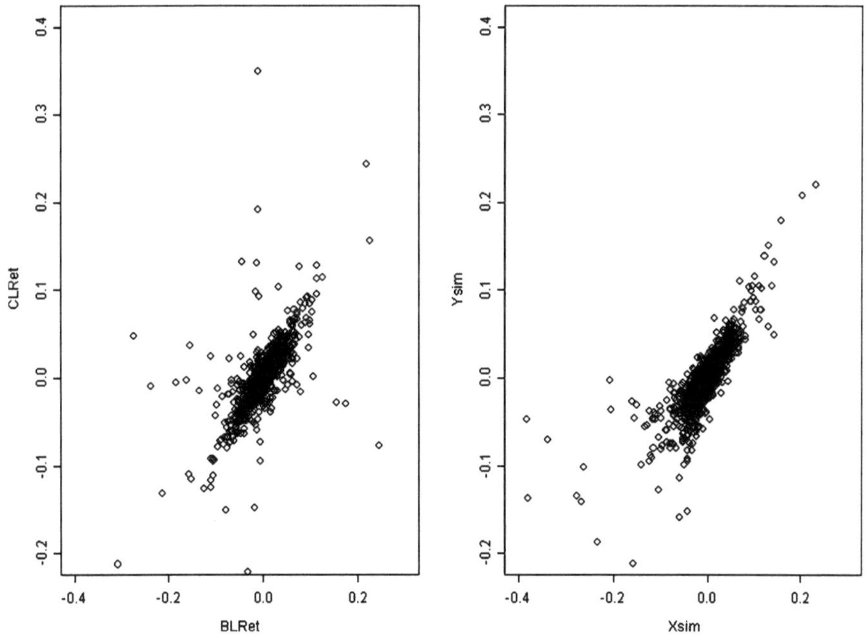

Fig. 3.20. Scatterplot of the Colombian coffee log-returns against the Brazilian ones (*left pane*), and scatterplot of the Monte Carlo samples produced with the dependence captured by the fitted copula (*right pane*)

```
> Kendalls.tau(ESTC, tol = 1e-2)
[1] 0.6654127
> Spearmans.rho(EMPCOP)
[1] 0.8356747
> Spearmans.rho(ESTC)
[1] 0.8477659
```

3.4.6 A Risk Management Example

Long before they were introduced in the valuation of baskets of debts and loan obligations, risk quantification for large portfolios of instruments with heavy tails was the main application of copulas in financial risk management. To this day, it remains the most important application of copulas.

We first describe the simple example of a portfolio of two instruments. Such an application is clearly limited in scope, but it should not be underestimated. We generalize it to arbitrary large portfolios at the end of the section. We chose to warm up with a simple structure for the ease of notation and for the freedom to use a large set of copula families.

The risk measures which we compute provide values which are orders of magnitude different from the values obtained from the theory of the Gaussian distribution. Relying on the Gaussian theory leads to overly optimistic figures ... and possibly to very bad surprises. Fitting heavy tail distributions and using copulas give more conservative (and presumably more realistic) quantifications of the risk carried by most portfolios.

In our simple model, the initial value of the portfolio is:

$$V_0 = n_1 S_1 + n_2 S_2$$

where n_1 and n_2 are the numbers of units of the two instruments, which are valued at S_1 and S_2 at the beginning of the period. Let us denote by S_1' and S_2' their values at the end of the period. Then the new value of the portfolio is:

$$V = n_1 S_1' + n_2 S_2'$$

and the raw return R over the period is

$$\begin{aligned}R = \frac{V - V_0}{V_0} &= \frac{n_1(S_1' - S_1) + n_2(S_2' - S_2)}{n_1 S_1 + n_2 S_2} \\ &= \frac{n_1 S_1}{n_1 S_1 + n_2 S_2} \frac{S_1' - S_1}{S_1} + \frac{n_2 S_2}{n_1 S_1 + n_2 S_2} \frac{S_2' - S_2}{S_2} = \lambda_1 X_1 + \lambda_2 X_2\end{aligned}$$

if we denote by λ_1 and λ_2 the fractions of the portfolio invested in the two instruments, and by $X_1 = (S_1' - S_1)/S_1$ and $X_2 = (S_2' - S_2)/S_2$ the raw returns on the individual instruments.

3.4.6.1 ***Value-at-Risk*** VaR_p

We now compute the value at risk of this portfolio. According to the discussion of Chap. 1, for a given level p, VaR_p was defined in relation to the capital needed to cover losses occurring with frequency p, and more precisely, VaR_p was defined as the $100p$-th percentile of the loss distribution. i.e. the solution r of the equation:

$$p = \mathbb{P}\{-R \geq r\} = \mathbb{P}\{R \leq -r\} = F_R(-r).$$

In order to solve for r in the above equation, we need to be able to compute the cdf of the log-return R. The latter can be expressed analytically as:

$$\begin{aligned}
\mathbb{P}\{-R \geq r\} &= \mathbb{P}\{\lambda_1 X_1 + \lambda_2 X_2 \leq -r\} \\
&= \int\int_{\{(x_1,x_2);\lambda_1 x_1+\lambda_2 x_2 \leq -r\}} f_{(X_1,X_2)}(x_1,x_2)\, dx_1 dx_2 \\
&= \int_{-\infty}^{+\infty} \left[\int_{-\infty}^{-r/\lambda_2-\lambda_1 x_1/\lambda_2} c(F_{X_1}(x_1), F_{X_2}(x_2)) f_{X_2}(x_2) dx_2\right] f_{X_1}(x_1) dx_1 \\
&= \int_0^1 \left[\int_0^{F_{X_2}(-r/\lambda_2-\lambda_1 F_{X_1}^{-1}(u)/\lambda_2)} c(u,v) dv\right] du \\
&= \int_0^1 du\, \frac{\partial}{\partial u} C(u,v)\Big|_{v=F_{X_2}(-r/\lambda_2-\lambda_1 F_{X_1}^{-1}(u)/\lambda_2)} du
\end{aligned}$$

where we used several substitutions to change variables in simple and double integrals. Despite all these efforts, and despite the fact that we managed to reduce the computation to the evaluation of a single integral, this computation cannot be pushed further in this generality. Even when we know more about the copula and the marginal distributions, this integral can very rarely be computed explicitly. We need to use numerical routines to compute this integral. In fact, we need to run these routines quite a lot of times to solve the equation giving the desired value of r.

3.4.6.2 ***Expected Shortfall*** $\mathbb{E}\{\Theta_p\}$

We now compute the other measure of risk which we introduced in the Appendix of Chap. 2. The analytic technicalities of the computation of the expected shortfall $\mathbb{E}\{\Theta_p\}$ are even more daunting than for the computation of the value at risk VaR_p. Just to give a flavor of these technical difficulties, we initialize the process by:

$$\begin{aligned}
\mathbb{E}\{\Theta_p\} &= \mathbb{E}\{-R| - R > VaR_p\} \\
&= \frac{1}{p}\int_{-\infty}^{-VaR_p} -rF_R(dr) \\
&= \frac{1}{p}\int_{-\infty}^{+\infty} dx_1\, f_{X_1}(x_1) \int_{-\infty}^{-VaR_p/\lambda_2-\lambda_1 x_1/\lambda_2} (\lambda_1 x_1 + \lambda_2 x_2) \\
&\qquad c(F_{X_1}(x_1), F_{X_2}(x_2))\, f_{X_2}(x_2) dx_2
\end{aligned}$$

where we have used the same notation as before. Unfortunately, it seems much more difficult to reduce this double integral to a single one, and it seems hopeless to try to derive reasonable approximations of the analytic expression of the expected shortfall which can be evaluated by tractable computations. Following this road, we ended up in a cul-de-sac.

Fortunately, random simulation of large samples from the joint distribution of (X, Y) and Monte Carlo computations can come to the rescue and save the day.

3.4.6.3 *Use of Monte Carlo Computations*

We illustrate the use of Monte Carlo techniques by computing the VaR_p and the expected shortfall $\mathbb{E}\{\Theta_p\}$ of a portfolio of Brazilian and Colombian coffee futures contracts. We solve the problem by simulation using historical data on the daily log-returns of the two assets. The strategy consists in generating a large sample from the joint distribution of X and Y as estimated from the historical data, and computing for each couple (x_i, y_i) in the sample, the value of R. Our estimate of the value at risk is simply given by the empirical quantile of the set of values of R thus obtained. We can now restrict ourselves to the values of R smaller than the negative of the VaR_p just computed, and the negative of the average of these R's gives the expected shortfall. This is implemented in the function `VaR.exp.sim` whose use we illustrate in the commands below.

```
> RES <- VaR.sim(n=10000, p=0.01, copula=ESTC, x.est=B.est,
        y.est=C.est, lambda1=0.7, lambda2=0.3)
> RES[1]
 Simulation size
        10000
> RES[2]
 VaR Q=0.01
 0.1073599
 > RES[3]
 ES Q=0.01
 0.1618901
```

which produce the value at risk and the expected shortfall over a one-period horizon of a unit portfolio with 70 % of Brazilian coffee and 30 % of Colombian coffee. Notice that the function `VaR.sim` returns a vector with three components. The first one is the number of Monte Carlo samples used in the computation, the second is the estimated value of the VaR while the third one is the estimated value of the expected shortfall. It is important to keep in mind that these numbers are the results of Monte Carlo computations, so they will be different each time the function is run, even if the parameters remain the same (recall the discussion in Chap. 1).

3.4.6.4 Comparison with the Results of the Gaussian Model

For the sake of comparison, we compute the same value at risk under the assumption that the joint distribution of the coffee log-returns is Gaussian. This assumption is implicit in most of the VaR computations done in everyday practice. Our goal here is to show how different the results are.

```
> Port <- c(.7,.3)
> MuP <- sum(Port*Mu)
> MuP
[1] -0.0007028017
> SigP <- sqrt(t(Port) %*% Sigma %*% Port)
> SigP
           [,1]
[1,] 0.03450331
>  - qnorm(p=.01,mean=MuP,sd=SigP)
[1] 0.08096951
```

For the given portfolio `Port`, we compute the mean `MuP` and the standard deviation `SigP` of the portfolio return, and VaR given by the model is the negative of the one percentile of the corresponding Gaussian distribution. We learn that only one percent of the time will the return be less than $8\,\%$ while the above computation was telling us that it should be expected to be less than $10\,\%$ with the same frequency. One cent on the dollar is not much, but for a large portfolio, ..., things add up!

3.4.7 A First Example from the Credit Markets

Our second illustration of the use of bivariate copulas and bivariate distributions will be developed in Sect. 3.4.9 into a full blown discussion of a realistic application to credit portfolios and Collateralized Debt Obligations (CDOs). For the time being, we assume τ_1 and τ_2 are the times of default of firms A_1 and A_2 respectively. The challenge is to compute probabilities of events such as "firm A_1 defaults before firm A_2", or "there are no default before a given time T", or any "probability or expectation involving both τ_1 and τ_2" when the values of firms A_1 and A_2 depend upon each other, and consequently, the random variables τ_1 and τ_2 are dependent. In order to tackle this problem, not only do we need to know the marginal distributions of τ_1 and τ_2, but we also need to capture the dependence between τ_1 ad τ_2. The knowledge of the correlation coefficient(s) is not enough and we need to use copulas.

In practice, the marginal distributions of the default times are estimated from historical data, or Credit Default Swaps (CDSs) depending upon the application at hand. For the sake of illustration, we assume that $\tau_1 \sim E(0.5)$ and $\tau_2 \sim E(0.9)$. The estimation of their copulas is a very touchy business (see the Notes and Complements for a discussion of its role in the collapse of the credit markets) but again, for the sake of illustration, we shall assume that it is a Gumbel copula with parameter $\delta = 1.5$. The computations of the probabilities mentioned earlier can be done in `R` as follows:

```
> N <- 5000
> SD <- rcopula(gumbel.copula(1.5),N)
> TAU1.sim <- qexp(0.5, SD$x)
> TAU2.sim <- qexp(0.9, SD$y)
> P1 <- mean(TAU1.sim < TAU2.sim)
[1] 0.9088
> TT <- 1
> P2 <- mean ( (TAU1.sim > TT) & (TAU2.sim > TT))
[1] 0.694
```

3.4.8 **Higher Dimensional Copulas**

We now extend our early discussion of bivariate copulas given starting Sect. 3.4.1, to the general multidimensional case. Unfortunately, a fair amount of repeats is necessary and we apologize for that.

A k-dimensional copula is the joint distribution of k random variables uniformly distributed over the unit interval $[0,1]$. So since we identify distributions and their cumulative distribution functions, copulas are the functions C from $[0,1]\times\cdots\times[0,1]$ into $[0,1]$ satisfying

$$C(u_1,\cdots,u_k) = \mathbb{P}\{U_1 \le u_1,\cdots,U_k \le u_k\}$$

for a set $U_1,\cdots,U_k$ of $U(0,1)$ random variables. An important result of the theory of copulas, known as Sklar's theorem, states that if $F = F_{\mathbf{X}}$ is the joint cdf of random variables $X_1,\cdots,X_k$ with marginals $F_1 = F_{X_1},\cdots,F_k = F_{X_k}$, then there is a copula C such that

$$F(x_1,\cdots,x_k) = C(F_1(x_1),\cdots,F_k(x_k)), \qquad (x_1,\cdots,x_k) \in \mathbb{R}^k. \tag{3.20}$$

Moreover, this copula is unique whenever the marginal cdfs are continuous. C is called the copula of F or the copula of the random variables $X_1,\cdots,X_k$. Conversely, given a k-dimensional copula C and k cdfs $F_1,\cdots,F_k$, formula (3.20) defines a distribution in $\mathbb{R}^k$ with marginals $F_1,\cdots,F_k$ and copula C. A simple consequence of Sklar's theorem is that the copula of a set of k random variables $X_1,\cdots,X_k$ is given by the formula

$$C(u_1,\cdots,u_k) = F(F_1^{-1}(u_1),\cdots,F_k^{-1}(u_k))$$

where F denotes the joint cdf of the random variables $X_1,\cdots,X_k$ whose quantile functions are $F_1^{-1},\cdots,F_k^{-1}$.

3.4.8.1 First Properties of Copulas

- $C(u_1,\cdots,u_k)$ is non-decreasing in each variable u_i;
- $C(1,\cdots,1,u_i,1,\cdots,1) = u_i$ for all i;

- If $u_i^1 \le u_i^2$ for all i, then:

$$\sum_{1\le i_1\le 2} \cdots \sum_{1\le i_k\le 2} (-1)^{i_1+\cdots+i_k} C(u_1^{i_1}, \cdots, u_k^{i_k}) \ge 0;$$

- The copula of k random variables is independent of their means (i.e. it does not change if one adds a number to each random variable);
- Copulas are scale invariant (i.e. they do not change if the individual random variables are multiplied by positive scalars);
- If $\psi_1, \cdots, \psi_k$ are monotone increasing functions, the copula of k random variables $X_1, \cdots, X_k$ is the same as the copula of the k random variables $\psi_1(X_1)$, $\cdots, \psi_k(X_k)$;
- If the random variables $X_1, \cdots, X_k$ have a joint density f, and if we denote by $f_1, \cdots, f_k$ the densities of $X_1, \cdots, X_k$ respectively, then the copula of $X_1, \cdots$, X_k has a density c which satisfies

$$f(x_1, \cdots, x_k) = c(F_1(x_1), \cdots, F_k(x_k)) \prod_{i=1}^{k} f_i(x_k), \qquad (x_1, \cdots, x_k) \in \mathbb{R}^k,$$

which can be rewritten as

$$c(u_1, \cdots, u_k) = \frac{f(F_1^{-1}(u_1), \cdots, F_k^{-1}(u_k))}{f_1(F_1^{-1}(u_1)) \cdots f_k(F_k^{-1}(u_k))}, \quad (u_1, \cdots, u_k) \in [0,1] \times \cdots \times [0,1].$$

The first property is an easy consequence of the fact that C is a cdf, the second property merely says that the marginals are uniformly distributed on $[0,1]$ while the next inequality (which is easily checked in the case $n = 2$) expresses the positivity of a probability distribution function. The next three claims are immediate consequences of simple properties of cdfs, the last of the three containing the previous two as particular cases. Finally, the last claim is a consequence of the fact that the density of a random vector is the multiple cross partial derivative of its distribution function. The following properties are geared toward future applications to Monte Carlo simulations with copulas. For the next two bullet points, we assume that the k uniform random variables on $[0,1]$, say $U_1, \cdots, U_k$, have copula C.

- If $1 \le h \le k$, the function $(u_1, \cdots, u_h) \hookrightarrow C(u_1, \cdots, u_h, 1, \cdots, 1)$ is the joint cdf of $(U_1, \cdots, U_h)$.
- The conditional cdf of U_h given that $U_1 = u_1, \cdots, U_{h-1} = u_{h-1}$, namely the function $u_h \hookrightarrow \mathbb{P}\{U_h \le u_h | U_1 = u_1, \cdots, U_{h-1} = u_{h-1}\}$ is given by

$$F_{U_h \,|\, U_1=u_1, \cdots, U_{h-1}=u_{h-1}}(u_h) = \frac{\partial^{h-1}_{u_1 \cdots u_{h-1}} C(u_1, \cdots, u_{h-1}, u_h, 1, \cdots, 1)}{\partial^{h-1}_{u_1 \cdots u_{h-1}} C(u_1, \cdots, u_{h-1}, 1, 1, \cdots, 1)}. \tag{3.21}$$

3.4.8.2 Examples of High Dimensional Copulas

1. The k-dimensional **independent copula** $C_{k,ind}$ is the copula of k independent uniform random variables. Is is given by:

$$C_{k,ind}(u_1, \cdots, u_n) = u_1 \cdots u_n$$

As in the bivariate case, this copula is highlighted for the sake of completeness as its practical use is minimal.

2. A k-dimensional **Gaussian copula** is the copula of k jointly Gaussian random variables. By translation and scale invariance, one can assume that these Gaussian random variables have mean zero and variance one, so that their joint distribution and hence their copula, are entirely determined by their correlation matrix, say $\boldsymbol{\Sigma}$. So the Gaussian copula with correlation matrix $\boldsymbol{\Sigma}$ is defined as

$$C_{Gauss,k,\boldsymbol{\Sigma}}(u_1, \cdots, u_k) = \Phi_{k,\mathbf{0},\boldsymbol{\Sigma}}(\Phi^{-1}(u_1), \cdots, \Phi^{-1}(u_k)), \tag{3.22}$$

for $(u_1, \cdots, u_k) \in [0,1] \times \cdots \times [0,1]$, where $\Phi_{k,\mathbf{0},\boldsymbol{\Sigma}}$ denotes the (joint) cdf of the k-dimensional Gaussian distribution with mean vector $\mathbf{0}$ and variance/covariance matrix $\boldsymbol{\Sigma}$, and as usual, $\Phi = \Phi_{1,0,1}$ denotes the cdf of the univariate standard Gaussian distribution. For $0 < \rho < 1$, we shall use the notation $C_{Gauss,k,\rho}$ for the k-dimensional Gaussian copula with correlation matrix $\Sigma_{k,\rho}$ which has ones on the diagonal and ρ everywhere else. We will see that this copula played a major role in the development of the market of Collateralized Debt Obligations (CDOs).

3. By analogy with the definition of the Gaussian copula introduced above, the k dimensional t **copula** is defined as the copula of a random vector with a k-dimensional t-distribution. It is defined as

$$C_{Student,k,\boldsymbol{\Sigma},\nu}(u_1, \cdots, u_k) = t_{\boldsymbol{\Sigma},\nu}(t_\nu^{-1}(u_1), \cdots, t_\nu^{-1}(u_k)), \tag{3.23}$$

for $(u_1, \cdots, u_k) \in [0,1] \times \cdots \times [0,1]$, where $t_{k,\boldsymbol{\Sigma},\nu}$ denotes the (joint) cdf of the k-dimensional Student distribution with variance/covariance matrix $\boldsymbol{\Sigma}$ and ν degrees of freedom, and as usual, t_ν denotes the cdf of the univariate standard Student distribution with ν degrees of freedom.

3.4.8.3 Archimedean Copulas

The class of Archimedean copulas is based on the notion of Archimedean copula generator defined as a strictly decreasing convex function $\psi : (0,1] \hookrightarrow [0,\infty)$ satisfying $\psi(1) = 0$. Its pseudo inverse is defined as

$$\psi^{[-1]}(x) = \begin{cases} \psi^{-1}(x) & \text{if } 0 < x < \psi(0) \\ 0 & \text{if } \psi(0) \le x < \infty \end{cases} \tag{3.24}$$

where $\psi(0)$ is defined as $\psi(0) = \lim_{t \searrow 0} \psi(t)$. If $\psi(0) = \infty$, then the pseudo inverse $\psi^{[-1]}$ is in fact a true inverse ψ^{-1} and the generator is said to be strict. The reason for the introduction of generator functions as defined above is given by a result due to Kimberlink stating that if ψ is an Archimedean copula generator, and if we define the function C by

$$C_\psi(u_1, \cdots, u_k) = \psi^{[-1]}(\psi(u_1) + \cdots + \psi(u_k)), \tag{3.25}$$

for $(u_1, \cdots, u_k) \in [0,1] \times \cdots \times [0,1]$, then C is a copula if and only if the function $\psi^{[-1]}$ is completely monotone on $[0, \infty)$ in the sense that it has derivatives of all orders which alternate in sign (even order derivatives are positive and odd order derivatives are negative).

The following are examples of the most commonly used classes of Archimedean copulas.

1. The k-dimensional **Gumbel copula** with parameter $\delta \in [1, \infty)$ is the Archimedean copula associated with the strict generator $\psi(u) = (-\log u)^\delta$. Such a copula is sometimes called the logistic copula.

$$C_{Gumbel,k,\delta}(u_1, \cdots, u_k) = \exp\big[-[(-\log u_1)^\delta + \cdots + (-\log u_k)^\delta]^{1/\delta}\big],$$

for $(u_1, \cdots, u_k) \in [0,1] \times \cdots \times [0,1]$.

2. The k-dimensional **Clayton copula** with parameter $\beta > 0$ is the Archimedean copula associated with the generator $\psi(u) = u^{-\beta} - 1$.

$$C_{Clayton,k,\beta}(u_1, \cdots, u_k) = \Big[\sum_{i=1}^k u_i^{-\beta} - k + 1\Big]^{-1/\beta},$$

$$(u_1, \cdots, u_k) \in [0,1] \times \cdots \times [0,1].$$

3. The k-dimensional **Frank copula** with parameter $\alpha \neq 0$ is the Archimedean copula associated with the generator rank

$$\psi(u) = -\log \frac{e^{-\alpha u} - 1}{e^{-\alpha} - 1}.$$

This generator is strict for $k \geq 3$ and $\alpha > 0$. The Frank copula is given by the formula

$$C_{F,k,\alpha}(u_1, \cdots, u_k) = -\frac{1}{\alpha} \log\Big[1 + \frac{\prod_{i=1}^k (e^{-u_i} - 1)}{(e^{-\alpha} - 1)^{k-1}}\Big].$$

3.4.8.4 Fitting a Copula to Data

In practice, one is given a sample $\mathbf{x}^{(1)} = (x_1^{(1)}, \cdots, x_k^{(1)})$, $\cdots$, $\mathbf{x}^{(n)} = (x_1^{(n)}, \cdots, x_k^{(n)})$ of size n from a k-dimensional distribution, and if the goal is to estimate the copula of the common distribution of these k dimensional vectors, it is quite straightforward to generalize to the present general set-up the strategy used in the bivariate example of the Brazilian and Colombian coffee prices:

- For $i = 1, \cdots, k$, we estimate the i-th marginal cdf F_i from the univariate sample $x_i^{(1)}, \cdots, x_i^{(n)}$ of size n, and denote by $\hat{F}_i$ the estimate.
- We create the k-dimensional sample $\boldsymbol{u}^{(1)} = (u_1^{(1)} = \hat{F}_1(x_1^{(1)}), \cdots, u_k^{(1)} = \hat{F}_k(x_k^{(1)}))$, $\cdots$, $\boldsymbol{u}^{(n)} = (u_1^{(n)} = \hat{F}_1(x_1^{(n)}), \cdots, u_k^{(n)} = \hat{F}_k(u_k^{(n)}))$ of size n. Notice that for each $i = 1, \cdots, k$, the univariate sample $u_i^{(1)}, \cdots, u_i^{(n)}$ should be a sample of size n from the uniform distribution $U(0, 1)$. The empirical distribution of the k-dimensional sample $\boldsymbol{u}^{(1)}$, $\cdots$, $\boldsymbol{u}^{(n)}$ is called the empirical copula of the original sample.
- Even though non-parametric methods (such as kernel density estimation) could be used at this stage, maximum likelihood estimates computed from a specific copula family are usually preferred, especially when Monte Carlo simulations from the fitted copula are needed. So the typical next step is to integrate the requirements of the analysis and choose a copula family (Gaussian, Gumbel, Clayton, ...), compute the likelihood of the empirical copula for the family in question, and compute the value of the parameter ($\hat{\rho}$, $\hat{\delta}$, $\hat{\alpha}$, $\cdots$) which maximizes this likelihood.

3.4.8.5 *Monte Carlo Simulations from Multivariate Distributions*

Let us assume that the goal is the generation of random samples $\mathbf{x}^{(1)}, \mathbf{x}^{(2)}, \cdots, \mathbf{x}^{(n)}$ from a given multivariate distribution. We denote the dimension of this distribution by k and by $x_j^{(i)}$ the j-th component of $\mathbf{x}^{(i)}$. Since the distribution in question is characterized by

- Its marginal cdfs F_1, $\cdots$, F_k;
- Its copula $C : (u_1, \cdots, u_k) \hookrightarrow C(u_1, \cdots, u_k)$,

a possible simulation algorithm goes as follows:

- Generate independent samples $\boldsymbol{u}^{(1)}, \boldsymbol{u}^{(2)}, \cdots, \boldsymbol{u}^{(n)}$ from the k-dimensional copula C
- Compute $x_j^{(i)} = F_j^{-1}(u_j^{(i)})$ for $i = 1, \cdots, n$ and $j = 1, \cdots, k$.

Such an algorithm relies on two prerequisites:

- Being able to evaluate the quantile functions F_j^{-1};
- Being able to generate random samples from the copula C.

Since we already addressed the first of these two bullet points, we concentrate on the second.

3.4.8.6 *Monte Carlo Simulations from Copulas*

The goal is thus to generate independent samples $\boldsymbol{u}^{(1)}, \boldsymbol{u}^{(2)}, \cdots, \boldsymbol{u}^{(n)}$ from a k-dimensional copula C. This is particularly easy when C is the copula of a multivariate distribution for which we have random vector generators. We illustrate this claim in the case of the Gaussian distribution. The case of the Student distribution can be treated in the same way.

In order to generate samples $\boldsymbol{u}^{(1)}, \boldsymbol{u}^{(2)}, \cdots, \boldsymbol{u}^{(n)}$ from the k-dimensional Gaussian copula with correlation matrix $\boldsymbol{\Sigma}$, we

- Generate independent samples $\mathbf{x}^{(1)}, \mathbf{x}^{(2)}, \cdots, \mathbf{x}^{(n)}$ from the k-dimensional Gaussian distribution $N_k(\mathbf{0}, \boldsymbol{\Sigma})$;
- Apply the standard Gaussian cdf Φ to each component $x_j^{(i)}$ of each of the $\mathbf{x}^{(i)}$'s, namely we compute $u_j^{(i)} = \Phi(x_j^{(i)})$.

It is also possible to generate samples for copulas by successive simulations of conditional densities. While theoretically simple, this strategy is typically unfeasible in practice. However, it happens to be manageable for most of the **Archimedean Copulas** when k is small ($k = 2$ or $k = 3$) and for general values of k when explicit formulas can be derived or numerical inversion of the marginal cdf's is feasible. This is indeed the case for the Clayton copula. We do not give the details as they are beyond the scope of the book. Finally, we notice that the use of the Laplace transform has been proposed as a tool for Monte Carlo sample generation from a copula. It was successfully implemented in the case of high dimensional Gumbel copulas.

3.4.8.7 ***VaR and Expected Short Fall Computations by Monte Carlo Simulations***

If the portfolio comprises a large number k of assets, if we denote by $X_1, \cdots, X_k$ the individual raw returns over a period of length Δt, and if we denote by $\lambda_1, \cdots, \lambda_k$ the relative weights of the individual stocks in the portfolio, then the Monte Carlo strategy can be implemented in the following way.

As explained in the previous subsection, success relies on two prerequisites. The first one is the estimation of the marginal distributions F_j of the k individual stocks (we denote by $\hat{F}_j$ the estimates), and the second one is the estimation of the copula C of the returns. So if the estimate $\hat{C}$ is from a family for which we have a random generator, then we can generate sample $\boldsymbol{u}^{(1)}, \boldsymbol{u}^{(2)}, \cdots, \boldsymbol{u}^{(n)}$ from the k-dimensional copula $\hat{C}$, and set $x_j^{(i)} = F_j^{-1}(u_j^{(i)})$ for $i = 1, \cdots, n$ and $j = 1, \cdots, k$. Monte Carlo samples $R^{(i)}$ of the portfolio return are then created via the formula

$$R^{(i)} = \lambda_1 x_1^{(i)} + \cdots + \lambda_k x_k^{(i)}, \qquad i = 1, \cdots, n.$$

The value at risk of the portfolio is given by the empirical quantile. More precisely, it is obtained by ordering the returns

$$R_{(1)} < R_{(2)} < \cdots < R_{(n)}$$

and by using $\widehat{VaR_p} = -R_{(1+[np])}$ as estimate. Here and in the following we use the notation $[x]$ for the integer part of x and we assumed $0 < p < 1$. Once the value at risk is estimated, the expected shortfall is estimated by replacing the conditional expectation entering in its definition by the corresponding empirical analog. In other words

$$\widehat{ES}_p = -\frac{1}{[np]}\sum_{i=1}^{[np]} R_{(i)}$$

3.4.8.8 ***Testing the Procedure with Real Data***

If `Ret` is a matrix of daily log returns with one row per day and one column per stock, the following commands can be used to compute the one-day value at risk at the level 0.01 by the Monte Carlo procedure described above.

```
> Nstock <- dim(Ret)[2]
> Port <- rep(1/Nstock, Nstock)
> Mu <- apply(Ret,2,mean)
> Sig <- var(Ret)
> Ret.est <- NULL
> for (I in 1:Nstock) Ret.est <-c(Ret.est,fit.gpd(Ret[,I]))
> SD <- rmvgaussian.copula(Nsim, Sigma = Sig)
> for (I in 1:Nstock) SD[,I] <- qgpd(Ret.est[[I]],SD[,I])
> Port.sim <- SD %*% Port
> VaR <- - quantile(Port.sim, probs=.01)
```

We can compute the expected shortfall with the command `ES <- - mean(Port.sim < - VaR)`. As before, we can compare the results with the results we would obtained under the assumption that the joint distribution of the daily log-returns is Gaussian, and as before, we would find that the VaR computed under the Gaussian assumption offers an overly optimistic picture of the risk of the portfolio as quantified by VaR.

3.4.9 **Multi Name Credit Derivatives and CDOs**

In this final subsection, we document the role of copulas (and especially the Gaussian copula) in the pricing of a special form of the (in)famous Collateralized Debt Obligations (CDOs for short). CDOs are financial contracts written on the cumulative losses of a portfolio of credit sensitive instruments (loans, mortgages, bonds, CDSs, ...). The premise of the creation of these new securities is the idea that aggregation and slicing into tranches lead to new securities with ratings and probabilities of default more favorable than from those of the original members of the pool. This securitization process was promoted as a clever way to produce parts whose values sum up to more than the value of the whole. Does this sound *too good to be true?* The collapse of the credit markets gives a clear answer.

In order to define precisely these securities, we introduce the notation we use for the major components of these new securities.

- First, the **maturity** T. For single tranche synthetic CDOs, T is typically 5, 7, or 10 years, $T = 5$ offering most liquidity.
- Then we denote by $\{1, 2, \cdots, k\}$ the k obligor names included in the portfolio and by $\{\tau_1, \tau_2, \cdots, \tau_k\}$ the default times of these obligors,
- We also denote by N_i the nominal and by R_i the recovery rate for obligor i, so that $(1 - R_i)N_i$ represents the loss given default of obligor i

We denote by $N(t)$ the total number of defaults occurring before or at time t and by $L(t)$ the cumulative losses at time t in the credit pool:

$$L(t) = \sum_{\tau_i \leq t} N_i(1 - R_i). \tag{3.26}$$

In most of the discussion that follows, we concentrate on homogeneous portfolios maintained by the Dow Jones, and we assume that the portfolio is homogeneous in the sense that all the nominals N_i, as well as the recovery rates, do not depend upon i. In such a particular case, both N_i and R_i can be removed from the definition of the cumulative loss, since for the sake of simplicity, we normalize the cumulative loss to a maximum of 1. Under such a simplifying assumption we have:

$$L(t) = \frac{N(t)}{k} \tag{3.27}$$

which gives the relative number of defaults occurring before time t.

Independently of the actual definition of the cumulative loss $L(t)$, the losses are organized in tranches defined by their end points called attachment points $0 < K_0 < K_1 < \cdots < K_M < 1$. Attachment points can be tailor made to the needs of investors. In the case of the Dow Jones indexes underlying the liquid single tranche synthetic CDOs, they are standardized as follows:

- $K_1 = 3\,\%$, $K_2 = 6\,\%$, $K_3 = 9\,\%$, $K_4 = 12\,\%$ and $K_5 = 22\,\%$ for the Dow Jones iTraxx Europ Index
- $K_1 = 3\,\%$, $K_2 = 7\,\%$, $K_3 = 10\,\%$, $K_4 = 15\,\%$ and $K_5 = 30\,\%$ for the Dow Jones CDX North American Index

Tranche names are mostly standard:

- *Equity Tranche* $0 = K_0 \leq L(T) \leq K_1$
- J-th *Mezzanine Tranche* $K_J \leq L(T) \leq K_{J+1}$
- *Senior Tranche* $K_{M-1} \leq L(T) < K_M$
- *Super Senior Tranche* $K_M \leq L(T)$

Originally, CDOs were written on customized portfolios chosen to match specific credit exposures, and one of the major problems was to choose the names, the attachment points K_j, and the spreads of each tranche to achieve desired ratings for the notes.

The following figure shows a typical sample realization of the change over time of the (relative) cumulative loss on a credit portfolio (Fig. 3.21).

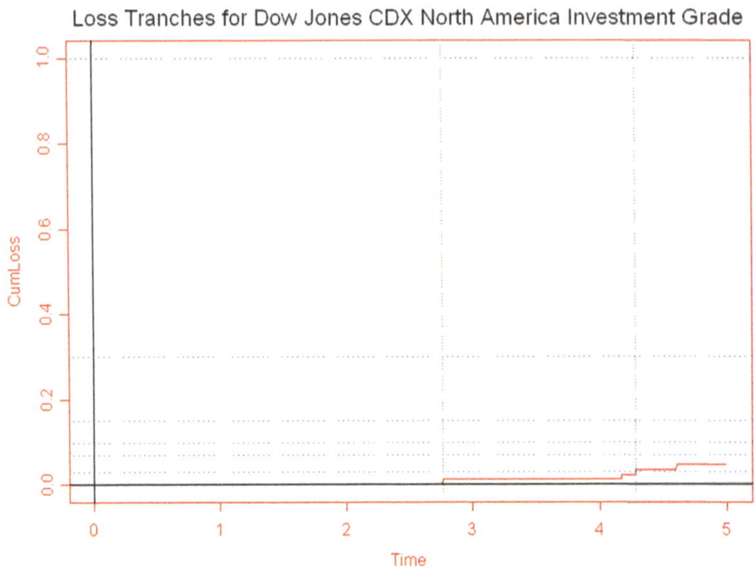

Fig. 3.21. Sample loss function of a credit portfolio

3.4.9.1 *Single Tranche Synthetic CDOs*

The first generation of CDOs gave a quantum leap to the popularity of instruments written on large credit portfolios. However, like CLOs and CBOs, their structures were quite complicated, often involving the creation of special entities like SPVs, and because of their customized nature, they did not offer much liquidity. The second generation of CDOs is based on indexes (typically the CDX indexes in the US and the ITRAX indexes in Europe). The indentures of the contracts are much simpler, and their liquidity is much improved. In this section, we concentrate on the single tranche synthetic CDOs.

3.4.9.2 *Single Tranche Synthetic CDO Mechanics*

The contract involves a protection buyer and a protection seller. As in the case of CDS contracts, the protection buyer pays regular coupons (typically every 3 months) until maturity or until the nominal of the tranche is wiped out by the losses incurred by the portfolio, whichever comes first. In exchange for these regular coupon payments, the protection seller will compensate the protection buyer for the losses of the tranche.

The following tables reproduce bid and ask quotes on the 4th and 5th series of the CDX-IG tranches on December 19, 2006.

IG4	0–3 %	3–7 %	7–10 %	10–15 %	15–30 %
5-year	38 1/4–39 1/4	106–112	26–32	11–16	6–7 1/2
7-year	51 3/8–52 1/8	244–254	47–54	26–32	8 1/2–11
10-year	57 1/2–59 1/8	598–617	118–126	58–66	16–22

IG5	0–3 %	3–7 %	7–10 %	10–15 %	15–30 %
5-year	41 1/4–42 1/4	107 1/4–112	26–29	11–14	6 1/2–9 1/2
7-year	54 3/4–55 5/8	290–300	45–51	27–31	7–10
10-year	61 3/4–62 3/4	685–705	118–124	61–66	17–21

The interpretation of these figures is the following.

We explain the meaning of these quotes by explaining in detail the cash flows associated with one of these tranches. We choose the super-senior tranche with attachment points 15 and 30 % on the 5-year CDX-IG index series 4. Let us assume that this tranche traded for 7 basis points. In this case, the protection buyer is to pay 0.07 % of the notional per year (in quarterly coupon payments made in arrear). In return, she will be compensated for any losses on the portfolio during the 5 years that are between 15 and 30 % of the principal. The losses are computed from the portfolio underlying the index at the original time of the trade.

The quotes for the all the other tranches are defined similarly except for the equity tranche for which the buyer of protection pays an upfront fee and a spread of 500 basis points per year. The published quotes give the bid and ask for the upfront fee expressed as a percent of the notional. The percent of the notional that the protection buyer of the equity-tranche has to pay on December 19, 2006 was between 38.25 and 39.25 % for 5-year protection.

The index is also quoted to indicate the cost of buying full protection against all $k = 125$ names.

3.4.9.3 *Stochastic Models*

As we already explained, we assume for the sake of simplicity that the recovery rates R_i are deterministic and equal to each other (this is a reasonable assumption for homogeneous credit baskets). So in order to model the cumulative losses of a credit portfolio, we only need to model the default times $\tau_1, \tau_2, \cdots, \tau_k$. These are non-negative random variables whose joint distribution can be given by the joint default cdf

$$F(t_1, t_2, \cdots, t_k) = \mathbb{P}\{\tau_1 \le t_1, \tau_2 \le t_2, \cdots, \tau_k \le t_k\}, \; (t_1, t_2, \cdots, t_k) \in [0, \infty)^k$$

or equivalently by the joint survival function

$$S(t_1, t_2, \cdots, t_k) = \mathbb{P}\{\tau_1 > t_1, \tau_2 > t_2, \cdots, \tau_k > t_k\}, \ (t_1, t_2, \cdots, t_k) \in [0, \infty)^k$$

One possible way to characterize this joint distribution is to first identify the marginal default cdfs or equivalently the marginal survival functions

$$F_i(t) = \mathbb{P}\{\tau_i \leq t\}, \qquad S_i(t) = \mathbb{P}\{\tau_i > t\}, \qquad t \in [0, \infty), \ i = 1, 2, \cdots, k$$

and put them together with the help of a copula $C(u_1, u_2, \cdots, u_k)$, in which case we get:

$$F(t_1, t_2, \cdots, t_k) = C(F_1(t_1), F_2(t_2), \cdots, F_k(t_k), \qquad (t_1, t_2, \cdots, t_k) \in [0, \infty)^k.$$

3.4.9.4 Default Models

A first approach to understand the statistics of the times of default τ_i is to view them as the first times the firms stop being solvent. Modeling the joint evolution of the k values of the firms and tracking these times of defaults go under the name of structural approach to default modeling. Figure 3.22 gives an example.

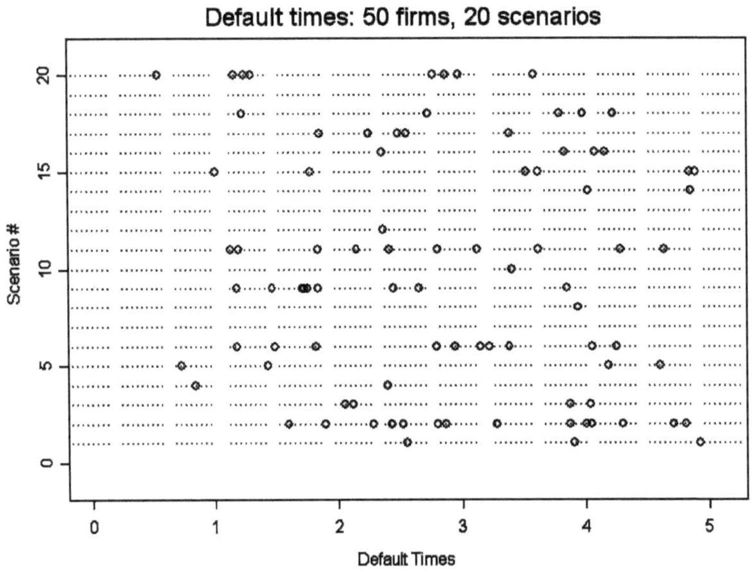

Fig. 3.22. Sample of 20 Monte Carlo scenarios of the default times of 50 firms in the structural approach

However, reduced form models based a simple factor models have gained popularity because of the ease with which they can be defined and calibrated. Typically, one chooses a vector $\mathbf{Y}$ of a small number of factors, and conditioned on the value of $\mathbf{Y}$, we assume that the default times τ_i are *independent*. So the dependencies between the defaults appear through the common factors $\mathbf{Y}$. The term structures of conditional *default* probabilities are then given by

$$F_{\mathbf{Y}}^{(i)}(t) = \mathbb{P}\{\tau_i \le t|\mathbf{Y}\}$$

We now describe the simplest possible way to model the state of the portfolio at maturity T by means of the Gaussian copula introduced by Dr Li. This approach is based on the knowledge of

- Marginal/individual default probabilities $\{p_1, p_2, \cdots, p_k\}$ by time T of the obligors;
- A copula $C(u_1, u_2, \cdots, u_k)$ for default dependency.

The individual probabilities of default

$$p_i = \mathbb{P}\{\tau_i \le T\}, \qquad i = 1, 2, \cdots, k$$

can be inferred from the CDS spread curves of the individual firms. CDS stands for Credit Default Swap. These instruments are liquidly traded and the estimation of the p_i's is not regarded as a serious obstacle. Once the copula is chosen, Monte Carlo simulations can be structured in the following way:

- Draw uniform samples $(U_1, U_2, \cdots, U_k)$ from the k-dimensional copula C;
- For each i, obligor i survives (i.e. does not default before T) if and only if $U_i \le p_i$

In particular,

- The probability of no default is given by

$$\mathbb{P}\{U_i \le p_i, \text{ for all } i \le k\} = C(p_1, p_2, \cdots, p_k)$$

- The probability that none of the first h obligors default

$$\mathbb{P}\{U_i \le p_i, \text{ for all } i \le h\} = C(p_1, \cdots, p_h, 1, \cdots, 1)$$

The rationale for the choice of a Gaussian copula for C is based on Merton's theory of default in which the logarithm of the value of the assets of a firm is assumed to be Gaussian, and default occurs by time T if the value of these assets drops below a solvency threshold. If these logarithms $\tilde{L}^{(i)}$ are modeled with a common factor Y and idiosyncratic noise components ϵ_i such that

$$\tilde{L}^{(i)} = \sqrt{\rho}Y + \sqrt{1-\rho}\epsilon_i, \qquad i = 1, 2, \cdots, k,$$

Y and ϵ_i $1 \le i \le k$ being independent $N(0,1)$ random variables, and ρ a positive number, then the correlation matrix $\boldsymbol{\Sigma}$ to be used in the copula has

- 1 **on** the diagonal
- ρ **off** the diagonal

in other words,

$$\boldsymbol{\Sigma} = \begin{bmatrix} 1 & \rho & \cdots & \cdots & \rho \\ \rho & 1 & \rho & \cdots & \rho \\ \vdots & & \ddots & & \vdots \\ \rho & \cdots & \cdots & 1 & \rho \\ \rho & \cdots & \cdots & \rho & 1 \end{bmatrix}$$

This Gaussian copula model was referred to as the *market standard* in the early 2000s and it was credited to Dr Li. Part of its popularity was due to the fact that it was depending upon a single parameter model ρ which could be implied for each tranche from market quotes, in complete analogy with the implied volatility of equity options. For this reason, these tranche dependent correlation parameters were called implied correlations, and the notion of correlation smile was observed and studied. As a result, trading CDOs was called *trading correlation*. We give a simple example for the purpose of illustration. We generate a sample of 5,000 Monte Carlo simulations of a credit portfolio of 125 firms and count, for each scenario, the number of defaults in the portfolio over a period of one year. We assume that the default times of the firms are all exponentially distributed with rate 0.3 and have a Gaussian copula with parameter $\rho = 0.7$. Our choices for the parameters were made for pedagogical reasons.

```
> Nsim <- 5000
> Nfirm <- 125
> LAMBDA <- rep(0.3,Nfirm)
> SD <- rmvgaussian.copula(Nsim,d=Nfirm,rho=0.7)
> TAU <- array(0,dim=c(Nsim,Nfirm))
> for(I in 1:Nfirm) TAU[,I] <- qexp(SD[,I],LAMBDA[I])
> NDefault <- apply(TAU<TT,1,sum)
> hist(NDefault,nclass=35, probability=TRUE, col="blue",
                               main="Number of Defaults")
> Prob <- mean(NDefault == 0)
[1] 0.1834
```

This computation shows that in one year (`TT=1`) one should expect no default in the portfolio with probability 18 %. The histogram of the Monte Carlo sample is reproduced in Fig. 3.23

Important Remark. The above discussion suggests that Monte Carlo simulation was the right approach to credit portfolio understanding and valuation. However, it is very important to emphasize that defaults are relatively rare (even in periods of economic downturn) and the probabilities of multiple defaults are small. Simulation of rare events (i.e. events with small probabilities) is a typical instance of situation where plain Monte Carlo simulations fail. Dedicated simulation methods involving importance sampling and other sophisticated re-weighting procedures need to be

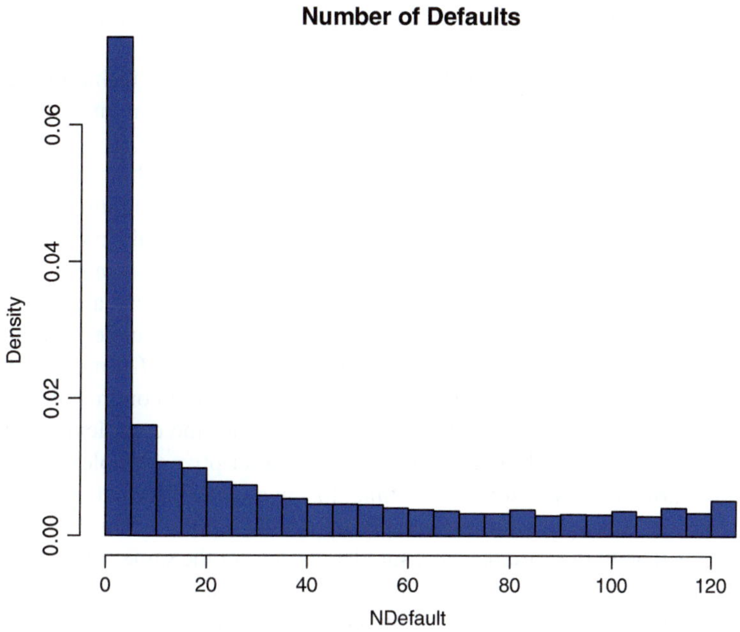

Fig. 3.23. Histogram of the Monte Carlo sample of numbers of defaults in the portfolio

used in order to reduce the variance of the Monte Carlo estimators. See the Notes & Complements at the end of the chapter for references on the subject.

3.5 Principal Component Analysis

Dimension reduction without significant loss of information is one of the main challenges of data analysis. The internet age has seen an exponential growth in the research efforts devoted to the design of efficient codes and compression algorithms. Whether the data are comprised of video streams, images, and/or speech signals, or financial data, finding a basis in which these data can be expressed with a small (or at least a smaller) number of coefficients is of crucial importance. Other important domains of applications are cursed by the high dimensionality of the data. Artificial intelligence applications, especially those involving machine learning and data mining, have the same dimension reduction problems. Pattern recognition problems are closer to the hearts of traditional statisticians. Indeed, regression and statistical classification problems have forced statisticians to face the curse of dimensionality, and to design systematic procedures to encapsulate the important features of high dimensional observations in a small number of variables. Principal component analysis as presented in this chapter, offers an optimal (in the least squares sense) solution to these dimension reduction issues.

3.5.1 Identification of the Principal Components of a Data Set

Principal component analysis (PCA, for short) is a data analysis technique designed for numerical data (as opposed to categorical data). The typical situation that we consider is where the data come in the form of a matrix $[x_{i,j}]_{i=1,\ldots,N,j=1,\ldots,M}$ of real numbers, the entry $x_{i,j}$ representing the value of the i-th observation of the j-th variable. As usual, we follow the convention of labeling the columns of the data matrix by the variables measured, and the rows by the individuals in the population under investigation. Examples are plentiful in most data analysis applications. We give below detailed analyses of several examples from the financial arena.

As we mentioned above, the N members of the population can be identified with the N rows of the data matrix, each one corresponding to an M-dimensional (row) vector of numbers giving the values of the variables measured on this individual. It is often desirable (especially when M is large) to reduce the complexity of the descriptions of the individuals and to replace the M descriptive variables by a smaller number of variables, while at the same time, losing as little information as possible. Let us consider a simple (and presumably naive) illustration of this idea. Imagine momentarily that all the variables measured are scores of the same nature (for examples they are all lengths expressed in the same unit, or they are all prices expressed in the same currency, ...) so that it would be conceivable to try to characterize each individual by the mean, and a few other numerical statistics computed on all the individual scores. The mean:

$$\overline{x_{i\cdot}} = \frac{x_{i1} + x_{i2} + \cdots + x_{iM}}{M}$$

can be viewed as a linear combination of the individual scores with coefficients $1/M$, $1/M, \ldots, 1/M$. Principal component analysis, is an attempt to describe the individual features in the population in terms of M linear combinations of the original features, as captured by the variables originally measured on the N individuals. The coefficients used in the example of the mean are all non-negative and sum up to one. Even though this convention is very attractive because of the probabilistic interpretation which can be given to the coefficients, we shall use another convention for the linear combinations. We shall allow the coefficients to be of any sign (positive as well as negative) and we normalize them so that the sum of their squares is equal to 1. So if we were to use the mean, we would use the normalized linear combination (NLC, for short) given by:

$$\widetilde{x_{i\cdot}} = \frac{1}{\sqrt{M}} x_{i1} + \frac{1}{\sqrt{M}} x_{i2} + \cdots + \frac{1}{\sqrt{M}} x_{iM}.$$

The goal of principal component analysis is to search for the main sources of variation in the M-dimensional row vectors by identifying M linearly independent and orthogonal NLC's in such a way that a small number of them capture most of the variation in the data. This is accomplished by identifying the eigenvectors and eigenvalues of the covariance matrix C_x of the M column variables. This covariance matrix is defined by:

$$C_x[j,j'] = \frac{1}{N}\sum_{i=1}^{N}(x_{ij} - \overline{x_{\cdot j}})(x_{ij'} - \overline{x_{\cdot j'}}), \qquad j,j' = 1,\dots,M, \tag{3.28}$$

where we used the standard notation:

$$\overline{x_{\cdot j}} = \frac{x_{1j} + x_{2j} + \cdots + x_{Nj}}{N}$$

for the mean of the j-th variable over the population of N individuals. It is easy to check that the matrix C_x is symmetric (hence diagonalizable) and non-negative definite (which implies that all the eigenvalues are non-negative). One usually rearranges the eigenvalues in decreasing order, say:

$$\lambda_1 \ge \lambda_2 \ge \cdots \ge \lambda_M \ge 0.$$

The corresponding eigenvectors are called the loadings. In practice we choose c_1 to be a normalized eigenvector corresponding to the eigenvalue λ_1, c_2 to be a normalized eigenvector corresponding to the eigenvalue λ_2, ..., and finally c_M to be a normalized eigenvector corresponding to the eigenvalue λ_M, and we make sure that all the vectors c_j are orthogonal to each other. This is automatically true when the eigenvalues λ_j are simple. See the discussion below for the general case. Recall that we say a vector is normalized if the sum of the squares of its components is equal to 1. If we denote by C the $M \times M$ matrix formed by the M column vectors containing the components of the vectors $c_1, c_2, \dots, c_M$ in this order, this matrix is orthogonal (since it is a matrix transforming one orthonormal basis into another) and it satisfies:

$$C_x = C^t D C$$

where we use the notation ^{t} to denote the transpose of a matrix or a vector, and where D is the $M \times M$ diagonal matrix with $\lambda_1, \lambda_2, \dots, \lambda_M$ on the diagonal. Notice the obvious lack of uniqueness of the above decomposition. In particular, if c_j is a normalized eigenvector associated to the eigenvalue λ_j, so is $-c_j$! This is something one should keep in mind when plotting the eigenvectors c_j, and when trying to find an intuitive interpretation for the features of the plots. However, this sign flip is easy to handle, and fortunately, it is the only form of non uniqueness when the eigenvalues are simple (i.e. nondegenerate). The ratio:

$$\frac{\lambda_j}{\sum_{j'=1}^{M}\lambda_{j'}}$$

of a given eigenvalue to the trace of C_x (i.e. the sum of its eigenvalues) has the interpretation of the proportion of the variation explained by the corresponding eigenvector, i.e. the loading c_j. In order to justify this statement, we appeal to the Raleigh-Ritz variational principle from linear algebra. Indeed, according to this principle, the eigenvalues and their corresponding eigenvectors can be characterized recursively in the following way. The largest eigenvalue λ_1 appears as the maximum:

$$\lambda_1 = \max_{x \in \mathbb{R}^M, \|x\|=1} x^t C_x x$$

while the corresponding eigenvector c_1 appears as the argument of this maximization problem:

$$c_1 = \arg \max_{x \in \mathbb{R}^M, \|x\|=1} x^t C_x x.$$

If we recall the fact that $x^t C_x x$ represents the quadratic variation (empirical variance) of the NLC's $x^t x_{1\cdot}$, $x^t x_{2\cdot}, \ldots, x^t x_{N\cdot}$, λ_1 can be interpreted as the maximal quadratic variation when we consider all the possible (normalized) linear combinations of the M original variables. Similarly, the corresponding (normalized) eigenvector has precisely the interpretation of this NLC which realizes the maximum variation.

As we have already pointed out, the first loading is uniquely determined up to a sign change if the eigenvalue λ_1 is simple. If this is not the case, and if we denote by m_1 the multiplicity of the eigenvalue λ_1, we can choose any orthonormal set $\{c_1, \cdots, c_{m_1}\}$ in the eigenspace of λ_1 and repeat the eigenvalue λ_1, m_1 times in the list of eigenvalues (and on the diagonal of D as well). This lack of uniqueness is not a mathematical difficulty, it is merely annoying. Fortunately, it seldom happens in practice! We shall assume that all the eigenvalues are simple (i.e. non-degenerate) for the remainder of this discussion. If they were not, we would have to repeat them according to their multiplicities.

Next, still according to the Raleigh-Ritz variation principle, the second eigenvalue λ_2 appears as the maximum:

$$\lambda_2 = \max_{x \in \mathbb{R}^M, \|x\|=1, x \perp c_1} x^t C_x x$$

while the corresponding eigenvector c_2 appears as the argument of this maximization problem:

$$c_2 = \arg \max_{x \in \mathbb{R}^M, \|x\|=1, x \perp c_1} x^t C_x x.$$

The interpretation of this statement is the following: if we avoid any overlap with the loading already identified (i.e. if we restrict ourselves to NLC's x which are orthogonal to c_1), then the maximum quadratic variation will be λ_2 and any NLC realizing this maximum variation can be used for c_2. We can go on and identify in this way all the eigenvalues λ_j (having to possibly repeat them according to their multiplicities) and the loadings c_j's.

Armed with a new basis of $\mathbb{R}^M$, the next step is to rewrite the data observations (i.e. the N rows of the data matrix) in this new basis. This is done by multiplying the data matrix by the *change of basis* matrix (i.e. the matrix whose columns are the eigenvectors identified earlier). The result is a new $N \times M$ matrix whose columns are called *principal components*. Their relative importance is given by the proportion of the variance explained by the loadings, and for that reason, one typically considers only the first few principal components, the remaining ones being ignored and/or treated as noise.

3.5.2 PCA with R

The principal component analysis of a data set is performed in R with the function princomp, which returns an object of class princomp that can be printed and plotted with generic methods. Illustrations of the calls to this function and of the interpretation of the results are given in the next subsections in which we discuss several financial applications of PCA.

3.5.3 Effective Dimension of the Space of Yield Curves

Our first application concerns the markets of fixed income securities which we will introduce in Sect. 4.8. The term structure of interest rates is conveniently captured by the daily changes in the yield curve. The dimension of the space of all possible yield curves is presumably very large, potentially infinite if we work in the idealized world of continuous-time finance. However, it is quite sensible to try to approximate these curves by functions from a class chosen in a parsimonious way. Without any a priori choice of the type of functions to be used to approximate the yield curve, PCA can be used to extract, one by one, the components responsible for the variations in the data.

3.5.3.1 PCA of the Yield Curve

For the purposes of illustration, we use data on the US yield curve as provided by the Bank of International Settlements (BIS, for short). These data are the result of a nonparametric processing (smoothing spline regression, to be specific) of the raw data. The details will be given in Sect. 5.4, but for the time being, we shall ignore the possible effects of this pre-processing of the raw data. The data are imported into an R-object named us.bis.yield which gives, for each of the 1,352 successive trading days following January 3rd 1995, the yields on the US Treasury bonds for times to maturity

$$x = 0, 1, 2, 3, 4, 5, 5.5, 6.5, 7.5, 8.5, 9.5 \quad \text{months.}$$

We run a PCA on these data with the following R commands:

```
> dim(us.bis.yield)
[1] 1352  11
> us.bis.yield.pca <- princomp(us.bis.yield)
> plot(us.bis.yield.pca,main="Proportions of the Variance
                              Explained by the Components")
```

The results are reproduced in Fig. 3.24 which gives the proportions of the variation explained by the various components. The first three eigenvectors of the covariance matrix (the so-called loadings) explain 99.9 % of the total variation in the data. This suggests that the effective dimension of the space of yield curves could be three. In other words, any of the yield curves from this period can be approximated by a linear

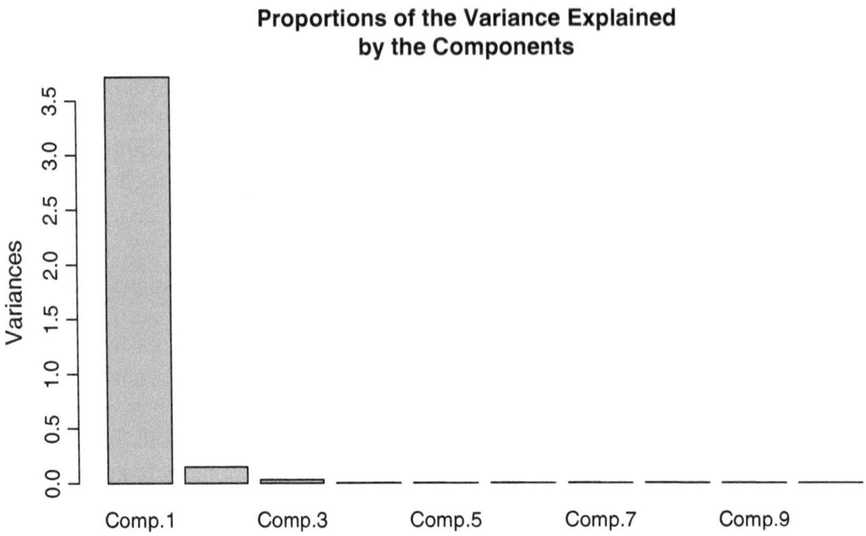

Fig. 3.24. Proportions of the variance explained by the components of the PCA of the daily US yield curves

combination of the first three loadings, the relative error being very small. In order to better understand the far reaching implications of this statement we plot the first four loadings.

```
> X <- c(0,1,2,3,4,5,5.5,6.5,7.5,8.5,9.5)
> par(mfrow=c(2,2))
> plot(X,us.bis.yield.pca$loadings[,1],ylim=c(-.7,.7))
> lines(X,us.bis.yield.pca$loadings[,1])
> plot(X,us.bis.yield.pca$loadings[,2],ylim=c(-.7,.7))
> lines(X,us.bis.yield.pca$loadings[,2])
> plot(X,us.bis.yield.pca$loadings[,3],ylim=c(-.7,.7))
> lines(X,us.bis.yield.pca$loadings[,3])
> plot(X,us.bis.yield.pca$loadings[,4],ylim=c(-.7,.7))
> lines(X,us.bis.yield.pca$loadings[,4])
> par(mfrow=c(1,1))
```

The results are reproduced in Fig. 3.25. The first loading is essentially flat, so a component on this loading will essentially represent the average yield over the maturities. Note that, contrary to what one could have expected, this first loading is always negative. But if we recall a remark we made on the non-uniqueness of the eigenvectors, a possible change in sign can be expected. Because of the monotone and increasing nature of the second loading, the second component measures the upward trend (if the component is positive, and the downward trend otherwise) in the yield. This second factor is interpreted as the tilt of the yield curve. The shape of the third loading suggests that the third component captures the curvature of the yield curve, whether positive or negative. Finally, the shape of the fourth loading does not seem to have

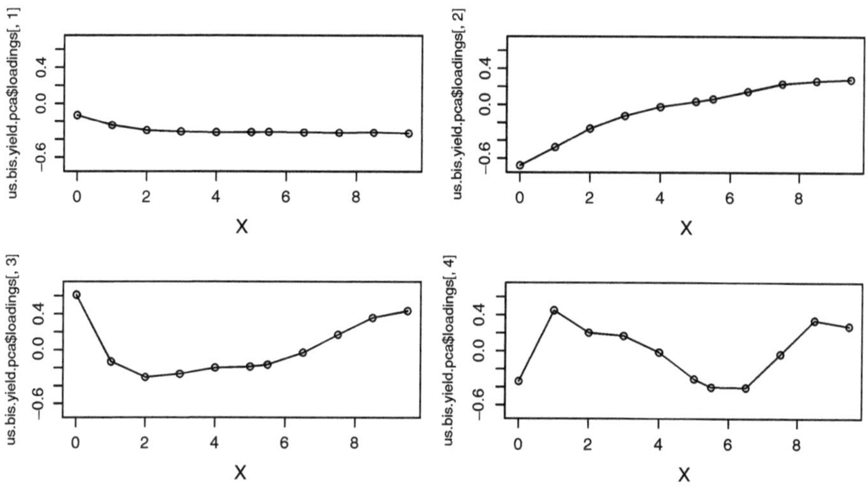

Fig. 3.25. From *left* to *right*, *top* to *bottom*, sequential plots of the first four US yield loadings

an obvious interpretation. It is mostly noise (remember that most of the variations in the yield curve are explained by the first three components). These features are very typical, and they should be expected in most PCA's of the term structure of interest rates.

The fact that the first three components capture so much of the yield curve may seem strange when compared to the fact that some estimation methods, which we discuss later in the book, use parametric families with more than three parameters! There is no contradiction there. Indeed, for the sake of illustration, we limited the analysis of this section to the first part of the yield curve. Restricting ourselves to short maturities makes it easier to capture all the features of the yield curve in a small number of functions with a clear interpretation.

Important Remarks.

1. PCA is most often used to study the daily changes in the yield curve, as opposed to the yield curves themselves. From a statistical point of view, working with differences over time helps resolve possible non-stationary issues which could plague the estimation of the covariance matrix given by its empirical counterpart (3.28). The next example gives an illustration of the procedure and of the interpretation of the results in terms of changes over time in the curves. In particular, even though the first component (i.e. the most important factor) will still be essentially flat, the interpretation will be that its impact on the actual yield curve is a parallel shift.
2. Most modern data sets provide measurements for a large number of variables M, and the order of magnitude of this number is often comparable to the number of observations N. In such situations, the estimation of the covariance matrix by the empirical covariance matrix (3.28) may be problematic given that the

number of parameters estimated is proportional to M^2, and a *modicum of care* is needed in the statistical interpretation of the results. Prompted by applications to genomics, a very active branch of statistical research has developed to provide robust statistical estimation procedures to handle situations when the number of variables is larger than the number of observations. We shall not dwell on this issue here, but the reader should be cautioned of the dangers to run PCA when $M \sim N$.

3.5.4 **Swap Rate Curves**

Swap contracts have been traded publicly since 1981. As of today, they are some of the most popular fixed income derivatives. Because of this popularity, the swap markets are extremely liquid, and as a consequence, they can be used to hedge interest-rate risk of fixed income portfolios at low cost. The estimation of the term-structure of swap rates is important in this respect and the PCA which we present below is the first step toward a better understanding of this term structure.

3.5.4.1 ***Swap Contracts and Swap Rates***

As implied by its name, a swap contract obligates two parties to exchange (or swap) some specified cash flows at agreed upon times. The most common swap contracts are interest rate swaps. In such a contract, one party, say counter-party A, agrees to make interest payments determined by an instrument P_A (say, a 10 year US Treasury bond rate), while the other party, say counter-party B, agrees to make interest payments determined by another instrument P_B (say, the London Interbank Offer Rate – LIBOR for short). Even though there are many variants of swap contracts, in a typical contract, the principal on which counter-party A makes interest payments is equal to the principal on which counterparty B makes interest payments. Also, the payment schedules are identical and periodic, the payment frequency being quarterly, semi-annually, ...

It is not difficult to infer from the above discussion that a swap contract is equivalent to a portfolio of forward contracts, but we shall not use this feature here. In this section, we shall restrict ourselves to the so-called plain vanilla contracts involving a fixed interest rate and the 3 or 6 months LIBOR rate.

We will not attempt to derive here a formula for the price of a swap contract, neither will we try to define rigorously the notion of swap rate. These derivations are beyond the scope of this book. See the Notes & Complements at the end of the chapter for references to appropriate sources. We shall use only the intuitive idea of the swap rate being a rate at which both parties will agree to enter into the swap contract.

3.5.4.2 PCA of the Swap Rate Daily Changes

Our second application of principal component analysis concerns the changes in the term structure of swap rates as given by the swap rate curves. As before, we denote by M the dimension of the vectors. We use data downloaded from `Data Stream`. It is quite likely that the raw data have been processed, but we are not quite sure what kind of manipulation is performed by `Data Stream`, so for the purposes of this illustration, we shall ignore the possible effects of the pre-processing of the data. In this example, the day t labels the rows of the data matrix. The latter has $M = 15$ columns, containing the swap rates with maturities T conveniently labeled by the times to maturity $x = T - t$, which have the values $1, 2, \ldots, 10, 12, 15, 20, 25, 30$ years in the present situation. We collected these data for each day t of the period from May 1998 to March 2000, and we rearranged the numerical values in a matrix $R = [r_{i,j}]_{i=1,\ldots,N,\ j=1,\ldots,M}$. Here, the index j stands for the time to maturity, while the index i codes the day the curve is observed.

The data are contained in the `R` object `swap`. We perform the PCA on the daily changes in the swap rate curve with the commands:

```
> dim(swap)
[1] 496  15
> DS <- diff(as.matrix(swap))
> swap.pca <- princomp(DS)
> plot(swap.pca, main= Proportions of the Variance
                            Explained by the Components")
> YEARS <- c(1,2,3,4,5,6,7,8,9,10,12,15,20,25,30)
> par(mfrow=c(2,2))
> plot(YEARS,swap.pca$loadings[,1],ylim=c(-.6,.6))
> lines(YEARS,swap.pca$loadings[,1])
> plot(YEARS,swap.pca$loadings[,2],ylim=c(-.6,.6))
> lines(YEARS,swap.pca$loadings[,2])
> plot(YEARS,swap.pca$loadings[,3],ylim=c(-.6,.6))
> lines(YEARS,swap.pca$loadings[,3])
> plot(YEARS,swap.pca$loadings[,4],ylim=c(-.6,.6))
> lines(YEARS,swap.pca$loadings[,4])
> par(mfrow=c(1,1))
```

The command `DS <- diff(as.matrix(swap))` turns the data frame `swap` into a matrix to which the `diff` operator can be apply to replace each column by the result of its differentiation to which the PCA is then applied. Figure 3.26 gives the proportions of the variation explained by the various components, while Fig. 3.27 gives the plots of the first four eigenvectors.

Looking at Fig. 3.27 one sees that the first and most important factor will create a parallel shift while the second one will produce a tilt of the curve, and the third factor will attempt to bend the curve to create a curvature.

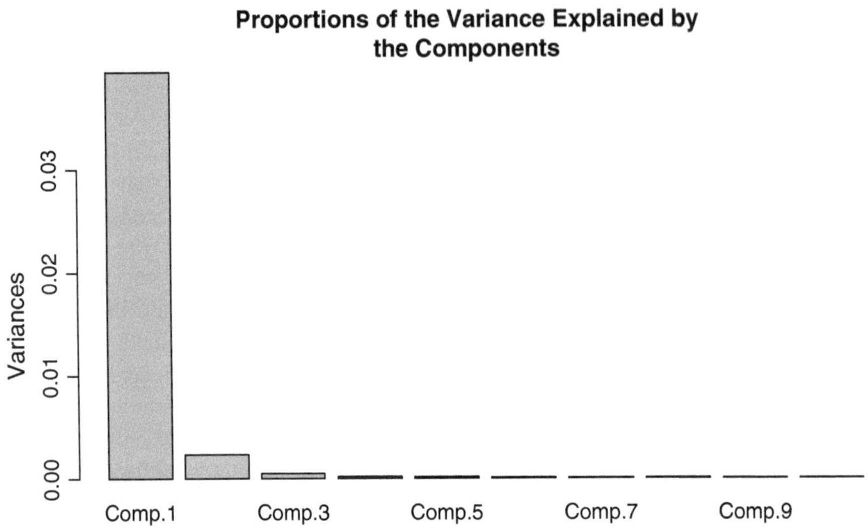

Fig. 3.26. Proportions of the variance explained by the components of the PCA of the daily changes in the swap rates for the period from May 1998 to March 2000

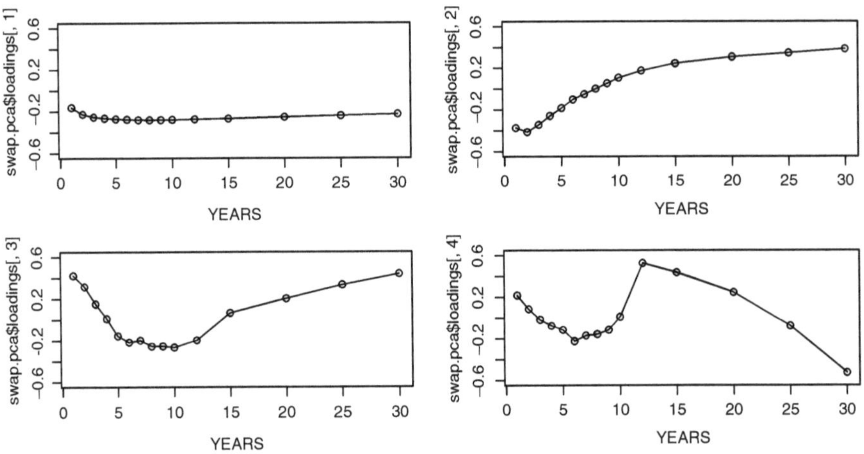

Fig. 3.27. From *left* to *right* and *top* to *bottom*, sequential plots of the eigenvectors (loadings) corresponding to the four largest eigenvalues. Notice that we changed the scale of the horizontal axis to reflect the actual times to maturity

APPENDIX 1: CALCULUS WITH RANDOM VECTORS AND MATRICES

The nature and technical constructs of this chapter justify our spending some time discussing the properties of random vectors (as opposed to random variables) and reviewing the fundamental results of the calculus of probability with random vectors.

Their definition is very natural: a random vector is a vector whose entries are random variables. With this definition in hand, it is easy to define the notion of expectation. The expectation of a random vector is the (deterministic) vector whose entries are the expectations of the entries of the original random vector. In other words,

$$\text{if} \qquad \boldsymbol{X} = \begin{bmatrix} X_1 \\ \vdots \\ X_n \end{bmatrix}, \qquad \text{then} \qquad \mathbb{E}\{\boldsymbol{X}\} = \begin{bmatrix} \mathbb{E}\{X_1\} \\ \vdots \\ \mathbb{E}\{X_n\} \end{bmatrix},$$

whenever these expectations exist. Notice that, if $\boldsymbol{B}$ is an n-dimensional (deterministic) vector then:

$$\mathbb{E}\{\boldsymbol{X} + \boldsymbol{B}\} = \mathbb{E}\{\boldsymbol{X}\} + \boldsymbol{B}. \tag{3.29}$$

Indeed:

$$\mathbb{E}\{\boldsymbol{X} + \boldsymbol{B}\} = \begin{bmatrix} \mathbb{E}\{X_1 + b_1\} \\ \vdots \\ \mathbb{E}\{X_n + b_n\} \end{bmatrix} = \begin{bmatrix} \mathbb{E}\{X_1\} + b_1 \\ \vdots \\ \mathbb{E}\{X_n\} + b_n \end{bmatrix} = \mathbb{E}\{\boldsymbol{X}\} + \boldsymbol{B}$$

where we used the notation b_i for the components of the vector $\boldsymbol{B}$. The notion of variance (or more generally of second moment) appears somehow less natural at first. We define the variance/covariance matrix of a random vector to be the (deterministic) matrix whose entries are the variances and covariances of the entries of the original random vector. More precisely, if $\boldsymbol{X}$ is a random vector as above, then its variance/covariance matrix is the matrix $\Sigma_{\boldsymbol{X}}$ defined by:

$$\Sigma_{\boldsymbol{X}} = \begin{bmatrix} \text{var}(X_1) & \text{cov}(X_1, X_2) & \cdots & \text{cov}(X_1, X_n) \\ \text{cov}(X_2, X_1) & \text{var}(X_2) & \cdots & \text{cov}(X_2, X_n) \\ \vdots & \vdots & \vdots & \vdots \\ \text{cov}(X_n, X_1) & \text{cov}(X_n, X_2) & \cdots & \text{var}(X_n) \end{bmatrix}, \tag{3.30}$$

In other words, on the i-th row and the j-th column of $\Sigma_{\boldsymbol{X}}$ we find the covariance of X_i and X_j. For the purposes of this appendix, we limit ourselves to random variables of order 2 (i.e. for which the first two moments exist) so that all the variances and covariances make perfectly good mathematical sense. Note that this is not the case for many generalized Pareto distributions, and especially for our good old friend the Cauchy distribution.

The best way to look at the variance/covariance matrix of a random vector is the following. Using the notation $\boldsymbol{Z}^t$ for the transpose of the vector or matrix $\boldsymbol{Z}$, we notice that:

$$[\boldsymbol{X} - \mathbb{E}\{\boldsymbol{X}\}][\boldsymbol{X} - \mathbb{E}\{\boldsymbol{X}\}]^t = \begin{bmatrix} X_1 - \mu_1 \\ \vdots \\ X_n - \mu_n \end{bmatrix} [X_1 - \mu_1, \ldots, X_n - \mu_n]$$

$$= \begin{bmatrix} (X_1 - \mu_1)^2 & (X_1 - \mu_1)(X_2 - \mu_2) & \cdots & (X_1 - \mu_1)(X_n - \mu_n) \\ (X_2 - \mu_2)(X_1 - \mu_1) & (X_2 - \mu_2)^2 & \cdots & (X_2 - \mu_2)(X_n - \mu_n) \\ \vdots & \vdots & \vdots & \vdots \\ (X_n - \mu_n)(X_1 - \mu_1) & (X_n - \mu_n)(X_2 - \mu_2) & \cdots & (X_n - \mu_n)^2 \end{bmatrix}$$

if we use the notation $\mu_j = \mathbb{E}\{X_j\}$ to shorten the typesetting of the formula. The variance/covariance matrix $\Sigma_{\boldsymbol{X}}$ is nothing more than the expectation of this random matrix, since the expectation of a random matrix is defined as the (deterministic) matrix whose entries are the expectations of the entries of the original random matrix. Consequently we have proven that, for any random vector $\boldsymbol{X}$ of order 2, the variance/covariance matrix $\Sigma_{\boldsymbol{X}}$ is given by the formula:

$$\Sigma_{\boldsymbol{X}} = \mathbb{E}\{[\boldsymbol{X} - \mathbb{E}\{\boldsymbol{X}\}][\boldsymbol{X} - \mathbb{E}\{\boldsymbol{X}\}]^t\}. \tag{3.31}$$

Notice that, if the components of a random vector are independent, then the variance/covariance matrix of this random vector is diagonal since all the entries off the diagonal must vanish due to the independence assumption.

Some Useful Formulae

If $\boldsymbol{X}$ is an n-dimensional random vector, $\boldsymbol{A}$ is an $m \times n$ deterministic matrix, and $\boldsymbol{B}$ is an m-dimensional deterministic vector, then:

$$\mathbb{E}\{\boldsymbol{AX} + \boldsymbol{B}\} = \boldsymbol{A}\mathbb{E}\{\boldsymbol{X}\} + \boldsymbol{B} \tag{3.32}$$

as can be checked by computing the various components of the m-dimensional vectors on both sides of the equality sign. Notice that formula (3.29) is merely a particular case of (3.32) when $m = n$ and $\boldsymbol{A}$ is the identity matrix. In fact, formula (3.32) remains true when $\boldsymbol{X}$ is an $n \times p$ random matrix and $\boldsymbol{B}$ is an $m \times p$ deterministic matrix. By transposition one gets that

$$\mathbb{E}\{\boldsymbol{XA} + \boldsymbol{B}\} = \mathbb{E}\{\boldsymbol{X}\}\boldsymbol{A} + \boldsymbol{B} \tag{3.33}$$

holds whenever $\boldsymbol{X}$ is an $n \times p$ random matrix and $\boldsymbol{A}$ and $\boldsymbol{B}$ are deterministic matrices with dimensions $p \times m$ and $n \times m$ respectively. Using (3.32) and (3.33) we get:

$$\Sigma_{\boldsymbol{AX}+\boldsymbol{B}} = \boldsymbol{A}\Sigma_{\boldsymbol{X}}\boldsymbol{A}^t \tag{3.34}$$

A proof of this formula goes as follows:

$$\begin{aligned} \Sigma_{\boldsymbol{AX}+\boldsymbol{B}} &= \mathbb{E}\{[\boldsymbol{AX} + \boldsymbol{B} - \mathbb{E}\{\boldsymbol{AX} + \boldsymbol{B}\}][\boldsymbol{AX} + \boldsymbol{B} - \mathbb{E}\{\boldsymbol{AX} + \boldsymbol{B}\}]^t\} \\ &= \mathbb{E}\{[\boldsymbol{A}(\boldsymbol{X} - \mathbb{E}\{\boldsymbol{X}\})][\boldsymbol{A}(\boldsymbol{X} - \mathbb{E}\{\boldsymbol{X}\})]^t\} \\ &= \mathbb{E}\{\boldsymbol{A}[\boldsymbol{X} - \mathbb{E}\{\boldsymbol{X}\}][\boldsymbol{X} - \mathbb{E}\{\boldsymbol{X}\}]^t\boldsymbol{A}^t\} \\ &= \boldsymbol{A}\mathbb{E}\{[\boldsymbol{X} - \mathbb{E}\{\boldsymbol{X}\}][\boldsymbol{X} - \mathbb{E}\{\boldsymbol{X}\}]^t\}\boldsymbol{A}^t \\ &= \boldsymbol{A}\Sigma_{\boldsymbol{X}}\boldsymbol{A}^t. \end{aligned}$$

Similar formulae can be proven for the variance/covariance matrix of expressions of the form $\boldsymbol{A}\boldsymbol{X}\boldsymbol{B}+\boldsymbol{C}$ when $\boldsymbol{X}$ is a random vector or a random matrix and when $\boldsymbol{A}$, $\boldsymbol{B}$ and $\boldsymbol{C}$ are deterministic matrices or vectors whose dimensions make the product meaningful.

Warning: Remember to be very cautious with the order in a product of matrices. Just because one can change the order in the product of numbers, does not mean that it is a good idea to do the same thing with a product of matrices, as the results are in general (very) different:

the product or matrices is not commutative!!!

Appendix 2: Families of Copulas

There are many parametric families of copulas, and new ones are created every month. For the sake of completeness, we review the families of bivariate copulas implemented in the library `Rsafd`. They can be organized in two main classes.

◇ **Extreme value copulas** are copulas of the form

$$C(x,y)=\exp\left[\log(xy)\;A\left(\frac{\log(x)}{\log(xy)}\right)\right],$$

where $A:[0,1]\to[0.5,1]$, is a convex function satisfying $\max(t,1-t)\le A(t)\le 1$ for all $t\in[0,1]$.

◇ **Archimedean copulas** are copulas of the form

$$C(x,y)=\phi^{-1}\left[(\phi(x)+\phi(y))\,A\left(\frac{\phi(x)}{\phi(x)+\phi(y)}\right)\right],$$

where $\phi(t)$ is convex and decreasing on $(0,1)$ (called the Archimedean generator), and A is as above.

We now list the most commonly used parametric families of copulas:

- **Bivarate Normal**, `"normal"`
`> normal.copula(rho)`

$$C(x,y)=\Phi_\rho\left(\Phi^{-1}(x),\Phi^{-1}(y)\right),$$

$0\le\rho\le 1$, where Φ^{-1} is the quantile function of the standard normal distribution, and Φ_ρ is the cdf of the joint distribution of two jointly Gaussian standard random variables with correlation $\rho>0$.

- **Frank Copula**, `"frank"`
  `> frank.copula(delta)`

$$C(x,y) = -\delta^{-1} \log \left(\frac{\eta - (1 - e^{-\delta x})(1 - e^{-\delta x})}{\eta} \right)$$

  $0 \le \delta < \infty$, and $\eta = 1 - e^{-\delta}$.
- **Kimeldorf–Sampson copula**, `"kimeldorf.sampson"`
  `> kimeldorf.sampson.copula(delta)`

$$C(x,y) = (x^{-\delta} + y^{-\delta} - 1)^{-1/\delta},$$

  $0 \le \delta < \infty$.
- **Gumbel copula**, `"gumbel"`
  `> gumbel.copula(delta)`

$$C(x,y) = \exp\left(- \left[(-\log x)^{\delta} + (-\log y)^{\delta} \right]^{1/\delta} \right),$$

  $1 \le \delta < \infty$. This is an extreme value copula with the dependence function

$$A(t) = (t^{\delta} + (1-t)^{\delta})^{1/\delta}.$$

- **Galambos**, `"galambos"`
  `> galambos.copula(delta)`

$$C(x,y) = xy \exp\left(\left[(-\log x)^{-\delta} + (-\log y)^{-\delta} \right]^{-1/\delta} \right),$$

  $0 \le \delta < \infty$. This is an extreme value copula with the dependence function

$$A(t) = 1 - (t^{-\delta} + (1-t)^{-\delta})^{-1/\delta}.$$

- **Hüsler–Reiss**, `"husler.reiss"`
  `> husler.reiss.copula(delta)`

$$C(x,y) = \exp\left(-\tilde{x}\, \Phi \left[\frac{1}{\delta} + \frac{1}{2} \delta \log \left(\frac{\tilde{x}}{\tilde{y}} \right) \right] - \tilde{y}\, \Phi \left[\frac{1}{\delta} + \frac{1}{2} \delta \log \left(\frac{\tilde{y}}{\tilde{x}} \right) \right] \right)$$

  $0 \le \delta < \infty$, $\tilde{x} = -\log x$, $\tilde{y} = -\log y$, and Φ is the cdf of the standard normal distribution. This is an extreme value copula with the dependence function

$$A(t) = t\, \Phi \left[\delta^{-1} + \frac{1}{2} \delta \log \left(\frac{t}{1-t} \right) \right] + (1-t)\, \Phi \left[\delta^{-1} - \frac{1}{2} \delta \log \left(\frac{t}{1-t} \right) \right]$$

- **Twan**, `"twan"`
  `> twan.copula(alpha, beta, r)`
  This is an extreme value copula with the dependence function

$$A(t) = 1 - \beta + (\beta - \alpha) + \{ \alpha^r t^r + \beta^r (1-t)^r \}^{1/r},$$

  $0 \le \alpha,\ \beta \le 1,\ 1 \le r < \infty$.

- **BB1**, `"bb1"`
  `> bb1.copula(theta, delta)`

$$C(x,y) = \left(1 + \left[(x^{-\theta} - 1)^{\delta} + (y^{-\theta} - 1)^{\delta}\right]^{1/\delta}\right)^{-1/\theta}$$

  $\theta > 0, \delta \geq 1$. This is an Archimedean copula with the Archimedean generator $\phi(t) = \left(t^{-\theta} - 1\right)^{\delta}$.
- **BB2**, `"bb2"`
  `> bb2.copula(theta, delta)`

$$C(x,y) = \left[1 + \delta^{-1} \log\left(e^{\delta(x^{-\theta})} + e^{\delta(y^{-\theta})} - 1\right)\right]^{1/\theta},$$

  $\theta > 0, \delta > 0$. This is an Archimedean copula with the Archimedean generator $\phi(t) = e^{\delta(t^{-\theta}-1)} - 1$.
- **BB3**, `"bb3"`
  `> bb3.copula(theta, delta)`

$$C(x,y) = \exp\left(-\left[\delta^{-1} \log\left(e^{\delta\tilde{x}^{\theta}} + e^{\delta\tilde{y}^{\theta}} - 1\right)\right]^{1/\theta}\right),$$

  $\theta \geq 1, \delta > 0, \tilde{x} = -\log x, \tilde{y} = -\log y$. This is an Archimedean copula with the Archimedean generator $\phi(t) = \exp\left\{\delta\left(-\log t\right)^{\theta}\right\} - 1$.
- **BB4**, `"bb4"`
  `> bb4.copula(theta, delta)`

$$C(x,y) = \left(x^{-\theta} + y^{-\theta} - 1 - \left[\left(x^{-\theta} - 1\right)^{-\delta} + \left(y^{-\theta} - 1\right)^{-\delta}\right]^{-\frac{1}{\delta}}\right)^{-\frac{1}{\theta}}$$

  $\theta \geq 0, \delta > 0$. This is an Archimedean copula with the Archimedean generator $\phi(t) = t^{-\theta} - 1$ and the dependence function $A(t) = 1 - (t^{-\delta} + (1-t)^{-\delta})^{-1/\delta}$ (same as for B7 family).
- **BB5**, `"bb5"`
  `> bb5.copula(theta, delta)`

$$C(x,y) = \exp\left(-\left[\tilde{x}^{\theta} + \tilde{y}^{\theta} - \left(\tilde{x}^{-\theta\delta} + \tilde{y}^{-\theta\delta}\right)^{-1/\delta}\right]^{1/\theta}\right),$$

  $\delta > 0, \theta \geq 1, \tilde{x} = -\log x, \tilde{y} = -\log y$. This is an extreme value copula with the dependence function

$$A(t) = \left[t^{\theta} + (1-t)^{\theta} - \left(t^{-\delta\theta} + (1-t)^{-\delta\theta}\right)^{-1/\delta}\right]^{1/\theta}.$$

- **BB6**, `"bb6"`
  `> bb6.copula(theta, delta)`
  This is an Archimedean copula with the generator
  $\phi(t) = \left[-\log\left(1 - (1-t)^{\theta}\right)\right]^{\delta} \qquad \theta \geq 1, \delta \geq 1.$

- **BB7**, "bb7"
  ```
  > bb7.copula(theta, delta)
  ```
  This is an Archimedean copula with the generator
  $\phi(t) = \left(1-(1-t)^{\theta}\right)^{-\delta} - 1, \qquad \theta \geq 1, \delta > 0.$
- **B1Mix**, "normal.mix"
  ```
  > normal.mix.copula(p, rho1, rho2)
  ```
  $$C(x,y) = p\, C_{Gauss,\rho_1}(x,y) + (1-p)\, C_{Gauss,\rho_2}(x,y),$$
  $0 < p, \rho_1, \rho_2 < 1.$

PROBLEMS

Ⓣ **Problem 3.1** *Assuming that you have access to a random generator for the multivariate Gaussian distribution, explain how you would generate Monte Carlo samples*

$$(x_1^{(1)}, x_1^{(2)}, x_1^{(3)}), \cdots, (x_n^{(1)}, x_n^{(2)}, x_n^{(3)})$$

from the 3-dimensional Gaussian copula with correlation matrix

$$\Sigma = \begin{bmatrix} 1 & \rho & \rho \\ \rho & 1 & \rho \\ \rho & \rho & 1 \end{bmatrix}$$

where $\rho \in (0,1)$ is given.

Ⓣ **Problem 3.2** *From the formula of the Gaussian copula $C_{Gauss,\rho}$ with parameter $\rho \in (0,1)$.*

1. *Compute the limit $C_0(u_1,u_2) = \lim_{\rho \searrow 0} C_{Gauss,\rho}(u_1,u_2)$*
2. *Say if $C_0(u,v)$ a copula and if yes, which one? Explain the intuition behind this result.*

Ⓣ **Problem 3.3** *Let us assume that X is a random variable uniformly distributed over the unit interval $[0,1]$ and let us define the random variable Y by:*

$$Y = 1 - |2X - 1|$$

1. *What is the distribution of Y?*
2. *Compute the covariance cov$\{X,Y\}$ of X and Y.*
3. *Are X and Y independent? Explain why.*

Ⓣ **Problem 3.4** *Let us assume that the random variables X_1 and X_2 represent the daily log-returns of two financial securities, and let us consider the log-return $X = 0.4X_1 + 0.6X_2$ of a portfolio invested 40/60 % in these securities. The goal of this problem is to compute the daily VaR of the portfolio at the level 1 %.*

1. *We first assume that we know the cumulative distribution function (c.d.f. for short) F_1 of X_1, the c.d.f. F_2 of X_2 and that we have a random generator for the copula C of X_1 and X_2. Saying that we know a c.d.f. means that we either have an analytic expression in closed form for its values, or that we have a program which can evaluate numerically*

values of this function. The goal of this first question is to give a procedure to compute a Monte Carlo approximation of the desired VaR based on a Monte Carlo sample of size $N = 10{,}000$*. Describe in a small number of short bullet points the steps you would take to do just that.*

NB*: No more than four to six bullet points of one or two sentences each suffice.*

2. *We now assume that* X_1 *and* X_2 *are jointly Gaussian and that we know the parameters* μ_1, μ_2, σ_1, σ_2 *and* ρ *of their joint distribution. Under this assumption, describe a shorter procedure to compute the exact value of the VaR without using Monte Carlo methods.*

Ⓣ **Problem 3.5** *1. Let* C *denote the 2-d cdf of the uniform distribution over the unit square* $[0, 1] \times [0, 1]$*. Is* C *a copula? If yes, which copula? We now assume that* X_1 *and* X_2 *are standard Gaussian random variables, i.e. each of them follows the standard normal distribution* $N(0, 1)$*, and we denote by* C *their copula.*

2. *Are* X_1 *and* X_2 *jointly Gaussian? Explain your answer.*
3. *Compute the* 10%*-tile of the random variable* $\max(X_1, X_2)$ *assuming that* $C(0.5, 0.5) = 0.1$.
4. *Compute the probability that* $X_2 \geq X_1$ *assuming that* C *is symmetric in the sense that* $C(u_1, u_2) = C(u_2, u_1)$ *for all* $u_1, u_2 \in [0, 1]$.

Ⓣ **Problem 3.6** *We assume that we have access to a random generator for the multivariate Gaussian distribution. Let us also assume that we are given a bivariate data sample*

$$(x_1^{(1)}, x_1^{(2)}), \cdots, (x_n^{(1)}, x_n^{(2)})$$

of observations of random variables X_1 *and* X_2.

1. *Assume that* X_1 *and* X_2 *are jointly Gaussian and describe in a small number of short bullet points the steps you would take to*
   1.1. *Estimate the joint distribution of* X_1 *and* X_2*;*
   1.2. *Compute analytically (i.e. with classical functions and the above estimate) an estimate of the* 5% *quantile of* $X_1 + X_2$*;*
   1.3. *Compute an estimate of the* 5% *quantile of* $X_1 + X_2$ *by a Monte Carlo computation.*
2. *Assume now that* X_1 *is Cauchy with location parameter* 0 *and scale parameter* 1*, that* X_2 *is exponential with rate* 1 *and that their copula is the Gaussian copula with parameter* $\rho = 0.7$*. Explain in detail how you would compute an estimate of the* 5% *quantile of* $X_1 + X_2$ *by a Monte Carlo computation. You can assume that functions computing the quantiles of the Cauchy and the exponential distributions are available, but you can also derive formulas for these functions.*

Ⓔ **Problem 3.7** *This problem is based on the data contained in the data set* `UTILITIES` *included in the library* `Rsafd`*. It is a matrix with two columns, each row corresponding to a given day. The first column gives the log of the weekly return on an index based on Southern Electric stock daily close and capitalization, (we'll call that variable* X*), and the second column gives, on the same day, the same quantity for Duke Energy (we'll call that variable* Y*), another large utility company.*

1. *Compute the means and the standard deviations of* X *and* Y*, and compute their correlation coefficients.*
2. *We first assume that* X *and* Y *are samples from a jointly Gaussian distribution whose parameters are equal to the estimates computed in question 1. Using the assumption of joint normality, compute the* 2*-percentiles of the variables* $X + Y$ *and* $X - Y$*, and compute the empirical estimates of these percentiles.*

3. *Fit generalized Pareto distributions (GPDs) to X and Y separately, and fit a copula of the Gumbel family to the data.*
4. *Generate a sample of size N (number of rows of the data matrix* `UTILITIES`*) from the joint distribution estimated in question 3.*
   4.1 *Use this sample to compute the same statistics as in question 1 (i.e. means and standard deviations of the columns, as well as their correlation coefficients), and compare the results to the numerical values obtained in question 1.*
   4.2 *Compute, still for this simulated sample, the two percentiles considered in question 2, compare the results, and comment.*

Ⓔ **Problem 3.8** *This problem is based on the data contained in the data set* `SPFUT`. *The first column gives, for each day, the log return of a futures contract which matures* 3 *weeks later, (we'll call that variable X), and the second column gives, on the same day, the log return of a futures contract which matures* 1 *week later (we'll call that variable Y).*

1. *Compute the means and the standard deviations of X and Y, and compute their correlation coefficients.*
2. *We first assume that X and Y are samples from a jointly Gaussian distribution with parameters computed in part 1. For each value $\alpha = 25\,\%$, $\alpha = 50\,\%$ and $\alpha = 75\,\%$, compute the q-percentile with $q = 2\,\%$ of the variable $\alpha X + (1-\alpha)Y$.*
3. *Fit a generalized Pareto distribution (GPD) to X and Y separately, and fit a copula of the Gumbel family to the empirical copula of the data.*
4. *Generate a sample of size N (where N is the number of rows of the data matrix) from the joint distribution estimated in question 3.*
   4.1. *Use this sample to compute the same statistics as in question 1 (i.e. means, standard deviations, as well as their correlation coefficients), and compare to the numerical values obtained in question 1.*
   4.2. *Compute, still for this simulated sample, the three percentiles considered in question 2, and compare the results.*

Ⓢ **Problem 3.9** *Let us assume that $(x_1, y_1), (x_2, y_2), \cdots\cdots, (x_n, y_n)$ represent the historical weekly log returns of two assets as measured over a period of* 5 *years. Let us also assume that you hold a portfolio invested in these two assets,* $40\,\%$ *of your portfolio being invested in the first asset, and* $60\,\%$ *in the second.*

1. *There are several ways to compute (from the data at hand) the $VaR_{0.01}$ of your portfolio over* 1 *week period. Describe in detail three possible ways, making sure that they are different, and explaining clearly the various steps of these three procedures.*
2. *If someone were to suggest that you estimate the mean $\hat{\mu}$ and the variance $\hat{\sigma}^2$ of the log return of your portfolio so that you could be sure that your log return belongs to the interval*
$$[\hat{\mu} - 3\hat{\sigma}, \hat{\mu} + 3\hat{\sigma}]$$
with probability at least $99\,\%$, *what kind of assumption would this someone be working under, and which of the three computations of question 1 would this someone be closest to?*

Ⓢ **Problem 3.10** 1. *Construct a vector of* 100 *increasing and regularly spaced numbers starting from* 0.1 *and ending at* 20. *Call it* `SIG2`. *Construct a vector of* 21 *increasing and regularly spaced numbers starting from* -1.0 *and ending at* 1.0. *Call it* `RHO`.
2. *For each entry σ^2 of* `SIG2` *and for each entry ρ of* `RHO`*:*

- *Generate a sample of size* $N = 500$ *from the distribution of a bivariate Gaussian vector* $Z = (X, Y)$*, where* $X \sim N(0, 1)$*, and* $Y \sim N(0, \sigma^2)$*, and the correlation coefficient of* X *and* Y *is* ρ*. Create a* 500×2 *matrix* `Z` *holding the values of the sample of* Z*;*
- *Create a* 500×2 *matrix* `EXPZ`*, with the exponentials of the entries of* `Z` *(the distributions of these columns are lognormal);*
- *Compute the correlation coefficient, call it* $\tilde{\rho}$*, of the two columns of* `EXPZ`

3. *Produce a scatterplot of all the points* $(\sigma^2, \tilde{\rho})$ *so obtained. Comment.*

Ⓣ **Problem 3.11** *This elementary exercise is intended to give an example showing that lack of correlation does not necessarily mean independence!*
Let us assume that $X \sim N(0, 1)$ *and let us define the random variable* Y *by:*

$$Y = \frac{1}{\sqrt{1 - 2/\pi}}(|X| - \sqrt{2/\pi})$$

1. *Compute* $\mathbb{E}\{|X|\}$
2. *Show that* Y *has mean zero, variance* 1*, and that it is uncorrelated with* X*.*

Ⓣ **Problem 3.12** *Let* X *and* Y *be continuous random variables with cdfs* F_X *and* F_Y *respectively, and with copula* C*. For each real number* t*, prove the following two equalities:*

1. $\mathbb{P}\{\max(X, Y) \le t\} = C(F_X(t), F_Y(t))$
2. $\mathbb{P}\{\min(X, Y) \le t\} = F_X(t) + F_Y(t) - C(F_X(t), F_Y(t))$

Ⓣ **Problem 3.13** *Suppose* X_1 *and* X_2 *are independent* $N(0, 1)$ *random variables and define* X_3 *by:*

$$X_3 = \begin{cases} |X_2|, & \text{if } X_1 > 0 \\ -|X_2|, & \text{if } X_1 < 0 \end{cases}$$

1. *Compute the cdf of* X_3*. Say if* X_3 *is Gaussian.*
2. *Compute* $\mathbb{P}\{X_2 + X_3 = 0\}$*.*
3. *Is* $X_2 + X_3$ *Gaussian? Is* (X_2, X_3) *jointly Gaussian?*

Ⓣ **Problem 3.14** *Let us assume that* X_1*,* X_2 *and* X_3 *are independent* $N(0, 1)$ *random variables and let us set*

$$Y_1 = \frac{X_1 + X_2 + X_3}{\sqrt{3}} \quad \text{and} \quad Y_2 = \frac{X_1 - X_2}{\sqrt{2}}$$

1. *Compute* $cov(Y_1, Y_2)$*.*
2. *Compute* $var(Y_1 Y_2)$*.*

Ⓣ **Problem 3.15** *Let us assume that* X *is a random variable uniformly distributed over the unit interval* $[0, 1]$ *and let us define the random variable* Y *by:*

$$Y = |2X - 1|$$

1. *What is the distribution of Y?*
2. *Compute the covariance* $cov\{X, Y\}$ *of* X *and* Y.
3. *Are* X *and* Y *independent? Explain why.*
4. *What can you say about the copula of* X *and* Y*? Describe its plot?*

Ⓣ **Problem 3.16** *The purpose of this problem is to provide another instance of the fact that lack of correlation does not imply independence, even when the two random variables are Gaussian!!! We assume that* X, ϵ_1 *and* ϵ_2 *are independent random variables, that* $X \sim N(0,1)$, *and that* $\mathbb{P}\{\epsilon_i = -1\} = \mathbb{P}\{\epsilon_i = +1\} = 1/2$ *for* $i = 1, 2$. *We define the random variable* X_1 *and* X_2 *by:*

$$X_1 = \epsilon_1 X, \qquad \textit{and} \qquad X_2 = \epsilon_2 X.$$

1. *Prove that* $X_1 \sim N(0,1)$, $X_2 \sim N(0,1)$ *and that* $\rho\{X_1, X_2\} = 0$.
2. *Show that* X_1 *and* X_2 *are not independent.*

Ⓣ **Problem 3.17** *The goal of this problem is to prove rigorously a couple of useful formulae for Gaussian random variables.*

1. *Show that, if* $Z \sim N(0,1)$, *if* $\sigma > 0$, *and if* f *is any function, then we have:*

$$\mathbb{E}\{f(Z)e^{\sigma Z}\} = e^{\sigma^2/2}\mathbb{E}\{f(Z+\sigma)\},$$

and use this formula to recover the well known fact, whenever $X \sim N(\mu, \sigma^2)$, *it holds:*

$$\mathbb{E}\{e^X\} = e^{\mu+\sigma^2/2}$$

2. *We now assume that* X *and* Y *are jointly-Gaussian mean-zero random variables and that* h *is any function. Prove that:*

$$\mathbb{E}\{e^X h(Y)\} = \mathbb{E}\{e^X\}\mathbb{E}\{h(Y + cov\{X, Y\})\}.$$

Ⓣ **Problem 3.18** *The goal of this problem is to prove rigorously the theoretical result illustrated by the simulations of Problem 3.10.*

1. *Compute the density of a random variable* X *whose logarithm* $\log X$ *is* $N(\mu, \sigma^2)$. *Such a random variable is called a lognormal random variable with mean* μ *and variance* σ^2. *Throughout the rest of the problem we assume that* X *is a lognormal random variable with parameters* 0 *and* 1 *(i.e.* X *is the exponential of an* $N(0,1)$ *random variable) and that* Y *is a lognormal random variable with parameters* 0 *and* σ^2 *(i.e.* Y *is the exponential of an* $N(0, \sigma^2)$ *random variable). We shall assume that* $(\log X, \log Y)$ *is jointly Gaussian even though the results remain true without this assumption. Finally, we use the notation* ρ_{min} *and* ρ_{max} *introduced in the last paragraph of Sect. 3.1.2.*
2. *Show that* $\rho_{min} = (e^{-\sigma} - 1)/\sqrt{(e-1)(e^{\sigma^2} - 1)}$.
3. *Show that* $\rho_{max} = (e^{\sigma} - 1)/\sqrt{(e-1)(e^{\sigma^2} - 1)}$.
4. *Check that* $\lim_{\sigma\to\infty} \rho_{min} = \lim_{\sigma\to\infty} \rho_{max} = 0$.

Ⓣ **Problem 3.19** *Let us assume that* X_1 *and* X_2 *are two positive random variables which we assume to be independent and having the same distribution. Say why the expectation*

$$\mathbb{E}\left\{\frac{X_1}{X_1 + X_2}\right\}$$

exists and compute its value.

Ⓣ **Problem 3.20** *In this problem, we consider the daily rates of exchange $p_{EUR/USD}$, $p_{GBP/USD}$ and $p_{EUR/GBP}$ between three currencies, the Euro (EUR), the US dollar (USD), and the British pound (GBP). The meaning of these spot rates of exchange is the following: $p_{XXX/YYY}$ gives the currency YYY value of one unit of currency XXX. Our goal is to study the joint distribution of the log-returns of these three rates.*

1. *Give an equation that $p_{EUR/USD}$, $p_{GBP/USD}$ and $p_{EUR/GBP}$ must satisfy if the model does not allow for arbitrage, by which we mean a way to make free money without risk. What is the corresponding no-arbitrage relationship between the log-returns which we denote by the letter r with the appropriate subscript?*
2. *Assuming that you are provided with daily observations of these three rates of exchange over the last $N = 500$ trading days, explain in a few bullet points how you would simulate $M = 1{,}000$ Monte Carlo samples from the joint distribution of the three log-returns?*
3. *We now consider the returns of the portfolio comprising a units of EUR/USD, b units of GBP/USD and c units of EUR/GBP.*
   3.1. *Write a formula for the expected return of the portfolio as a function of a,b,c, $p_{EUR/USD}$, $p_{GBP/USD}$, $r_{EUR/USD}$ and $r_{GBP/USD}$.*
   3.2. *If you had to ask for a single number to capture the correlation structure of the random variables, which one would it be? Remember that all you want to estimate is the expected return of the portfolio.*

Ⓣ **Problem 3.21** *If C denotes the copula of the two random variables X and Y, prove that the Spearman correlation coefficient $\rho_S(X,Y)$ is given by the formula*

$$\rho_S(X,Y) = 12 \int_0^1 \int_0^1 uvC(u,v)\, dudv - 3.$$

One will assume that the cdfs are continuous and strictly increasing.

Ⓔ **Problem 3.22** *Assuming that you are given the bivariate sample*

$$(x_1^{(1)}, x_1^{(2)}), \cdots\cdots, (x_n^{(1)}, x_n^{(2)})$$

of size n from a given bivariate distribution, describe in a few bullet points how you would generate $N = 10{,}000$ Monte Carlo samples from the 2-dimensional distribution which has the same copula as the data, the exponential distribution with rate 1 for its first marginal, and the classical Pareto distribution with shape parameter $\xi = 0.5$, location parameter $m = 0$, and scale parameter $\lambda = 1$ for its second marginal.

Ⓢ **Problem 3.23** *The data needed for this problem are contained in the data frame* `TC` *comprising two columns named* `X` *and* `Y` *respectively. You do not have to know where these data come from, but for the sake of definiteness, you can think of* `X` *as giving the temperature fluctuation around the average temperature at a given location, and* `Y` *as giving the price of a physical commodity for delivery near this location.*

1. *Say if the tails of the distribution of* `X` *are heavy (explain your answer) and describe which steps you would take if you were asked to fit a distribution to the data in the vector* `X`. *NB: you do not have to actually* take these steps, *just describe what you would do.*
2. *Same question for the vector* `Y`.
3. *Compute the Pearson, Kendall and Spearman correlation coefficients between* `X` *and* `Y` *and comment on the numerical values so obtained.*

4. *Create a new data set in the form of a two-column matrix with the rows of the original data matrix* `TC` *for which the value of* `X` *is smaller than* -5.48, *i.e. for which* $X < -5.48$, *and compute the same three correlation coefficients as before for the columns of this new data set.*
5. *Compare the two sets of correlation coefficients and comment.*

Ⓢ **Problem 3.24** *1. For each* $\rho = 0, 0.05, 0.1, 0.15, \ldots, 0.9, 0.95, 1.0$, *generate a sample of size* $N = 2{,}000$ *from the bivariate Gaussian copula with parameter* ρ, *and call this sample* `SD`. *Each such sample can be viewed as a* `Nx2` *matrix, the first column being called* `SD$x` *and the second column* `SD$y`. *For each of these samples:*

1.1. *Transform the columns* `SD$x` *and* `SD$y` *in such a way that they become samples from the standard normal distribution.*
1.2. *Compute the Pearson, Kendall and Spearman correlation coefficients of the samples so obtained, and store these values.*
1.3. *Give a plot of the Pearson correlations coefficients against the values of* ρ *and explain what you find.*
1.4. *Give plots of the Spearman and Kendall correlation coefficients against the corresponding values of* ρ. *Comment your results.*
1.5. *Redo the same thing by transforming the columns* `SD$x` *and* `SD$y` *in such a way that they become samples from the standard Cauchy distribution, recompute all the correlation coefficients as before, produce the same plots, and explain the differences.*

2. *For each* $\beta = 1, 1.5, 2, 2.5, \ldots, 19, 19.5, 20$, *generate a sample of size* $N = 2{,}000$ *from the bivariate Gumbel copula with parameter* β, *and call this sample* `SD`. *As before, each such sample can be viewed as a* `Nx2` *matrix, the first column being called* `SD$x` *and the second column* `SD$y`. *For each of these samples:*

2.1. *Transform the columns* `SD$x` *and* `SD$y` *in such a way that they become samples from the standard normal distribution.*
2.2. *Compute the Pearson, Kendall and Spearman correlation coefficients of the samples so obtained, and store these values.*
2.3. *Give a plot of the Pearson correlations coefficients against the values of* β *and comment.*
2.4. *Give plots of the Spearman and Kendall correlation coefficients against the corresponding values of* ρ. *Comment your results.*
2.5. *Redo the same thing by transforming the columns* `SD$x` *and* `SD$y` *in such a way that they become samples from the standard Cauchy distribution, recompute all the correlation coefficients as before, produce the same plots, and explain the differences.*

ⓈⓉ **Problem 3.25** *Let* a, b, *and* c *be real numbers satisfying* $a < c < b$. *We say that a random variable* X *has the triangular distribution with range parameters* a *and* b *and mode* c, *fact which we denote by* $X \sim \Delta(a, b, c)$, *if it has the distribution function*

$$F_X(x) = \begin{cases} 0 & \text{if } x < a \\ \frac{(x-a)^2}{(b-a)(c-a)} & \text{if } a \le x \le c \\ 1 - \frac{(b-x)^2}{(b-a)(b-c)} & \text{if } c < x \le b \\ 1 & \text{if } b < x \end{cases}.$$

In this problem we assume that $X \sim \Delta(a, b, c)$ *and we set* $\beta = (c - a)/(b - a)$.

1. *Compute the density* f_X *and sketch its graph.*
2. *Compute the quantile function,* F_X^{-1}.
3. *Construct an algorithm that generates samples from the* $\Delta(a, b, c)$ *distribution.*
4. *We now assume that* $Z \sim \Delta(0, 1, \beta)$ *and we define the random variable* Y *by* $Y = a + (b - a)Z$. *Show that* $Y \sim \Delta(a, b, c)$.
5. *Set* $\alpha = 1\%$, *let* β_1 *and* β_2 *be real numbers satisfying* $0 < \beta_1 < \beta_2 < 1$, *and let* $Z_1 \sim \Delta(0, 1, \beta_1)$ *and* $Z_2 \sim \Delta(0, 1, \beta_2)$. *Which is larger,* $VaR_\alpha(Z_1)$ *or* $VaR_\alpha(Z_2)$?
6. *Recall that the (bivariate) Gumbel copula with parameter* δ *is given by*

$$C_\delta(x, z) = \exp\left(-\left[(-\log(x))^\delta + (-\log(z))^\delta\right]^{1/\delta}\right),$$

where $1 \le \delta < \infty$. *Let us assume that the two random variables* X *and* Z *considered above have a Gumbel copula with parameter* δ, *and define the random variable* W *by*

$$W = \frac{F_X(Y) + F_Y(X)}{2}.$$

6.1. *Identify the Gumbel copula* C_1 *with parameter* $\delta = 1$.
6.2. *What can you say about the dependence between* X *and* Y *in this case?*
6.3. *Show that for* $\delta = 1$ *the random variable* W *has a triangular distribution and give its parameters.*

Ⓔ **Problem 3.26** *The data to be used for this problem are contained in the data set* `Power08`. *It is a numeric matrix with* 366 *rows and* 2 *columns, the names of the rows being the days of the quotes. Each column gives the price of* 1 *MWhr of electricity to be delivered during a specific hour of the day: the first column gives the price for a* 5–6 *pm delivery and the second column for the* 12–1 *pm delivery. The data is viewed as a sample of* $n = 366$ *independent observations from a bivariate distribution, the joint distribution of the prices of* P_1 *and* P_2 *of* 1 *MWhr during these two periods of the day.*

1. *Fit a bivariate distribution to the data. Explain the steps you take and why you take them.*
2. *Assume that an oracle tells you that the distribution of* P_1 *is log-normal and the distribution of* P_2 *is exponential. Under these conditions*
   2.1. *Estimate the parameters of these marginal distributions*
   2.2. *Provide a new estimate of the joint distribution of* P_1 *and* P_2 *taking the information given by the oracle into account.*
3. *Assuming that you need to purchase a fixed quantity of electricity every day, either between* 12 *and* 1 *pm, or between* 5 *and* 6 *pm, and that you are indifferent in which of these two periods you buy it, an energy provider offers you the option to acquire electricity at the lower price each day. To be specific, for each MWhr, you purchase the electricity at the price* P_2 *and buy from the energy provider an option which pays* $(P_2 - P_1)^+$. *Before showing some interest in the deal (and negotiating the premium) you want to compute (a) the probability that on any given day, the option will be exercised; (b) the expected minimum price* $\mathbb{E}\{P_1 \wedge P_2\}$ *per MWhr purchased; (c) the expected pay-off* $\mathbb{E}\{(P_2 - P_1)^+\}$ *per MWhr purchased. For each of the three quantities a), b) and c) above, propose three estimates based on*

3.1. The price data contained in `Power08`*;*
3.2. The bivariate distribution fitted in question 1 above;
3.3. The bivariate distribution fitted in question 2 above.

***NB:** The notation x^+ stands for the maximum of x and 0 while $x \wedge y$ stands for the minimum of the numbers x and y.*

Ⓢ Ⓣ **Problem 3.27** *The first question concerns the computation in* `R` *of the square root of a symmetric nonnegative-definite square matrix.*

1. *Write an* `R`*-function, call it* `msqrt`*, with argument A which:*
   - *Checks that A is a square matrix and exits if not;*
   - *Checks that A is symmetric and exits if not;*
   - *Diagonalizes the matrix by computing the eigenvalues and the matrix of eigenvectors (hint: check out the help file of the function* `eigen` *if you are not sure how to proceed);*
   - *Checks that all the eigenvalues are nonnegative and exits, if not;*
   - *Returns a symmetric matrix of the same size as A, with the same eigenvectors, the eigenvalue corresponding to a given eigenvector being the square root of the corresponding eigenvalue of A.*

   The matrix returned by such a function `msqrt` *is called the square root of the matrix A and it will be denoted by $A^{1/2}$.*

   The second question concerns the generation of Gaussian random vectors in `R`*. In other words, we write an* `R` *function to play the role of the function* `rnorm` *in the case of multi-dimensional random vectors. Such a function does exist in the* `R` *distribution. It is called* `mvrnorm`*. The goal of this second question is to understand how such a generation method works.*
2. *Write an* `R`*-function, call it* `vnorm`*, with arguments* `Mu`*,* `Sigma` *and* `N` *which:*
   - *Checks that* `Mu` *is a vector, exits if not, and otherwise reads its dimension, say* `L`*;*
   - *Checks that* `Sigma` *is an* `LxL` *symmetric matrix with nonnegative eigenvalues and exits, if not;*
   - *Creates a numeric array with* `N` *rows and* `L` *columns and fills it with independent random numbers with the standard normal distribution $N(0, 1)$;*
   - *Treats each row of this array as a vector, and multiplies it by the square root of the matrix* `Sigma` *(as computed in question 1 above) and adds the vector* `Mu` *to the result;*
   - *Returns the random array modified in this way.*

   The array produced by the function `vnorm` *is a sample of size* `N` *of* `L`*-dimensional random vectors (arranged as rows of the matrix outputted by* `vnorm`*) with the Gaussian distribution with mean* `Mu` *and variance/covariance matrix* `Sigma`*. Indeed, this function implements the following simple fact reviewed during the lectures.*

   If X is an L-dimensional Gaussian vector with mean 0 and covariance matrix given by the $L \times L$ identity matrix (i.e. if all the L entries of X are independent $N(0, 1)$ random variables), then:

$$Y = \mu + \boldsymbol{\Sigma}^{1/2} X$$

   is an L-dimensional Gaussian vector with mean μ and variance/covariance matrix $\boldsymbol{\Sigma}$.

Notes & Complements

This chapter concentrated on multivariate distributions and on dependence between random variables. The discussion of the various correlation coefficients is modelled after the standard treatments which can be found in most multivariate statistics books. The originality of this chapter lies in the statistical analysis of the notion of dependence by way of copulas. The latter are especially important when the marginal distributions have heavy tails, which is the case in most financial applications as we saw in Chap. 2. The recent renewal of interest in the notion of copula prompted a rash of books on the subject. We shall mention for example the monograph [73] of R.B. Nelsen, or the book of D. Drouet Mari and S. Kotz [68]. We refer the interested reader to their bibliographies for further references on the various notions of dependence and copulas.

The `R` methods used in this chapter to estimate copulas and generate random samples from multivariate distributions identified by their copulas are from the library `Rsafd`. They were originally developed for the library `EVANESCE` [15] by J. Morrisson and the author. This library was included in the `Splus` public-domain library `safd`, and the `S+FinMetrics` module of the commercial distribution of `SPlus`.

Comprehensive presentations of the credit markets and the mechanics of CDOs before 2005 can be found in the textbooks [27, 59, 86]. The popular press has offered analyses of the role of the Gaussian copula in the pricing of CDOs and the misunderstanding of its limitations. From the realistic journalistic accounts we selected Felix Salmon's WIRED Magazine article *Recipe for Disaster: the Formula that Killed Wall Street* and the Financial Times piece *The Formula that felled Wall Street* by Sam Jones. Dedicated Monte Carlo algorithms for the computation of rare events involving multiple defaults are developed in the articles [16, 17].

To the best of my knowledge, the first attempt to apply principal component analysis to the yield curve is due to Litterman and Scheinkmann [60]. Rebonato's book [77], especially the short second chapter, and the book [2] by Anderson, Breedon, Deacon, Derry, and Murphy, are good sources of information on the statistical properties of the yield curve. Discussions of interest rate swap contracts and their derivatives can also be found in these books. The reader interested in a discussion of the mathematical models of the fixed income markets with developments in stochastic analysis including pricing and hedging can consult the book by Carmona and Tehranchi [18].

An application of PCA to variable selection in a regression model is given in Problem 5.25.

The decomposition of a data set into its principal components is known in signal analysis as the Karhunen-Loève decomposition, and the orthonormal basis of principal components is called the Karhunen-Loève basis. This basis was identified as optimal for compression purposes. Indeed, once a signal is decomposed on this basis, most of the coefficients are zero or small enough to be discarded without significantly affecting the information contained in the signal. Not surprisingly, the optimality criterion is based on a form of the entropy of the set of coefficients. PCA is most useful for checking that data do contain features which are suspected to be present. For this reason, some authors suggest to remove by regression the gross features identified by a first PCA run, and to then run PCA on the residuals. PCA has been successfully used in many applications, especially in signal and image analysis.

Part II

REGRESSION

4

PARAMETRIC REGRESSION

This chapter provides an introduction to several types of regression analysis: simple and multiple linear, as well as simple polynomial and nonlinear. In all cases we identify the regression function in a parametric family, hence the title of the chapter. We introduce the idea of robustness, and we illustrate the concept with a parallel comparison of the least squares and the least absolute deviations regression methods. Even though we introduce regression from a data smoothing point of view, we systematically interpret the results in terms of statistical models, and we derive the statistical inference and diagnostic tools provided by the theory underpinning the statistical models. The chapter ends with a thorough discussion of the parametric estimation of the term structure of interest rates based on the Nelson-Siegel and Swensson families. As before, we try to work from examples, introducing theoretical results as they become relevant to the discussions of the analyzes which are used as illustrations.

4.1 SIMPLE LINEAR REGRESSION

Although multiple linear regression is ubiquitous in economic and in econometric applications, simple linear regression does not play a very central role in quantitative finance. Nevertheless, its mathematical theory and inferential results are important enough to compel us to present them, even if the conclusions drawn from its use in practical financial applications are not always earth-shattering. As in earlier chapters, we choose particular data sets to illustrate the statistical concepts and techniques which we introduce. In the first part of this chapter we choose to work with the values of an energy index, as they relate to the values of several utilities, but as a disclaimer, it is important to emphasize that this choice was made for illustration purposes only. Most financial data come naturally in the form of time series, and the serial dependencies contained in the data may not be appropriate for some forms of regression analysis. With this in mind, the reader should be prepared to have a critical attitude toward the results produced by the algorithms introduced in this chapter. Case in point, after examining residual patterns, we abandon the original form of our

R. Carmona, *Statistical Analysis of Financial Data in R*, Springer Texts in Statistics, DOI 10.1007/978-1-4614-8788-3_4, © Springer Science+Business Media New York 2014

first regression, and switch to a form of the data more amenable to regression. The reason for our choice is to understand the theoretical underpinnings of a regression model, so it is purely didactic.

The original index data were computed following the methodology and data used by Dow Jones to produce its Total Market Index. Our goal is to investigate to which extent two large companies can influence the index of an entire economic sector. We chose the energy/utility sector because of its tremendous growth in the second half of the 1990s. However for obvious reasons, we will stay away from what happened during the California energy crisis and after ENRON's bankruptcy.

4.1.1 **Getting the Data**

A very convenient way to store data in an `R` object is to use the structure of `data frame`. Whether the data set is already part of the `R` distribution or it has been read from a disk file, or downloaded from the internet (see the introduction to `R` at the end of the book for details), one way to make a data set available to the current `R` session is to attach the `data frame` using the `R` function `attach` whose main purpose is to make sure that the values stored in the columns of the data frame can be manipulated by `R` commands using their names without having to refer to the name of the data frame. We use the data set `UTIL.index` included in the library `Rsafd`.

```
> head(UTIL.index)
           ENRON.index DUKE.index UTILITY.index
01/04/1993    135.0000   104.2857      99.94170
01/05/1993    135.3714   103.5714      99.49463
01/06/1993    132.8571   104.2857      99.86034
01/07/1993    130.7143   103.5714      98.70023
01/08/1993    126.8000   101.8000      97.93630
01/11/1993    127.5143   101.8000      98.69736
```

As always in a `data.frame` object, the first row contains the names of the column variables, which in the case at hand, are the indexes computed from the share values and the capitalizations of ENRON and DUKE Energy, and a utility index computed from all the publicly traded large utilities. The left most column contains the names of the row variables, which in the present situation, are the dates of the quotes. As we can see, they cover the year 1993. I chose this year in order to stay away from the speculative period of the late 1990s during which the energy sector heated up to unhealthy levels. The dimensions (i.e. the number of rows and the number of columns) of the data frame can be obtained with the `R` generic command `dim`. For example, the result of the command:

```
> dim(UTIL.index)
[1] 260   3
```

tells us that the data frame UTIL.index has 260 rows and 3 columns. Manipulating the numerical vectors forming the rows and the columns of the data frame can be done by subscripting as we did before. For example, the command UTIL.index[,3] gives the numeric vector with length 260 formed by the entries of the third column of the data frame. Since this column has a name, we may want to use its name UTILITY.index in order to make the text of the commands more meaningful. This can be done as UTIL.index$UTILITY.index is a valid alternative to UTIL.index[,3]. However, this method of extracting columns leads to lengthy and cumbersome commands. As explained above, a solution to this quandary is offered by the command attach. After running the command

```
> attach(UTIL.index)
[1] 260   3
```

using UTILITY.index will not trigger and error message, and the three columns of the data frame UTIL.index can be accessed by their names.

4.1.2 First Plots

It is always a good idea to look at graphical representations of the data whenever possible. In the present situation one can split the graphic window in two columns and place a scatterplot of the UTILITY.index variable against the ENRON.index variable on the left entry of this 1 by 2 matrix of plots, and the scatterplot of the DUKE.index and the UTILITY.index variables on the right. This is done with the commands:

```
> par(mfrow=c(1,2))
> plot(ENRON.index,UTILITY.index)
> plot(DUKE.index,UTILITY.index)
> par(mfrow=c(1,1))
```

The last command resets the graphics window to a 1 by 1 matrix of plots for subsequent use. The results are shown in Fig. 4.1.

Plots of this type (one variable against another) are called *scatterplots.* They are very convenient for getting a feeling of the dependence/independence of two variables, as we saw in Chap. 3. When the data frame contains more than two variables, it is possible to get all these 2 by 2 plots simultaneously with the command pairs

```
> pairs(UTIL.index)
```

The result is shown in Fig. 4.2. The plots on the diagonal entries of this matrix of plots are not given for an obvious reason: they would be un-informative since all the points would have to be on the main diagonal.

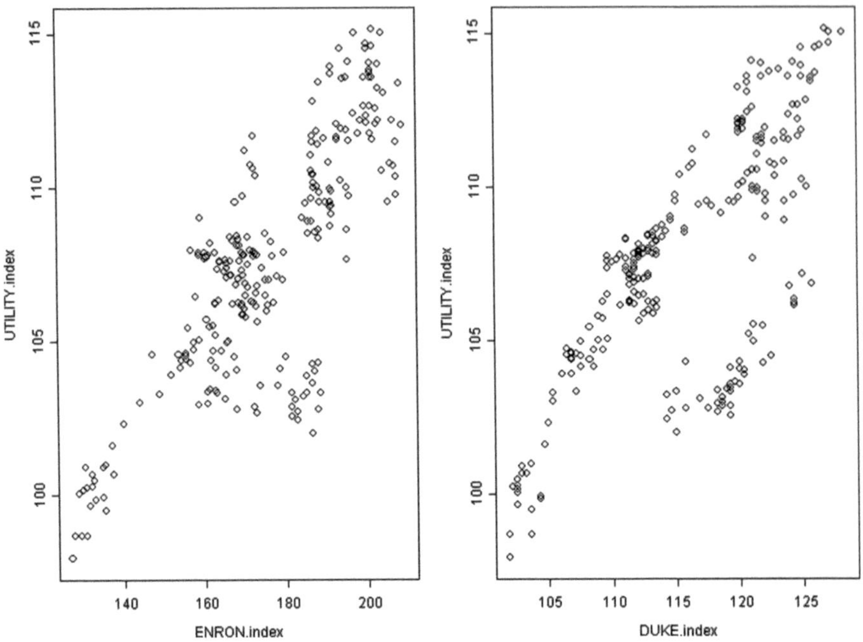

Fig. 4.1. Scatterplot of `UTILITY.index` against `ENRON.index` (*left*) and of `UTILITY.index` against `DUKE.index` (*right*) of the `UTIL.index` data set

4.1.3 **Regression Set-Up**

As expected, the scatterplots give strong evidence of relationships between the values of the utility index and the two other variables. Indeed, the points would be scattered all over the plotting area if it weren't for these relationships. Regression is a way to capture these dependencies, and we proceed to fix the notation which we will use when setting up a regression problem.

The general form of the (simple) regression set-up is given by observations

$$y_1, y_2, \cdots\cdots, y_n$$

of one variable, which we call the *response* variable (think for example of the $n = 260$ values of the utility index in the case discussed in this section), and of observations:

$$x_1, x_2, \cdots\cdots, x_n$$

of an explanatory variable, which we call the *regressor.* For the sake of definiteness, we first consider the case of the $n = 260$ values of the index computed from the values of ENRON shares and capitalization. Regression is the search for a functional relationship of the form:

$$y \approx \varphi(x)$$

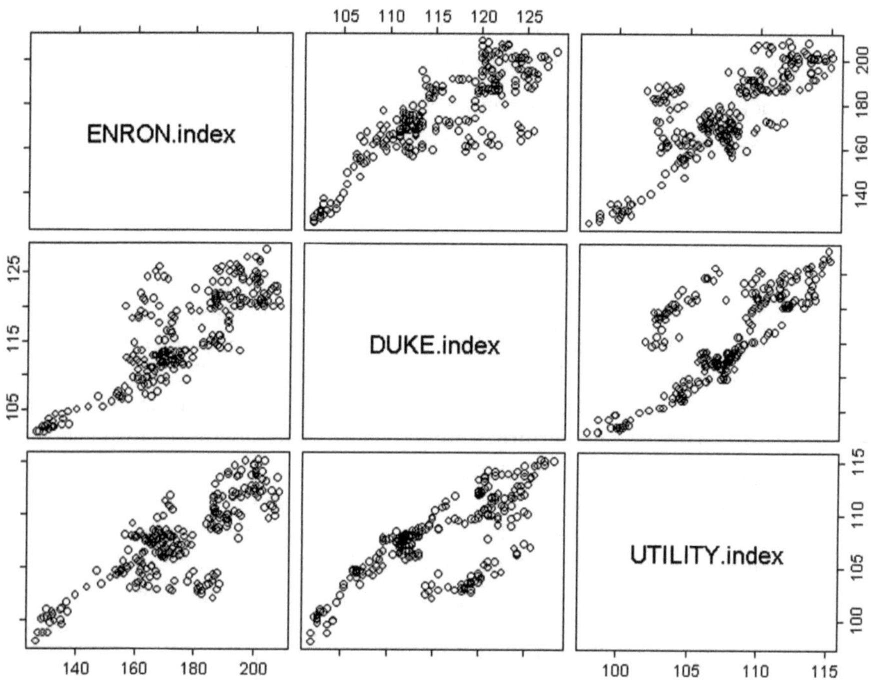

Fig. 4.2. Matrix plot of the pair scatterplots of the variables in the `UTIL.index` data frame

so that the actual observations satisfy:

$$y_i = \varphi(x_i) + \text{error term}, \qquad i = 1, \ldots, n$$

where the error term is hopefully small, but most importantly unpredictable in a sense which we will try to make clear later on.

Warning. The regressor variable is sometimes called the *independent variable* to be consistent with the terminology *dependent variable* often used for the response. We believe that this terminology is very misleading and we shall refrain from using it.

The regression terminology is well established and quite extensive, and we shall conform to the common usage. For example, we talk of

- *Simple linear regression* when we have only one regressor and when the dependence is given by an affine function of the form $\varphi(x) = \beta_0 + \beta_1 x$. The regression problem is then to estimate the parameters β_1 giving the slope of the regression line, and β_0 giving the intercept, and possibly some statistical characteristics of the error terms (sometimes call noise) such as its variance for example;
- *Simple nonlinear regression* when we have only one regressor and when the dependence is given by a general (typically nonlinear) function φ;
- *Simple spline regression* when we have only one regressor and when the dependence is given by a function φ constructed from spline functions;

- *Multiple regression* (whether linear or nonlinear) when we have several regressors which we usually bind in a vector of *explanatory variables.*

Coming back to our utility index example, and trying to explain the values of the utility index (representing the entire energy sector) from the values of the ENRON and DUKE indexes, we can choose the entries of column variable `UTILITY.index` for the values of y_i and

- Looking at the scatterplot in the left pane of Fig. 4.1, we can decide to choose the entries of `ENRON.index` for the values of x_i and perform a simple linear regression of `UTILITY.index` against `ENRON.index`, searching for real numbers β_0 and β_1 and a regression function φ of the form $\varphi(x) = \beta_0 + \beta_1 x$;
- Alternatively, looking at the scatterplot in the right pane of Fig. 4.1, we can decide that the variable `DUKE.index` better explains the changes in the response `UTILITY.index`, and perform a simple linear regression of the response variable `UTILITY.index` against the regressor `DUKE.index`;
- Finally, suspecting that a cleverly chosen combination of the values of the variables `ENRON.index` and `DUKE.index` could provide a better predictor of the values of `UTILITY.index`, we could decide to opt for a multiple linear regression of the response variable `UTILITY.index` against both variables `ENRON.index` and `DUKE.index`. In this case, we would choose to view the entries of the variable `ENRON.index` as observations of a first explanatory variable $x^{(1)}$, the entries of `DUKE.index` as observations of a second explanatory variable $x^{(2)}$, bundle these two explanatory variables into a bivariate vector $\boldsymbol{x} = (x^{(1)}, x^{(2)})$ and search for real numbers β_0, β_1 and β_2 and a function φ of the form:
$$y = \varphi(x^{(1)}, x^{(2)}) = \beta_0 + \beta_1 x^{(1)} + \beta_2 x^{(2)}.$$

We do just that in this chapter. Notice that we purposely restricted ourselves to linear regressions. Indeed, the shapes of the clouds of points appearing in the scatterplots of Figs. 4.1 and 4.2 are screaming for linear regressions, and it does not seem reasonable to embark on a search for nonlinear regression functions for the data at hand.

Going through the program outlined by the first set of bullets will keep us busy for the next two chapters. Despite the classification introduced by this terminology, regressions are most often differentiated by the specifics of the algorithms involved, and regression methods are frequently organized according to the dichotomy *parametric* versus *nonparametric regression methods* which we shall define later on.

Coming back to the general set-up introduced in this subsection, we outline search strategies for the regression function φ. A standard approach is to associate a *cost* to each admissible candidate φ, and to choose the candidate (hopefully it will exist and be defined unambiguously) which minimizes this cost. Even though there are many possible choices we shall concentrate on the two most common ones:

$$\mathcal{L}_2(\varphi) = \sum_{j=1}^{n} [y_j - \varphi(x_j)]^2 \tag{4.1}$$

and:

$$\mathcal{L}_1(\varphi) = \sum_{j=1}^{n} |y_j - \varphi(x_j)|. \tag{4.2}$$

Since both cost functions are based on the sizes of the differences $y_i - \varphi(x_i)$, the resulting function φ provides the best possible fit *at the observations* x_i. Given a family $\boldsymbol{\Phi}$ of functions φ, and given the data, the function $\varphi \in \boldsymbol{\Phi}$ which minimizes $\mathcal{L}_2(\varphi)$ over $\boldsymbol{\Phi}$ is called the least squares regression over $\boldsymbol{\Phi}$, and the function which minimizes $\mathcal{L}_1(\varphi)$ over $\boldsymbol{\Phi}$ is called the least absolute deviations regression over $\boldsymbol{\Phi}$. We shall often use the abbreviations L2 and L1 regression respectively, but the reader should be aware of the fact that the alternative abbreviations LS and LAD are also used frequently in the literature, and we may be using them from time to time. We first consider cases for which $\boldsymbol{\Phi}$ is a set of linear functions.

Remark. But first, it is important to emphasize that this L2/L1 duality is not new. We already encountered it in introductory statistics for the solution of the simpler problem of the statistical estimation of the location of a data sample. Indeed, given a sample $x_1, x_2, \ldots, x_n$ of real numbers, the classical estimates of the location given by the sample mean $\overline{x}$ and the sample median $\hat{\pi}_{0.5}$ are known to be the solutions of the minimization problems:

$$\overline{x} = \arg\min_m \sum_{i=1}^{n} |x_i - m|^2$$

and

$$\hat{\pi}_{0.5} = \arg\min_m \sum_{i=1}^{n} |x_i - m|$$

respectively. As we are about to see, least squares simple linear regression and least absolute deviations simple linear regression are mere generalizations of the two location estimation problems just mentioned. We shall revisit this example in Sect. 4.2.2 below.

4.1.4 Simple Linear Regression

As explained in the introduction, the set-up is given by input data of the form

$$(x_1, y_1), \ldots, (x_n, y_n)$$

where:

- n denotes the sample size,
- The x_i's denote the observations of the explanatory variable
- The y_i's denote the observations of the response variable,

and the problem is to find a straight line (whose equation will be written as $y = \beta_0 + \beta_1 x$) summarizing the data as faithfully as possible. For the purposes of the present discussion we limit ourselves to the two cost functions introduced earlier.

4.1.4.1 ***Least Squares (LS) Simple Regression***

We first consider the case of the least squares regression and we illustrate its use in the case of the utility indexes data. We shall see later that R has powerful methods for performing least squares linear regression, but for the time being, we shall restrict ourselves to the function `lsfit` in order to emphasize the parallel with the least absolute deviations regression method `l1fit`.

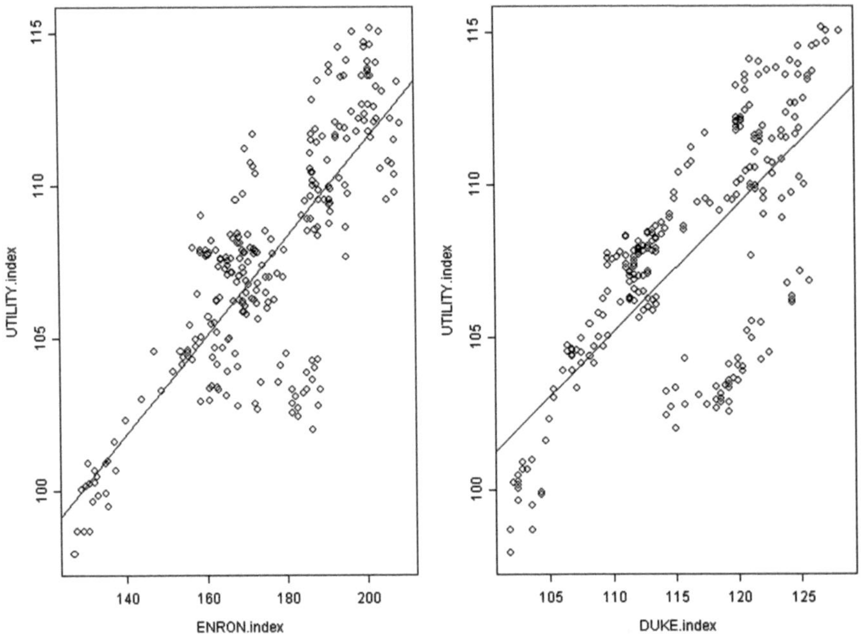

Fig. 4.3. Least squares regression of the utility index against the ENRON index (*left*) and the DUKE index (*right*)

The plots of Fig. 4.3 were obtained with the commands:

```
> UEl2 <- lsfit(ENRON.index,UTILITY.index)
> UDl2 <- lsfit(DUKE.index,UTILITY.index)
> par(mfrow=c(1,2))
> plot(ENRON.index,UTILITY.index)
> abline(UEl2)
> plot(DUKE.index,UTILITY.index)
> abline(UDl2)
> par(mfrow=c(1,1))
```

The function `lsfit` produces objects of class `lsfit` containing all the information necessary to process the results of the regression. The function `abline` adds a line to an existing plot. It is a very convenient way to add a regression line to a scatterplot. In particular, the intercept β_0 and slope β_1 of the regression lines can be extracted from the object produced by `lsfit` by using the extension `$coef`.

```
> UEl2$coef
 Intercept          X
  79.22646 0.1614243
 > UDl2$coef
 Intercept          X
 59.00217 0.4194389
```

As already explained, we chose to use the R function `lsfit` to emphasize the parallel with the least absolute deviations regression and the function `l1fit` provided in the library `Rsafd` for that purpose. The use of `lsfit` will eventually be abandoned when we come across the more powerful method `lm` designed to fit more general linear models.

4.1.4.2 ***Least Absolute Deviations (LAD) Simple Regression***

For the sake of comparison, we produce the results of the least absolute deviations regression in the same format. We shall often use the abbreviation `LAD` or `L1` for *least absolute deviations* while *least squares* is usually abbreviated as `LS` or L2. The plots of Fig. 4.4 were obtained with the commands:

```
> UEl1 <- l1fit(ENRON.index,UTILITY.index)
> UDl1 <- l1fit(DUKE.index,UTILITY.index)
> par(mfrow=c(1,2))
> plot(ENRON.index,UTILITY.index)
> abline(UEl1)
> plot(DUKE.index,UTILITY.index)
> abline(UDl1)
> par(mfrow=c(1,1))
```

The graphical results seem to be very similar, and we will need to investigate further to understand the differences between these two regression methods. As before we can print the coefficients of the regression lines:

```
> UEl1$coef
Intercept          X
  77.92709 0.1706265
 > UDl1$coef
 Intercept          X
  53.98615 0.4706897
```

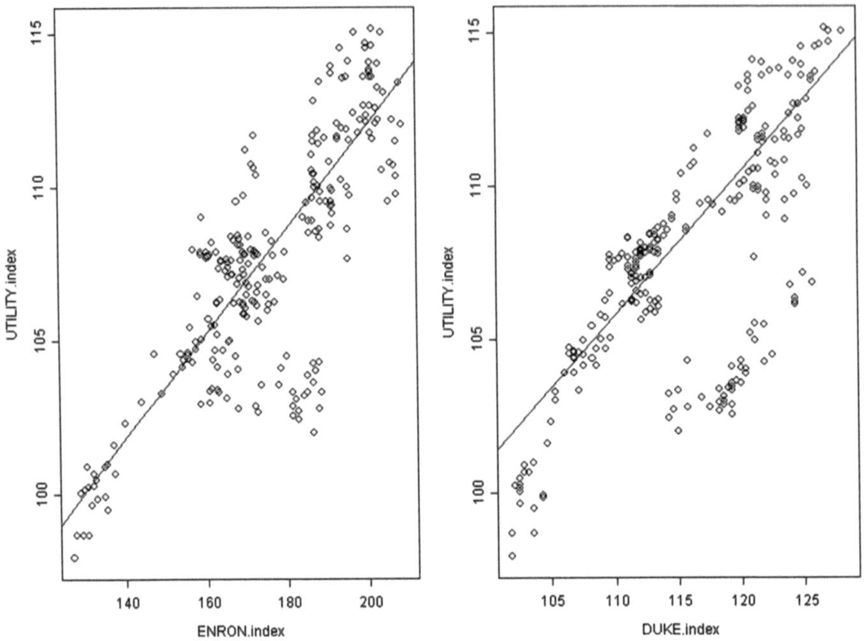

Fig. 4.4. Simple least absolute deviations regressions of the utility index on the ENRON index (*left*) and on the DUKE index (*right*)

We notice that the coefficients produced by the least absolute deviations regression are different from those produced by the least squares regression. Nevertheless, it is difficult at this stage to assess how serious these differences are, and how statistically significant they may be. More on that later.

4.1.5 **Cost Minimizations**

In a simple linear regression problem, the least squares regression line is given by its slope $\hat{\beta}_1$ and intercept $\hat{\beta}_0$ which minimize the function:

$$(\beta_0, \beta_1) \hookrightarrow \mathcal{L}_2(\beta_0, \beta_1) = \sum_{i=1}^{n} (y_i - \beta_0 - \beta_1 x_i)^2. \qquad (4.3)$$

This function is quadratic in β_0 and β_1, so it is easily minimized. Indeed, one can compute explicitly (in terms of the values of $x_1, \ldots, x_n$ and $y_1, \ldots, y_n$) the partial derivatives of $\mathcal{L}_2(\beta_0, \beta_1)$ with respect to β_0 and β_1, and setting these derivatives to zero gives a system of two equations with two unknowns (often called the first order conditions in the jargon of optimization theory) which can be solved. The solution is given by the formulae:

$$\hat{\beta}_1 = \frac{\text{cov}(x, y)}{\sigma_x^2} \qquad \text{and} \qquad \hat{\beta}_0 = \overline{y} - \frac{\text{cov}(x, y)}{\sigma_x^2}\overline{x}. \qquad (4.4)$$

Recall that the empirical means $\overline{x}$ and $\overline{y}$ were already defined in (3.10) of Chap. 3, as well as $\text{cov}(x, y)$ and σ_x which were defined in (3.11). This is in contrast with the LAD regression for which we cannot find formulae for the optimal slope and intercept parameters. The least absolute deviations regression line is determined by the values of the intercept $\hat{\beta}_0$ and the slope $\hat{\beta}_1$ which minimize the function:

$$(\beta_0, \beta_1) \hookrightarrow \mathcal{L}_1(\beta_0, \beta_1) = \sum_{i=1}^{n} |y_i - \beta_0 - \beta_1 x_i|. \tag{4.5}$$

Unfortunately, one cannot expand the absolute value as one does for the squares, and solving for vanishing derivatives cannot be done by closed-form formulae in this case. Nevertheless, it is relatively easy to show that the function $(\beta_0, \beta_1) \hookrightarrow \mathcal{L}_1(\beta_0, \beta_1)$ is convex and hence, that it has at least one minimum. But because it is not strictly convex, uniqueness of this minimum is not guaranteed, and as in the classical case of the median, we may have entire intervals of minima. Efficient algorithms exist to compute the minima of $\mathcal{L}_1(\beta_0, \beta_1)$. They are based on a reduction of the problem to a classical linear programming problem. But the lack of uniqueness and the lack of explicit formulae is still a major impediment to widespread use of the L1 method.

4.1.6 Regression as a Minimization Problem

This subsection does not contain any new technical material, and as a consequence, it can be skipped on a first reading. Its prose is intended to shed some light on the importance of optimization in statistical estimation, and to use this opportunity to highlight some of the nagging problems coming with the territory.

The approach to regression which we advocate in this section is based on the idea of *smoothing* of a cloud of points into a curve that captures the main features of the data. In this spirit, a simple regression problem can be formulated in the following way. We start from observations:

$$(x_1, y_1), \ldots\ldots, (x_n, y_n),$$

and we try to find a function $x \hookrightarrow \varphi(x)$ which minimizes a loss or penalty, say $\mathcal{L}(\varphi)$, associated to each specific choice of the candidate φ. This candidate can be picked in a specific class of functions, say $\boldsymbol{\Phi}$, e.g. the set of affine functions of x when we consider linear regression, the set of polynomials when we deal with a polynomial regression problem later in this chapter, ... What distinguishes parametric regression from nonparametric regression is the fact that the class $\boldsymbol{\Phi}$ can be described in terms of a small number of parameters. For example, in the case of simple linear regression, the parameters are usually chosen to be the slope and the intercept, while in the case of polynomial regression the parameters are most often chosen to be the coefficients of the polynomial. See nevertheless Sect. 4.6.3 later in this chapter. We shall use the notation θ for the parameter used to label the candidate (i.e. the set

$\Theta = \{\theta\}$ replaces the set $\boldsymbol{\Phi} = \{\varphi\}$ via a correspondence of the type $\theta \leftrightarrow \varphi_\theta$. In the case of simple linear regression this amounts to setting:

$$\theta = (\beta_0, \beta_1) \qquad \text{and} \qquad \varphi_\theta(x) = \beta_0 + \beta_1 x.$$

Notice that the present discussion is not limited to the case of a single scalar regressor variable. Indeed, the regression variables x_i can be multivariate as when each observed value of x_i is of the form $\boldsymbol{x}_i = (x_{i,1}, \ldots, x_{i,p})$. As explained earlier, the parameter θ is likely to be multivariate. In the discussion above the regression problem always reduces to minimization of a function of the form:

$$\theta \hookrightarrow \mathcal{L}(\theta). \tag{4.6}$$

This problem does not have a clear-cut answer in general. We say that the problem is *well posed* if there exists at least one value, say $\theta_0 \in \Theta$, such that:

$$\mathcal{L}(\theta_0) = \inf_{\theta \in \Theta} \mathcal{L}(\theta). \tag{4.7}$$

When (as is often the case) Θ is a subset of a vector space, then the problem is often well posed when the loss function (4.6) is convex. Moreover, there is a unique value of θ realizing the minimum (4.7) whenever the loss function (4.6) is actually strictly convex.

4.1.6.1 ***The Search for a Minimum***

As we saw in the case of least squares regression, the search for an optimal value of the parameter θ is usually replaced by the search for values of θ at which all the partial derivatives of the function $\mathcal{L}$ vanish. In other words one looks for solutions of the equation:

$$\nabla \mathcal{L}(\theta) = 0 \tag{4.8}$$

where the notation ∇ is used for the gradient, i.e. the vector of partial derivatives. Notice that this equation is in fact a system of k-equations when θ is k-dimensional, since in this case the gradient is made of k partial derivatives and Eq. (4.8) says that all of them should be set to 0. A solution of (4.8) is called a *critical point.* Such a strategy for the search of a minimum is reasonable because, if

$$\theta_0 = \arg \inf_{\theta \in \Theta} \mathcal{L}(\theta)$$

is a point at which the minimum is attained, and if the function $\theta \hookrightarrow \mathcal{L}(\theta)$ is differentiable, then all the partial derivatives of $\mathcal{L}$ vanish for $\theta = \theta_0$. Unfortunately, the converse is not true in the sense that there might be critical points which are not solutions of the minimization problem. Indeed, the gradient of $\mathcal{L}$ vanishes at any local minimum, (or even at any local maximum for that matter) even if it is not a global minimum, and this is a source of serious problems, for there is no good way to find out if a critical point is in fact a global minimum, or even to find out how good or bad a proxy it can be for such a global minimum.

To make matters worse, one is very often incapable of solving Eq. (4.8). Indeed except for the very special case of least squares linear regression (see the previous subsection for a re-cap of all the formulae derived in this case), it is not possible to find closed form formulae for a solution, and one is led to compute numerical approximations by iterative procedures. As is often the case, stability and convergence problems plague such a strategy.

4.2 Regression for Prediction & Sensitivities

Regression is often performed to explain specific features of the existing (i.e. already collected) data. This involves looking at the values of the response variables given by the evaluation of the regression function $\varphi(x)$ for values of the regressor(s) contained in the data set, and searching for an interpretation of the differences (also called residuals) between the observed values y_i and the values $\varphi(x_i)$ provided by the regression function. However, regression is also often used for prediction purposes. The functional relationship between the response and the regressor variables identified by the regression function φ, is taken advantage of to predict what the response would be for values of the regressor(s) for which the responses have not yet been observed. Filling in missing data appears as an intermediate situation between these two uses of regression.

4.2.1 **Prediction**

We give a first informal presentation of the notion of prediction operator. We shall make this concept more precise later in the book. The purpose of a prediction operator is to produce the best *reasonable* guess for a random quantity. It is a good idea to have in mind the least squares error criterion as a measure of how *reasonable* a guess can be. So such a prediction operator, say P, will take a random variable, say Y, into a number $P(Y)$ which serves as its best guess. In the absence of any other information, and if one uses the least squares criterion, the operator P is given by the expectation operator in the sense that $P(Y) = \mathbb{E}\{Y\}$. That is, the number $m = \mathbb{E}\{Y\}$ is the number for which the criterion $\mathbb{E}\{|Y - m|^2\}$ is minimum. Other criteria lead to other prediction operators. It is a very intuitive fact (with which we have been familiar since our first introductory statistics class) that, in the absence of any extra information, the best (in the least squares sense) predictor of a random quantity is its expected value. We shall revisit (and prove) this statement in the next subsection.

We are interested in prediction when partial information is available. Apart from the fact that we will thus work with conditional expectations instead of the usual expectations, nothing should be different. If the information x is available, we shall denote a prediction operator by P_x. As explained above, we should think of $P_x(Y)$ as the best predictor of the random variable Y in the presence of the information x. The following properties are imposed on such an operator:

- $P_x(Y)$ should be a *linear* function of Y
- If Y is not really random, and it is known that $Y = y$ for a given number y, then one should have $P_x(Y) = y$.

Obviously, the expected valued (conditioned on the information x) is a prediction operator which satisfies these properties. But many other operators are possible. Very often, we will also demand that $P_x(Y) = 0$ whenever Y represents a noise term, and this is often modeled as Y having mean zero. For most of the applications discussed in this book, the prediction operators will be given by expectations and conditional expectations.

We will still use the data on the utility indexes for illustration purposes, but we will concentrate for the next little while, on the analysis of the dependence of `UTILITY.index` upon `DUKE.index`. So for the time being, we might as well forget that we also have the variable `ENRON.index` to explain the values of the overall utility index.

For an example of prediction based on regression results, let us imagine that back in January 1994, we looked into our crystal ball, and we discovered that the stock of DUKE energy was going to appreciate in a dramatic fashion over the next 2 years. The DUKE index would increase accordingly and this would presumably imply a significant increase in the utility index as well. But how could we quantify these qualitative statements. To be more specific, could we predict the value of the utility index if we knew that the value of the DUKE index was 150?

Using the results of the regressions performed earlier, one needs only to compute the value of $\beta_0 + \beta_1 x$ for $x = 150$ and β_0 and β_1 determined by each regression. We compute predictions both for the least squares and the least absolute deviations regressions.

```
> NEWDUKE <- 150
> PRED2 <- UDl2$coef[1]+UDl2$coef[2]*NEWDUKE
> PRED2
  121.918
> PRED1 <- UDl1$coef[1]+UDl1$coef[2]*NEWDUKE
> PRED1
  124.5896
```

One can see that the two predictions are different and deciding which one to trust is a touchy business. We cannot resolve this issue at this stage, especially since we cannot talk about confidence intervals yet. Nevertheless, even though we do not have the tools to justify the claims we are about to make, and in order to shed some light on the reasons why the results are different, we do venture the following explanations: The value of the explanatory variable for which we seek a value of the response variable, i.e. 150, is far from the bulk of the data from which the regression lines were constructed. Indeed the values of the explanatory variable ranged between 101 and 128 during the year 1993. Predictions that far from the available data can be very unstable.

4.2.2 Introductory Discussion of Sensitivity and Robustness

Before relying on the results of a regression for decision making, it is a good idea to understand how stable and/or reliable the coefficients of a regression are. In each case (i.e. for the least squares and least absolute deviations regressions) we investigate the sensitivity of the values of the slope and the intercept when variations in the observations are present (or introduced). We shall also illustrate the sensitivity of our predictions.

A (simple) linear regression is said to be significant when the results of the regression confirm the influence of the explanatory variable on the outcome of the response, in other words when the slope is determined to be nonzero. When this is not the case, the regression line is horizontal and the value of the explanatory variable has no effect on the response, the latter being merely explained by the measure of location given by the intercept. We shall introduce the notion of robustness by first discussing this case.

For a set of n real numbers $y_1, \ldots, y_n$, the most common measure of location is the sample mean:

$$\overline{y} = \frac{1}{n}\sum_{i=1}^{n} y_i.$$

As we already pointed out in Sect. 4.1.5, the sample mean $\overline{y}$ is the solution of a least squares minimization problem since:

$$\overline{y} = \arg\min_m \sum_{i=1}^{n} |y_i - m|^2.$$

In other words, the sample mean is to the location problem what the slope and the intercept of the least squares regression line are to the (simple) linear regression problem.

Next to the sample mean, the median also enjoys great popularity. Bearing some annoying non-uniqueness problems when n is even (problems which are solved by agreeing on a specific convention), the median $\hat{\pi}_{0.5}$ is an element of the data set which splits the data into two subsets of approximately the same size. Like the mean, it also appears as the solution of a minimization problem since:

$$\hat{\pi}_{0.5} \in \arg\min_m \sum_{i=1}^{n} |y_i - m|.$$

In other words, the sample median is to the location problem what the slope and the intercept of the least absolute deviations regression line are to the (simple) linear regression problem. We claim that the median is much less sensitive than the mean to perturbations or errors in the data, and we illustrate this fact on the following simple example. Let us consider the data:

$$y_1 = 1.5 \;\; y_2 = 0.7 \;\; y_3 = 5.1, \;\; y_4 = 2.3, \;\; y_5 = 3.4.$$

The mean of this sample is:

$$\overline{y} = (y_1 + y_2 + y_3 + y_4 + y_5)/5 = (1.5 + 0.7 + 5.1 + 2.3 + 3.4)/5 = 13/5 = 2.6$$

while the median is obtained by first ordering the data:

$$y_{(1)} = 0.7 < y_{(2)} = 1.5 < y_{(3)} = 2.3 < y_{(4)} = 3.4 < y_{(5)} = 5.1$$

and then picking the *mid-value* $\hat{\pi}_{0.5} = 2.3$. Let us now assume the value $y_5 = 3.4$ is erroneously recorded as $y_5 = 13.4$. In this case the new mean is increased by 2 since:

$$\overline{y} = 1.5 + 0.7 + 5.1 + 2.3 + 13.4)/5 = 23/5 = 4.6$$

while the *value of the median does not change!* In fact the median will not change as long as the changes do not affect the number of values greater than the median. This feature is extremely useful in the case of undesirable erroneous measurements, and/or uncontrollable perturbations of the data. Indeed, it is plain to see that the mean can be made arbitrarily large or small by appropriately changing a single measurement! With this introductory example in mind, we compare the robustness of the least squares regression to the least absolute deviations regression.

4.2.3 **Comparing L2 and L1 Regressions**

As in the case of the discussion of the robustness of the measures of location (mean and median) we try to quantify (or at least visualize) the effects that perturbations of the data have on the regression lines.

In order to do so, we create a new data set, say `NEWUTIL.index`, by modifying the first 20 entries of `UTILITY.index`. We then perform simple least squares and least absolute deviations regressions of this new index against the DUKE index, and we compare the resulting regression lines to the lines obtained with the original data. The results are given in Fig. 4.5. They were produced using the following commands:

```
> GOOD <- 21:260
> NEWUTIL.index <- c(UTILITY.index[-GOOD]-10,
                          UTILITY.index[GOOD])
> NUDl2 <- lsfit(DUKE.index,NEWUTIL.index)
> NUDl1 <- l1fit(DUKE.index,NEWUTIL.index)
> par(mfrow=c(1,2))
> plot(DUKE.index,NEWUTIL.index)
> abline(UDl2,lty=4)
> abline(NUDl2)
> plot(DUKE.index,NEWUTIL.index)
> abline(UDl1,lty=4)
> abline(NUDl1)
> par(mfrow=c(1,1))
```

The first command defines the row numbers for which we are not going to change the value of `UTILITY.index`. The second command actually creates the new utility index by concatenating two vectors with the `R` function `c`. The first vector is formed by the first 20 entries of `UTILITY.index` from which we subtract 10, while the second vector is formed by the remaining entries of `UTILITY.index` which we leave untouched. Notice how we used the subscript `-GOOD` to extract the entries whose row numbers are not in `GOOD`. The next two commands perform the least squares and the least absolute deviations regressions of the new utility index against the old DUKE index, and the remaining commands produce the plots of Fig. 4.5. We used the parameter `lty` to draw the original L2 and L1 regressions as dash lines.

The differences between the two regression methods are clearly illustrated on these plots. The L2 line changed dramatically while the L1 line remained the same. As in the case of the median, changing a small number of values did not affect the result of the L1 fitting/estimation procedure. This robustness of the least absolute deviations regression can be extremely useful. Indeed, there are times when one wants the estimations and predictions to change with changing data, however, with noisy data, it is generally not a good idea to use estimation and prediction procedures which are too sensitive to small changes, mostly because the latter are very likely due to the noise, and for this reason, they should not have an overly dramatic impact on the outcome.

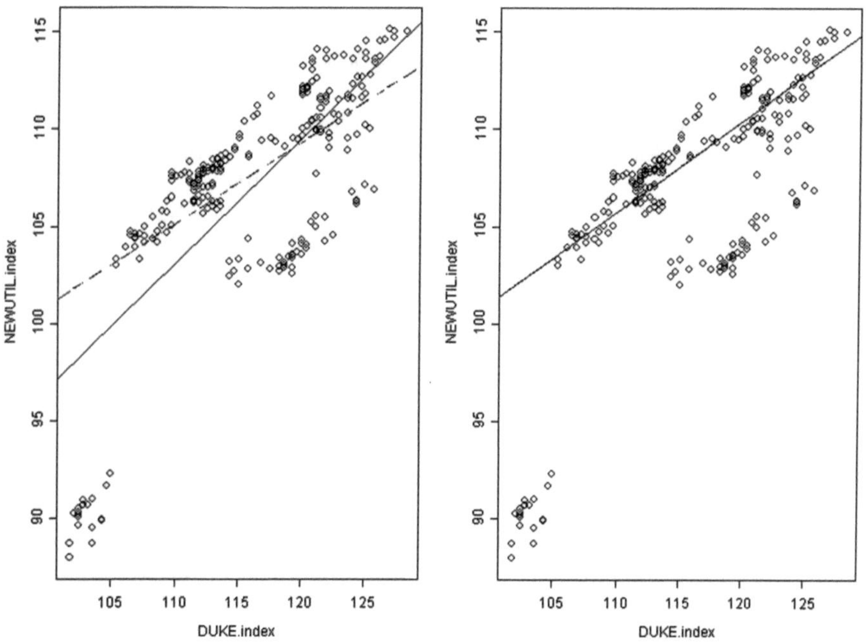

Fig. 4.5. *Left*: simple least squares regression of the modified utility index against the DUKE index, with the original least squares regression line superimposed as a *dashed line*. The two lines are very different. *Right*: same thing for the least absolute deviations regressions. The two lines are identical!

We can also compare the effects of data perturbations on the performance of the predictions produced by the regressions. If we revisit the prediction of the value of the utility index when the DUKE index reaches the value 150, we now find:

```
> NEWDUKE <- 150
> NPRED2 <- NUD12$coef[1]+NUD12$coef[2]*NEWDUKE
> NPRED2
    128.8155
> NPRED1 <- NUD11$coef[1]+NUD11$coef[2]*NEWDUKE
> NPRED1
    124.5896
```

from which we see that the L1 prediction does not change while the L2 prediction goes from 121.9 to 128.8. As we already explained, it is difficult to assess how bad such a fluctuation is, but at a very intuitive level, it may be comforting to see that some prediction systems do not jump all over the place when a few data points change! More on that later when we discuss statistical inference issues.

We now proceed to illustrate the differences between L1 and L2 regression methods with the coffee data. See also the Notes & Complements at the end of the chapter for references to textbooks devoted to the important aspects of robustness in statistical inference.

4.2.4 **Taking Another Look at the Coffee Data**

We revisit the case of the coffee daily price data already considered in the previous chapters. A scatterplot of the daily log-returns of the two commodities shows that there is a strong dependence between the two, and that a linear relation may hold. We compute the least squares and the least absolute deviations regressions of the Brazilian coffee daily log-returns against the Colombian ones and plot the regression lines so-obtained.

```
> plot(CLRet,BLRet)
> BCL2 <- lsfit(CLRet,BLRet)
> BCL1 <- l1fit(CLRet,BLRet)
> abline(BCL2)
> abline(BCL1,lty=3)
```

We give the results in Fig. 4.6. The least absolute deviations regression captures the upward trend of the cloud of points more faithfully than the least squares regression. The few low values of the Brazilian coffee seen on days the Colombian coffee had a large return (i.e. the few points on the middle right part of the scatterplot) are influencing the results of the least squares regression, pulling the regression line down. On the other hand, these points do not seem to have much influence on the least absolute deviations regression line. We shall explain this extra sensitivity of the least squares regression and the robustness of the least absolute deviations regression later when we revisit this issue in the context of the statistical distribution theories associated with these two different types of regression, and when we bring into the picture the tails of the distributions of the variables involved in the regression.

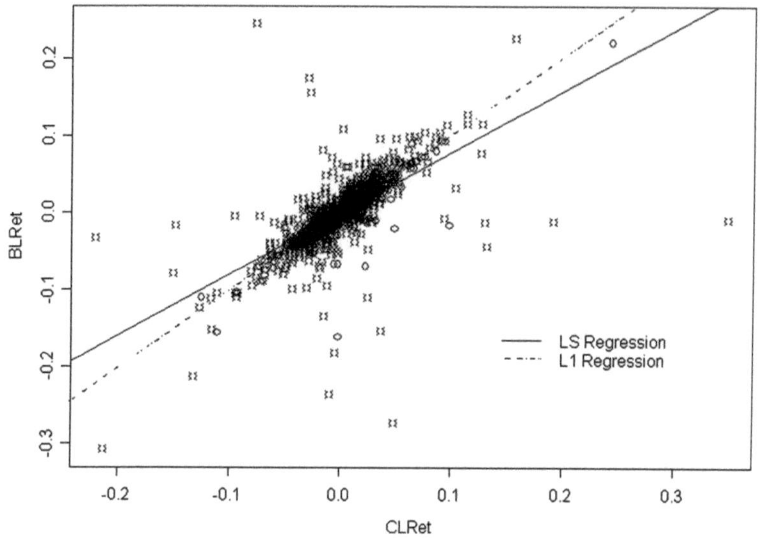

Fig. 4.6. Least squares and least absolute deviations regressions of the Brazilian daily log-returns against the Colombian daily log-returns

Very Important Remark. It is crucial to understand the differences between the regressions performed on the utility indexes and the regressions of this subsection. The latter are computed from log-returns, and the features of these transformed data are very different from the statistical features of the raw price data. Indeed, these raw prices show strong dependencies among themselves, say from day to day, and regression with this type of data is a very *touchy business*. We warn the reader against the pitfalls of this trade. These regressions will be called time series regressions later. This important remark will be revisited in Sect. 4.3.2 below.

4.3 Smoothing Versus Distribution Theory

So far, our approach to regression was based on a *smoothing* philosophy, and our discussion did not rely upon any statistical assumption or principle. For this reason, it belongs more in a course on computer graphics, or on function approximation, than a statistics course. To remedy this, we assume the existence of a statistical model for the roughness and/or uncertainty in the data. More precisely we assume that the values $x_1, \dots, x_n$ and $y_1, \dots, y_n$ are observed realizations of n random variables $X_1, \dots, X_n$ and $Y_1, \dots, Y_n$. The main assumption of a regression model is not about the overall distributions of the response variables $Y_1, \dots, Y_n$, but rather, their conditional distributions with respect to the explanatory variables $X_1, \dots, X_n$, hence their names.

4.3.1 Regression and Conditional Expectation

Since it happens very often in practice that the observations $x_1, \ldots, x_n$ are in fact realizations of random variables $X_1, \ldots, X_n$, it is more convenient to consider the observations (x_i, y_i) as realizations of couples (X_i, Y_i) of random variables having the same joint distribution. The conditional expectation of any of the random variables Y_i given the value of the corresponding X_i becomes a deterministic function of X_i. In other words, if one knows that the random variable X_i has the value x, then the conditional expectation of Y_i given $X_i = x$ is a function of x which we could denote by $\varphi(x)$. Notice that this function of x does not depend upon the index i, since we assume that the joint distributions of all the couples (X_i, Y_i) are the same. This function is called the *regression function* of Y against X, and the values $y_1, \ldots, y_n$ can be viewed as (noisy) observations of the (expected) values $\varphi(x_1), \ldots, \varphi(x_n)$.

It is reasonable to expect the function φ to be nonlinear in general. Nevertheless, when the joint distribution of the (X_i, Y_i) is Gaussian, the function φ is linear (affine, to be precise) and linear regression is fully justified in this case.

As we have already seen, the goal of regression analysis is to quantify the way the values of the response variable are influenced by the values of the explanatory variables (if there is any influence at all). In other words, given the values $x_1, \ldots, x_n$ of the explanatory (random) variables, we assume that the means of the respective response variables are of the form $\varphi(x_1), \ldots, \varphi(x_n)$. More precisely, we assume:

$$\mathbb{E}\{Y_i | X_i = x_i\} = \varphi(x_i), \qquad i = 1, \ldots, n.$$

Still another form of the same assumption is to postulate that for $i = 1, \ldots, n$, given that $X_i = x_i$, we have:

$$Y_i = \varphi(x_i) + \epsilon_i, \tag{4.9}$$

for some mean-zero random variable ϵ_i. We shall further assume that all the random variables ϵ_i have the same variance, say σ^2, and that they are uncorrelated. As we have seen in the R tutorial, such a sequence $\{\epsilon_i\}_{i=1,\ldots,n}$ is called a white noise. In fact, more than merely assuming that the ϵ_i's are uncorrelated, we shall assume most of the time that they are *independent.* Recall that

- Choosing the number of explanatory variables determines if the regression is called a *simple regression* or a *multiple regression*;
- Choosing if the function φ is to be restricted to a limited class of functions which are determined by the choice of a small number of parameters as opposed to be allowed to be of a general type, determines if the regression is called a *parametric regression* or a *nonparametric regression*;
- In the case of parametric regression, deciding whether or not the function φ should be linear determines if the regression is called a *linear regression*.

Once these choices have been made, choosing the distribution of the *noise terms* ϵ_i completely determines the statistical model and makes statistical inference possible.

As before, we shall use the comparison of the least squares (simple linear) regression to the least absolute deviations (simple linear) regression as an illustrative example, but one should keep in mind that the conclusions of this discussion are not restricted to these two particular regressions.

4.3.2 Maximum Likelihood Approach

Specifying the common distribution of the noise terms ϵ_i makes it possible to write down the joint distribution of the response variables $Y_1, \ldots, Y_n$. Recall that we assume that the ϵ_i are independent and that consequently, the response variables $Y_1, \ldots, Y_n$ are, conditionally on the knowledge of the values $X_i = x_i$ of the explanatory variables, independent, with means $\varphi(x_i)$, and with the same variance σ^2 as the ϵ_i's. Given the fact that, in the statistical jargon, the likelihood function of the model, say $\mathcal{L}(y_1, \ldots, y_n)$, is nothing but the joint density $f_{(Y_1,\ldots,Y_n)}(y_1, \ldots, y_n)$ of the observed responses $Y_1, \ldots, Y_n$, specifying the distribution of the noise terms ϵ_i determines the likelihood function of the model. Notice that the likelihood function also depends upon the values of the means $\varphi(x_i)$ and the variance σ^2. These dependencies will be emphasized or de-emphasized depending on the goal of the computation.

Next, we re-derive the least squares and the least absolute deviations regression procedures as instances of the general maximum likelihood approach.

4.3.2.1 *Simple Least Squares Regression Revisited*

In this subsection we assume that the noise terms ϵ_i are independent and normally distributed:

$$\epsilon_i \sim N(0, \sigma^2)$$

where the common variance σ^2 is assumed to be unknown. In this case we have:

$$\begin{aligned}
\mathcal{L}_2(\beta_0, \beta_1, \sigma^2) &= \mathcal{L}(\beta_0, \beta_1, \sigma^2, y_1, \ldots, y_n) \\
&= f_{(Y_1,\ldots,Y_n)}(y_1, \ldots, y_n) \\
&= f_{Y_1}(y_1) \cdots f_{Y_n}(y_n) \\
&= \frac{1}{\sqrt{2\pi\sigma^2}} e^{-[y_1-(\beta_0+\beta_1 x_1)]^2/2\sigma^2} \cdots \frac{1}{\sqrt{2\pi\sigma^2}} e^{-[y_n-(\beta_0+\beta_1 x_n)]^2/2\sigma^2} \\
&= \frac{1}{\sqrt{2\pi}^n} \sigma^{-n} e^{-\frac{1}{2\sigma^2}\sum_{i=1}^n [y_i-(\beta_0+\beta_1 x_i)]^2}
\end{aligned}$$

where we used the independence of the observation variables Y_i to claim that the joint density of the Y_i's was the product of the individual densities, and we used the specific form of the Gaussian density. From this expression it is plain to see (even without computing the partial derivatives of the objective function) that maximizing the likelihood of the observations $y_1, \ldots, y_n$ given the values $x_1, \ldots, x_n$ of the explanatory variables, is equivalent to minimizing the sum of square errors. So the slope $\hat{\beta}_1$ and the intercept $\hat{\beta}_0$ of the least squares regression line appear as the maximum likelihood estimators of the parameters β_1 and β_0 of the model. As a consequence, results from the inference theory of normal families can be applied. In particular one obtains:

- Confidence intervals for the true slope (from which we can test if the slope is zero or not, and in so doing, assess the significance of the regression);
- Confidence intervals for the true intercept;
- Joint confidence regions (typically ellipsoids) for the couple (β_0, β_1) of parameters;
- Maximum likelihood and unbiased estimates of the noise variance σ^2 and corresponding confidence intervals;
- Coefficient of determination R^2 giving the proportion of the variation explained by the regression,

and much more. All the statistical properties of normal families can be used for inferential purposes. This abundance of tools is due to the fact that we assumed that the noise terms ϵ_i were mean zero, i.i.d. and normally distributed. Some of the statistical diagnostics mentioned above can be obtained from the output of the function `ls.diag` which takes as argument any object created with the function `lsfit`. We shall see more of the statistical inference tools provided by `R` when we discuss linear models and the function `lm`.

4.3.2.2 *Simple Least Absolute Deviations Regression Revisited*

We now assume that the noise terms ϵ_i have a double exponential distribution (also called Laplace distribution) given by the density:

$$f_{\epsilon_i}(x) = \frac{1}{2\lambda} e^{-|x|/\lambda}$$

for some variance-like scale parameter $\lambda > 0$ which is assumed to be unknown. A computation similar to the computation done earlier in the Gaussian case gives:

$$\begin{aligned}
\mathcal{L}_1(\beta_0, \beta_1, \lambda) &= \mathcal{L}(\beta_0, \beta_1, \lambda, y_1, \ldots, y_n) \\
&= f_{(Y_1, \ldots, Y_n)}(y_1, \ldots, y_n) \\
&= f_{Y_1}(y_1) \cdots f_{Y_n}(y_n) \\
&= \frac{1}{2\lambda} e^{-(1/\lambda)|y_1 - (\beta_0 + \beta_1 x_1)|} \cdots \frac{1}{2\lambda} e^{-(1/\lambda)|y_n - (\beta_0 + \beta_1 x_n)|} \\
&= \frac{1}{2^n} \lambda^{-n} e^{-(1/\lambda) \sum_{i=1}^n |y_i - (\beta_0 + \beta_1 x_i)|}
\end{aligned}$$

where as before, we used the independence of the variables Y_i to claim that the joint density of the Y_i's was the product of the individual densities, and the specific form of the density of the double exponential distribution. From this expression we see that, maximizing the likelihood of the observations $y_1, \ldots, y_n$ given the values $x_1, \ldots, x_n$ of the explanatory variables, is equivalent to minimizing the sum of absolute deviations. This implies that the slope $\hat{\beta}_1$ and the intercept $\hat{\beta}_0$ of the least absolute deviations regression line appear as the maximum likelihood estimators of the parameters β_1 and β_0 of the model in which the noise terms have a double exponential distribution.

Unfortunately, the usefulness of this result is limited. Indeed, there are no closed formulae for the estimators, and double exponential families do not have a distribution theory as developed as that of the normal families. As a consequence, we need to use approximations each time we need a confidence interval, a test, ... These approximations are usually derived theoretically from asymptotic results or from Monte Carlo simulations. The latter can be computer intensive and in any case, it is difficult if not impossible to control the extend of the errors produced by the approximations.

Nevertheless, this result sheds light on the robustness of the least absolute deviations regression as compared to the least squares regression. Indeed, the Laplace distribution of the noise has thicker tails (since for large values of $|x|$, the exponential of a negative multiple of $|x|$ is significantly larger than the exponential of a negative multiple of x^2) and consequently, the model allows for larger values of the error terms ϵ_i. In other words, the LAD regression will be more tolerant of points far away from the regression curve $\varphi(x)$, while the LS regression will try harder to get the regression curve $\varphi(x)$ to be closer to these points.

4.3.2.3 ***When Can or Should We Perform a Regression?***

In this subsection, we elaborate on the Very Important Remark at the end of Sect. 4.2. In order to allow for statistical inference, the (simple) regression set-up requires that the data $(x_1, y_1), \ldots, (x_n, y_n)$ form a sample of observations from identically distributed random couples $(X_1, Y_1), \ldots, (X_n, Y_n)$ having the same joint distribution. To be more specific, the statistical models made explicit in this section in order to derive expressions for the likelihood function are based on the following premises:

> *the residuals $r_i = y_i - \varphi(x_i)$ given by the differences between the observations y_i of the responses and the values $\varphi(x_i)$ of the regression function at the observed explanatory variables, are realizations of independent identically distributed random variables.*

The top pane of Fig. 4.7 gives the sequential plots of the residuals of the least squares regression of `ENRON.index` against `DUKE.index` while the bottom pane gives the sequential plots of the residuals of the least squares regression of `BLRet` against `CLRet`. These sequential plots were produced with the commands:

```
> plot(EDL2$residuals, type="l")
> plot(BCL2$residuals, type="l")
```

Both residual sequences have mean zero by construction. However, our experience with i.i.d. sequences of mean zero random variables (recall the Introductory Session to R reproduced in Appendix) tells us that if the bottom plot could be accepted as the plot of a white noise (this is the terminology we use for i.i.d. sequences of mean zero random variables), this is certainly not the case for the plot on the top! In fact, if we were to compute the auto-correlation function (acf for short) of the utility index residuals, we would see strong correlations which are inconsistent with the assumptions of the regression set-up. At this stage, the reader should bare with us since we are slightly ahead of ourselves. Indeed, the notion of acf will be introduced and explained in Chap. 6, while regression diagnostics will be presented later in this chapter.

The statistical model (4.9) is based on a sequence $(\epsilon_i)_i$ of uncorrelated noise terms. Since the residuals can be considered as proxies for these noise terms, we could expect them to be uncorrelated as well. We shall see later in this chapter that this is not the case, and we shall compute explicitly the Pearson correlation coefficient between residuals. While the pattern of the residuals of the coffee log-returns regression appears to be normal (though one can suspect the presence of heavy tails), the serial dependence of the residuals of the top pane of Fig. 4.7 is not consistent with the premises of the statistical set-up of a regression problem as restated above. Not only does the distribution of the couple of random variables `DUKE.index` and `ENRON.index` change over time, but the couples actually observed day after day are not independent. These two facts were part of the reasons which pushed us into studying the log-returns of the coffee data instead of the original raw data of the

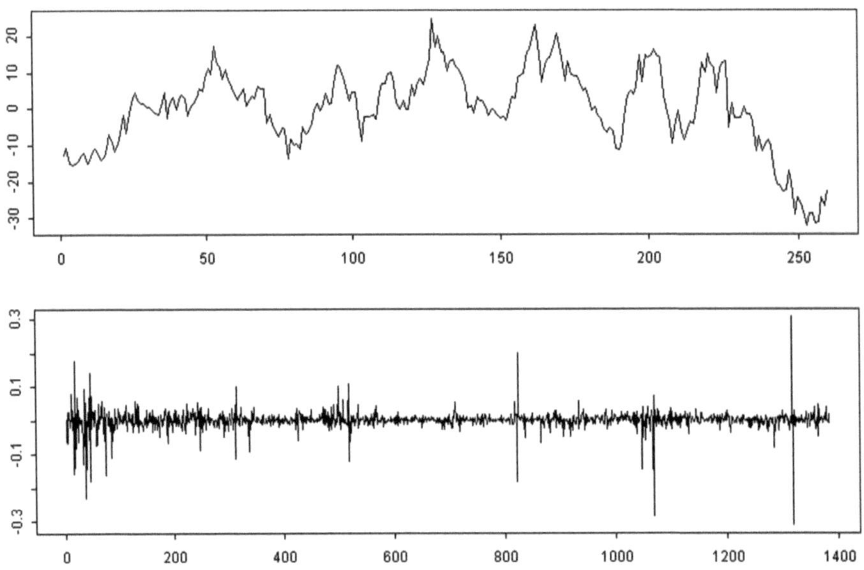

Fig. 4.7. Sequential plot of the residuals of the simple least squares regression of ENRON's index against DUKE's index (*top*) and of the Brazilian coffee daily log-returns against the Colombian coffee daily log-returns (*bottom*)

index values. So as announced earlier, we stop using the utility data in its original form. From now on, for the purposes of regression analysis of the energy indexes, we replace the actual indexes by their log-returns.

```
> UtilLRet <- diff(log(UTILITY.index))
> EnronLRet <- diff(log(ENRON.index))
> DukeLRet <- diff(log(DUKE.index))
```

Recall that as explained earlier, the difference can be undone by a cumulative sum, and the logarithm can be inverted with an exponential function. In this way, conclusions, estimations, predictions, etc., concerning the log-returns can be re-interpreted as conclusions, estimations, predictions, etc. for the original index values. But before we switch gear and embark on the multiple regression journey, it is important to summarize the main differences between the two regression procedures we discussed so far.

4.3.2.4 ***When Should We Use Least Absolute Deviations?***

Advantages of the least squares regression:

- Existence and uniqueness of the estimators $\hat{\beta}_0$ and $\hat{\beta}_1$;
- Existence of explicit formulae for the estimators $\hat{\beta}_0$ and $\hat{\beta}_1$;
- Fast and reliable computations;
- Existence of a distribution theory leading to convenient statistical inferences, e.g. exact confidence intervals, tests.

Drawbacks of the least squares regression:

- Sensitivity to outliers and extreme observations.

Advantages of the least absolute deviations regression:

- Existence of the estimators $\hat{\beta}_0$ and $\hat{\beta}_1$ and reasonable computational algorithms;
- Extreme robustness to one type of outlier.

Drawbacks of the least absolute deviations regression:

- Lack of uniqueness of the estimators $\hat{\beta}_0$ and $\hat{\beta}_1$;
- Lack of a convenient distribution theory: the estimators, the tests, the confidence intervals, etc., have too often to be computed by lengthy Monte Carlo methods.

So to summarize:

- If the complexity of the computations and/or the computing time is an issue one may want to use least squares regression;
- On the other hand if robustness is important then it is likely that the least absolute deviations regression will give more satisfactory results.

4.4 Multiple Regression

Back to the analysis of our utility data. If our goal is to explain the overall utility index `UTILITY.index` using all the information at our disposal, we may want to use both individual indexes, namely `ENRON.index` and `DUKE.index`, together in the same formula. As explained earlier, from now on we work with the log-returns instead of the original time series data, and restricting ourselves to linear – affine to be more specific – functions, we may seek a relationship of the form

$$\texttt{UtilLRet} = \beta_0 + \beta_1 * \texttt{EnronLRet} + \beta_2 * \texttt{DukeLRet} + \texttt{noise}.$$

This is the epitome of a multiple linear regression model.

4.4.1 **Notation**

As before, we denote by n the sample size and we assume that the observations come as n pairs

$$(\boldsymbol{x}_1, y_1), \ldots\ldots, (\boldsymbol{x}_n, y_n),$$

where the last component y_i is the response variable whose value we try to explain from the values of the explanatory variables, i.e. the components of $\boldsymbol{x}_i$. The main difference is that we now allow the explanatory variable $\boldsymbol{x}_i$ to be a vector of p different scalar explanatory variables $x_{i,1}, \ldots, x_{i,p}$. In the case of the utility data, the response variable should be $y = \texttt{UtilLRet}$, and the bivariate explanatory variable (so $p = 2$) is now $\boldsymbol{x} = (\texttt{EnronLRet}, \texttt{DukeLRet})$.

As in the case of simple linear regression, the `R` functions `lsfit` and `l1fit` can still be used to perform the regression in the sense of least squares and in the sense of least absolute deviations respectively. One simply needs to bind the p explanatory "column" variables into an $n \times p$ matrix with the command `cbind`. For example, the following commands:

```
> UtilLRetS <- cbind(EnronLRet,DukeLRet)
> UEDls <- lsfit(UtilLRetS,UtilLRet)
> UEDl1 <- l1fit(UtilLRetS,UtilLRet)
```

perform the least squares and the least absolute deviations linear regression of the utility index log-return against the ENRON and DUKE log-returns. As before, one can argue that these two regressions are the results of maximum likelihood estimations of the coefficients of the regression "planes" when the noise terms are assumed to be normally and double-exponentially distributed, respectively. Again, inferential tools are not as developed in the case of the least absolute deviations regression, and for this reason, we shall mostly concentrate on the least squares method. See nevertheless the Notes & Complements at the end of this chapter for further information. Although diagnostic tools have been added to the function `lsfit`, statistical inference is best done with the powerful function `lm` which we now introduce.

4.4.2 The R Function `lm`

As explained in our comparison of the least squares and least absolute deviations regressions, linear models come with a distribution theory providing exact tests of significance, confidence intervals, ..., when the error terms are assumed to be independent and identically distributed (i.i.d. for short) and normally distributed. Because of its weak distribution theory, we shall momentarily refrain from using the least absolute deviations regression, and even though statistical inference can be performed with the function `lsfit`, we shall start using the more powerful function `lm` provided by R. The regression objects produced by `lm` are the result of least squares regression, and the statistical estimates, confidence intervals, p-values, ... are based on the assumption that the error terms ϵ_i are i.i.d. $N(0, \sigma^2)$, for some unknown $\sigma > 0$. The command `lm` gives an implementation of the most general linear models defined below in Sect. 4.5, but it is still worth using it even in the restrictive framework of plain linear regression.

1. We shall give many examples of uses of the method `lm` and the reader is invited to carefully read the help file. Resuming the analysis of the utility data, we first perform the least squares regression of the utility index daily log-return against the ENRON daily log-return. This could have been done with the command:

   ```
   > UeL2 <- lsfit(EnronLRet,UtilLRet)
   ```

   but we do it now with the command:

   ```
   > Ue <- lm(UtilLRet ~ EnronLRet)
   ```

   Notice that the argument of the function `lm` is a formula stating that the variable `UtilLRet` which appears on the left of the tilde, has to be expressed as a function of the variable `EnronLRet` which appears on its right. The function `lm` has an optional argument which can be set by the parameter `data`. It is very useful when the variables are columns in a data frame. It gives the name of the data frame containing the variables. The object produced by such a command is of class `lm`. Contained in the object `Ue` which we just created, we find the estimated slope 0.0784, and the estimated intercept 0.0001, values which could have been obtained with the function `lsfit`. The numerical results extracted from the `lm` object `Ue` by the command `summary(Ue)` are very detailed. They contain the coefficients of the simple linear model:

   $$\texttt{UtilLRet} = 0.0001 + 0.0784 \times \texttt{EnronLRet} + \epsilon,$$

   but also extra information on the model. In particular, they include estimates of the variances of the slope and the intercept (which can be used to compute confidence intervals for these parameters), and a p-value for tests that the parameters are significantly different from 0 (i.e. tests of significance of the regression). We discuss the well known R^2 coefficient in the next subsection.
2. Similarly, instead of using the command:

   ```
   > UdL2 <- lsfit(DukeLRet,UtilLRet)
   ```

one now use the R command:

```
> Ud <- lm(UtilLRet ~ DukeLRet)
```

in order to perform the simple least squares linear regression of UtilLRet against DukeLRet. In this case the linear model is:

$$\texttt{UilLRet} = -0.0001 + 0.5090 \times \texttt{DukeLRet} + \epsilon$$

as we can see by reading the values of estimated slope and estimated intercept from the summary of the lm object Ud.

3. To perform the multiple least squares linear regression of UtilLRet against the variables EnronLRet and DukeLRet together, we use the function lm in the following way:

```
> Ued <- lm(UtilLRet~EnronLRet+DukeLRet)
```

The plus sign "+" in the above formula should not be understood as a sum, but rather as a way to get the regression to use both variables EnronLRet and DukeLRet. The estimated coefficients can be read off the summary of the lm object Ued. The fitted model is now:

$$\texttt{UtilLRet} = -0.0001 + 0.0305 \times \texttt{EnronLRet} + 0.4964 \times \texttt{DukeLRet} + \epsilon.$$

4.4.3 R^2 as a Regression Diagnostic

We will now discuss the well known R^2 coefficient which is used to quantify the quality of a regression. This number gives the proportion of the variance explained by the regression. In the case of a simple least squares regression, it is defined by the formula:

$$R^2 = 1 - \frac{SSE}{TSS} \tag{4.10}$$

where

$$SSE = \sum_{i=1}^{n}(y_i - \hat{y}_i)^2 = \sum_{i=1}^{n}(y_i - \hat{\beta}_0 - \hat{\beta}_1 x_i)^2 \tag{4.11}$$

is the sum of squared residuals from the regression, called the sum of squared errors, and where:

$$TSS = \sum_{i=1}^{n}(y_i - \overline{y})^2 \tag{4.12}$$

represents the total sum of squares. Notice that SSE is always smaller than TSS – hence their ratio is always between 0 and 1 – since SSE is the smallest sum of squared errors when we try all the possible lines while TSS is the smallest sum of squared errors when we only try horizontal lines!

The closer R^2 to 1, the better the regression. R^2 is often called the coefficient of determination of the regression. It is not difficult to imagine possible generalizations of this coefficient to the case of least absolute deviations regression. We shall consider some of these generalizations in the problems at the end of the chapter.

In the three least squares regressions performed earlier, the values of the R^2 coefficients were

- 0.0606 in the case of `UtilLRet` against `EnronLRet`;
- 0.5964 in the case of `UtilLRet` against `DukeLRet`;
- 0.6052 in the case of `UtilLRet` against both variables `EnronLRet` and `DukeLRet`.

By comparing the R^2 values in the cases of the first and second bullets, one could conclude that the variable `EnronLRet` is a far worse predictor than `DukeLRet` when it comes to explaining the overall sector log-returns `UtilLRet`. Indeed, the former has a much smaller value of R^2 (0.0606) than the latter (0.5964). In our zealous attempt to show that the raw index data were inappropriate for direct regression analysis, we departed from our tradition, and we performed the regression analysis of the log-returns without plotting the data first. It is time to change that. The scatterplots given in Fig. 4.8 will help us understand the values of the R^2. The striking differences between the scatterplots explain the difference between the two R^2 scores. Indeed, a quick look at the scatterplot in the left pane of Fig. 4.8 shows that there does not seem to be any functional relation between the values of the utility index log-returns and ENRON's log-returns. On the other hand, the scatterplot contained in the right pane of the figure where the dependence of `UtilLRet` upon `DukeLRet` is visualized, shows that the data is reasonably well suited for a linear regression.

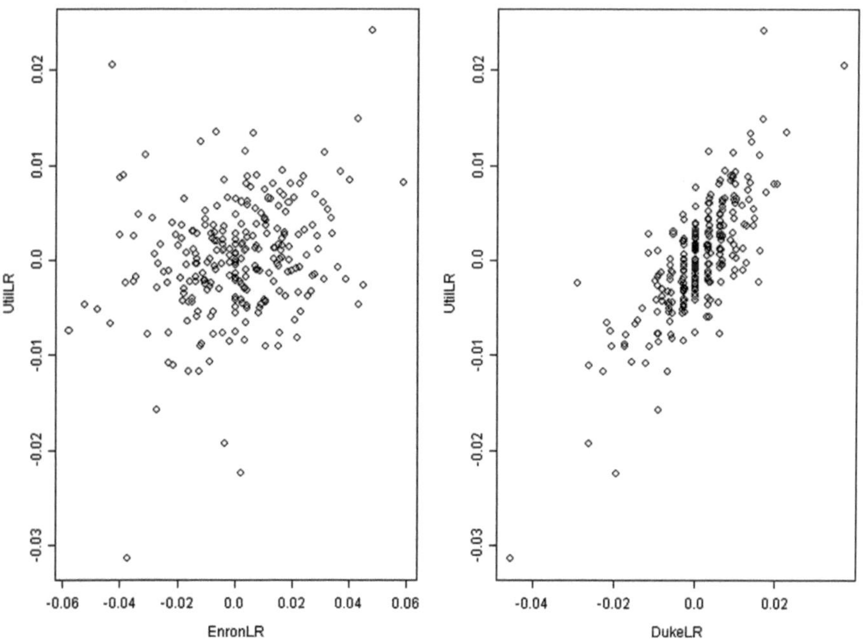

Fig. 4.8. Scatterplot of the utility index daily log-returns against ENRON's daily log-returns (*left*) and DUKE's daily log-returns (*right*)

However, the most interesting remark prompted by the numerical values of R^2 is the following. Despite the fact that, as expected, the R^2 value of 0.6052 obtained with multiple regression is larger than the values of R^2 obtained with simple regressions, the improvement does not seem to be significant. It seems that adding an extra explanatory variable does not help much with explaining the response. This is presumably due to the fact that when it comes to explaining the fluctuations of `UtilLRet`, the information carried by `EnronLRet` is most likely already contained in `DukeLRet` for the most part. We shall discuss this issue in more details later.

4.5 Matrix Formulation and Linear Models

We will now expand on the notion of linear model alluded to in our discussion of the `R` function `lm`. The statistical formulation of the simple linear regression problem is based on the model:

$$y_i = \beta_0 + \beta_1 x_i + \epsilon_i, \qquad i = 1, \ldots, n,$$

which can be rewritten in matrix notation as:

$$\mathbf{Y} = \mathbf{X}\boldsymbol{\beta} + \boldsymbol{\epsilon} \tag{4.13}$$

provided we set:

$$\mathbf{Y} = \begin{bmatrix} y_1 \\ \vdots \\ y_n \end{bmatrix}, \quad \mathbf{X} = \begin{bmatrix} 1 & x_1 \\ \vdots & \vdots \\ 1 & x_n \end{bmatrix}, \quad \boldsymbol{\epsilon} = \begin{bmatrix} \epsilon_1 \\ \vdots \\ \epsilon_n \end{bmatrix}, \quad \text{and} \quad \boldsymbol{\beta} = \begin{bmatrix} \beta_0 \\ \beta_1 \end{bmatrix}.$$

Similarly, the multiple linear regression model:

$$y_i = \beta_0 + \beta_1 x_{i,1} + \cdots + \beta_p x_{i,p} + \epsilon_i, \qquad i = 1, \ldots, n$$

can be recast in the same matrix formulation (4.13) provided we set:

$$\mathbf{Y} = \begin{bmatrix} y_1 \\ \vdots \\ y_n \end{bmatrix}, \quad \mathbf{X} = \begin{bmatrix} 1 & x_{1,1} & \cdots & x_{1,p} \\ \vdots & \vdots & & \vdots \\ 1 & x_{n,1} & \cdots & x_{n,p} \end{bmatrix}, \quad \boldsymbol{\epsilon} = \begin{bmatrix} \epsilon_1 \\ \vdots \\ \epsilon_n \end{bmatrix}, \quad \text{and} \quad \boldsymbol{\beta} = \begin{bmatrix} \beta_0 \\ \beta_1 \\ \vdots \\ \beta_p \end{bmatrix}.$$

4.5.1 **Linear Models**

We promote the notation introduced above to the rank of definition by stating that we have a *linear model* when:

$$\mathbf{Y} = \mathbf{X}\beta + \epsilon \tag{4.14}$$

where:

- $\mathbf{Y}$ is a (random) vector of dimension n;
- $\mathbf{X}$ is an $n \times (p+1)$ matrix (called the *design matrix*);
- $\boldsymbol{\beta}$ is an $(p+1)$ – dimensional vector of parameters;
- ϵ is a mean zero random vector of uncorrelated errors with common (unknown) variance σ^2.

From now on, we will assume that the design matrix $\mathbf{X}$ has full rank. Since in all practical applications we have more observations than explanatory variables (i.e. $n > p$) this assumption merely says that the rank of X is $p+1$.

Normal Linear Models: When the individual error terms ϵ_i are assumed to be jointly normal we have:

$$\epsilon \sim N_n(0, \sigma^2 \mathbf{I}_n)$$

and consequently:

$$\mathbf{Y} \sim N_n(\mathbf{X}\boldsymbol{\beta}, \sigma^2 \mathbf{I}_n).$$

Since the noise term of a linear model is of mean 0, one has:

$$\mathbb{E}\{\mathbf{Y}\} = \mathbf{X}\boldsymbol{\beta}. \tag{4.15}$$

since $\mathbb{E}\{\epsilon\} = 0$. Notice also that, since all the components ϵ_i are assumed to have the same variance σ^2, the variance/covariance matrix of ϵ is in fact σ^2 times the $n \times n$ identity matrix $\mathbf{I}_n$. Finally, since changing the expectation of a random variable does not change its variance, we conclude that the variance/covariance matrix of the observation vector $\mathbf{Y}$ is also $\sigma^2 \mathbf{I}_n$. Hence:

$$\boldsymbol{\Sigma}_{\mathbf{Y}} = \boldsymbol{\Sigma}_{\epsilon} = \sigma^2 \mathbf{I}_n. \tag{4.16}$$

4.5.2 Least Squares (Linear) Regression Revisited

The purpose of least squares linear regression is to find the value of the $(p+1)$-dimensional vector β of parameters minimizing the loss function:

$$\begin{aligned}
\mathcal{L}_2(\beta) &= \sum_{i=1}^{n} [y_i - \beta_0 - \beta_1 x_{i,1} - \cdots - \beta_p x_{i,p}]^2 \\
&= \sum_{i=1}^{n} [y_i - {}^t\mathbf{X}_i \boldsymbol{\beta}]^2 \\
&= \|\mathbf{Y} - \mathbf{X}\boldsymbol{\beta}\|^2
\end{aligned} \tag{4.17}$$

where we use the notation $\|\cdot\|$ for the Euclidean norm of an n-dimensional vector, i.e. $\|\boldsymbol{y}\| = \left(\sum_{i=1}^{n} y_i^2\right)^{1/2}$. Notice that we are now using the notation $\boldsymbol{\beta}$ for the parameter which we called θ in our general discussion of Sect. 4.1.6. As announced earlier, there is a unique solution to the above problem (although we need to use the full rank assumption to rigorously justify this statement) and this solution is given by an explicit formula which we give here without proof:

$$\hat{\boldsymbol{\beta}} = [\mathbf{X}^t\mathbf{X}]^{-1}\mathbf{X}^t\mathbf{Y}. \tag{4.18}$$

A two-line proof of this result can be given if one is familiar with vector differential calculus. Indeed, the minimum $\hat{\boldsymbol{\beta}}$ can be obtained by solving for the zeroes of the gradient vector of the loss function $\mathcal{L}_2(\boldsymbol{\beta})$ as given by (4.17), and this can be done in a straightforward manner. The interested reader is invited to consult the references given in the Notes & Complements at the end of this chapter.

4.5.2.1 **Properties of $\hat{\boldsymbol{\beta}}$**

The following list summarizes some of the most important properties of the least squares estimator $\hat{\boldsymbol{\beta}}$. The first property is a simple remark on formula (4.18), while the other ones rely on the distribution properties of the linear model (4.14).

- $\hat{\boldsymbol{\beta}}$ is *linear* in the sense that it is a linear function of the observation vector $\mathbf{Y}$;
- $\hat{\boldsymbol{\beta}}$ is *unbiased* in the sense that it is equal on the average to the true (and unknown) value of $\boldsymbol{\beta}$, whatever this value is. Mathematically this is expressed by the formula $\mathbb{E}\{\hat{\boldsymbol{\beta}}\} = \boldsymbol{\beta}$;
- The variance/covariance matrix of this estimator can be computed explicitly. It is given by the formula $\boldsymbol{\Sigma}_{\hat{\boldsymbol{\beta}}} = \sigma^2[\mathbf{X}^t\mathbf{X}]^{-1}$;
- $\hat{\boldsymbol{\beta}}$ is optimal in the sense that it has *minimum variance* among all the linear unbiased estimators of $\boldsymbol{\beta}$.

4.5.2.2 **Properties of the Fitted Values $\hat{\mathbf{Y}}$**

The formula giving $\hat{\boldsymbol{\beta}}$ implies that the fitted values are given by:

$$\hat{\mathbf{Y}} = \mathbf{X}\hat{\boldsymbol{\beta}} = \mathbf{X}[\mathbf{X}^t\mathbf{X}]^{-1}\mathbf{X}^t\mathbf{Y} = \boldsymbol{H}\mathbf{Y}$$

if we introduce the notation

$$\boldsymbol{H} = \mathbf{X}[\mathbf{X}^t\mathbf{X}]^{-1}\mathbf{X}^t.$$

As for the parameter estimate(s), the fitted value(s) are linear in the observations, since the vector $\hat{\mathbf{Y}}$ of fitted values is obtained by multiplying a matrix and the vector $\mathbf{Y}$ of observations of the response variable. The matrix $\boldsymbol{H}$ plays an important role in the analysis of the properties of least squares regression. It is called the *hat matrix* or the *prediction matrix* since the formula $\hat{\mathbf{Y}} = \boldsymbol{H}\mathbf{Y}$ tells us how to transform the observations $\mathbf{Y}$ into the values $\hat{\mathbf{Y}}$ predicted by the model. We shall see below that the diagonal elements $h_{i,i}$ enter in an explicit way into the variance of the raw residuals, but for the time being we shall stress that the $h_{i,i}$'s measure the *influence* of the corresponding observation: a large value of $h_{i,i}$ (since the $h_{i,i}$ are never greater than 1, a large value means a value close to 1) indicates that the corresponding observation plays a crucial role in the computation of the resulting value of $\hat{\beta}$ and consequently, greatly influences the interpretation of the results. This notion of influential observation has to be carefully distinguished from the notion of outlying observation discussed below. Indeed, there is nothing wrong with an influential observation, it should not be disregarded, but should simply be looked at carefully in order to understand why some of the results of the regression are what they are.

We display the following formula for future reference.

$$\hat{\mathbf{Y}} = \mathbf{X}\hat{\boldsymbol{\beta}} = \boldsymbol{H}\mathbf{Y} \qquad \hat{\epsilon} = \mathbf{Y} - \hat{\mathbf{Y}} = [\mathbf{I}_n - \boldsymbol{H}]\mathbf{Y}. \tag{4.19}$$

In order to differentiate them from the modified residuals which we introduce later on, the components $\hat{\epsilon}_i$ of the vector $\hat{\epsilon}$ defined above are called the *raw residuals* since:

$$\hat{\epsilon}_i = \hat{y}_i - y_i. \tag{4.20}$$

They were merely called residuals up to now.

4.5.2.3 *Properties of the Residuals*

The properties of the raw residuals $\hat{\epsilon}_i$ are summarized in the following bullet points:

- The $\hat{\epsilon}_i$'s are mean zero, $\mathbb{E}\{\hat{\epsilon}\} = 0$;
- Their variance/covariance matrix is given by the formula $\boldsymbol{\Sigma}_{\hat{\epsilon}} = \sigma^2[\mathbf{I}_n - \boldsymbol{H}]$;
- In particular
  - The $\hat{\epsilon}_i$'s are correlated;
  - They do not have the same variance since $\sigma_{\hat{\epsilon}_i} = \sigma\sqrt{1 - h_{i,i}}$.

We use the notation $h_{i,j}$ for the (i, j)-th entry of the hat matrix $\boldsymbol{H}$. It is important to reflect on the meaning of these statements. Indeed, given the fact that the $\hat{\epsilon}_i$ come as candidates for realizations of the actual error terms ϵ_i, one could expect that they form a white noise. But we just learned that this is not the case. This fact was already mentioned when we compared the residuals of the time series regression of the utility index with the residuals of the regression of the coffee log-returns. Indeed, there are at least two good reasons why the plot of the raw residuals should not look like a white noise! First the variance of $\hat{\epsilon}_i$ changes with i, and second, the $\hat{\epsilon}_i$'s are correlated. Any plot of the raw residuals should show these facts.

Notice that even though the dependence of the $\hat{\epsilon}_i$ may look shocking at first, it should not be surprising since after all, the $\hat{\epsilon}_i$'s are computed from $\hat{\boldsymbol{\beta}}$ instead of from the (deterministic) true value $\boldsymbol{\beta}$ which is not available, and also the estimator $\hat{\boldsymbol{\beta}}$ is a function of all the observations y_j, and consequently of all the error terms ϵ_j. This is the source of this unexpected correlation between the residuals.

4.5.2.4 **Estimator of the Variance** σ^2

We first introduce the notation:

$$R_0^2 = \|\mathbf{Y} - \mathbf{X}\hat{\boldsymbol{\beta}}\|^2 = \sum_{i=1}^{n} [y_i - \hat{\beta}_0 - \hat{\beta}_1 x_{i,1} - \cdots - \hat{\beta}_p x_{i,p}]^2 \tag{4.21}$$

for the minimum sum of squared residuals (i.e. the minimum value of the loss function $\mathcal{L}_2(\boldsymbol{\beta})$). In the notation used in our discussion of the simple least squares linear regression, this quantity was denoted by `SSE` and called the sum of squared errors. The variance parameter σ^2 is estimated by:

$$\begin{aligned}
\hat{\sigma}^2 &= \frac{1}{n-p-1} R_0^2 \\
&= \frac{1}{n-p-1} \|\mathbf{Y} - \mathbf{X}\hat{\boldsymbol{\beta}}\|^2 \\
&= \frac{1}{n-p-1} \sum_{i=1}^{n} [y_i - \hat{\beta}_0 - \hat{\beta}_1 x_{i,1} - \cdots - \hat{\beta}_p x_{i,p}]^2.
\end{aligned}$$

As usual, the normalization of R_0^2 is done by division after subtracting the number of parameters ($p+1$ in our case) from the number of observations. This correction is included to make sure that $\hat{\sigma}^2$ is an unbiased estimator of the variance σ^2.

4.5.2.5 **Standardized Residuals** $\hat{\epsilon}'_i$

In order to resolve the issue of heteroskedasticity (i.e. fluctuations in the variance) we would like to divide the raw residuals by their standard deviations. In other words, we would like to use:

$$\frac{\hat{\epsilon}_i}{\sigma\sqrt{1-h_{i,i}}},$$

but since we do not know σ we instead use:

$$\hat{\epsilon}'_i = \frac{\hat{\epsilon}_i}{\hat{\sigma}\sqrt{1-h_{i,i}}} \tag{4.22}$$

The $\hat{\epsilon}'_i$ so defined are called the *standardized residuals.* Obviously, they are mean zero and have unit variance, but there is still a lot of correlation between them.

4.5.2.6 **Studentized Residuals** $\hat{\epsilon}^*_i$

In order to circumvent the problem caused by the serial correlation of the residuals considered so far, another type of residual was proposed. The $\hat{\epsilon}^*_i$ defined below are called the *studentized residuals.* For each $i = 1, \ldots, n$, we

- Remove the observation (x_i, y_i);
- Fit a linear model (by least squares) to the remaining $(n-1)$ observations;
- Call $\hat{y}_{(i)}$ the prediction of the response for the value x_i of the regressor.

Notice that the value of the error term ϵ_i does not enter into the computation of the new *residual* $y_i - \hat{y}_{(i)}$, so we would hope that these new residuals are de-correlated. In any case, we set:

$$\hat{\epsilon}_i^* = \frac{y_i - \hat{y}_{(i)}}{\sqrt{\text{var}\{y_i - \hat{y}_{(i)}\}}}. \tag{4.23}$$

These studentized residuals should look like white noise and a reasonable diagnostic for the fit of a linear model is to check how this materializes. As defined, the studentized residuals require enormous computer resources, since the computation of each single $\hat{y}^{(i)}$ requires a linear regression! Fortunately, the set of all the studentized residuals can be computed from one single linear regression and a few updates. Indeed, simple arithmetic can be used to derive the following formula:

$$\hat{\epsilon}_i^* = \sqrt{\frac{n-p-1}{n-p-\hat{\epsilon}_i'^2}}\hat{\epsilon}_i'$$

linking the standardized residuals to the studentized ones. Despite the existence of a function `ls.diag`, which can be used with an object created by the function `lsfit`, there is no convenient way to produce the residual diagnostics from an object created by the function `lm`. For this reason, we provided a simple wrapper called `lm.diag`, which does just that. Its use is illustrated in Sect. 4.5.2.8 below.

4.5.2.7 More Residual Diagnostics

A standard way to gauge the goodness of fit of a regression model is to use graphical diagnostics. For example if a plot of the standardized or studentized residuals (whether it is a sequential plot, or a plot against the fitted values or the observations) shows values outside the range $[-2, +2]$, one should suspect that something is wrong. Either the model is not appropriate for the data, or if the model is reasonable, the observations responsible for these extreme residual values do not belong. These observations are usually called *outliers*. It is a good practice to try to identify the potential outliers and if one has faith in the model, to re-fit a regression after the outlying observations have been removed.

4.5.2.8 Revisiting the Analysis of the Utility Indexes

We use the utility data to illustrate how to produce the suite of regression diagnostics in R. They are created with the commands

```
> Ue.diag <- lm.diag(Ue)
> Ud.diag <- lm.diag(Ud)
```

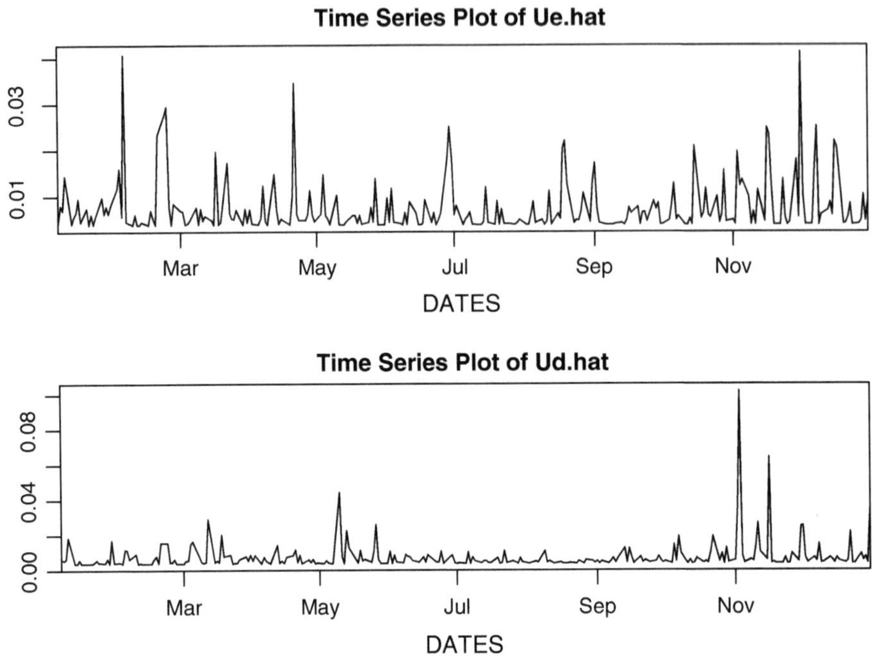

Fig. 4.9. Sequential plot of the $h_{i,i}$ in the case of the simple regression of utility index daily log-returns against ENRON's daily log-returns (*top*) and against DUKE's daily log-returns (*bottom*)

Figure 4.9 gives the plots of the influence measures (i.e. the diagonal entries of the hat matrix) in the case of the simple regression of `UtilLRet` against `EnronLRet` and `DukeLRet`, respectively. These numerical vectors are extracted from the diagnostic objects by means of the extension ··· `$hat`. In preparing Fig. 4.9, we did not rely on the usual `plot` function to produce the plots of the vectors `Ue.diag$hat` and `Ud.diag$hat`. Instead, we used time series plots in order to have the dates appear on the horizontal axes. We shall learn how to do that in `R` when we introduce the `timeSeries` objects in Chap. 6. It is clear from this plot that in both cases, the Winter season and the period spanning the end of Summer and early Fall are most influential. Notice nevertheless that the scales of the vertical axes are not the same, and taking this fact into account, one sees that a few days in November are extremely influential in the regression of the utility index log-returns against DUKE's log-returns. As we shall see in the later part of the book, departure from normality can be an indication of the existence of significant dependencies among the residuals, so it is always a good idea to check the Q-Q plots of the standardized and the studentized residuals against the quantiles of the normal distribution. We give these Q-Q plots in Fig. 4.10. Significant departure from normality seems to be present. The numerical vectors containing the standardized residuals and the studentized residuals can be extracted from the `R` diagnostic objects by means of the extensions ··· `$stdres`

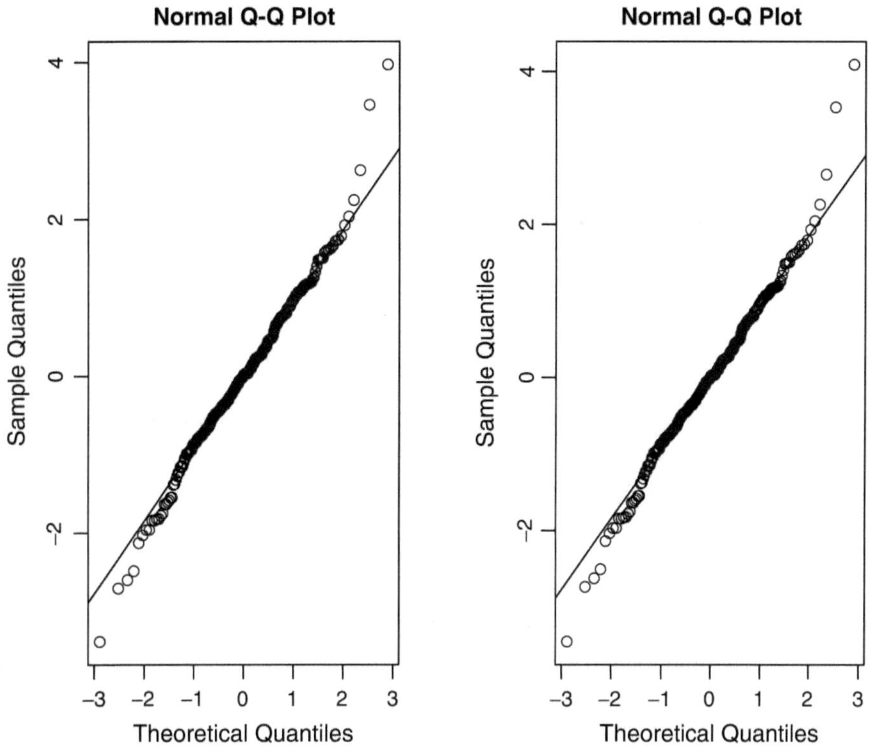

Fig. 4.10. Q-Q plots of the standardized residuals (*left*) and the studentized residuals (*right*) of the multiple least squares regression of the utility daily log-returns against both ENRON and DUKE daily log-returns

and $\cdots$ `$studres` respectively. The Q-Q plots of Fig. 4.10 were created with the commands:

```
> qqnorm(Ued.diag$stdres)
> qqnorm(Ued.diag$studres)
```

This plot shows that the residuals exhibit heavy tails.

4.5.3 Confidence and Prediction Intervals

4.5.3.1 Confidence Interval

In statistics, we typically *estimate* parameters, and *predict* (future) random outcomes. In the set up of simple least squares linear regression from a data set $(x_1, y_1), \cdots, (x_n, y_n)$, if we consider the prediction of the response variable for a new value x_0 of the explanatory variable, the number $\mu = \beta_0 + \beta_1 x_0$ can be considered as a *new parameter* of the model, and the inferential theory of Gaussian families gives a *confidence interval* for μ. For example, the end points of a $100\,\alpha\,\%$ confidence interval for μ are given by

$$\hat{\mu} \pm t_{n-2,\frac{1+\alpha}{2}} \sqrt{\hat{\sigma}^2 \left(\frac{1}{n} + \frac{(x_0 - \bar{x})^2}{\sum_{i=1}^n (x_i - \bar{x})^2} \right)},$$

where $t_{k,\gamma}$ denotes the γ-quantile of the t-distribution with k degrees of freedom, and

$$\bar{x} = \frac{1}{n} \sum_{i=1}^n x_i \quad \text{and} \quad \hat{\sigma}^2 = \frac{1}{n-2} \sum_{i=1}^n (y_i - \hat{y}_i)^2.$$

Remember that when n is large, say $n \geq 50$, the t-distribution with n degrees of freedom is very close to the standard Gaussian distribution and using quantiles of the Gaussian distribution would provide reasonable approximate confidence intervals. Notice that the estimate

$$\hat{\mu} = \hat{\beta}_0 + \hat{\beta}_1 x_0.$$

of the parameter is the same as the least squares prediction $\hat{Y}_0$ of the random variable Y_0 whose conditional expectation is given by

$$\hat{Y}_0 = \mathbb{E}\{Y | X = x_0\} = \hat{\beta}_0 + \hat{\beta}_1 x_0$$

4.5.3.2 ***Prediction Interval***

However, the random variable Y_0 we are trying to predict is of the form

$$Y_0 = \mu + \epsilon_0$$

ϵ_0 being a random variable independent of $\hat{\beta}_0$ and $\hat{\beta}_1$, providing an extra source of noise. Consequently, an interval covering the true value of Y_0 with probability at least α should be larger than the confidence interval for μ, leading to the notion of *prediction interval* for this prediction

$$\hat{Y}_0 \pm t_{n-2,\frac{1+\alpha}{2}} \sqrt{\hat{\sigma}^2 (1 + \frac{1}{n} + \frac{(x_0 - \bar{x})^2}{\sum_{i=1}^n (x_i - \bar{x})^2})},$$

4.5.4 **First Extensions**

The formalism of linear models is so general that the theory can be applied to many diverse situations. We consider two such applications in this subsection. Many more will be discussed later in the chapter. Since the applications considered in this section will not be used in the sequel, this section can be skipped in a first reading.

4.5.4.1 *Weighted Regression*

Each diagonal entry of the hat matrix $\boldsymbol{H}$ gives an indication of the influence of the corresponding measurement. It is sometime desirable to decrease (resp. increase) the influence of a given measurement on the final regression result. This can be done by means of a minor change in the way each individual measurement actually contributes to the value of the loss function. Let us consider for example the weighted sum of squares:

$$\mathcal{L}_2^{(w)}\varphi = \sum_{i=1}^{n} w_i|y_i - \varphi(x_i)|^2. \tag{4.24}$$

for some predetermined set of nonnegative weights w_i. In order to avoid large contributions to the overall loss, any function φ minimizing this loss function will try very hard to have a value $\varphi(x_i)$ very close to y_i for the indices i for which the weight w_i is large. On the other hand, if a weight w_i is smaller than the bulk of the other weights, then the fit may be loose without causing much damage to the overall value of the loss.

The computations involved in the minimization of the weighted loss function $\mathcal{L}_2^{(w)}$ are of the same complexity as the computation of the minimum of the corresponding unweighted least squares loss function $\mathcal{L}_2$. For this reason, it is possible in R to specify a set of weights and expect as result the regression function φ minimizing the weighted loss function $\mathcal{L}_2^{(w)}$.

Example. Let us imagine that one tries to explain the variable `UtilLRet` by the variable `DukeLRet` by giving more importance to the days when the ENRON index is larger than the DUKE index. One can use simple least squares linear regression using the variable `ENRON.index/DUKE.index` as a set of weights. This would be accomplished by the R commands:

```
> WEIGHTS <- (ENRON.index/DUKE.index)[-1]
> TstUtil.lm <- lm(UtilLRet ~ DukeLRet, weights=WEIGHTS)
```

Note the `[-1]` in the first command. It is there because of a fact we encountered each time we computed returns: the vectors `UtilLRet` and `DukeLRet` are one unit shorter than the vectors `ENRON.index` and `DUKE.index` as the computation of the log-returns was not possible for the first entry of these vectors.

Remark. Because of the parallel that we drew between the least squares regression and its robust cousin, the least absolute deviations regression, it would be natural to consider a weighted loss function of the form:

$$\mathcal{L}_1^{(w)}\varphi = \sum_{i=1}^{n} w_i|y_i - \varphi(x_i)|, \tag{4.25}$$

and by analogy with the weighted least squares regression to have the weighted least absolute deviations regression function be the result of the search for the function φ

minimizing $\mathcal{L}_1^{(w)}$. Unfortunately, this minimization is quite involved and weighted least absolute deviations regression is rarely an option: the function `l1fit` does not offer the possibility to include weights in the regression.

4.5.4.2 Seemingly Unrelated Regressions (SUR)

Let us assume that for each $j = 1, \ldots, J$ we fit a multiple linear regression model

$$\mathbf{Y}^{(j)} = \mathbf{X}^{(j)} \boldsymbol{\beta}^{(j)} + \boldsymbol{\epsilon}^{(j)}$$

to the sample

$$(\boldsymbol{x}_1^{(j)}, y_1^{(j)}), \ldots, (\boldsymbol{x}_n^{(j)}, y_n^{(j)})$$

where the explanatory variables $\boldsymbol{x}_i^{(j)}$ all have the same dimension $k = p+1$. In such a situation, one performs the J regressions in sequence, independently of each other. However, it happens in many financial applications that the explanatory variables are the same for all the regressions. In this case the model can be rewritten as:

$$\mathbf{Y}^{(j)} = \mathbf{X} \boldsymbol{\beta}^{(j)} + \boldsymbol{\epsilon}^{(j)}$$

and it can be ran in `R` by binding all the response vectors into a $n \times J$ response matrix. According to the theory developed earlier, each parameter vector $\boldsymbol{\beta}^{(j)}$ can be estimated by ordinary least squares, and the results are given in matrix form by:

$$\hat{\boldsymbol{\beta}}^{(j)} = [\mathbf{X}^t \mathbf{X}]^{-1} \mathbf{X}^t \mathbf{Y}^{(j)}.$$

Now comes the interesting part of this subsection. Even though the response variables $Y_i^{(j)}$ and $Y_{i'}^{(j')}$ are uncorrelated if they come from two different observations (i.e. if $i \neq i'$), we now assume that, the responses associated to the same observation are correlated. In other words, we assume the existence of a variance/covariance matrix $\Gamma = [\gamma_{j,j'}]_{j,j'=1,\ldots,J}$ such that:

$$\text{cov}\{Y_i^{(j)}, Y_i^{(j')}\} = \gamma_{j,j'}, \qquad j, j' = 1, \ldots, J, \; i = 1, \ldots, n, \tag{4.26}$$

or in vector form, $\boldsymbol{\Sigma}_{\mathbf{Y}_i} = \Gamma$ for all $i = 1, \ldots, n$. As we already pointed out, this assumption is quite realistic in many financial applications. For example it holds when the response variables $Y_i^{(j)}$ are the log-returns of J stocks in the same economic sector, and when the explanatory vectors $\boldsymbol{x}_i$ are observations of vectors of economic factors driving the dynamics of the share prices of the public companies in the sector. A linear model specified this way is called a set of seemingly unrelated regressions, or SUR for short. The usual ordinary least squares approach is not appropriate because of the dependence among the response variables. In order to understand why, we rewrite the model in the standard form of a linear model with a scalar response variable. In order to do so, we bind the response vectors $\mathbf{Y}^{(j)}$ into a $(nJ) \times 1$ column vector $\tilde{\mathbf{Y}}$, the noise vectors $\boldsymbol{\epsilon}^{(j)}$ into a $(nJ) \times 1$ column vector $\tilde{\boldsymbol{\epsilon}}$, the parameters $\boldsymbol{\beta}^{(j)}$

into a $(kJ) \times 1$ column vector $\tilde{\boldsymbol{\beta}}$, and we create the new design matrix $\tilde{\mathbf{X}}$ as the $(nJ) \times (kJ)$ matrix with J copies of $\mathbf{X}$ on the diagonal:

$$\tilde{\mathbf{Y}} = \begin{bmatrix} \mathbf{Y}^{(1)} \\ \vdots \\ \mathbf{Y}^{(J)} \end{bmatrix}, \quad \tilde{\epsilon} = \begin{bmatrix} \epsilon^{(1)} \\ \vdots \\ \epsilon^{(J)} \end{bmatrix}, \quad \tilde{\boldsymbol{\beta}} = \begin{bmatrix} \boldsymbol{\beta}^{(1)} \\ \vdots \\ \boldsymbol{\beta}^{(k)} \end{bmatrix}, \quad \tilde{\mathbf{X}} = \begin{bmatrix} \mathbf{X} & \mathbf{0} & \cdots & \mathbf{0} \\ \mathbf{0} & \mathbf{X} & \cdots & \mathbf{0} \\ \vdots & \vdots & & \vdots \\ \mathbf{0} & \mathbf{0} & \cdots & \mathbf{X} \end{bmatrix}.$$

With these notation, the SUR model reads:

$$\tilde{\mathbf{Y}} = \tilde{\mathbf{X}}\tilde{\boldsymbol{\beta}} + \tilde{\epsilon}$$

but the ordinary least squares estimate

$$\hat{\tilde{\boldsymbol{\beta}}}^{(j)} = [\tilde{\mathbf{X}}^t\tilde{\mathbf{X}}]^{-1}\tilde{\mathbf{X}}^t\tilde{\mathbf{Y}}$$

is presumably not the best estimate. Indeed, even when the noise terms are Gaussian, this estimate does not coincide with the maximum likelihood estimator. This is due to the fact that the above least squares estimate is based on the assumption that the variance/covariance matrix of the noise vector $\tilde{\epsilon}$ is a multiple of the $(nJ) \times (nJ)$ identity matrix I_{nJ}, and this is not the case in the present situation. Indeed, this variance/covariance matrix is not diagonal as a simple computation shows. It is in fact equal to what is called the Kroenecker product of the matrix Γ and the $n \times n$ identity matrix I_n. We shall not pursue the analysis any further. We refer the interested reader to the Notes & Complements section at the end of the chapter for references.

4.5.5 Testing the CAPM

We now present the Capital Asset Pricing Model (CAPM for short) as an illustration of the linear regression techniques introduced in this chapter.

We consider a market economy containing N financial assets, and we denote by R_{jt} the log-return on the j-th asset on day t. For the purposes of the present discussion, we assume the existence of a market portfolio, and we denote by $R_t^{(m)}$ its log-return on day t. Finally, we also assume the existence of lending and borrowing at the same risk-free rate which we denote by r, and which we assume to be deterministic. The Sharpe-Lintner version of the CAPM states that the excess return (over the risk-free rate) of each asset j is, up to noise, a linear function of the excess return of the market portfolio. In other words, for each j:

$$\widetilde{R_{jt}} = \alpha_j + \beta_j \widetilde{R_t^{(m)}} + \epsilon_{jt}$$

where the noise sequence $\{\epsilon_{jt}\}_t$ is uncorrelated with the market portfolio return. We use a tilde to denote the returns in excess of the risk-free rate:

$$\widetilde{R_{jt}} = R_{jt} - r \quad j = 1, 2, \ldots \qquad \text{and} \qquad \widetilde{R_t^{(m)}} = R_t^{(m)} - r. \tag{4.27}$$

One of the claims of the CAPM theory is that, if one uses excess returns, the intercept α_j appearing in the linear regression equation (4.27) should be zero. In other words, the regression lines should all go through the origin, whatever the choice of the asset. Notice that the same model can be used for small portfolio returns R_{jt} instead of individual stock returns.

For each stock, a least squares regression will provide estimates for the slope β_j and the intercept α_j. The estimate $\hat{\beta}_j$ is given by a normalized form of the covariance between the j-th asset (or the j-th portfolio) and the market portfolio: this is known as the investment beta for the j-th asset. It measures the sensitivity of the return to variations in the returns of the market portfolio. So assets (or portfolios) with a $\hat{\beta}_j$ greater than one are regarded as risky, while those with $\hat{\beta}_j$ smaller than one are much less sensitive to market fluctuations. The validity of CAPM depends upon

- The existence of a significant linear relationship (so we shall look at the R^2);
- A zero intercept (so we shall test the hypothesis that $\alpha_j = 0$);
- Uncorrelated normally distributed error terms (which we shall check by means of a residual analysis);
- All of this being stable (which we shall check by testing the constancy of the beta's over time).

4.5.5.1 ***Empirical Tests***

In order to test the CAPM, we consider the daily returns on five stocks of one of the utility subindexes of the Dow Jones Industrial Average. We use the S&P 500 index as a proxy for the market portfolio, and we use the yield on the 13 weeks T-bill (see Sect. 4.8 for the definition of this instrument) as a proxy for the risk-free rate of borrowing from which the excess returns are computed. In the late 1990s, gas and electricity trading became an important source of revenues. Deregulation raised high expectations, and speculative trading overshadowed hedging and risk management. Energy companies experienced growth throughout this period. Things changed dramatically in 2000. The California crisis and Enron's bankruptcy ignited a sudden reversal in trading activity, and a spectacular downfall for the sector. So we divided the data into the period starting 01/01/1995 and ending 01/01/1999, and the period starting 01/01/1999 and ending 01/01/2003. Figure 4.11 shows the results of simple linear regressions of the American Electric Power (AEP) weekly excess returns against the market excess returns over the first period on the left, and over the second period on the right. In each case we plotted both the least squares and least absolute deviations regression lines. The beta of the second period is much bigger than the beta of the first period, which is consistent with our interpretation of a risky stock in terms of the size of its beta. Only looking at this plot may not show this fact. Indeed it is partially masked by the fact that the scales on the vertical axes are not the same. See Tables 4.1 and 4.2 for the exact values of the least squares estimates of the betas. In the case of the first period, the shape of the cloud of points is different, and the robust estimate of β_{AEP} is greater.

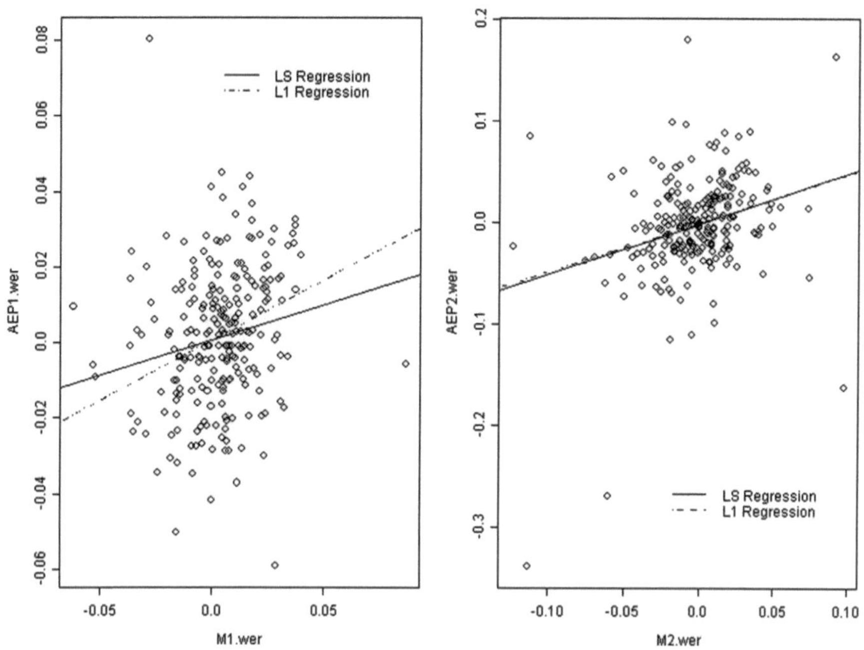

Fig. 4.11. Least squares (*solid line*) and least absolute deviations (*dashed line*) regressions of the American Electric Power (AEP) weekly excess returns against the market excess returns for the period starting 01/01/1995 and ending 01/01/1999 (*left*) and for the period starting 01/01/1999 and ending 01/01/2003 (*right*)

	AEP Estimate (p-value)	**DUKE** Estimate (p-value)	**PCG** Estimate (p-value)	**SO** Estimate (p-value)	**TXU** Estimate (p-value)
Intercept	0.0009(0.4604)	0.0015(0.2826)	0.0011(0.5233)	0.0008(0.5691)	0.0011(0.4286)
Slope	0.1861(0.0050)	0.1563(0.0326)	0.0127(0.8893)	0.2520(0.0010)	0.1762(0.0176)

Table 4.1. Intercept and slope estimates, and corresponding p-values (in parentheses) for the least squares regression tests of the CAPM for a set of five electric companies over the period starting 01/01/1995 and ending 01/01/1999

Figure 4.12 shows the results of the same linear regression analysis in the case of Texas Utilities (TXU). The results are qualitatively the same.

Next we test the null hypothesis of a zero intercept in the case of five of the largest electric companies in the Dow Jones Utility Index. We reproduce the intercept estimates and the p-values of the t-test in Tables 4.1 and 4.2 for the two periods. In all cases, the test statistic could not reject the null hypothesis of a zero intercept. However, looking carefully at the estimates of the betas shows clearly that these

	AEP Estimate (p-value)	DUKE Estimate (p-value)	PCG Estimate (p-value)	SO Estimate (p-value)	TXU Estimate (p-value)
Intercept	−0.0009	−0.0010	−0.0029	0.0021	−0.0027
	(0.7621)	(0.7307)	(0.5813)	(0.3492)	(0.5840)
Slope	0.4982	0.1563	0.4294	−0.0184	0.4834
	(0.0000)	(0.0000)	(0.0180)	(0.8074)	(0.0052)

Table 4.2. Intercept and slope estimates, and corresponding p-values (in parentheses) for the least squares regression tests of the CAPM for a set of five electric companies over the period starting 01/01/1999 and ending 01/01/2003

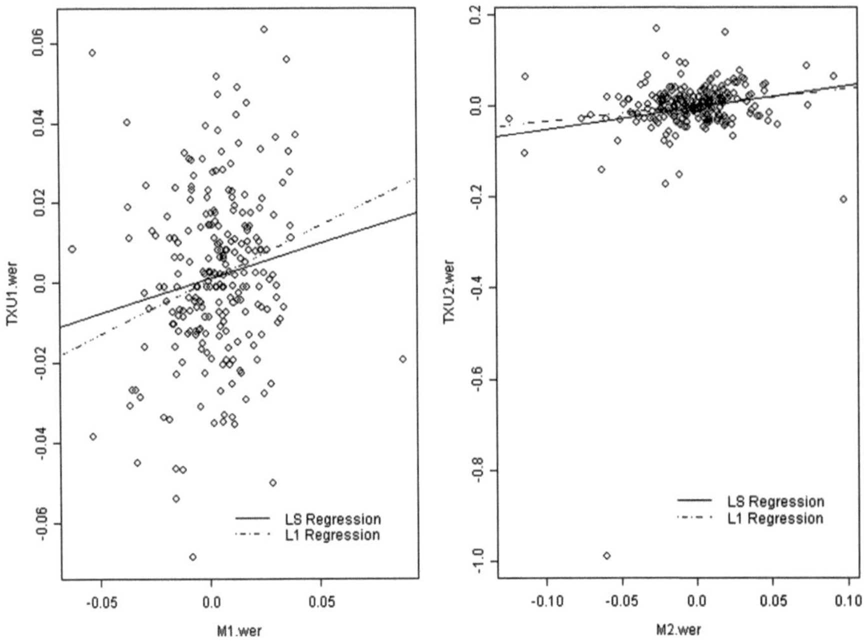

Fig. 4.12. Least squares (*solid line*) and least absolute deviations (*dashed line*) regressions of the Texas Utilities (TXU) weekly excess returns against the market excess returns for the period starting 01/01/1995 and ending 01/01/1999 (*left*) and for the period starting 01/01/1999 and ending 01/01/2003 (*right*)

betas change from one period to the next. But the academic literature is populated with many papers arguing that the model should be rejected on the basis of empirical tests of this type. See the Notes & Complements for further discussion of the issue. In any case, our numerical results show that the stability of the betas over time (and consequently the choice of the periods over which one should estimate them) can become a serious issue. In order to avoid this difficult challenge, and in order to satisfy the pundits debating the validity of the model, we generalize the CAPM to time varying betas in Chap. 7, and we apply the filtering theory developed in that chapter to estimate these *changing betas*.

4.6 Polynomial Regression

After a short excursion into the realm of multivariate regression, we turn our attention back to simple regression (i.e. one real-valued explanatory variable x) and we force the regression function φ to be a polynomial. In other words, we restrict ourselves to regression functions of the form:

$$\varphi(x) = \beta_0 + \beta_1 x + \beta_2 x^2 + \cdots + \beta_p x^p \tag{4.28}$$

for some integer $p \geq 1$ and unknown parameters β_0, β_1, ..., β_p. This type of regression generalizes the simple linear regression which we saw earlier, since the latter corresponds to the case $p = 1$. As such it appears as a very attractive way to resolve some of the shortcomings we noticed. Unfortunately, polynomial regression can be very poor, especially when it comes to prediction of the response for values of the explanatory variables outside the range of the data. So our advice is to avoid its use unless one has VERY GOOD REASONS to believe that the model (i.e. the true regression function φ) is indeed a polynomial.

4.6.1 **Polynomial Regression as a Linear Model**

Even though a polynomial of degree p is a nonlinear function of the variable x when $p > 1$, it can be regarded as a linear function (affine to be specific) of the variables x, x^2, ..., and x^p. So, we could fit a linear model to the data as long as the observations x_1, x_2, ..., x_n on the single univariate explanatory variable x are replaced by n observations of x and its powers which we collect together in a design matrix:

$$\mathbf{X} = \begin{bmatrix} 1 & x_1 & x_1^2 & \cdots & x_1^p \\ \vdots & \vdots & \vdots & & \vdots \\ 1 & x_n & x_n^2 & \cdots & x_n^p \end{bmatrix}$$

But we should not have to do that, `R` should do it for us. The `R` command used for polynomial regression is:

```
lm(response ~ poly(regressor,degree))
```

to which we may want to add the `data.frame` by setting the parameter `data`.

4.6.2 **Example of `R` Commands**

We use the data set `FRWRD` containing the values in US \$ of the 36 Henry Hub natural gas forward contracts traded on March 23rd 1998. We use those prices as observations of our response variable, and we use the index of the delivery month of the forward contract as explanatory variable. For the sake of simplicity we use the integers $1, 2, \ldots, 36$. This example may not be representative of most regression problems, for, in general, the explanatory variable does not take values in a regular deterministic grid. We chose it for the sake of illustration: natural gas forward curves have a strong seasonal component which distinguishes them from most of the other commodity forward curves. Figure 4.13 gives the results of several polynomial regressions of the forward prices against the month indexes. We used degrees 3, 6 and 8 respectively. The plots were produced with the R commands:

```
> plot(1:36,FRWRD,main="Polynomial Gas Forward Curves")
> lines(1:36,fitted(lm(FRWRD~poly(1:36,3))))
> lines(1:36,fitted(lm(FRWRD~poly(1:36,6))),lty=3)
> lines(1:36,fitted(lm(FRWRD~poly(1:36,8))),lty=6)
> legend(locator(1),c("Degree=3","Degree=6","Degree=8"),
                                      lty=c(1,3,6),bty="n")
```

The estimates fitted to the response variable are extracted from the `lm` object by the function `fitted`. The points corresponding to the fitted values have been joined by straight line segments by the R function `lines`. This produces a continuous curve and gives the impression that the plot of the polynomial is actually given, even though we only plotted broken lines between the fitted points. More on the properties of this function `lines` later at the end of this chapter.

None of the polynomial fits given in Fig. 4.13 is very good. Indeed, the forward prices do not seem to be a polynomial function of the month of maturity, and it would take a much higher degree to get a satisfactory fit throughout the 36 months period.

4.6.3 **Important Remark**

Polynomial regression, as we just introduced it, is a particular case of the estimation of the regression function φ when the latter is known to belong to a specific vector space. In the present situation the vector space is the vector space of polynomials of degree at most p. This space has dimension $p + 1$. Indeed, the special polynomials 1, x, x^2, ..., x^p form a basis for this space, since any polynomial of degree at most p can be decomposed in a unique way as a linear combination of these particular $(p + 1)$ polynomials. So any element of this vector space is entirely determined by $(p + 1)$ numbers, the coefficients of the decomposition in this basis. This is why polynomial regression is a particular case of parametric statistics, and this is how we recast this seemingly nonlinear simple regression into the framework of a linear multivariate regression. But vector spaces have many different bases. In fact, one of the very nice features of vector algebra is the option to change basis. Indeed, re-expressing a problem in a different basis may sometimes make it easier to understand and to handle.

Since what matters to us is the function φ and not so much the way in which it is parameterized, or the specific basis in which one chooses to decompose it, it should not matter how R handles the linear model set up to perform a polynomial regression. In fact, even though we like to think of the particular basis $\{1, x, x^2, \ldots, x^p\}$ when we introduce a polynomial of degree at most p, there is no reason why R should work with this particular basis. In other words, there is no reason why the R function `poly` could not create the design matrix $\mathbf{X}$ by expressing the polynomial candidate for the regression function in a different basis of the same vector space of polynomials. In fact, this is exactly what it does. This is R internal politics which we should not have to be concerned with. More on this subject later in the appendix at the end of this chapter when we discuss R's idiosyncrasies.

4.6.4 **Prediction with Polynomial Regression**

Whether it is for speculative reasons or for hedging future price exposures, owners and operators of gas fired power plants are often involved in long term contracts with maturities going far beyond the longest maturity appearing in the publicly available forward curve data. They are not the only ones to do that, and some amateurs can be burned at this game. The California electricity crisis is full of instances of such mishaps. In any case, the prediction of the forward prices for very long maturities is a challenge, and we show why one should not use polynomial regression in the case of natural gas. Having fitted a polynomial to the existing curve, it is straightforward to compute the value of this fitted polynomial for any time to maturity. Let us assume for example that we want to compute the prediction of the price of a 4 year time-to-maturity forward contract (i.e. a forward contract maturing in 48 months) or 6 year

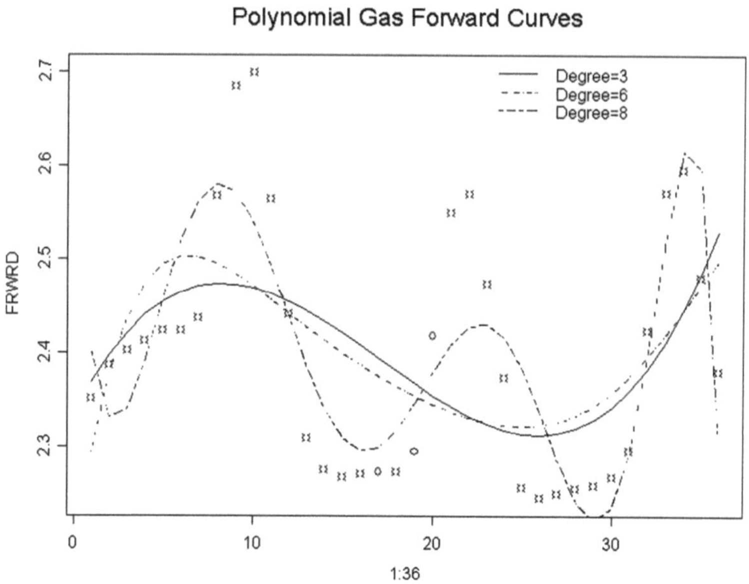

Fig. 4.13. Scatterplot of the prices on March 23rd, 1998 of the 36 traded natural gas forward contracts, together with the results of three different polynomial regressions

time-to-maturity forward contract (i.e. a forward contract maturing in 72 months). We create the object FPOLY of class lm by running the polynomial regression as a linear model as we did earlier, and we use the generic method predict to get the actual predictions. The values of the explanatory variable(s) for which the predictions are computed are passed to the function predict through the parameter newdata which needs to be an object of class data.frame. The following commands give a first example of the steps which need to be taken to do just that.

```
> x <- 1:36
> FPOLY <- lm(FRWRD~poly(x,8))
> F48PRED <- predict(FPOLY, newdata=data.frame(x=c(48)))
> F48PRED
[1] -271.0561
> F72PRED <- predict(FPOLY, newdata=data.frame(x=c(72)))
> F72PRED
[1] -42482.44
```

Something is obviously wrong, these numbers should not be negative, and in any case, they should not be so large. In order to illustrate how bad the situation is, we compute the values of the fitted polynomial for all the monthly maturities ranging from 0 to 48 months, and we superimpose these values on the scatterplot of the values of the forward prices for the first 36 months.

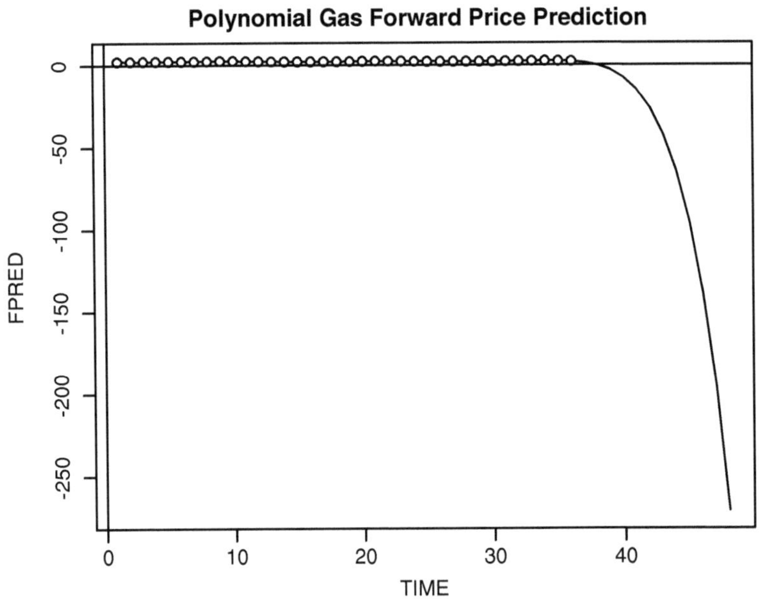

Fig. 4.14. Prediction of the price of a 48 month natural gas forward contract using the best of the polynomial regressions done on March 23rd, 1998

```
> TIME <- 1:48
> FPRED <-  predict(FPOLY, newdata=data.frame(x=1:48))
> plot(TIME,FPRED, type="l",
      main="Polynomial Gas Forward Price Prediction")
> points(x,FRWRD)
> abline(h=0,v=0)
```

The resulting plot is reproduced in Fig. 4.14. The scale of the values for the times to maturity larger than 3 years is so large that the original part of the forward curve is dwarfed to the point of looking like a horizontal straight line. We use this example to emphasize one more time how inappropriate it can be to try to predict the response for values of the explanatory variable(s) too far outside the range of the data.

Back to the Very Important Remark. Before we close this subsection, we come back to one of the important features of the prediction procedure used above. We used the generic method `predict` which was able to extract all the necessary information contained in the object `FPOLY` of class `lm` and compute the desired predictions. This is very simple and very convenient, but unfortunately, the prediction was done in a *black box*, without any indication on how the prediction was actually computed. This is somehow in contradiction with the spirit of our approach to data analysis.

As we saw earlier when we computed predictions from linear regression models, predictions can be obtained by plugging the values of the explanatory variable(s) for which the predictions are desired, into the formula giving the regression function estimate $\hat{\varphi}$. In the present situation, a command of the form `sum(COEFS*(48^(0:8)))` should do just that for a time to maturity of 48 months, provided the 9 coefficients of the polynomial are stored in the components of the vector `COEFS`. Indeed we have:

```
sum(COEFS*(48^(0:8)))=COEFS[1]*48^0+COEFS}[2]*48^1
                                    +....+COEFS[9]*48^8
```

which is the desired evaluation of the value of the polynomial for the value 48 of the explanatory variable x. A naive attempt to compute the predictions $\hat{\varphi}(48)$ and $\hat{\varphi}(72)$ with this plugging strategy would give:

```
> COEFS <- coef(FPOLY)
> NPRED48 <- sum(COEFS*(48^(0:8)))
> NPRED48
[1] -4.461495e+12
> NPRED72 <- sum(COEFS*(72^(0:8)))
> NPRED72
[1] -1.118850e+14
```

Clearly, these naive predictions are *way off*. Something went wrong without our noticing it. We revisit this issue in the appendix at the end of the chapter, and we give detailed explanations for the reasons of this unexpected behavior of the coefficients of the regression. The guilty party is the function `poly` used by `lm` to transform the (nonlinear) polynomial regression into a linear regression by replacing a polynomial by the sequence of its coefficients. Indeed, instead of using the basis 1, x, x^2, $\cdots$, x^8, the function poly chose to decompose polynomials of degree at most 8 in a different basis. For reasons which we do not want to get into, R chose to construct the design matrix $\mathbf{X}$ in a special basis of orthogonal polynomials instead of the canonical basis $1, x, x^2, \cdots, x^8$. Indeed such a design matrix is better conditioned for numerical computations, inversions, ..., and more efficient and robust algorithms can be used to compute the elements of the linear model so designed. However, at the risk of facing unstable or inefficient computations, the user can always force R and the function poly to work with the canonical polynomial basis. This can be done by setting the parameter raw to TRUE as illustrated below.

```
> FPOLY <- lm(FRWRD~poly(x,8, raw=TRUE))
> COEFS <- coef(FPOLY)
> SPRED48 <- sum(COEFS*(48^(0:8)))
> SPRED48
[1] -271.0561
> SPRED72 <- sum(COEFS*(72^(0:8)))
> SPRED72
[1] -42482.44
```

As expected, we recovered the predictions computed by the *black box* method predict.

4.6.5 Choice of the Degree p

The choice of degree p is a particular case of the delicate problem of the choice of the dimension of a model. A larger degree (in general a large number of explanatory variables) leads to a smaller sum of squares errors and a larger R^2, which is desirable. The case of polynomial regression is a good example with which to illustrate this fact. If the sample is of size n, for most data sets, it is possible to find a polynomial of degree $n+1$ going through all the points and, consequently, providing a ZERO sum of squares!!! But if the data is noisy (i.e. if the variance σ^2 of the noise is not 0) the fit provided by this polynomial of degree $n+1$ is very unsatisfactory: we are fitting the noise, and this is not desirable.

In fact, too large of a degree will produce absurd predictions, especially if the model is used to predict values of the response variable for values of the regressor outside the range of the data. The general principle is:

BE PARSIMONIOUS.

Criteria (mostly of an intuitive nature) have been designed to help with the choice of dimension of a linear model (and in particular with the degree of a polynomial regression). The most popular of them seem to be based on principles of information theory and entropy measures. R provides several of them, but to our astonishment, the most popular seems to be the (in)famous AIC which stands for Akaike Information Criterion. Its use is widespread despite well-documented flaws. We will reluctantly conform to the common practice and shall mention its use throughout.

4.7 NONLINEAR REGRESSION

We have already encountered several examples of simple regression for which the regression function $\varphi(x)$ was a nonlinear function of the explanatory variable x. However, in all these cases, the estimation of φ was based on the solution of a linear model. We now consider examples of models which cannot be reduced to linear models. We shall apply the techniques developed in this section to the constructions of yield curves used by the various Central Banks all over the world.

We present the main ideas of nonlinear regression on an example, and we use the same type of illustration as in our discussion of polynomial regression: as before we use the analysis of commodity forward curves as a test bed. The explanatory variable is again the index of the month of delivery for a forward contract, and for the sake of simplicity we use the integers $1, 2, \ldots, 18$ which we put in a vector x. The response variable y is now the forward price of a crude oil contract. The two specific examples we analyze below are from a period (pre-2000) when the price of a barrel of crude oil was much smaller than in the late 2000s! The main difference with the natural gas forward curve considered earlier is the lack of seasonality. Typical crude oil forward curves can be either increasing (we say that they are in contango) or decreasing (in which case we say that they are in backwardation). But in any case, they are almost always monotone with significant curvature.

4.7.1 **A First Model**

We propose to fit the forward quotes with a function of the form:

$$y = \varphi_\theta(x) = F_\infty \frac{x+K}{x+1}, \tag{4.29}$$

where the constant F_∞ has the interpretation of an asymptotic forward price while K is a positive constant. Economic theory justifies the existence of a limiting price F_∞ representing the cost of production. We use the notation $\theta = (F_\infty, K)$ for the parametrization of the regression function φ_θ. We choose the parametric form (4.29) because the forward curve is in contango when $K < 1$ and in backwardation when $K > 1$, and obviously flat when $K = 1$. Least squares regression can be used to determine the values of the parameter $\theta = (F_\infty, K)$ which minimize the sum of squares, giving:

$$\hat{\theta} = \arg\inf_\theta \sum_i |y_i - \varphi_\theta(x_i)|^2.$$

Obviously, these functions φ_θ are nonlinear. But they form a specific family parameterized by the two-dimensional parameter θ, and given the observations (x_i, y_i) contained in the data, one can try to use a minimization procedure to look for the optimal θ. This can be done using the `R` function `nls` whose call is similar to a call to `lm` in the sense that a formula and possibly a reference to the data frame used are required. In the present situation the formula will have to be nonlinear. It should read:

```
y ~ FINF*(x+K)/(x+1)
```

if we use the R objects FINF and K for the parameters F_∞ and K, respectively. We perform the nonlinear regression with the command:

```
> Ffit <- nls(y ~ FINF*(x+K)/(x+1), start=c(FINF=17,K=.1))
```

Should the explanatory and response variables be columns of a data frame, the name of this data frame should be passed to the function nls by setting the parameter data. The function nls relies on an iterative optimization routine, and this makes it somewhat unpredictable. So, it is always a good idea to initialize the optimization by providing reasonable values for the parameters. The initial values for the parameters are passed with the argument start. We chose to initialize F_∞ to 17 because looking at the scatter plot of (x, y), it appears that this value is in the range of the asymptotic value of the forward price. Also, since the forward curve seems to be in contango, we start K with a value smaller than 1. The numerical results produced by the command above can be viewed using the command summary

```
> summary(Ffit)
    Formula: y ~ (FINF * (x + K))/(x + 1)
  Parameters:
            Value Std. Error  t value
FINF 17.631500  0.0338120 521.4560
   K  0.790918  0.0139659  56.6321
Residual standard error: 0.143233 on 32 degrees of freedom
  Correlation of Parameter Estimates:
    FINF
K -0.672
```

or they can be plotted with the commands:

```
> plot(x,y,main="Crude Oil Forward Curve")
> lines(x, fitted(Ffit))
```

As before, we use the function fitted to extract the fitted values from the nls object Ffit. The results are given in the left pane of Fig. 4.15.

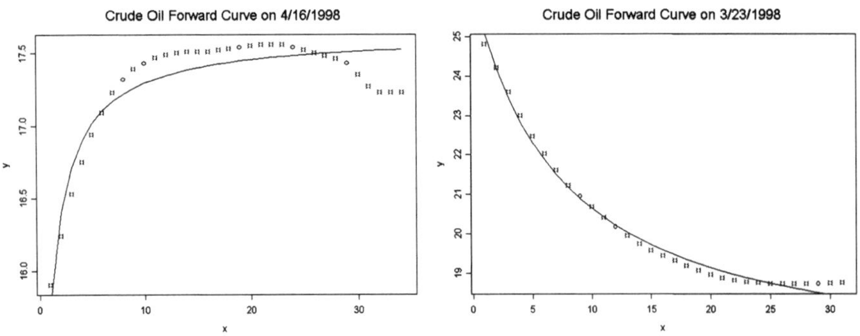

Fig. 4.15. Crude oil forward curves produced by nonlinear regression

4.7.2 **Transformation of the Variables**

As we are about to see in the next subsection, the above model will be abandoned because it cannot give a reasonable account of some of the forward curve shapes which appear on a regular basis. We kept it anyway because of its simplicity and because of the following thought-provoking remark. Notice that:

$$\varphi_\theta(x) = F_\infty \frac{x+K}{x+1} = F_\infty \left(1 + (K-1)\frac{1}{x+1}\right),$$

so that if we set $z = 1/(x+1)$, then the model (4.29) can be rewritten as:

$$y = \beta_0 + \beta_1 z$$

if we set $\beta_0 = F_\infty$ and $\beta_1 = F_\infty(K-1)$. So it should be possible to obtain the results of the nonlinear regression used earlier with a simple linear regression of the response variable y against the explanatory variable z, and a simple transformation of the estimates of the slope and intercept to get back to estimates of the original parameters F_∞ and K. Recall that maximum likelihood estimates of transformed parameters can be obtained by transforming the maximum likelihood estimates of the parameters, so our strategy is consistent when our nonlinear least squares estimates are maximum likelihood estimates, and this typically is the case when the noise terms form a Gaussian white noise. This transformation idea can be implemented in R in the following way:

```
> z <- 1/(x+1)
> LFit <- lm(y~z)
> LFINF <- coef(LFit)[1]
> LFINF
  17.63148
> LK <-1+ coef(LFit)[2]/LFINF
> LK
  0.7909179
```

So the linear regression provides exactly the same estimates for the parameters F_∞ and K. This is a beautiful example of a situation where a clever transformation of one of the variables reduces the complexity of the problem. However, it is important to keep in mind that recovering the same values for the estimates is not the whole story. Indeed, both the function `nls` and the function `lm` provide a suite of test statistics, confidence intervals and regions, and regression diagnostic tests, and the correspondence between these goodies is not so clear any more. Let us give an example for the sake of illustration. The function `lm` returns the correlation coefficient between the estimates of the slope β_1 and the intercept β_0. However, the significance of this number for the correlation between the parameters of interest F_∞ and K, is anyone's guess, because these parameters are nonlinear functions of β_0 and β_1.

4.7.3 A Second Model

The right pane of Fig. 4.15 gives an example of crude oil forward curve in backwardation, also produced by nonlinear regression. The data are from April 16, 1998. We originally tried to fit a curve from the parametric family of functions φ_θ given by (4.29) but the results were very poor. Because it can only play with two parameters, the fitting procedure could not match the curvature of the forward curve on that day. So we decided to introduce a third parameter to give the fitting procedure more leeway. Implementing the nonlinear regression from the parametric family:

$$y = \varphi_\theta(x) = F_\infty \frac{x + K_1}{x + K_2}$$

with the command:

```
> Ffit<-nls(y~FINF*(x+K1)/(x+K2),start=c(FINF=17,K1=2,K2=1))
```

we obtained the results reproduced in the right pane of Fig. 4.15. This confirms the general philosophy that more parameters produce a better fit. But as we explained earlier, following this trend too systematically can lead to disaster, especially in the presence of random noise (which is not obvious in the present situation).

Nonlinear regression is a very touchy business, and it should not be practiced without a license!

4.8 Term Structure of Interest Rates: A Crash Course

The size and level of sophistication of the market of fixed income instruments increased dramatically over the last 20 years, and it has become a prime test bed for financial institutions and academic research. The fundamental object to model is the term structure of interest rates. We approach it via the prices of treasury bond issues. Models for these prices are crucial for pricing the important swap contracts and liquid derivatives such as caplets, swaptions, etc., quantifying and managing financial risk, and setting monetary policy. In this section, we restrict ourselves to Treasury issues to avoid having to deal with the possibility of default. The highly publicized defaults of counties (such as the bankruptcy of Orange County in 1994), of sovereigns (like Russia defaulting on its bonds in 1998) and the ensuing ripple effects on worldwide markets have brought the issue of credit risk to the forefront. We discuss some of the models and instruments of the credit markets in our discussion of the credit markets and CDOs in Chap. 3.

In order to be in a position to tackle some of the fundamental statistical issues of the bond markets, we need a crash course on the mechanics of interest rates and the fixed income securities. However, in order to make our life easier, we assume that all the bonds used are default free, that there have no embedded options, no call or convertibility features, and we ignore the effects of taxes and transaction costs.

4.8.1 Zero Coupon Bonds

We first introduce the *time value of money* by valuing the simplest possible fixed income instrument. It is a financial instrument providing cash flow with a single payment of a fixed amount (the principal X) at a given date in the future. This date is called the maturity date. If the time to maturity is exactly n years, the present value of this instrument is:

$$P_X(n) = \frac{1}{(1+r)^n} X. \tag{4.30}$$

So this formula gives the present value of a nominal amount X due in n years time. Such an instrument is called a *discount bond* or a *zero coupon bond.* The positive number r is referred to as the (yearly) *discount rate* or *spot interest rate* for time to maturity n, since it is the interest rate which is applicable today (hence the terminology *spot*) on an n-year loan. Formula (4.30) gives a one-to-one correspondence between bond prices and interest rates. More generally, at any given time t, we denote by $P(t, t+m)$ the price of a zero coupon bond with unit principal and time to maturity m, or maturity date $T = t+m$. This price can be expressed in terms of an interest by the formula:

$$P(t, t+m) = \frac{1}{(1 + r(t, t+m))^m}, \tag{4.31}$$

where $r(t, t+m)$ is the yearly spot interest rate prevailing at time t for time of maturity $T = t+m$. We assumed implicitly that the time to maturity $\tau = T - t$ is a whole number of years. This definition can be rewritten in the form:

$$\log(1 + r(t, t+\tau)) = -\frac{1}{\tau} \log P(t, t+\tau)$$

and considering the fact that $\log(1+x) \sim x$ when x is small, the same definition gives the approximate identity:

$$r(t, t+\tau) \sim -\frac{1}{\tau} \log P(t, t+\tau)$$

which becomes an exact equality if we use continuous compounding. We use the Greek letter τ for the time to maturity $\tau = T - t$. This formula justifies the terminology discount rate for r. Considering payments occurring in m years time, the spot rate $r(t, t+\tau)$ is the single rate of return used to discount all the cash flows for the discrete period from time t to time $t+m$. As such, it appears as some sort of composite of interest rates applicable over shorter periods. Moreover, this formula offers a natural generalization to continuous time models with continuous compounding of the interest. In this case, it reads:

$$P(t, T) = e^{-(T-t)r(t,T)}. \tag{4.32}$$

4.8.2 Coupon Bearing Bonds

Treasury Bills are the perfect example of zero coupon bonds. They are issued and sold at auctions on a regular basis. Their maturities are shorter than 1 year, and they may not be very useful when it comes to understanding bonds with much longer maturities. Most Treasury bonds carry coupons. In general, what is called a bond (or a coupon bearing bond), is a regular stream of future cash flows. To be more specific, a *coupon bond* is a series of payments amounting to C_1, C_2, ..., C_m, at times T_1, T_2, ..., T_m, and a nominal payment X at the maturity date T_m. X is also called the face value, or principal value of the bond. The bond price at time t should be given by the formula:

$$P(t) = \sum_{j=1}^{m} C_j P(t, T_j) + XP(t, T_m). \tag{4.33}$$

This all purpose formula can be specialized advantageously, for in most cases, the payments C_j's are coupon payments made at regular time intervals. Coupon payments C_j are most often quoted as a percentage c of the face value X of the bond. In other words, $C_j = cX$. This percentage is given as an annual rate, even though payments are usually made every 6 months in the US, or different frequencies depending upon the country. It is convenient to introduce a special notation, say n_y, for the number of coupon payments per year. For example, $n_y = 2$ for coupons paid semi-annually. If we denote by r_1, r_2, ..., r_m the interest rates for the m periods ending with coupon payment dates T_1, T_2, ..., T_m, then the present value of the bond cash flow is given by the formula:

$$\begin{aligned} P &= \frac{C_1}{1 + r_1/n_y} + \frac{C_2}{(1 + r_2/n_y)^2} + \cdots + \frac{C_m}{(1 + r_m/n_y)^m} \\ &= \frac{cX}{n_y(1 + r_1/n_y)} + \frac{cX}{n_y(1 + r_2/n_y)^2} + \cdots + \frac{cX + X}{n_y(1 + r_m/n_y)^m}. \end{aligned} \tag{4.34}$$

Note that we divided the rates r_n by the frequency n_y because the rates are usually quoted in years. Formulae (4.33) and (4.34) are often referred to as the *bond price equations.* An important consequence of these formulae is the fact that on any given day, the value of a bond is entirely determined by the discount curve (i.e. the available sequence of discrete observations of the function $T \hookrightarrow P(t, T)$ on that day).

Remarks.

1. Reference to the present date t will often be dropped from the notation when no confusion is possible. Moreover, instead of working with the absolute dates T_1, T_2, ..., T_m, which can represent coupon payment dates as well as maturity dates of various bonds, it will be often more convenient to work with the times to maturities, which we denote by $\tau_1 = T_1 - t$, $\tau_2 = T_2 - t$, ..., $\tau_m = T_m - t$. We will use whatever notation is more convenient for the discussion at hand.
2. Unfortunately for us, bond prices are not quoted as a single number. Instead, they are given by a *bid-ask* interval. We ignore the existence of this bid-ask spread in

most of the discussion that follows, collapsing this interval to a single value by considering its midpoint only. We shall re-instate the bid-ask spread in the next section when we discuss the actual statistical estimation procedures.

3. Formula (4.33) shows that a coupon bearing bond can be viewed as a composite instrument comprising a zero coupon bond with the same maturity T_m and face value $(1+c)X/n_y$, and a set of zero coupon bonds whose maturity dates are the coupon payment dates T_j for $1 \le j < m$ and face value cX/n_y. This remark is much more than a mere mathematical curiosity. Indeed, the principal and the interest components of some US Treasury bonds have been traded separately under the Treasury STRIPS (Separate Trading of Registered Interest and Principal Securities) program since 1985.

4.8.3 Constructing the Term Structure by Linear Regression?

In principle, formula (4.33) above can be used to recover the prices of the zero coupon bonds in terms of the coupon bonds quoted on the market. Indeed, let us assume for example that on a given day t, we have access to the prices $P_i(t)$ of n instruments whose future cash flows are given by payments $C_{i,j}$ at the times $T_1 < T_2 < \cdots < T_m$. Notice that we assume that the payments are made at the same time T_j (instead of the auction anniversary dates). In this case, formula (4.33) can be rewritten in the form:

$$P_i(t) = \sum_{j=1}^{m} C_{i,j} P(t, T_j),$$

which shows that the vector $\boldsymbol{P}(t, \cdot\,)$ of the prices of the zero coupon bonds can be recovered from the vector $\boldsymbol{P}(t)$ of the quoted coupon bonds $P_i(t)$ when the matrix $C = [C_{i,j}]_{i,j}$ is invertible, in which case we have:

$$\boldsymbol{P}(t, \cdot\,) = C^{-1}\boldsymbol{P}(t).$$

Once the zero coupons are determined for $T = T_j$, one produces a full term structure $T \hookrightarrow P(t,T)$ by mere linear interpolation. Unfortunately, the coupon payments of the coupon bearing bonds priced on the market do not take place on the same dates, and the resulting matrix C is practically never invertible. A way to overcome this problem is to assume that the prices observed on the market are in fact noisy perturbations, and to assume that in fact:

$$P_i(t) = \sum_{j=1}^{m} C_{i,j} P(t, T_j) + \epsilon_i, \qquad i = 1, \ldots, n$$

and to extract the coefficients $P(t, T_j)$ by an ordinary least squares multiple regression. This procedure is hardly ever used in practice and we shall not discuss it any further.

4.8.4 Clean Prices & Duration

Formulae (4.33) and (4.34) implicitly assumed that t was the time of a coupon payment, and consequently, that the time to maturity was an integer multiple of the time separating two successive coupon payments. Because of the nature of the coupon payments occurring at isolated times, the prices given by the bond pricing formula (4.33) are discontinuous, in the sense that they jump at the times the coupons are paid. This is regarded as an undesirable feature, and systematic price corrections are routinely implemented to remedy the jumps. Since the bond price jumps by the amount cX/n_y at the times T_j of the coupon payments, the most natural way to smooth the discontinuities is to adjust the bond price for the *accrued interest* earned by the bond holder since the time of the last coupon payment. This notion of accrued interest is quantified in the following way. If the last coupon payment (before the present time t) was made on date T_n, then the accrued interest is defined as the quantity:

$$AI(T_n, t) = \frac{t - T_n}{T_{n+1} - T_n} \frac{cX}{n_y}, \tag{4.35}$$

and the *clean price* of the bond is defined by the requirement that the transaction price be equal to the clean price plus the accrued interest. In other words, if $T_n \leq t < T_{n+1}$, the clean price $CP(t, T_m)$ is defined as:

$$CP(t, T_m) = P_{X,C}(t, T_m) - AI(t, T_n)$$

where $P_{X,C}(t, T_m)$ is the transaction price given by (4.33) with the summation starting with $j = n + 1$. Notice that in all cases, the price of a bond appears as a multiple of its nominal value X. Since the role of the latter is merely a multiplicative factor, we shall assume, without any loss of generality that $X = 1$ from now on.

The maturity of a zero coupon bond measures the length of time the bond holder has invested his money, but it is desirable to have an analog for the case of coupon bearing bonds. Since such a bond can be viewed as a sequence of individual payments, a natural proxy could be the expected maturity of all these payments. This is the concept of duration proposed by Macaulay. For the sake of simplicity, we give its definition when the maturity is equal to an integer multiple of the length of time between two consecutive coupon payments. The duration at time t of a bond with annual coupon payment C, nominal value $X = 1$, and time to maturity m is given by the formula:

$$D_C(t, m) = \frac{C}{1 + Y_C(t, m)} + 2\frac{C}{(1 + Y_C(t, m))^2} + \cdots + m\frac{1 + C}{(1 + Y_C(t, m))^m}, \tag{4.36}$$

where $Y_C(t, m)$ is the yield of the bond, i.e. the number such that we can write the price $P_C(t, m)$ of the bond in the form:

$$P_C(t, m) = \frac{1}{(1 + Y_C(t, m))^m}. \tag{4.37}$$

The following two properties of the Macaulay duration fit well with the intuition behind the above definition. The duration of a bond is always smaller than its actual maturity. Moreover, the duration of a zero coupon bond (corresponding to the case $C = 0$) is equal to the actual maturity of the bond.

The concept of duration plays an important role in the immunization of fixed income portfolios, but a discussion of the details would take us far beyond the scope of this book. We shall limit the use of the duration to the weighting of various bond prices in the least squares term structure estimation procedures which we discuss below.

4.8.5 The Three Different Forms of Term Structure

The above discussion was aimed at identifying the one-to-one correspondence between the prices of discount bonds and some interest rates. Recall that we assume that $X = 1$. We also hinted at the fact that, when the compounding frequency was increasing without bound, the interest compounding formula

$$P(t, T = t + m) = \frac{1}{(1 + r(t, t + m)/n)^{nm}}$$

converged toward its continuously compounded analog

$$P(t, T = t + m) = e^{-Y(t,t+m)(T-t)}.$$

This formula can easily be inverted to give:

$$Y(t, T) = -\frac{1}{T - t} \log P(t, T).$$

The continuously compounded interest rate $Y(t, T)$ prevailing at time t for the maturity T is sometimes called the yield, and the curve $t \hookrightarrow Y(t, T)$ is called the yield curve. Practitioners use still a third way to capture the term structure of interest rates. It requires the notion of instantaneous forward rate which we shall only define in the continuous case. The instantaneous forward rate $f(t, T)$ prevailing at time t for the date of maturity T is defined by:

$$f(t, T) = -\frac{\frac{d}{dT} P(t, T)}{P(t, T)} = -\frac{d}{dT} \log P(t, T). \tag{4.38}$$

This definition implies that:

$$P(t, t + \tau) = e^{-\int_0^\tau f(t,t+u)du} \tag{4.39}$$

and in terms of the spot rate or yield:

$$Y(t, t + \tau) = -\frac{1}{\tau} \int_0^\tau f(t, t + u)du. \tag{4.40}$$

Even though we shall not use this fact, it is easy to see that this relation can be inverted to express the forward rates as a function of the spot rates:

$$f(t,T) = r(t,T) + (T-t)r'(t,T). \tag{4.41}$$

So on any given day t, the term structure of interest rate can be given by any one of the following three curves:

$$x \hookrightarrow P(t,t+x)$$
$$x \hookrightarrow Y(t,t+x)$$
$$x \hookrightarrow f(t,t+x)$$

We shall take advantage of this convenience when it comes to estimating the term structure from available data.

4.9 Parametric Yield Curve Estimation

This section reviews the methods of yield curve estimation used by some of the central banks which report to the Bank for International Settlements (BIS for short). Apart from the U.S. and Japan, most central banks use parametric estimation methods to infer smooth curves from daily quotes of prices of bonds and other liquid interest rate derivatives. Caplets and swaptions are most frequently used, but for the sake of simplicity, we shall limit ourselves to bond price data.

We postpone the discussion of nonparametric methods to the next chapter. The use of parametric estimation methods is justified by the principal component analysis performed in Chap. 3. There, we showed that the effective dimension of the space of yield curves is low, and consequently, a small number of parameters should be enough to describe the elements of this space. Moreover, another advantage of the parametric approach is the fact that one can estimate the term structure of interest rates by choosing to estimate first the yield curves, or the forward curves, or even the zero coupon curves as functions of the maturity. Indeed, which one of these quantities is estimated first is irrelevant: once the choice of a parametric family of curves and of their parametrization has been made, the parameters estimated from the observations, together with the functional form of the curves, can be used to derive estimates of the other sets of curves. We shall most often parameterize the set of forward rate curves, and derive formulae for the other curves (yields curves and discount bond curves) by means of the relationships made explicit earlier in formulae (4.38), (4.40) and (4.41).

On any given day, say t, one uses the available values of the discount factors to produce a curve $x \hookrightarrow f(t,x)$ for the instantaneous forward rates as functions of the time to maturity x. For the sake of notational convenience, we shall drop the reference to the present t in most of our discussions below. In this section, we limit ourselves to the fitting of a parametric family of curves to the data. We shall revisit this problem in Sect. 5.3 of Chap. 5 when we discuss nonparametric smoothing techniques based on splines.

4.9.1 Estimation Procedures

In this section we restrict ourselves to the two most commonly used curve families: the Nelson-Siegel and the Swensson families. We refer to Problem 4.12 for an illustration of the use of a third exponential family. For the sake of simpler notation, we drop the current date t from the notation.

4.9.1.1 The Nelson-Siegel Family

This family is parameterized by a 4-dimensional parameter $\boldsymbol{\theta} = (\theta_1, \theta_2, \theta_3, \theta_4)$. It is defined by:

$$f_{NS}(x, \boldsymbol{\theta}) = \theta_1 + (\theta_2 + \theta_3 x)e^{-x/\theta_4} \tag{4.42}$$

where θ_4 is assumed to be strictly positive, and as a consequence, the parameter θ_1, which is also assumed to be strictly positive, gives the asymptotic value of the forward rate. The value $\theta_1 + \theta_2$ gives the forward rate today, i.e. the starting value of the forward curve. Since this value has the interpretation of the instantaneous (short) interest rate r_t prevailing at time t, it is also required to be positive. The remaining parameters θ_3 and θ_4 are responsible for the so-called *hump*. This hump does exist when $\theta_3 > 0$, however, it is a dip when $\theta_3 < 0$. The magnitude of this hump/dip is a function of the size of the absolute value of θ_3, while θ_3 and θ_4 govern the location, along the maturity axis, of this hump/dip. Once the four parameters have been estimated, a formula for the zero-coupon yield can be obtained by plain integration from formula (4.40). We get:

$$Y_{NS}(x, \boldsymbol{\theta}) = \theta_1 + (1 - e^{-x/\theta_4}) - \theta_3\theta_4 e^{-x/\theta_4}. \tag{4.43}$$

A formula for the discount factor can be deduced by injecting the yield given by this formula into $P_{NS}(x) = e^{-xY_{NS}(x)}$. The Nelson-Siegel family and these formulae are used in countries such as Finland and Italy to produce yield curves.

4.9.1.2 The Swensson Family

To improve the flexibility of the curves and the fit, Swensson proposed a natural extension to the Nelson-Siegel's family by adding an extra exponential term which can produce a second hump/dip. This extra flexibility comes at the cost of two extra parameters which have to be estimated. The Swensson family is generated by mixtures of exponential functions of the Nelson-Siegel type. To be specific, the Swensson family is parameterized by a 6-dimensional parameter θ, and defined by:

$$f_S(x, \boldsymbol{\theta}) = \theta_1 + (\theta_2 + \theta_3 x)e^{-x/\theta_4} + \theta_5 x e^{-x/\theta_6}. \tag{4.44}$$

As before, once the parameters are estimated, the yield curve can be estimated by plain integration of (4.44). We get:

$$Y_S(x,\boldsymbol{\theta}) = \theta_1 - \frac{\theta_2\theta_4}{x}(1-e^{-x/\theta_4}) + \theta_3\theta_4\left[\frac{\theta_4}{x}(1-e^{-x/\theta_4}) - e^{-x/\theta_4}\right]$$
$$+\,\theta_5\theta_6\left[\frac{\theta_6}{x}(1-e^{-x/\theta_6}) - e^{-x/\theta_6}\right]. \tag{4.45}$$

The Swensson family is used by the Central Banks of many countries including Canada, Germany, France and the UK.

4.9.2 **Practical Implementation**

Enough talk, let's see how all these ideas work in practice.

4.9.2.1 Description of the Available Data

On any given day t, financial data services provide, for a certain number of bond issues, the times to maturity $x_j = T_j - t$, the coupon payments and their frequencies, and various pre-computed quantities. We used data from `Data Stream`. For the purposes of illustration, we chose to collect data on German bonds. These instruments are very liquid and according to BIS, the Deutsche Bundesbank uses the Swensson's extension of the Nelson-Siegel family to produce yield curves. As an added bonus, the coupons on the instruments we chose are paid annually, which makes numerical computations easier.

4.9.2.2 The Actual Fitting Procedure

Let B_j be the bond prices available on a given day t. Let $B_j(\theta)$ be the prices one would get using formula (4.33), with zero coupon values computed from formula (4.39) when the forward curve is given by the element of the parametric family of forward curves determined by the parameter $\boldsymbol{\theta}$. Then our estimate of the term structure of interest rates is given by the zero-coupon/yield/forward curve corresponding to the (vector) parameter $\hat{\boldsymbol{\theta}}^*$ which minimizes the quadratic loss function:

$$\mathcal{L}(\boldsymbol{\theta}) = \sum_j w_j |B_j - B_j(\boldsymbol{\theta})|^2 \tag{4.46}$$

for a given set of weights w_j which are usually chosen as functions of the duration (4.36) and the yields to maturity. The dependence of the loss function upon the parameters $\boldsymbol{\theta}$ appears to be complex and extremely nonlinear. In any case, this least squares estimation procedure is very much in the spirit of the nonlinear regression discussed in Sect. 4.7. As explained earlier, fitting the parameters depends upon delicate optimization procedures which can be very unstable and computer intensive. For this reason we will give only a limited sample of results below.

Remarks.

1. Many Central Banks do not use the full spectrum of available times to maturity. Indeed, the prices of many short-term bonds are very often influenced by liquidity problems. For this reason, they are often excluded from the computation of the parameters. For example each of the Bank of Canada, the Bank of England and the Deutsche Bundesbank consider only those bonds with a remaining time-to-maturity above 3 months. The French Central Bank also filters out the short term instruments.
2. Even though it appears less general, the Nelson-Siegel family is often preferred to its Swensson relative. Being of a smaller dimension, the model is more robust and less unstable. This is especially true for countries with a relatively small number of issues. Finland is one of them. Spain and Italy are other countries using the original Nelson-Siegel family for stability reasons.
3. The bid-ask spread is another form of illiquidity. Most central banks choose the mid-point of the bid-ask interval for the value of B_j. The Banque de France does this for most of the bonds, but it also uses the last quote for some of them. Suspicious that the influence of the bid-ask spread could overwhelm the estimation procedure, the Finnish Central Bank uses a loss function which is equal to the sum of squares of errors where the individual errors are defined as the distance from $B(j, \boldsymbol{\theta})$ to the bid-ask interval (this error being obviously 0 when $B(j, \boldsymbol{\theta})$ is inside this interval).
4. It is fair to assume that most Central Banks use accrued interests and clean prices to fit a curve to the bond prices. This practice is advocated in official documents of the Bank of England and the US Treasury.
5. Some of the countries relying on the Swensson family, first fit a Nelson-Siegel family to their data. Once this 4-dimensional optimization problem is solved, they use the argument they found, together with two other values for θ_5 and θ_6 (often 0 and 1), as initial values for the minimization of the loss function for the Swensson family. Even then, these banks opt for the Swensson family, only when the final θ_5 is significantly different from 0 and θ_6 is not too large! This mixed procedure is used by Belgium, Canada and France.

4.9.3 R Experiments

The following R code can be used to compute values of the forward rate function in the Nelson-Siegel model. The graphs of Fig. 4.16 were produced with the commands:

```
> THETA <- c(.07,-.03,.1,2.5)
> XX <- seq(from=0,to=30,by=1/12)
> FORW <- fns(XX,THETA)
> YIELD <- yns(XX,THETA)
> par(mfrow=c(1,2))
> plot(XX,FORW,type="l",ylim=c(0,.25))
> title("Example of a Nelson Siegel Forward Curve")
> plot(XX,YIELD,type="l",ylim=c(0,.25))
```

```
> title("Example of a Nelson Siegel Yield Curve")
> par(mfrow=c(1,1))
```

Finally, the price $B_{NS}(\boldsymbol{\theta})$ of a coupon bond can be computed from its coupon rate COUPON, the accrued interest AI, the time to maturity LIFE given in years, and the parameters THETA of the Nelson-Siegel family, using the function bns, whose code implements the various formulae derived in this subsection. In order to illustrate the use of this function we chose the example of the German bond quotes available on May 17, 2000. After some minor reformatting and editing to remove the incomplete records, we created the data frame GermanB041700 included in the library Rsafd. Its main purpose is to offer a testbed for the nonlinear fit of the Nelson-Siegel family.

```
> head(GermanB041700)
 Issue Coupon Maturity  Price Intrst.Yield Redemp.Yield Accrud.Intrst  Life
1    1992    8.00     2002 105.28         7.60         5.202          7.67  2.04
3    1993    6.75     2003 104.37         6.47         5.158          6.47  3.04
5    1999    3.75     2009  90.43         4.15         5.135          1.07  8.72
7      G3    3.00     2010  77.00         3.90         6.080          0.12 10.46
9      G4    3.00     2010  77.00         3.90         6.080          0.12 10.46
11     G5    3.00     2010  77.00         3.90         6.080          0.12 10.46
```

We compute and minimize the loss (4.46) in the case of the Nelson-Siegel parametrization. We shall denote it by $\mathcal{L}_{NS}(\theta)$. It is given by the formula:

$$\mathcal{L}_{NS}(\boldsymbol{\theta}) = \sum_j w_j |B_j - B_{NS}(j,\boldsymbol{\theta})|^2. \tag{4.47}$$

The parameters THETA[j] are obtained by minimizing the sum of square deviations (4.47). Since we do not know what kind of weights (if any) are used by the German Central Bank, we set $w_j = 1$. We use the R function optima to perform the minimization of the sum of square errors which we compute for each candidate parameter set THETA with the homegrown function BondSSE defined below.

```
> BondSSE <- function(x)
{
  sum((Price-bns(COUPON=Coupon,AI=Accrud.Intrst,LIFE=Life,THETA=x)$price)^2)
}
> GB.fit <- optim(par=as.numeric(c(0.07,-0.03, 0.10,2.50)),
     fn=BondSSE, gr=NULL, method = "L-BFGS-B",
           lower=c(0,-Inf,-Inf,0), upper=c(Inf,Inf,Inf,Inf))
> GB.fit
$par
[1]  0.00000000 -0.03701645 -0.01441902  9.00320236
$value
[1] 221.5779
$counts
function gradient
     117      117
$convergence
[1] 0
```

We visualize the quality of the resulting term structure of bond prices with the plots in Fig. 4.17.

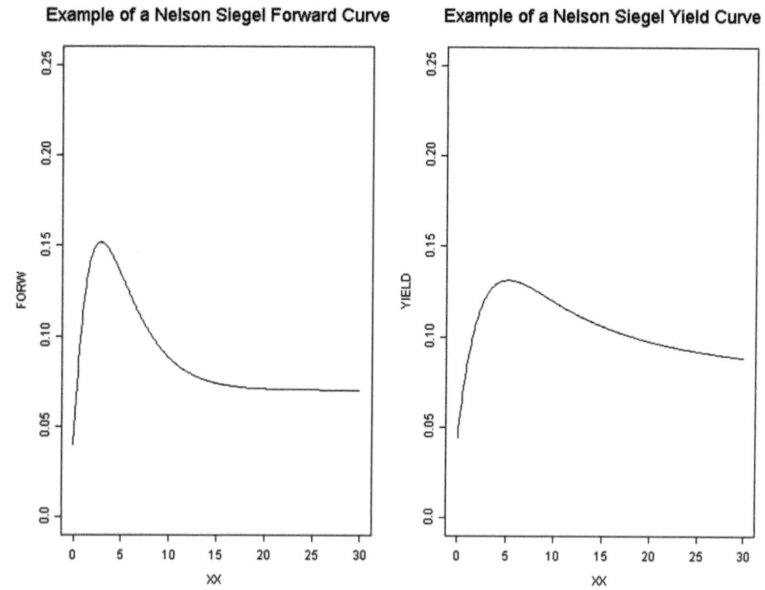

Fig. 4.16. Example of a forward curve (*left*) and a yield curve from the Nelson-Siegel family with parameters $\theta_1 = 0.07$, $\theta_2 = -0.03$, $\theta_3 = 0.1$ and $\theta_4 = 2.5$

The plot in the right pane of Fig. 4.17 was produced with the commands:

```
> plot(Price,bns(COUPON=Coupon,AI=Accrud.Intrst,LIFE=Life,
                  THETA=GB.fit$par)$price,ylab="")
> title("Nelson-Siegel fitted Bond Prices against actual Prices")
> abline(0,1)
```

The results are surprisingly good given the poor nature of the data as exemplified by the scatterplot of the bond prices against the times to maturity given by the variable `Life` reproduced in the left pane of Fig. 4.17.

4.9.4 Concluding Remarks

The results reported above show an extreme variability in the estimates at the *short end of the curve.* This confirms the widely-admitted fact that the term structure of interest rates is more difficult to estimate for short maturities. This is one of the reasons why many central banks do not provide estimates of the term structure for the left hand of the maturity spectrum.

All in all it seems clear that the various estimates are stable and reliable in the maturity range from 1 to 10 years.

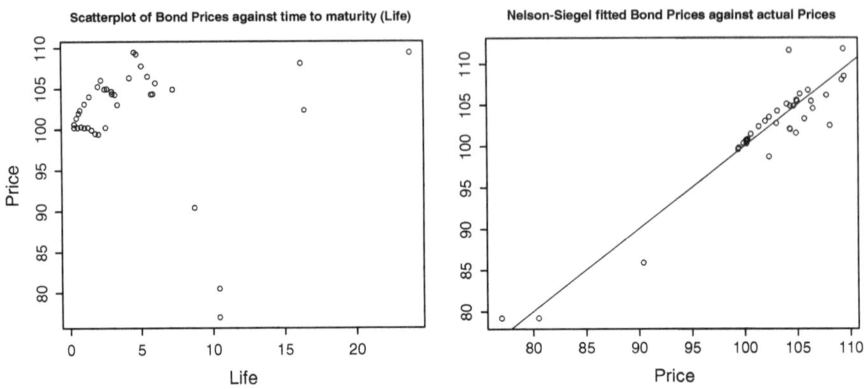

Fig. 4.17. Scatterplot of the bond prices against time to maturity (*left*) and comparison with the prices from a fitted bond curve from the Nelson-Siegel parametric family (*right*)

APPENDIX: CAUTIONARY NOTES ON R IDIOSYNCRACIES

The specific features of R which we describe in this appendix can be sources of headaches for first time users.

R and Polynomial Regression

We use a controlled experiment to illustrate an important point made in the text.

The Experiment

We generate a set of observations (x_i, y_i) for which the values of x are uniformly distributed between 0 and 100 and the values of y are, up to a noise which is normally distributed with variance one, given by the polynomial function:

$$y = \varphi(x) = 50 - 43x + 31x^2 - 2x^3.$$

The R commands to do that are:

```
> x <- runif(100,0,100)
> y <- 50 - 43*x + 31*x^2 -2*x^3 + rnorm(100)
```

Given the construction of x and y, we expect that a polynomial regression (using a polynomial of degree 3) of the variable y against the variable x will recover the exact coefficients we started with. Well, let's see:

```
> xylm <- lm(y ~ poly(x,3))
> xylm
Call:
lm(formula = y ~ poly(x, 3))
```

```
Coefficients:
 (Intercept) poly(x, 3)1 poly(x, 3)2 poly(x, 3)3
   -465738.7    -4798444    -2052599   -358566.4
```

Obviously, the coefficients given in the R summary are not what we expected!

So What Is Wrong?

It turns out that the estimated coefficients printed out by the program are the coefficients obtained in the decomposition of the function φ in a basis of 4 orthogonal polynomials which are **NOT** the polynomials $1 = x^0$, x, x^2 and x^3. As we explained in the section on polynomial regression, the internal manipulations of the program can be done in any basis. It turns out that, for reasons of numerical efficiency and stability, R chose to work with a basis of orthogonal polynomials different from the natural basis $\{1, x, x^2, x^3\}$ which is often called the canonical basis. As explained in the text, it is always possible to force R to construct the design matrix and compute the regression coefficients in the canonical basis, by using the function `poly` with the parameter `raw` set to `TRUE`. In the present situation, this gives

```
> xylmT <- lm(y ~ poly(x,3,raw=TRUE))
> xylmT
Call:
lm(formula = y ~ poly(x, 3, raw=TRUE))
Coefficients
(Intercept) poly(x,3,raw=TRUE)1 poly(x,3,raw=TRUE)2 poly(x,3,raw=TRUE)3
 49.95       -42.99                31.00                -2.00
```

which is what we expected!

Is There Something Wrong with the Function `lines`?

The function `lines` is a convenient graphical tool which can be used to produce a continuous curve by joining points on a graph. But as we are about to see, a careless call to this function can produce unexpected results. We use the `ethanol` data set from the R distribution to make our point. These data, though non-financial, are very well suited for the illustration we have in mind. They contain 88 measurements of the variable NOx giving the concentration of Nitric Oxide in a series of tests of a single-cylinder automobile engine. The regressor variables used to predict the response are the compression ratio C and the equivalence ratio E. We use the following commands to produce a scatterplot together with a smooth curve giving the general trend in the data. We shall come back to the function `loess.smooth` in the next chapter on nonparametric regression. It does not play a significant role here.

```
> attach(ethanol)
> dim(ethanol)
88  3
> names(ethanol)
"NOx" "C"   "E"
> plot(NOx~E)
> lines(loess.smooth(E,NOx))
```

Notice that, in order to illustrate the various forms of the various ways scatterplots can be performed, we fed the formula `NOx~E` to the generic function `plot`. This form `plot(y~x)` corresponds to the wording *plot y against x*. The result is shown in Fig. 4.18. Now, from the appearance of the result, one might think that a similar result could be obtained with a polynomial regression of high enough degree. We could do that by first computing the fitted values and then by joining these fitted values by straight line segments. After all, we have used this trick several times already. Using the commands:

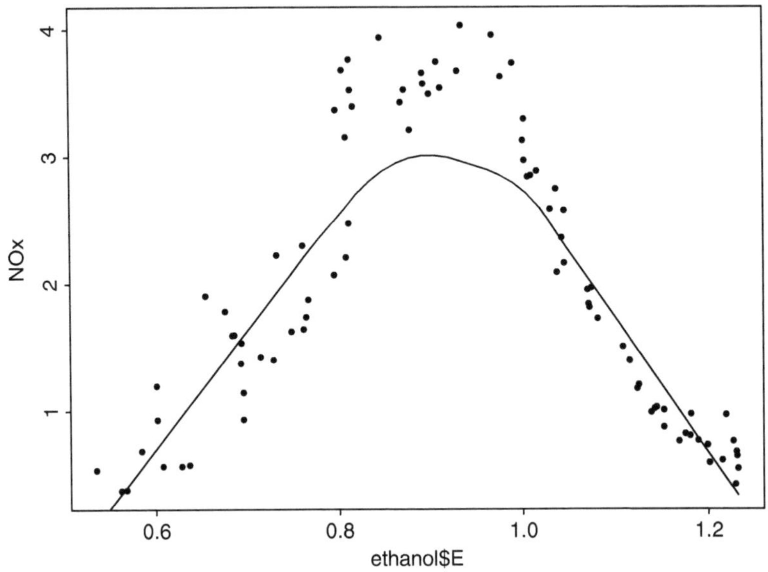

Fig. 4.18. Scatterplot of the ethanol data set

```
> plot(E,NOx)
> lines(E, fitted(lm(NOx ~ poly(E,8))))
```

we get the results reproduced in Fig. 4.19.

Obviously, something went wrong. The results are nothing near what we expected! To explain this apparent anomaly, we notice that, in our previous calls to the function `lines`, the values of the first components of the points to be joined were always in increasing order. On the other hand, if we look at the values of the current explanatory variable `E` we see that this is not the case.

```
> E
  1     2     3     4     5     6     7     8     9
0.907 0.761 1.108 1.016 1.189 1.001 1.231 1.123 1.042
.....................................................
.....................................................
 79    80    81    82    83    84    85    86    87    88
1.18 0.795 0.99 1.201 0.629 0.608 0.584 0.562 0.535 0.655
```

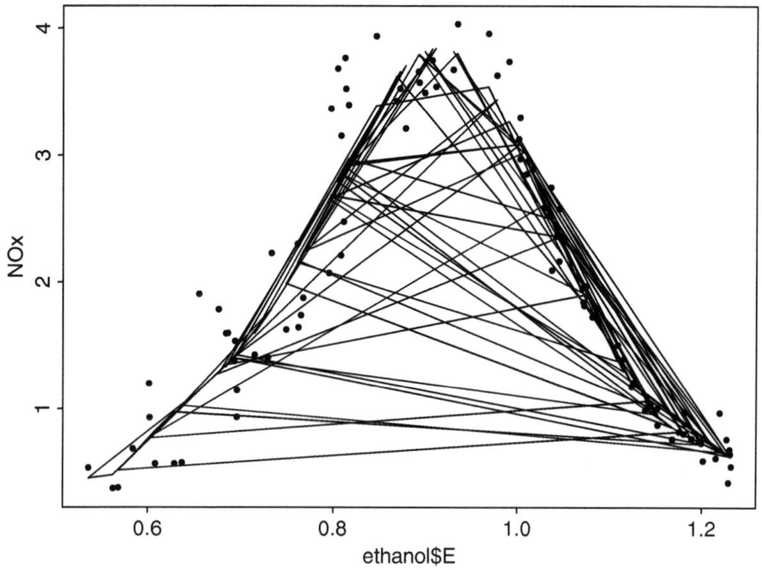

Fig. 4.19. Example of the WRONG use of the function `lines`

Now that we understand what is going wrong, can we find a solution to this problem? The answer is YES of course! We simply need to re-order the points so that the `E` – values are in increasing order. We can use the `R` function `order` to find the order of the entries of the vector `E`, and we can use this order to subscript the vectors containing the coordinates of the points. The following commands show how one can do all this.

```
> RKS <- order(E)
> EE <- E[RKS]
> NN <- NOx[RKS]

> par(mfrow=c(1,2))
> plot(E,NOx)
> plot(EE,NN)
> par(mfrow=c(1,1))
```

The plot is reproduced in Fig. 4.20. We can now try again to smooth the data cloud and use the function `lines` once more.

```
> plot(EE,NN)
> lines(EE, fitted(lm(NN ~ poly(EE,8))))
```

The results are reproduced in Fig. 4.21, and this time, our expectations are met.

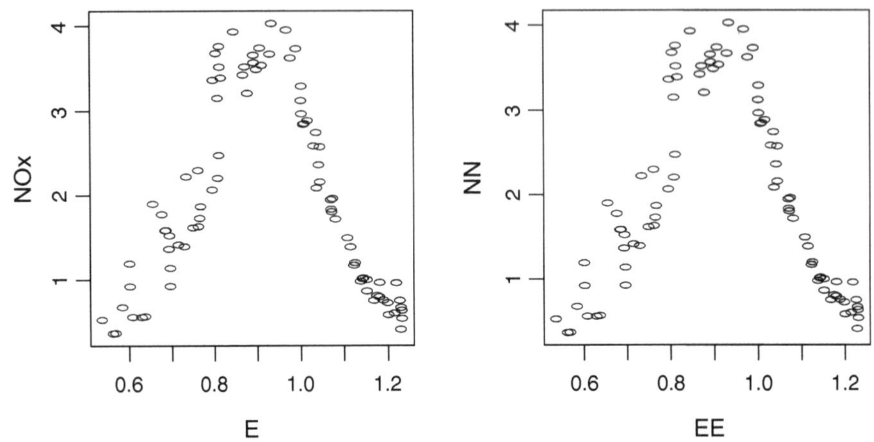

Fig. 4.20. Clearly, ordering the data does not affect the scatterplot

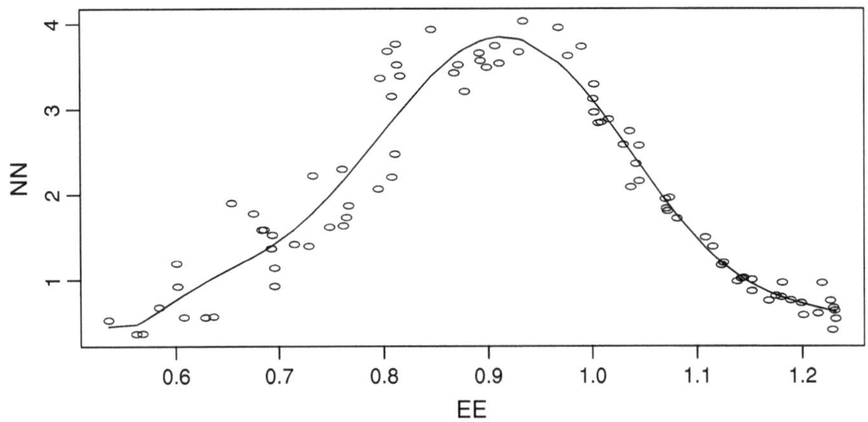

Fig. 4.21. Correct use of the function `lines` on the polynomial regression of degree 8 of the ethanol data

PROBLEMS

Ⓔ **Problem 4.1** *1. Can the plot appearing in the left pane of Fig. 4.22 be the plot of the raw residuals of a linear least squares regression? Explain why.*

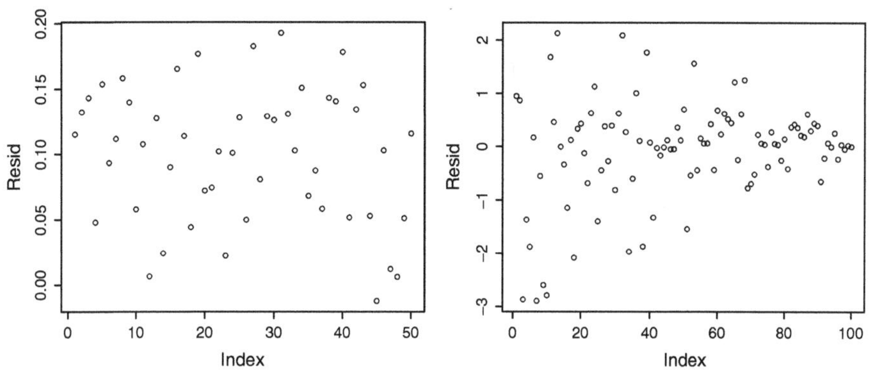

Fig. 4.22. Possible residuals from least squares simple linear regressions

2. *The right pane of Fig. 4.22 gives the sequential plot of the raw residuals of a multiple least squares linear regression. What can you say about the diagonal entries* $h_{i,i}$ *of the* hat matrix*?*
3. *Say if the line in Fig. 4.23 is a least squares regression line or a least absolute deviations regression line and explain why.*

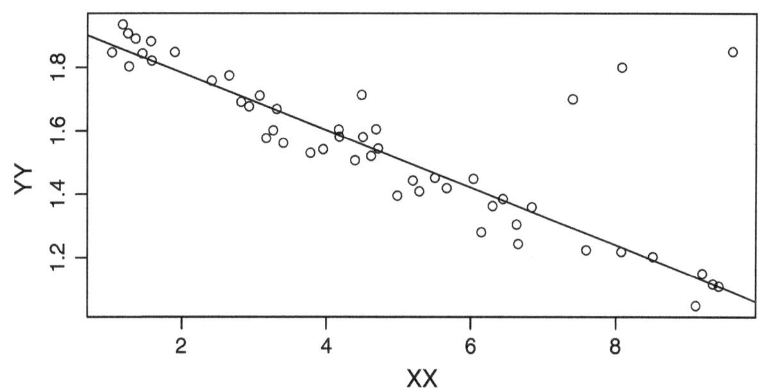

Fig. 4.23. Possible least squares simple linear regression

4. *The same scatterplot appears on both sides of the Fig. 4.24. The response variable appears on the vertical axis and the explanatory variable on the horizontal axis. Sketch (your best guess of) the graph of the least squares polynomial regression of degree* 2 *on the left, and of the least absolute deviations polynomial regression of degree* 2 *on the right. Explain the differences between the two sketches if any, and explain why you did what you did.*

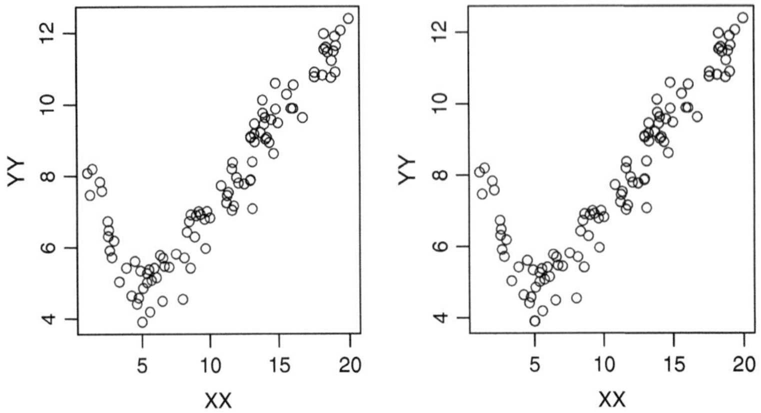

Fig. 4.24. Scatterplots of the same sample of points

Ⓔ **Problem 4.2** *1. Can the plot appearing below be the plot of the raw residuals of a linear least squares regression? Explain your answer.*

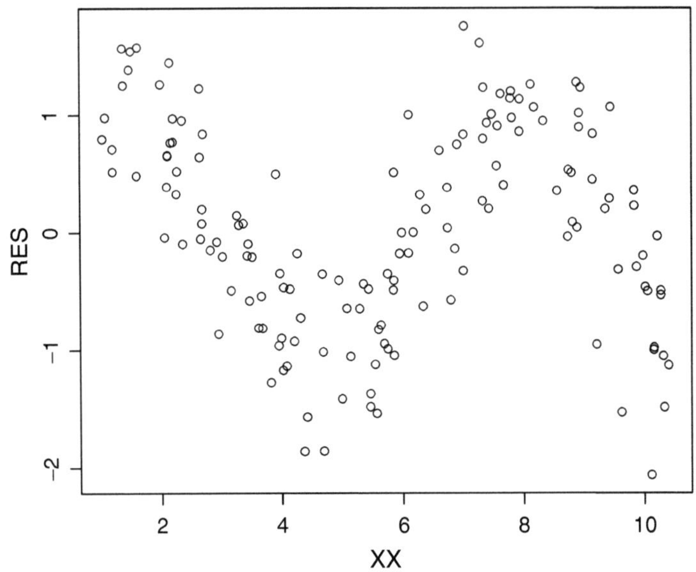

Assuming they are raw residuals of a linear least squares regression, say which feature of the data could have produced such residuals, and propose a fitting procedure which, when applied to the residual data, would produce residuals with smaller variance and more in line with what is expected from the residuals of a linear least squares regression.

2. *Assuming now that* (x_1, y_1), $\cdots$, (x_n, y_n) *is a bivariate sample and* $\hat{\beta}_0$, $\hat{\beta}_1$, *and* $\hat{\sigma}^2$ *are the maximum likelihood estimators of the intercept, slope and noise variance of a least squares simple linear regression performed on these data,*

2.1 Explain what is understood by "the regression is significant".
2.2 Recall the definition of the R^2 and explain (at least qualitatively) how its value relates to the significance of the regression.

Ⓣ **Problem 4.3** *A ski resort operator has determined that on any given year, the P&L P_i at the end of the winter season (April 15) is, up to an additive noise term, a linear function of the snow pack, say S_i as measured in Snow Water Equivalent (SWE) on December 31 of the previous year. In other word, it is given by a relation of the form:*

$$P_i = \beta_0 + \beta_1 S_i + \epsilon_i$$

where the ϵ_i are independent identically distributed mean zero Gaussian random variables with common variance σ^2, and β_0, β_1 and σ^2 are constants. We assume that the last 15 years for which the resort operator has data are coded by $i = 1, 2, \cdots, 15$. The least squares simple linear regression on the 15 years worth of data gave the estimates

$$\hat{\beta}_0 = 1.5, \qquad \hat{\beta}_1 = -0.1, \qquad \text{and} \qquad \hat{\sigma}^2 = 0.12.$$

1. *What should the resort operator expect for the P&L of the season if on December 31 the Natural Resource Conservation Service announces that the snow pack measurement was $S = 3.6$.*
2. *Explain how you would determine the end points of an interval which would contain the future value of the P&L with probability 95 %. Should this interval be computed as a confidence interval or a prediction interval? Explain the difference, and the reasons of your choice.*

Ⓣ **Problem 4.4** *According to the CAPM theory, the daily excess return R on a given stock is related to the excess return $R^{(m)}$ of the market portfolio by a simple linear model of the form*

$$R = \alpha + \beta R^{(m)} + \epsilon$$

where ϵ is a mean zero Gaussian random variable with variance σ^2 and α, β and σ^2 are constants.

1. *We first assume that an oracle announced that the true values of the parameters are*

$$\alpha = 0.05, \qquad \beta = -0.1, \qquad \text{and} \qquad \sigma^2 = 0.09.$$

Assuming that, on a given day, $R^{(m)} = 1.5$,
1.1. What would be your estimate of R?
1.2. For such an estimate, should you compute a confidence interval *or a* prediction interval*? Explain your answer. How would you compute such a 98 % interval for the true value of R?*
1.3. How would you compute the daily 1 % VaR?

2. *Having no luck getting an oracle to help, you perform a simple linear least squares regression from data collected over the last $n = 252$ days in order to obtain estimates of the same 1 % VaR. Let*

$$(r_1^{(m)}, r_1), (r_2^{(m)}, r_2), \cdots\cdots, (r_n^{(m)}, r_n)$$

be the values of the excess returns of the market portfolio and the stock obtained over this period, denote by $\hat{\alpha}$, $\hat{\beta}$ and $\hat{\sigma}^2$ the estimates obtained for the parameters of the model, and by $e_1 = r_1 - \hat{\alpha} - \hat{\beta} r_1^{(m)}, \cdots, e_n = r_n - \hat{\alpha} - \hat{\beta} r_n^{(m)}$ the corresponding raw residuals.

2.1. *How could you get the variance/covariance of the raw residuals?*
2.2. *How would you change the computation of the interval of question 1.2? How does the width of the interval change compared to question 1.2?*
2.3. *How would you estimate the daily 1 % VaR in the present situation?*

Ⓣ **Problem 4.5** *Assume that a simple linear model of the form*

$$Y = \beta_0 + \beta_1 X + \epsilon$$

is used to explain the P&L (profits and losses) Y from the value of an index X. As usual, ϵ is a mean zero Gaussian random variable with variance σ^2, and β_0, β_1 and σ^2 are constants.

1. *We first assume that an oracle announced that the true values of the parameters are*

$$\beta_0 = 0.5, \qquad \beta_1 = -0.05, \qquad \text{and} \qquad \sigma^2 = 0.09.$$

Given that $X = 5$ on a given day:
1.1. *What would be your estimate of Y?*
1.2. *For such an estimate, should you compute a* confidence interval *or a* prediction interval? *How would you compute such a 98 percentile interval for the true value of Y? The 99 % quantile of standard normal distribution is 2.326.*
1.3. *How would you compute the daily 1 % VaR?*

2. *Let us now assume that having no luck trying to access an oracle, we perform a simple linear least squares regression from data collected over the last $n = 252$ days. Let*

$$(x_1, y_1), (x_2, y_2), \cdots\cdots, (x_n, y_n)$$

be the values of the indexes and subsequent P&Ls, denote by $\hat\beta_0$, $\hat\beta_1$ and $\hat\sigma^2$ the least squares estimates of the parameters of the model, and by $r_1 = y_1 - \hat\beta_0 - \hat\beta_1 x_1, \cdots, r_n = y_n - \hat\beta_0 - \hat\beta_1 x_n$ the corresponding residuals.
2.1. *How could you get the variance/covariance of the residuals?*
2.2. *How would you change the computation of the interval of question 1.2? How does the width of the interval change compared to question 1.2?*
2.3. *How would you estimate the daily 1 % VaR in the present situation?*

Ⓔ Ⓣ **Problem 4.6** *Each year, the first quarter (Q1) profit and loss (P&L) in millions of local currency, say Y, of a Northern Europe aluminum producer is given by a formula of the form*

$$Y = \beta_0 + \beta_1 X + \epsilon$$

where X is the price on the first trading day of the month of January of that year of the forward Natural Gas contract with April delivery at the nearest city gate, ϵ is a mean zero Gaussian random variable with variance σ^2, and β_0, β_1 and σ^2 are constants.

1. *We first assume that the Chief Risk Officer (CRO) of the company has access to an oracle who tells her that the true values of the parameters are*

$$\beta_0 = 1, \qquad \beta_1 = -0.125, \qquad \sigma^2 = 0.36.$$

We shall consider two scenarios:
- *The January price of the natural gas contract for April delivery is $x_0 = 4.1$*
- *Due to damage to a pipeline, the price of the same contract is $x_0 = 16.7$*

How should the CRO compute the 1 % VaR for the Q1 P&L at the end of the first trading day of January in each of these two scenarios?

2. *Let us now assume that the CRO does not have access to the oracle. She performs a simple linear regression from data* $(x_1, y_1), (x_2, y_2), \cdots\cdots, (x_n, y_n)$ *collected over the last* $n = 15$ *years, and let us assume that the minimum of the observed* x_i*'s is* 2.5 *while the maximum is* 12.8*. The CRO performs a simple linear regression and obtains estimates* $\hat{\beta}_0$, $\hat{\beta}_1$ *and* $\hat{\sigma}^2$ *for the parameters of the model.*
   2.1. *Given the assumption stated above, which regression method should the CRO use?*
   2.2. *Explain how in each of the two scenarios considered above, the VaR should change.*

Ⓔ **Problem 4.7** *The purpose of this problem is to analyze the data contained in the data set* `STRENGTH`*. The first column gives the fracture strength* `X` *(as a percentage of ultimate tensile strength) and the second column gives the attenuation* `Y` *(the decrease in amplitude of the stress wave in neper/cm) in an experiment on fiberglass re-inforced polyester composites.*

1. *Is a linear regression model reasonable to describe the data?*
2. *Perform a linear least squares regression of* `attenuation` *against* `strength`*, determine the estimated intercept, the estimated slope and estimate the variance of the noise.*
3. *Plot the fitted values of* `attenuation` *against the corresponding actual values of* `attenuation`*, compute the coefficient of determination (the famous* R^2*), and assess the relevance of the linear model. Compute the predicted values of the variable* `attenuation` *for the values* $x = 20$, $x = 50$ *and* $x = 75$ *of* `strength`*.*
4. *Compute the raw residuals and plot them against the fitted values of the* `attenuation` *together with a* `qqnorm` *plot of these raw residuals. Comment on the results.*
5. *Same question as in 2 and 3 (with the exception of the computation of* R^2*) for the least absolute deviations regression instead of the least squares regression.*

Ⓔ **Problem 4.8** *The purpose of this problem is to perform a regression analysis of the* 35 *Scottish hill races data contained in the data set* `hills`*.*

1. *Produce the scatterplot of the variable* `time` *against the variable* `climb`*, and superimpose the least squares regression line of* `time` *against the variable* `climb`*. Repeat the same thing for* `time` *against* `dist`*. For each of these regressions, compute the* R^2*, and compare their values. Compute the raw residuals and produce their normal Q-Q plot. Comment.*
2. *Go through the same steps as in the question above using the least absolute deviations method instead, and an analog of* R^2 *appropriate for this type of regression. Compare the results so obtained to the results of the least squares regression.*
3. *Using the result of the regression of* `time` *against* `dist`*, compute both in the case of the least squares regression and in the case of the absolute deviations regression, the predictions of the record times for a marathon (i.e. the value* 26.2 *for the variable* `dist`*).*
4. *For this question, we suppose that the information for one of the races was entered incorrectly. Create a new data set, call it* `Thills`*, with an erroneous information for the run* `Lairig Ghru`*. Use:*
   `Lairig Ghru 28.0 2100 92.667`
   instead of
   `Lairig Ghru 28.0 2100 192.667` *In other words, create a new vector* `Tdist` *identical to* `dist` *but change the 11-th entry of the vector* `time` *and create a new vector* `Ttime` *with the new value* 92.667*. On the scatterplot of* `time` *against* `dist`*, superimpose the least squares regression line of* `time` *against* `dist`*, and of* `Ttime` *against*

`Tdist`. *Create a new scatterplot of* `time` *against* `dist`, *and superimpose on this new scatterplot the least absolute deviations regression line of* `time` *against* `dist`, *and of* `Ttime` *against* `Tdist`. *Explain what you see, compare the regression coefficients, and explain.*

5. *Perform the multiple least squares regression of* `time` *against* `dist`, *and* `climb`, *compute the* R^2, *and compare its value to the values of* R^2 *found for the simple least squares regressions of* `time` *against* `dist`, *and* `time` *against* `climb`. *Explain the results. Compute the raw residuals and produce their normal Q-Q plot. As in the case of* R^2, *compare the result to the Q-Q plots of the simple least squares regressions.*

Ⓔ **Problem 4.9** *The purpose of this problem is to perform a first analysis of a classical data set contained in the data file* `BASKETBALL`. *The intended analysis is very much in the spirit of the previous problem.*

1. *We first use simple regression techniques.*
   1.1. *Use least squares simple linear regression to explain the number of points scored by a player as given by the variable* `points` *in terms of the variable* `height`.
   1.2. *Similarly, use least squares simple linear regression to explain* `points` *in terms of the variable* `minutes` *giving the time played per game.*
   1.3. *Use least squares simple linear regression to explain* `points` *in terms of the values of* `age` *obviously giving the age of the player.*
   1.4. *Use regression diagnostics to compare these regressions and rank the variables* `height`, `minutes` *and* `age` *according to their ability to predict the variable* `points`.
2. *Use a least squares multiple regression to explain the values of the variable* `points` *in terms of* `height` *and* `minutes`. *Similarly use a least squares multiple regression to explain* `points` *in terms of* `height`, `minutes` *and* `age`. *Does the inclusion of* `age` *in the regression improve the performance of the model?*
3. *Redo question 1 with least absolute deviations regression instead of least squares regression. Compare the results and comment on the (possible) differences.*

Ⓔ **Problem 4.10** *The goal of this problem is to analyze the data set* `MID1`. *Denote the first and second columns by* X *and* Y *respectively. We want to compare the least squares and least absolute deviations polynomial regressions of degree 3 of the values of* Y *against the corresponding values of* X. *In other words, we want to compare the real numbers* β_0, β_1, β_2 *and* β_3, *which minimize the criterion:*

$$C_2(\beta_0, \beta_1, \beta_2, \beta_3) = \sum_{j=1}^{134} |Y[j] - \beta_0 - \beta_1 X[j] - \beta_2 X[j]^2 - \beta_3 X[j]^3|^2$$

as well as a measure of the goodness of the fit of your choice to those obtained from the minimization of the criterion:

$$C_1(\beta_0, \beta_1, \beta_2, \beta_3) = \sum_{j=1}^{134} |Y[j] - \beta_0 - \beta_1 X[j] - \beta_2 X[j]^2 - \beta_3 X[j]^3|.$$

In this problem, one is expected to use the `R` *functions* `lsfit` *and* `l1fit`, ***not*** *the function* `lm`.

1. *Give the commands needed to perform these two regressions, run these commands and give the resulting coefficients β_0, β_1, β_2 and β_3 and the optimal values of the criteria in both cases. In particular, explain how you construct the design matrix and compute the fitted values in both cases.*
2. *Give, on the same plot, the scatter plot of the values of* `X` *and* `Y`*, the graph of the least squares polynomial regression and the graph of the absolute deviations polynomial. Say which of these two regressions looks better to you and explain why the method you prefer performed better.*

Ⓔ **Problem 4.11** *The purpose of this problem is to analyze a classical example of nonlinear regression which can be found in most textbooks discussing the subject.*

The data, which are not of a financial nature, are contained in the data frame `Puromycin` *included in the* `R` *distribution. They contain three variables pertaining to a specific set of biomedical experiments on cells which were either* `treated` *or* `untreated` *with the drug Puromycin (this information is given by the third variable* `state`*). The response variable y is the initial velocity* `vel` *of the reaction while the explanatory variable x is the enzyme concentration given in the first column* `conc`*. It is expected that the regression function $\varphi(x)$ will be given by the so-called Michaelis-Menten relationship:*

$$y = \varphi(x) = V_a \frac{x}{x+K},$$

where the constant V_a has the interpretation of an asymptotic velocity while K is a constant.

1. *Attach the data frame* `Puromycin` *to your* `R` *session and extract the rows corresponding to the treated cases. Perform a nonlinear regression of the velocity y against the enzyme concentration x, starting with initial values $V_a = 200$ and $K = 0.1$ for the parameters.*
2. *Give the values of the estimates obtained for the parameters V_a and K, and plot, on the same graph, the original data points of the "treated" sub-sample together with a broken line going through the points fitted by the nonlinear regression.*

Ⓔ Ⓣ **Problem 4.12** *In this problem we introduce a new exponential family (called the generalized Vasicek family) to parameterize the term structure of interest rates. For each value of the 4-dimensional parameter $\boldsymbol{\theta} = (\theta_1, \theta_2, \theta_3, \theta_4)$ we define the function $Y_{GV}(x, \boldsymbol{\theta})$ by:*

$$Y_{GV}(x, \boldsymbol{\theta}) = \theta_1 - \theta_2\theta_4 \frac{1 - e^{-x/\theta_4}}{x} + \theta_3\theta_4 \frac{(1 - e^{-x/\theta_4})^2}{4x} \tag{4.48}$$

where θ_4 is assumed to be strictly positive.

1. *Derive an analytic formula for the instantaneous forward rates and write an* `R` *function* `fgv` *to compute their values. Mimic what was done in the text in the case of the Nelson-Siegel family to give an interpretation to the roles of the parameters θ_i's, and plot the graphs of three of these forward curves for three values of the parameter θ which you will choose to illustrate your comments on their interpretation.*
2. *Derive an analytic formula for the prices of the zero coupon bonds when the term structure of interest rates is given by the yield curve (4.48) and write an* `R` *function* `bgv` *to compute the values of the zero coupon bonds. Plot the zero coupon curves corresponding to the three values of $\boldsymbol{\theta}$ chosen above in question 1.*
3. *Using this new function family, estimate the term structure of interest rates (as given by the zero coupon curve, the forward curve and the yield curve) for the German bond data used in the text.*

Notes & Complements

The least squares method is a classic of statistical theory and practice. Its sensitivity to extreme features of the data is well documented and the need for robust alternatives is widely accepted. The least absolute deviations method is the simplest example of a robust method of statistical estimation. Understanding the sensitivity of the estimation procedures to extreme measurements or *outliers* is a very important part of statistical inference, and an industry was developed on this basis. We refer the interested reader to the books of P.J. Huber [49] and Rousseeuw and Leroy [82] and the references therein. Realizing the importance of robust methods of estimation, `R` proposes a library of functions which can be used with the commercial distribution.

Complements to the tools of multivariate regression analysis and the theory of linear models, as introduced in this chapter, can be found in most multivariate analysis statistical textbooks. We refer the interested reader to the classical books of Montgomery and Peck [69], Mardia, Kent, and Bibby [67], or Chap. 14 of Rice's book [54] for an elementary exposition. Examples of `R` analyses with the function `lm` are given in [94].

Complements on seemingly unrelated regressions (SUR) can be found for example in the introductory econometric textbook of Ruud [84]. The CAPM was developed by Sharpe and Lintner after the pioneering work of Markowitz on the solution of the mean-variance optimal portfolio selection problem. An exposé of this theory can be found in any textbook on financial econometrics. We refer the interested reader to, for example, [42] or [13]. Most empirical tests of the model rely on market indexes as proxies for the market portfolio. The composition and the weights used in the computations of the market indexes have many features which vary over time (membership, capitalization of the members, ...), and such features have been argued to be the main reason for the empirical rejection of CAPM. See for example the enlightening discussion by Roll in [80].

Our presentation of nonlinear regression is pretty elementary and it remains at an intuitive level. Nonlinear regression can be very technical, and the reader interested in mathematical developments, and especially the differential geometric aspects of likelihood maximization, is referred to Amari's book [1] or to the monograph [3] written in French by Antoniadis, Berruyer and Carmona. Nonlinear regression with `R` is explained in detail in Chap. 10 of Chambers' book [21], from which we borrow the classic example treated in Problem 4.11.

The Bank of International Settlements (BIS for short) provides information on the methodologies used by the major central banks to estimate the term structure of interest rates in their respective countries. It also provides data and samples of curve estimates upon request. The Nelson-Siegel family was introduced by Nelson and Siegel in [74] and the Swensson's generalization was proposed by Swensson in [91]. The use of cubic splines was proposed by Vasicek and Fong in [93]. We learned of the estimation of the term structure of interest rates using the Vasicek exponential family proposed in Problem 4.12 from Nicole El Karoui in a private conversation. The methods reviewed in this chapter are used by Central Banks for econometric analysis and policy making. Fixed income desks of investment banks are more secretive about the methods they use to estimate and calibrate the term structure of interest rates.

5

LOCAL AND NONPARAMETRIC REGRESSION

This chapter is devoted to a class of regression procedures based on a new paradigm. Instead of searching for regression functions in a space whose elements are determined by a (small) finite number of parameters, we derive the values of the regression function from local properties of the data. As we shall see, the resulting functions are given by computational algorithms instead of formulae in closed forms. As before, we emphasize practical implementations over theoretical issues, and we demonstrate the properties of these regression techniques on financial applications: we revisit the construction of yield curves, and we study a non-parametric regression alternative to the Black-Scholes formula for the pricing of liquid options.

5.1 Review of the Regression Setup

Although we have already worked in the regression framework for an entire chapter, we thought it would be useful to review once more the general setup of a regression problem, together with the notation used to formalize it. This will give us a chance to stress the main differences between the parametric point of view of Chap. 4 and the nonparametric approach of this chapter.

The general setup is the same as in our discussion of multiple linear regression. We have a sample of n observations

$$(\boldsymbol{x}_1, y_1), \ldots\ldots, (\boldsymbol{x}_n, y_n)$$

where for each $i = 1, 2, \ldots, n$, $\boldsymbol{x}_i$ is a vector of p numerical components which we arrange in a row $\boldsymbol{x}_i = [x_{i,1}, x_{i,2}, \ldots, x_{i,p}]$, and y_i is a real number. The components of the $\boldsymbol{x}_i$'s are the observed values of the p scalar explanatory variables, while the y_i's are the observed values of the corresponding response variable.

The $\boldsymbol{x}_i$'s can be deterministic: this happens when they are chosen by design. In this case, we assume that the y_i's are noisy observations of the values of a deterministic function $\boldsymbol{x} \hookrightarrow \varphi(\boldsymbol{x})$ which is to be estimated from the observations. But the $\boldsymbol{x}_i$'s can also be realizations of random vectors $\boldsymbol{X}_i$ (which are usually assumed

R. Carmona, *Statistical Analysis of Financial Data in R*, Springer Texts in Statistics,
DOI 10.1007/978-1-4614-8788-3_5,

to be independent), in which case the y_i's are interpreted as realizations of random variables Y_i so that the $(\boldsymbol{x}_1, y_1), \ldots, (\boldsymbol{x}_n, y_n)$ appear as a sample of realizations of random couples $(\boldsymbol{X}_1, Y_1), \ldots, (\boldsymbol{X}_n, Y_n)$. For the sake of notation, we will use the notation $(\boldsymbol{X}, Y)$ for a generic couple (random vector, random variable) with the same joint distribution. The statistical dependence of the Y-component upon the $\boldsymbol{X}$-component is determined by the knowledge of the (common) joint distribution of the couples $(\boldsymbol{X}_i, Y_i)$. This distribution *sits* in $(p+1)$ dimensions, and it can be a very complicated object. It is determined by the marginal distribution of the p-variate random vector $\boldsymbol{X}$ (which sits in p dimensions), and the conditional distribution of Y which gives, for each possible value of $\boldsymbol{X}$, say $\boldsymbol{x}$, the conditional distribution of Y given that $\boldsymbol{X} = \boldsymbol{x}$. Instead of considering the whole conditional distribution, one may want to consider first the conditional mean (i.e. the expected value of this conditional distribution) and this leads to the analysis of the function:

$$\boldsymbol{x} \hookrightarrow \varphi(\boldsymbol{x}) = \mathbb{E}\{Y|\boldsymbol{X} = \boldsymbol{x}\} \tag{5.1}$$

which is called the regression function of Y against $\boldsymbol{X}$. The graph of the function $\boldsymbol{x} \hookrightarrow y = \varphi(\boldsymbol{x})$ captures graphically the properties of such a regression. It is a one-dimensional curve when $p = 1$, a 2-dimensional surface when $p = 2$, and it becomes a hypersurface more difficult to visualize for larger values of the number p of explanatory variables.

In this chapter, except possibly for requiring that the function φ is (at least piecewise) smooth, nothing is assumed on the structure of φ (as opposed to the linearity, or polynomial character assumed in Chap. 4). In particular, we do not restrict the function φ to families of functions which can be characterized by a small number of parameters. This is what differentiates nonparametric regression from the parametric regression procedures seen in the previous chapter.

We first consider the univariate case $p = 1$. In this case, one can view the search for the regression function φ, as a search for a graphical summary of the dependence between the two coordinates of the couples $(x_1, y_1), \ldots, (x_n, y_n)$ which are usually visualized as a cloud of points in the plane. In this way, the graph of φ appears as a scatterplot smoother. We review the most frequently used nonparametric scatterplot smoothers in Sect. 5.3, and we illustrate their use in the construction of yield curves in Sect. 5.4. When we compare nonparametric smoothing to the parametric techniques introduced in Chap. 4, the main novelty is the notion of local averages. Indeed, in all the regression procedures considered so far, including polynomial regression, when computing the value $\varphi(x)$ of the regression function for a given value of x, each single observation (x_i, y_i) contributes to the result, whether or not x_i is close to x. This lack of localization in the x variable when averaging the y_i's is a source of many difficulties which the nonparametric smoothers are able to avoid.

This shortcoming due to the *lack of localization* in the x-variable is not universally shared by all parametric regression techniques. Natural splines regression is a case in point. Regression with natural splines is very popular in computer graphics. It is an integral part of parametric regression because it can be recast in the framework of the linear models of Chap. 4. However, because of their local character, natural splines share many of the properties of the nonparametric scatterplot smoothers, and we decided to include them in this chapter. We briefly explain how to use them in Sect. 5.2 below.

We illustrate the use of natural splines and nonparametric smoothers on the same FRWRD data set on which we tested polynomial regression in Chap. 4. Recall Fig. 4.13. The last three sections of the chapter are devoted to the analysis of the multivariate case $p > 1$. We first introduce the kernel method which is recommended for small values of p. Then we discuss projection pursuit and other additive procedures designed to overcome the curse of dimensionality. For lack of time and space, we chose to ignore the tree-based regression methods, referring the interested reader to the references in the Notes & Complements at the end of the chapter.

5.2 Basis Expansion Regression

The contents of this section could be part of a discussion of parametric as well as nonparametric regression.

5.2.1 Natural Splines as Local Smoothers

This subsection deals only with the univariate case $p = 1$. While still an integral part of parametric regression, natural splines offer a regression procedure satisfying the localization requirement formulated above. A spline of order $m + 1$ is a function constructed from a subdivision of the range of x values. The points of this subdivision are called knots. This function is equal to a polynomial of degree $m + 1$ in between two consecutive points of the subdivision, and at each knot, all the left and right derivatives of order up to m match. This guarantees that the function is in fact m times continuously differentiable, discontinuities occurring possibly in the $(m + 1)$-th derivative, and only at the knots, but nowhere else. In most cases, this cannot be detected by the human eye. Given the knots, the splines of a given order form a finite dimensional vector space. Consequently, if we choose a basis for this linear space, each such spline function can be decomposed on this basis, and hence, it is entirely determined by a finite number of parameters, namely the coefficients in such a decomposition. It should now be clear that the discussion of polynomial regression given in Chap. 4 applies, and one can recast this type of regression within the class of linear models.

We refrain from discussing the details of the constructions of natural spline bases, restricting ourselves to a couple of illustrative examples. R has a fast implementation of a cubic natural splines regression. The typical form of the R command needed to run a natural splines regression is:

```
ns.fit <- lm(response ~ ns(regressor, df=DF), data=DATA)
```

As in the case of polynomial regression, natural splines regression is performed by the generic lm method, after the data have been massaged by the specific function `ns`. Completely in analogy with the function `poly`, the function `ns` generates a design matrix from the basis of cubic splines associated with the specified sequence of knots and boundary conditions. In the case of natural cubic splines, the functions are taken to be linear (polynomial of degree 1) to the left of the smallest knot, and to the right of the largest knot. When the parameter `df` is supplied, the function `ns` chooses (`df` $-$ 1) knots at suitably chosen empirical quantiles of the x_i's. The knots can also be supplied with the optional parameter `knots` if the user so desires.

We illustrate the use of natural splines on the gas forward data used earlier to test polynomial regression. Recall Fig. 4.13 in Chap. 4. The results are given in Fig. 5.1. This plot was produced by the following R commands:

```
> x <- 1:36
> plot(x,FRWRD,main="Natural Splines")
> lines(x,fitted(lm(FRWRD~ns(x,df=5))))
> lines(x,fitted(lm(FRWRD~ns(x,df=8))), lty=3)
> lines(x,fitted(lm(FRWRD~ns(x,df=15))), lty=6)
```

As seen from Fig. 5.1, the natural splines regression is much more local than a plain polynomial regression. Indeed, with a large enough number of knots, it does a reasonable job of staying very close to the data throughout. Also, it is clear that the higher the number of knots, the tighter the fit, even at the price of losing some smoothness in the regression function.

Like all regression methods, natural splines can be used for prediction purposes. Indeed, the linear model used to fit a natural splines regression model, can also be used to predict the response for new values of the explanatory variable. But like in the case of polynomial regression, this application needs to be restricted to new values of the explanatory variable which are within the range of the data, for the same disastrous results can be expected outside this range.

5.2.2 Feature Function Expansions

We now consider a natural generalization of polynomial and natural spline regression. It is based on the same rationale, and as we are about to see, its implementation is the same, irrespective of the dimension d of the explanatory variable $\mathbf{x}$. The premise of the method is a finite set $\{\varphi_0, \varphi_1, \cdots, \varphi_p\}$ of functions of the explanatory variable $\boldsymbol{x}$ which are called feature functions, and the ansatz that the regression function φ is of the form

$$\varphi(\mathbf{x}) = \sum_{j=0}^{p} \beta_j \varphi_j(\boldsymbol{x}) \tag{5.2}$$

for a set of coefficients $\beta_0, \beta_1, \cdots, \beta_p$. The set of feature functions $\{\varphi_0, \varphi_1, \cdots, \varphi_p\}$ is often chosen to be a basis of a vector space of functions, hence the alternative terminology basis functions for the φ_is, in which case the unknown parameters β_j are the coefficients of the decomposition of the regression function φ onto this basis.

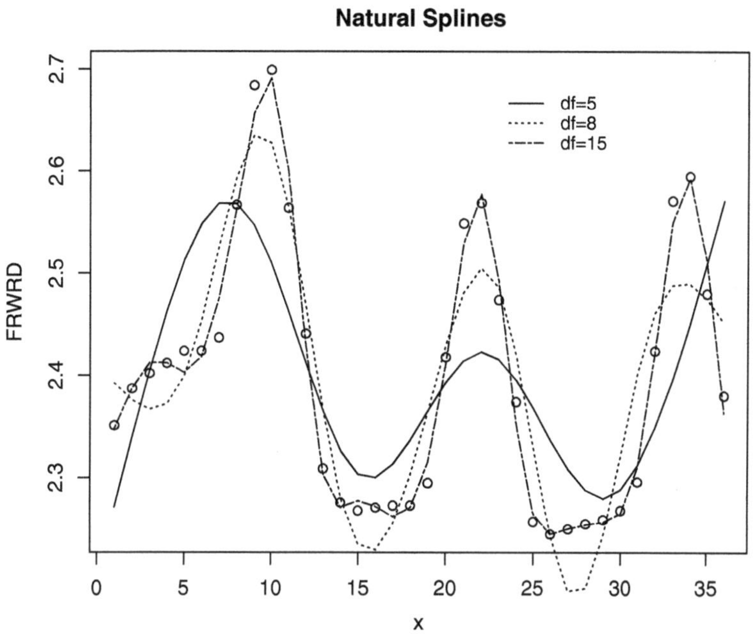

Fig. 5.1. Natural spline regressions with 5, 8 and 15 degrees of freedom, for the natural gas forward data which were used to illustrate polynomial regression in Chap. 4

For example, the set of feature functions $\{\varphi_0, \varphi_1, \cdots, \varphi_p\}$ happens to be the natural basis of polynomials of degree at most p in the case of polynomial regression since $\varphi_i(x) = x^i$, while even though we did not give explicit formulas for the functions φ_i in the subsection above, we explained that it was a basis of natural spline functions constructed from the choice of the knots in the case of natural splines regression. We now explain how the power of the linear models (and of the function `lm`) can be unleashed to determine least square estimates of the parameters β_j.

In order to do so, we assume that we are given a sample of n observations

$$(\boldsymbol{x}_1, y_1), (\boldsymbol{x}_2, y_2), \ldots\ldots, (\boldsymbol{x}_n, y_n)$$

where the *explanatory variables* $\boldsymbol{x}_1, \boldsymbol{x}_2, \ldots, \boldsymbol{x}_n$ are d-dimensional. In other words, each $\boldsymbol{x}_i$ is a d-vector of the form $\boldsymbol{x}_i = (x_{i,1}, \ldots, x_{i,d})$ while the response variables $y_1, y_2, \ldots, y_n$ are univariate (i.e. scalar). As usual, we use bold face letters $\boldsymbol{x}$ to emphasize that we are dealing with a *multivariate* explanatory variable which we sometimes call an explanatory vector to emphasize that its dimension d can be greater than one. We can then form the $n \times (p+1)$ matrix $\boldsymbol{X}$

$$X_{i,j} = \varphi_j(\boldsymbol{x}_i), \qquad i = 1, \cdots, n, \; j = 0, \cdots, p \tag{5.3}$$

by evaluating the feature functions at the values of the observations of the explanatory variables. Since the regression function is expected to satisfy $y_i = \varphi(\mathbf{x}_i) + \epsilon_i$ for realizations of independent (or at least uncorrelated) noise terms ϵ_i, and since

$$\varphi(\boldsymbol{x}_i) = \sum_{j=0}^{p} \beta_j \varphi_j(\boldsymbol{x}_i) = \sum_{j=0}^{p} X_{i,j}\beta_j, \qquad i = 1, \cdots, n$$

the matrix $\boldsymbol{X}$ appears as the design matrix of the linear model whose solution provides estimators $\hat{\beta}_0, \hat{\beta}_1, \cdots, \hat{\beta}_p$ minimizing the least squares criterion

$$\sum_{i=1}^{n} \left| y_i - \sum_{j=0}^{p} \beta_j \varphi_j(\boldsymbol{x}_i) \right|^2 .$$

So computing the design matrix $\boldsymbol{X}$ given by (5.3) and running a linear model (for example using the `R` function `lm`), we obtain estimates $\hat{\beta}_0, \hat{\beta}_1, \cdots, \hat{\beta}_p$ of the coefficients and as a consequence, the estimate

$$\boldsymbol{x} \hookrightarrow \hat{\varphi}(\boldsymbol{x}) = \sum_{j=0}^{p} \hat{\beta}_j \varphi_j(\boldsymbol{x})$$

of the regression function φ. The broad appeal of this regression method has many sources: it benefits from an intuitive rationale, the power of the numerical implementations of linear models, and the flexibility provided by the choice of the feature functions. While they may still suffer from the curse of dimensionality discussed in detail later in the chapter, Monte Carlo versions have been proposed for the pricing of American options on large baskets of interests, and became very popular in the financial industry.

Basis function expansions are very popular methods of regression at the boundary between parametric and non-parametric regression. The most ubiquitous are the expansions in natural splines discussed in the previous subsection, the Fourier expansions and the wavelet expansions. See the Notes & Complements at the end of the chapter for references.

5.2.2.1 *Control of the Error*

Assuming that the regression function φ belongs to a specific function space, and assuming that the feature functions $\varphi_0, \varphi_1, \cdots, \varphi_p$ have been chosen, the best approximation of φ by a linear combination of the φ_js, in the least squares sense, is the orthogonal projection of φ onto the linear subspace generated by the φ_js. Estimating the error incurred by replacing the unknown function φ by its orthogonal projection is a well studied problem in approximation theory and numerical analysis. Rates of convergence have been derived, especially when the φ_j are part of a complete orthonormal system of the function space. Coefficients $\hat{\beta}_0, \hat{\beta}_1, \cdots, \hat{\beta}_p$ determining the orthogonal projection of φ can be obtained in algorithmic fashion in theory, and the estimation of these coefficients from a sample $(\boldsymbol{x}_1, y_1), (\boldsymbol{x}_2, y_2), \ldots\ldots, (\boldsymbol{x}_n, y_n)$ of observations is the only source of statistical errors. The theory of linear models presented in Chap. 4 provides controls for the expected mean squared error.

Remark 1. The strategy described in this subsection is tailored to situations when we do not have a priori information quantifying the relative sizes of the observation errors. If we know (or have reasonable guesses for) the measurement errors, we can include weights in the above least squares regression and choose the coefficients β_0, $\beta_1, \cdots, \beta_p$ as minimizers of the weighted least squares criterion:

$$\sum_{i=1}^{n} w_i \left| y_i - \sum_{j=0}^{p} \beta_j \varphi_j(\boldsymbol{x}_i) \right|^2$$

5.3 Nonparametric Scatterplot Smoothers

As the use of the word scatterplot suggest, we are now back to the univariate case $p = 1$. Scatterplot smoothers can be viewed as nonparametric regression procedures for univariate explanatory variables, the idea being to represent a cloud of points $(x_1, y_1), \ldots, (x_n, y_n)$ by the graph of a function $x \hookrightarrow y = \varphi(x)$. As explained earlier, the terminology nonparametric is justified by the fact that the function φ is not expected to be determined by a small number of parameters. In fact, except for a mild smoothness requirement, this regression function will not be restricted to any specific class.

We review some of the most commonly used scatterplot smoothers, and we give the precise definitions of those implemented in R. We restrict ourselves to the smoothers used in applications discussed in this book and the problem sets. For the sake of illustration, we compare their performance on data already used in our discussions of the polynomial and natural spline regressions.

5.3.1 **Smoothing Splines**

We begin our review of scatterplot smoothers with smoothing splines. This choice was made to emphasize the similarities and differences with the natural splines discussed above. The idea behind the smoothing splines procedure is common to many applications in signal and image processing: it relies on regularization. To be specific, the scatterplot smoother φ is obtained by minimizing the objective function:

$$\mathcal{L}(\varphi) = \sum_{i=1}^{n} w_i |y_i - \varphi(x_i)|^2 + \lambda \int |\varphi^{(m)}(x)|^2 \, dx \qquad (5.4)$$

for some constant $\lambda > 0$ called the smoothing parameter, for an integer m giving the order of derivation, and for a set of weights w_i which are most often taken to be unity. As usual, we use the notation $\varphi^{(m)}(x)$ to denote the m-th derivative of the function φ. The desired scatterplot smoother is the function $x \hookrightarrow \varphi(x)$ which minimizes the cost function $\mathcal{L}(\varphi)$, i.e. the argument of the minimization problem:

$$\varphi = \arg\min_f \mathcal{L}(f).$$

The goal of the first term in the objective function (5.4) is to guarantee the fit to the data, while the second term tries to ensure that the resulting scatterplot smoother is indeed smooth. The order of derivation m has to be chosen in advance, and the parameter λ balances the relative contributions, to the overall cost, of the lack of fit and the possible lack of smoothness. As defined, the minimizing function φ does not seem to have any thing to do with splines. But surprisingly enough, it turns out that the solution of the optimization problem (5.4) is in fact a spline of order $m+1$, i.e. an m times continuously differentiable function which is equal to a polynomial of degree $m+1$ on each subinterval of a subdivision of the range of the explanatory variable x. In particular, it has the same local properties as the regression by natural splines. R gives an implementation of this scatterplot smoother for $m = 2$, in which case the resulting function φ is a cubic spline. The name of the R function is `smooth.spline`. The smoothing parameter λ appearing in formula (5.4) can be specified through the value of an optional smoothing parameter `spar` in the interval $(0,1]$. When the parameter `spar` is specified, the actual value of λ used in (5.4) is $\lambda = r256^{3\texttt{spar}-1}$ where r is a constant computed from the values of the explanatory variables x_i's. Alternatively, the balance between the fit and smoothness terms of formula (5.4) can be controlled by setting the parameter `df` which stands for number of degrees of freedom, and which is essentially given by the integer part of the real number:

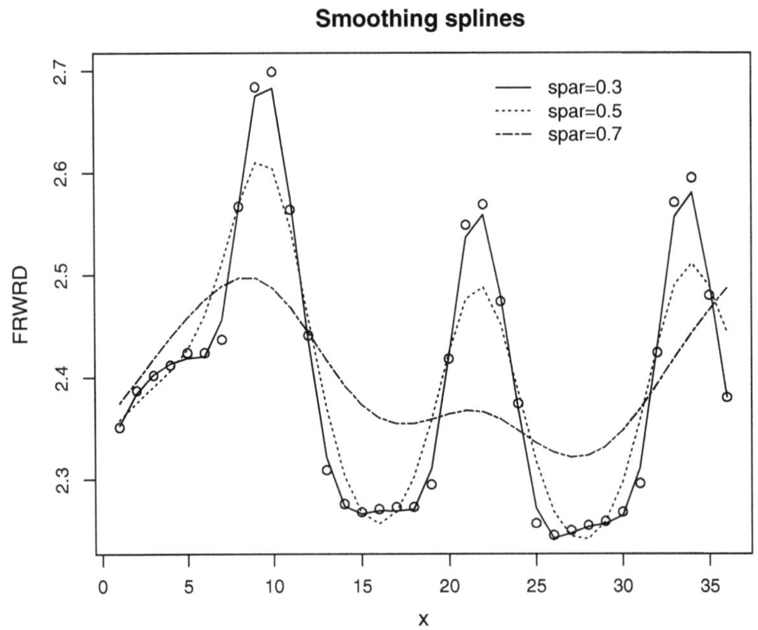

Fig. 5.2. Examples of smoothing splines with smoothing parameters $\lambda = 0.001$, $\lambda = 0.0001$ and $\lambda = 0.00001$

$$n\frac{(\max x_j - \min x_j)^3}{\lambda}$$

(recall that n stands for the sample size). If neither of these parameters is specified, the function `smooth.spline` chooses λ by cross validation. The plots appearing in Fig. 5.2 were created with the commands:

```
> x <- 1:36
> plot(x,FRWRD,main="Smoothing splines")
> lines(smooth.spline(x,FRWRD,spar=.3))
> lines(smooth.spline(x,FRWRD,spar=.5),lty=3)
> lines(smooth.spline(x,FRWRD,spar=.7),lty=6)
```

The value of the smoothing parameter was determined from the numerical value passed to the argument `spar` from the rules detailed above. Notice that the smaller this parameter, the closer the scatterplot smoother is to the points (since the fit contribution is more important) and that the larger this parameter, the smoother the resulting curve (since the smoothing contribution dominates the objective function).

5.3.2 Locally Weighted Regression

A quick review of the least squares paradigm for linear regression shows that the values of $\varphi(x)$ depend linearly on the observed responses y_i. Indeed, the optimal function for the least squares criterion happens to be of the form:

$$\varphi(x) = \sum_{i=1}^{n} w_i(x) y_i,$$

where each weight $w_i(x)$ depends upon all values x_j of the explanatory variable. As explained in the introduction, one of the goals of nonparametric smoothers is to keep this form of $\varphi(x)$ as a weighted average of the observed responses y_i, while at the same time choosing the weights $w_i(x)$ to emphasize the contributions of the y_i's corresponding to x_i's which are near the value x at which the function φ is being computed.

A first implementation of this idea is given by the R function `loess`. This scatterplot smoother depends upon a parameter called `span` which gives the proportion of observations included in the neighborhood of each point. In practice, reasonable `span` values range from 0.3 to 0.5. Let us denote by k this number of points. For each x at which one wishes to compute the function $\varphi(x)$, we denote by $N(x)$ the set of the k nearest values x_i of the explanatory variable, and by $d(x)$ the largest distance between x and the points of $N(x)$. In other words:

$$d(x) = \max_{x_i \in N(x)} |x - x_i|.$$

To each $x_i \in N(x)$ we associate the weight

$$w_x(x_i) = W\left(\frac{|x - x_i|}{d(x)}\right) \qquad (5.5)$$

where the weight function W is defined by:

$$W(u) = \begin{cases} (1-u^3)^3 & \text{if } 0 \le u \le 1 \\ 0 & \text{otherwise,} \end{cases}$$

and we choose for $\varphi(x)$ the value of the weighted least squares regression line for the k points (x_j, y_j) for which $x_j \in N(x)$ and for the weights $w_x(x_j)$ defined in (5.5).

We illustrate the use of the `loess` function on the gas forward data on which we already tested several regression procedures. The results are given in the left pane of Fig. 5.3. This plot was produced with the following `R` commands:

```
> plot(x,FRWRD,main="Loess Regression")
> lines(x,fitted(loess(FRWRD~x,span=.4)))
> lines(x, fitted(loess(FRWRD~x)),lty=3)
```

We set the value of the smoothing parameter `span` to 0.4 for the sake of definiteness, but as before, varying this value leads to plots similar to those of Figs. 5.1 and 5.2, the smaller the value of `span` the closer the loess regression to the original data points, and the larger the value of `span` the smoother the loess regression curve. As in most cases, the smoothing parameter has a default value which is used by the program when the value of the parameter is not set in the call to the function. However, the results can be pretty bad when using the default value of the smoothing parameter and experimenting with several values of the smoothing parameter is always a useful *sanity check...*

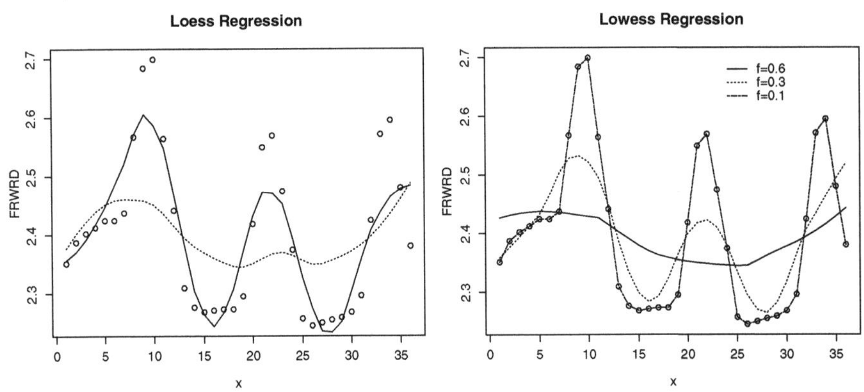

Fig. 5.3. *Left*: `loess` regression using 0.4 (*solid line*) and the default value of the parameter `span` (*dotted line*). *Right*: robust smooth given by `lowess`

5.3.3 A Robust Smoother

There is still another scatterplot smoother based on the same idea of local linear regression. It is a robust form of the weighted local linear regression given by the function `loess` described above. It is called `lowess`. Given a number of neighboring data points, this function performs locally a robust regression to determine the value of the function $\varphi(x)$. The use of this robust scatterplot smoother is illustrated in the right pane of Fig. 5.3. We produced this plot with the following R commands:

```
> plot(x,FRWRD,main="Lowess Regression")
> lines(lowess(x,FRWRD,f=.6))
> lines(lowess(x,FRWRD,f=.3),lty=3)
> lines(lowess(x,FRWRD,f=.1),lty=6)
```

The optional parameter f gives the proportion of neighboring points used in the local robust regression. We used three values, f$= 0.6$, f$= 0.3$ and f$= 0.1$. The function lowess does not return an object of a specific class from which we extract the fitted values with the generic function fitted. It merely returns, in the old-fashioned way, a vector of x-values containing the values of the explanatory variable, and a vector of y-values containing the values fitted to the response variable. As before, we use the function lines to produce a continuous graph from discrete isolated points. Here, each of the values which are joined by straight lines to produce these three curves were obtained by averaging 12, 10 and 3 neighboring values of the response variable, respectively. This explains why these curves change from very smooth (and not fitting the data very well) to rather ragged, and why they give the impression of a broken line. We included the present discussion of the function lowess for the sake of completeness only. Indeed, since robust regressions are more computer intensive than least squares regressions, the lowess procedure can be very slow when the sample size n is large. For this reason, we do not recommend its use for large data sets.

5.3.4 The Super Smoother

The super smoother function supsmu is our last example of a scatterplot smoother before we start our long discussion of the kernel method. It is based on the same idea as the function loess, but its implementation has been optimized for computational speed. The main difference is the fact that the span (or equivalently the size k of the neighborhood $N(x_j)$) is now chosen by a form of local *cross validation* as a function of each x_j. Because of its computational speed, it is part of the projection pursuit algorithm which we present in detail later in this chapter. Calls to the function supsmu, and objects returned by these calls, have the structure described above in the case of the function lowess, so we refrain from illustrating them with examples.

5.3.5 The Kernel Smoother

As with the other scatterplot smoothers, the idea of the kernel smoother is to rely on the observed responses to neighboring values of x to predict the response $\varphi(x)$. The only difference is that, instead of relying on a limited number of observations y_i of the response, the local character of the averaging is realized by a weighted average of all the observed values y_i, the weights being decreased with the distance between x and the corresponding value x_i of the explanatory variable. To be more specific, the weights are computed by means of a *kernel* function $x \hookrightarrow K(x)$, and our good old enemy, the smoothing parameter. The latter is called *bandwidth* in the case of the kernel method and is denoted by $b > 0$. By now, we should be familiar with the terminology and the notation associated with the kernel method. Indeed, we already introduced them in our discussion of kernel density estimation. We give a lucid account of the relationship between the applications of the kernel method to density estimation and to regression in the Appendix at the end of the chapter. The actual formula giving the kernel scatterplot smoother $\varphi(x)$ is:

$$\varphi(x) = \varphi_{b,K}(x) = \frac{\sum_{i=1}^{n} y_i K\left(\frac{x-x_i}{b}\right)}{\sum_{j=1}^{n} K\left(\frac{x-x_j}{b}\right)}. \tag{5.6}$$

Notice that the formula giving $\varphi(x)$ can be rewritten in the form:

$$\varphi(x) = \sum_{i=1}^{n} w_i(x) y_i \tag{5.7}$$

provided we define the weights $w_i(x)$ by the formula:

$$w_i(x) = \frac{K\left(\frac{x-x_i}{b}\right)}{\sum_{j=1}^{n} K\left(\frac{x-x_j}{b}\right)}. \tag{5.8}$$

Understanding the properties of these weights is crucial to understanding the very nature of kernel regression. These properties will be clear once we define what we mean by a kernel function. Recall that a nonnegative function $x \hookrightarrow K(x)$ is called a kernel function if it is integrable and if its integral is equal to 1, i.e. if it satisfies:

$$\int_{-\infty}^{+\infty} K(x)\, dx = 1,$$

in other words, if K is a probability density. The fact that the integral of $K(x)$ is equal to one is merely a normalization condition useful in applications to density estimation. It will not be of any consequence in the case of applications to regression since K always appear simultaneously in the numerator and the denominator: indeed, as one can easily see from formulae (5.7) and (5.8), multiplying K by a constant does not change the value of the regression function φ as defined in (5.6). But in order to be useful, the kernel $K(x)$ has to take relatively large values for small values of x, and relatively small values for large values of x. In fact, it is also often assumed that K is symmetric in the sense that:

$$K(-x) = K(x)$$

and that $K(x)$ decreases as x goes from 0 to $+\infty$. The above symmetry condition implies that:

Kernel function	Formula
`box`	$K_{box}(x) = \begin{cases} 1, \text{ if } \lvert x\rvert \le 0.5 \\ 0 \ \text{ otherwise} \end{cases}$
`triangle`	$K_{triangle}(x) = \begin{cases} 1 - \lvert x\rvert, \text{ if } \lvert x\rvert \le 1 \\ 0 \qquad \text{otherwise} \end{cases}$
`parzen`	$K_{parzen}(x) = \begin{cases} (9/8) - (3/2)\lvert x\rvert + x^2/2, & \text{if } 1/2 \le \lvert x\rvert \le 3/2 \\ 3/4 - x^2, & \text{if } \lvert x\rvert \le 1/2 \\ 0 & \text{otherwise} \end{cases}$
`normal`	$K_{normal}(x) = \sqrt{2\pi}^{-1} e^{-x^2/2}$

Table 5.1. Table of the four kernel functions used by the `R` function `ksmooth`

$$\int_{-\infty}^{+\infty} xK(x)\, dx = 0 \qquad (5.9)$$

which will be used in our discussion of the connection with kernel density estimation in the Appendix. Recall Fig. 1.19 from Chap. 1 which gives the graphs of the four kernel functions used by the `R` density estimation method. They are also some of the most commonly used kernel functions when it comes to regression. Notice that the first three of them vanish outside a finite interval, while the fourth one (theoretically) never vanishes. Nevertheless, since its computation involves evaluating exponentials, it will not come as a surprise that such a kernel can be (numerically) zero because of the evaluation of exponentials of large negative numbers: indeed for all practical purposes, there is no significant difference between e^{-60} and 0, and exponents which are that negative can appear very often! The `R` function `ksmooth` gives an implementation of the univariate kernel scatterplot smoother. The value of b is set by the parameter `bandwidth`. The kernel function K is determined by the choice of the parameter `kernel`. Four possible choices are available for the parameter `kernel`, `box` being the default. Explicit formulae are given in the Table 5.1. The option `triangle` gives the triangular function found in Fig. 1.19, the `parzen` option gives a function proposed by Manny Parzen, who was one of the early proponents of the kernel method. Notice that this kernel replaces the cosine kernel used in density estimation. Finally the choice `normal` selects the Gaussian density function. We give the formulae for these functions in Table 5.1.

Except for the kernel function `box`, which leads to the crudest results, the other three kernels give essentially the same results in most applications. The situation is different when it comes to the choice of bandwidth parameter b. Indeed, the choice of the bandwidth is the *Achilles heel* of kernel regression. This choice can have an enormous impact, and the results can vary dramatically: small values of b give rough graphs which fit the data too closely, while too large a value of b produces a flatter graph. By experimenting with the choice of bandwidth, one can easily see that as b

tends to ∞, the graph of the kernel smoother converges toward the horizontal straight line with intercept at the level of the mean $\overline{y}$ of the observed responses y_j. A rigorous proof of this is the subject of Problem 5.3. As we explained earlier, the regression becomes meaningless in this limiting case since the explanatory variable does not have any influence on the value of the prediction of the response variable.

We conclude this subsection with a short discussion of the R implementation of the univariate kernel smoother. As already mentioned during our discussion of histograms and density estimators, both R and S have a command ksmooth for the implementation of univariate kernel regression. They take the same parameters, and when they do work, they produce the same results. However, because the R implementation was optimized for numerical speed, it is based on a different algorithm (presumably a fast Fourier transform) which at times, typically for extremely small values of the bandwidth, gives NAs where the algorithm used in the S implementation still gives finite numbers. In particular, if we run the commands

```
> plot(x,FRWRD,main="1-D Gaussian Kernel Regression")
> lines(ksmooth(x,FRWRD,"normal",bandwidth=1))
> lines(ksmooth(x,FRWRD,"normal",bandwidth=3),lty=3)
> lines(ksmooth(x,FRWRD,"normal",bandwidth=10),lty=5)
```

in R and in S-Plus, the plot corresponding to the value 0.01 of the bandwidth will not appear in R as the values of the kernel smoother are mostly NAs. To illustrate this claim, Fig. 5.4 provides the results obtained in R (left) and S-Plus (right).

More on the kernel scatterplot smoother later in Sect. 5.5 when we discuss the multivariate case and the kernel regression method.

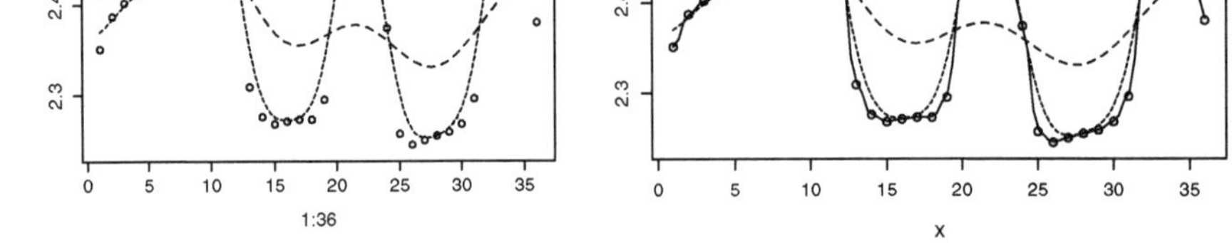

Fig. 5.4. Effect of the choice of bandwidth on the result of a kernel smoother in R (*left*) and in S-Plus (*right*)

5.4 More Yield Curve Estimation

Given our newly acquired knowledge of scatterplot smoothers and nonparametric curve estimation, we revisit the problem of estimation of the term structure of interest rates, as given for example by the instantaneous forward interest rate curves, which was originally tackled in Sect. 4.9 by means of parametric methods.

5.4.1 A First Estimation Method

The first procedure we present was called *iterative extraction* by its inventors, but it is known *on the street* as the *bootstrap method.* We warn the reader that this use of the word bootstrap is more in line with the everyday use of the word bootstrap than with the standard statistical terminology.

We assume that the data at hand consist of coupon bearing bonds with maturity dates $T_1 < T_2 < \cdots < T_m$ and prices $B_1 < B_2 < \cdots < B_m$. The so-called bootstrap method seeks a forward curve which is constant on each of the intervals $[T_j, T_{j+1})$. For the sake of simplicity, we assume that $t = 0$. In other words, we postulate that:

$$f(0,T) = f_j \qquad \text{for} \qquad T_j \le T < T_{j+1} \qquad j = 1, \ldots, m-1$$

for a sequence $\{f_j\}_j$ of deterministic rates to be determined recursively by calibration to the observed bond prices. Let us assume momentarily that $f_1, \ldots, f_j$ have already been determined, and let us describe the procedure for identifying f_{j+1}. If we denote by X_{j+1} the principal of the $(j+1)$-th bond, by $\{t_{j+1,i}\}_i$ the sequence of coupon payment times, and by $C_{j+1,i} = c_{j+1}/n_y$ the corresponding payment amounts (recall that we use the notation c_j for the annual coupon rate, and n_y for the number of coupon payments per year), then the bond's price at time $t = 0$ can be obtained by discounting all the future cash flows associated with this bond:

$$B_{j+1} = \sum\nolimits_{t_{j+1,i} < T_j} P(0, t_{j+1,i}) \frac{c_{j+1} X_{j+1}}{n_y} \tag{5.10}$$

$$+P(0,T_j) \left(\sum\nolimits_{T_j < t_{j+1,i} \le T_{j+1}} e^{-(t_{j+1,i} - T_j) f_{j+1}} \frac{c_{j+1} X_{j+1}}{n_y} + e^{-(t_{j+1,i} - T_j) f_{j+1}} X_{j+1} \right).$$

Notice that all the discount factors appearing in this formula are known since, for $T_k \le t < T_{k+1}$ we have:

$$P(0,t) = \exp\left[-\sum_{h=1}^{k} (T_h - T_{h-1}) f_h - (t - T_k) f_{k+1} \right]$$

(recall formula (4.39) linking the price of the zero coupon bond to the instantaneous forward rate) and all the forward rates are known if $k \le j$. Consequently, rewriting (5.10) as:

$$\frac{B_{j+1} - \sum_{t_{j+1,i} \le T_j} P(0, t_{j+1,i}) \frac{c_{j+1} X_{j+1}}{n_y}}{P(0, T_j)}$$
$$= \frac{c_{j+1} X_{j+1}}{n_y} \sum_{T_j < t_{j+1,i} \le T_{j+1}} e^{-(t_{j+1,i} - T_j) f_{j+1}} + e^{-(t_{j+1,i} - T_j) f_{j+1}},$$

and noticing that the left hand side can be computed, and that the unknown forward rate f_{j+1} appears only in the right hand side, this equation can be used to determine f_{j+1} from the previously evaluated values f_k for $k \le j$.

Remark. Obviously, the forward curve produced by the bootstrapping method is discontinuous, since by construction, it jumps at all the input maturity dates. These jumps are the source of an artificial volatility: this is the main shortcoming of this method of estimation. Several remedies have been proposed to alleviate this problem. The simplest one is to artificially increase the number of maturity dates T_j to interpolate between the observed bond (or swap) prices. Another proposal is to add a smoothness penalty which will force the estimated curve to avoid jumps. This last method is in the spirit of the smoothing spline estimation method which we discuss now.

5.4.2 A Direct Application of Smoothing Splines

For the purposes of this subsection we use the data contained in the `R data frame` `USBN041700`. These data comprise the quotes on April 17, 2000 of the outstanding US Treasury Notes and Bonds. Figure 5.5 gives the plot of the redemption yield as a function of the time to maturity, together with the plot of the smoothing spline. This plot was created with the following commands:

```
> plot(LIFE,INT.YIELD,main="Smoothing Spline .... Yields")
> lines(smooth.spline(LIFE,INT.YIELD))
```

The (smooth) yield curve plotted in Fig. 5.5 is unusual because it has two humps, but despite this unusual feature, it can still be regarded as a reasonable yield curve.

5.4.3 US and Japanese Instantaneous Forward Rates

Even though we stated in Sect. 4.9 that the yield curves and forward rate curves published by the US Federal Reserve and the Bank of Japan were computed using smoothing (cubic) splines, they are not produced in the simplistic approach described above. The instantaneous forward rate curve produced on a given day t is the function $x \hookrightarrow \varphi(x)$ which minimizes the loss function:

$$\mathcal{L}_{JUS}(\varphi) = \sum_{i=1}^{n} w_i |P_i - P_i(\varphi)|^2 + \lambda \int |\varphi''(x)|^2 \, dx, \tag{5.11}$$

where $\varphi''(t)$ stands for the second derivative of $\varphi(t)$, the P_i's are the prices quoted for the outstanding bonds and notes available on day t, and the $P_i(\varphi)$'s are the prices

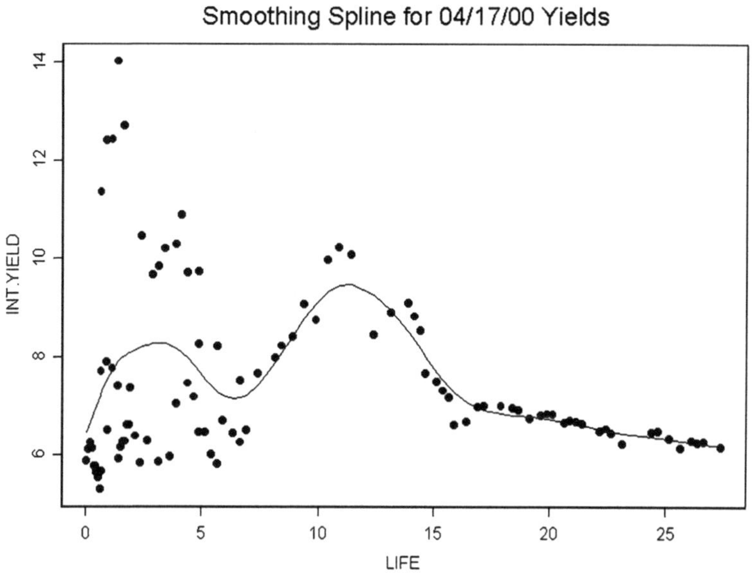

Fig. 5.5. Plot of the US Treasury notes and bonds redemption yields on April 17, 2000 together with the smoothing spline

one would get from pricing the bonds and notes on the forward curve given by φ using the fundamental pricing formula (4.33) in which the discount factors are computed from the forward curve given by φ. As in the parametric case, the weights w_i are chosen as functions of the duration of the i-th bond. See Sect. 4.9 for details.

The curve construction based on the minimization of the objective function $\mathcal{L}_{JUS}(\varphi)$ defined in (5.11) is reminiscent of smoothing splines regression. The difference lies in the fitting part $|P_i - P_i(\varphi)|^2$, which replaces the usual $|y_i - \varphi(x_i)|^2$. Instead of directly comparing the observation y_i to the values $\varphi(x_i)$ of the regression function, we compare the price P_i of a bond to the theoretical price $P_i(\varphi)$ that it would have if the function φ gives the true values of the instantaneous forward rate curve. This difference is enough to prevent us from directly using the `R` function `smooth.splines`, and it explains why the example we gave in Sect. 5.4.2 was for the construction of a yield curve $x \hookrightarrow \varphi(x)$ from observations y_i of the yields $\varphi(x_i)$.

5.5 Multivariate Kernel Regression

Multivariate kernel regression is a typical example of multivariate nonparametric nonlinear regression, but it can also be viewed as a high dimensional generalization of the procedure described in the subsection on the kernel scatterplot smoother, and especially the discussion of the function `ksmooth`. Indeed, most of what we said then, can be generalized to the case where the dimension p of the explanatory variables is not necessarily equal to 1. Indeed, formula (5.6) can be used in the form:

$$\varphi(\boldsymbol{x}) = \varphi_{b,K}(\boldsymbol{x}) = \frac{\sum_{j=1}^{n} y_j K\left(\frac{\boldsymbol{x}-\boldsymbol{x}_j}{b}\right)}{\sum_{j=1}^{n} K\left(\frac{\boldsymbol{x}-\boldsymbol{x}_j}{b}\right)}, \tag{5.12}$$

provided the function $\boldsymbol{x} \hookrightarrow K(\boldsymbol{x})$ is a kernel function in p dimensions, in the sense that it is a nonnegative function of p variables which integrates to one.

The simplest example of a p-dimensional kernel function is given by a function of the form:

$$K(\boldsymbol{x}) = k(\text{dist}(\boldsymbol{x}, 0)) \tag{5.13}$$

for some nonnegative and non-increasing function $d \hookrightarrow k(d)$ of one variable, and some choice of a notion of distance from the origin in p dimensions. Possible choices for this notion of distance include the usual Euclidean norms in $\mathbb{R}^p$:

$$\text{dist}(\boldsymbol{x}, 0) = \left(\sum_{j=1}^{p} x_j^2\right)^{1/2} \qquad \text{or} \qquad \text{dist}(\boldsymbol{x}, 0) = \left(\sum_{j=1}^{p} w_j x_j^2\right)^{1/2}$$

or non-Euclidean norms such as:

$$\text{dist}(\boldsymbol{x}, 0) = \sum_{j=1}^{p} |x_j| \qquad \text{or} \qquad \text{dist}(\boldsymbol{x}, 0) = \sup_{j=1,\ldots,p} |x_j|.$$

These choices are popular because of their convenient scaling properties. With the exception of the Euclidean distance computed with different weights w_j for the different components x_j of the explanatory vector $\boldsymbol{x}$, all these kernel functions share the same shortcoming: all the components of the explanatory vector are treated equally, and this may be very inappropriate if the numerical values are on different scales. Indeed, in such a case, the value of the distance is influenced mostly (if not exclusively) by the variables having the largest values. We illustrate this point with a short discussion of an example which we will study in detail in the later part of this chapter. Let us imagine, for example, that the first explanatory variable is an annualized interest rate. Its values are typically of the order of a few percentage points. Let us also imagine that the second explanatory variable is a time to maturity. If for some strange reason this second variable is expressed in days instead of years, its values will quite often be in the hundreds, and a distance of the type given above will ignore the small changes in interest rate, and be sensitive only on the differences in maturity. A change in unit in one of the variables can dramatically change the qualitative properties measured by these notions of distance, and consequently strongly affect the results of the kernel regression. This effect is highly undesirable. We discuss below alternative choices of kernel functions which can overcome this difficulty, as well as a standardization procedure which re-scales all the explanatory variables in an attempt to balance their relative contributions to the regression results.

Another very popular class of kernel functions is given by direct products (sometimes called tensor products) of one dimensional kernel functions. Indeed, if K_1, K_2, ... , K_p are one-dimensional kernel functions (possibly equal to each other), then the function:

$$K(\boldsymbol{x}) = K(x_1, x_2, \ldots, x_p) = K_1(x_1)K_2(x_2)\cdots K_p(x_p) \qquad (5.14)$$

is obviously a p-dimensional kernel function. For these kernel functions, the weight multiplying the i-th response y_i is proportional to:

$$K\left(\frac{\boldsymbol{x}-\boldsymbol{x}_i}{b}\right) = K_1\left(\frac{x_1 - x_{i,1}}{b}\right) K_2\left(\frac{x_2 - x_{i,2}}{b}\right) \cdots K_p\left(\frac{x_p - x_{i,p}}{b}\right)$$

and from this expression one sees that there is no harm in choosing different values for the p occurrences of the bandwidth b in the right hand side. In other words, it is possible to choose p different bandwidths b_1, b_2, ..., b_p, one for each component of the explanatory variable. This feature of the direct product kernels makes them very attractive. In some sense, normalizing the scalar explanatory variables and using one single bandwidth amounts to the same thing as using different bandwidths for the components of the explanatory vector. See Sect. 5.5.2 for an example of standardization before running a kernel regression.

We now recast some of the most important properties of kernel regression as elementary remarks which also apply to the one dimensional case of the kernel scatterplot smoother `ksmooth` discussed earlier.

- The kernel regression estimate $\varphi(\boldsymbol{x})$ is a linear function of the observations. Indeed, the definition formula (5.12) can be rewritten in the form:

$$\varphi(\boldsymbol{x}) = \sum_{i=1}^{n} w_i(\boldsymbol{x}) y_i,$$

where the weights $w_i(\boldsymbol{x})$ are defined by:

$$w_i(\boldsymbol{x}) = \frac{K\left(\frac{\boldsymbol{x}-\boldsymbol{x}_i}{b}\right)}{\sum_{j=1}^{n} K\left(\frac{\boldsymbol{x}-\boldsymbol{x}_j}{b}\right)}.$$

Notice that these weights are nonnegative and sum up to one. Because the kernel function is typically very small when its argument is large and relatively large when its argument is small, the weight $w_i(\boldsymbol{x})$ is (relatively) large when $\boldsymbol{x}$ is close (i.e. similar) to the observation $\boldsymbol{x}_i$ and small otherwise. This shows that the kernel regression function φ given by (5.12) is a weighted average of the observed values y_i of the response (and hence it is linear in the y_i's) with weights which favor the responses to the values $\boldsymbol{x}_i$ close to the value $\boldsymbol{x}$ of the explanatory variables at which we try to compute the regression function.

- The choice of bandwidth is a very touchy business. Many proposals have been made for an automatic (i.e. data driven) choice of this smoothing parameter. Whether one uses the results of difficult asymptotic analyses to implement bootstrap or cross validation procedures or simple rules of thumb, our advice is to be wise and to rely on experience to detect distortions due to a poor choice of bandwidth.
- As explained earlier, it is tempting to use a separate bandwidth for each explanatory variable. This is especially the case when the kernel function is of the product type given in (5.14) and when the dynamic ranges of these variables are very different. For example, if a variable is expressed in a physical unit, changing the unit system may dramatically change the range of the actual values of the measurements, and small numbers can suddenly become very large as a consequence of the change of units. Accordingly, the influence of this variable on the computation of the kernel regression can increase dramatically. This undesirable effect is often overcome by normalizing the variables. See details in the discussions of the practical examples presented in Sect. 5.5.2 below.
- The sample observations of the explanatory vector form a cloud of points in the p-dimensional Euclidean space $\mathbb{R}^p$. The larger the dimension p, the further apart these points appear. Filling up space with points is more difficult in higher dimensions, and in any given neighborhood of a point $\boldsymbol{x} \in \mathbb{R}^p$, we are less likely to find points from the cloud of sample observations when p is large. This fact is known as Bellman's *curse of dimensionality.* When the number n of observations is not excessively large, the kernel regression has proven to be very powerful when the number of explanatory variables (i.e. the number p) is reasonably small, typically 2 or 3. How small this number should be obviously depends upon the sample size n, and the more observations we have, the larger the number of explanatory variables we may include. This form of Bellman's *curse of dimensionality* can easily be illustrated by heuristic arguments, but it can also be quantified by rigorous asymptotic results which show that n should grow exponentially with p. This is a serious hindrance.

5.5.1 **Running the Kernel in** R

When $p = 1$ the kernel regression is implemented by the scatterplot smoother `ksmooth` described in the Sect. 5.3.5. Unfortunately, there is no R function implementing the multivariate kernel regression. We propose to use the home-grown function `kreg` for the purposes of this book.

Since this function can be called quite often in a typical application, the user needs to be aware of the fact that computer times can be unexpectedly long when the sample size n and/or the dimension p of the explanatory variables are large. The output of the function `kreg` is a list containing the input variables and a variable `ypred`. A call to the function `kreg` of the form

```
> PRED <- kreg(X,Y,kernel=triangle,b=.4)
```

returns a variable `PRED$ypred` containing the values $\hat{y}_i = \varphi(\boldsymbol{x}_i)$ of the fitted values for the values of the regressor variable in the set of observations if the argument `xpred` is missing. When this argument is set to a vector of possible values of the explanatory variable/vector, as in the example:

```
> PRED <- kreg(X,Y,xpred=GRID,kernel=triangle,b=.4)
```

then `PRED$ypred` contains the values of this regression function φ for the specific values of the explanatory variables/vectors contained in the argument `xpred`. In other words, the latter is equal to the observations by default.

In the two dimensional case, one may be interested in computing φ over a grid of points for plotting purposes. One can generate such a surface plot by setting the parameter `PLOT` to `TRUE` by adding `PLOT=T` in the call to the function `kreg`. The surface is computed by default over a grid of 256×256 regularly spaced points between the minima and the maxima of the two explanatory variables. This grid can also be user specified.

5.5.2 An Example Involving the June 1998 S&P Futures Contract

This experiment is based on high frequency data on the S&P 500 index introduced and first manipulated in Chap. 6. For the sake of definiteness, we chose to work with the June 1998 futures contract for which we collected ALL the transaction records. Then for each of the 59 trading days between `March 15, 1998` and `June 8, 1998` we computed

- Six indicators at 12:00 pm (noon) for the morning transactions;
- The same six indicators for the afternoon transactions;

the goal of the experiment being to predict the values of the afternoon indicators from the knowledge of the morning ones. The 6 morning indicators are stored in a 59×6 matrix which we call `MORN.mat` and the corresponding afternoon values are stored in another `59 x 6` matrix which we call `AFT.mat`. These indicators were computed from the high frequency tick-by-tick data of all the quotes taking place in the morning and afternoon sessions, respectively. We shall describe later in Sect. 6.1.6 of Chap. 6 some of the tools we used to compute these indicators. For the time being, a quick explanation of what these indicators are will suffice. The first indicator is called `range`. It represents the difference between the highest quote and the lowest quote of the morning. The next indicator is called `nbticks`. It gives the number of transactions during the session, whether it is the morning or afternoon session. The third indicator gives the standard deviation of the log-returns between two successive transactions (and computed ignoring the length of the time interval separating these transactions), while the next two indicators give the volatility ratio `volratio` and the quantile slope `l2slope` which we will define rigorously in the Notes and Complements at the end of the chapter. Finally, the last indicator gives the mean separation time between two consecutive transactions. It is called `ticksep`. Printing the first five rows of the data matrix of our six morning indicators, we get:

```
> MORN.mat[1:5, ]
     range nbticks         vol  volratio l2slope ticksep
 [1,]   4.8     199 9.68800e-08 0.0584073 298.328 7.12836
 [2,]   6.5     199 1.01720e-07 0.0763445 293.569 6.36461
 [3,]   5.1     200 6.59284e-08 0.0206114 299.409 7.00117
 [4,]   4.5     200 7.13838e-08 0.0376023 317.831 6.59341
 [5,]   8.8     207 1.06649e-07 0.0456482 334.903 6.62047
```

Obviously:

the six indicators **are not on the same scale**.

As a solution we propose to

standardize the explanatory variables.

As we already pointed out, this procedure is ubiquitous in nonlinear and nonparametric regression, and in classification. The R-function `scale` does just that. We shall use it in two different ways. First, with only one single parameter the numeric matrix `MORN.mat`. In this case, the function `scale` returns a matrix, say `NORMMORN.mat` obtained by subtracting the column means from their corresponding columns, and by dividing the (centered) columns so-obtained by their root-mean-square. This is exactly *what the doctor ordered*!

```
> NORMMORN.mat <- scale(MORN.mat)
```

Since the column means and standard deviations which are used by the function `scale` to center and re-scale the columns will be needed in the sequel, we show how they can be computed in R. We do this with the function `apply`.

```
> MEANMORN <- apply(MORN.mat,2,mean)
> MEANMORN
    range nbticks         vol  volratio l2slope ticksep
  8.68474 203.389 1.46168e-07 0.0707302 309.809 5.74796
```

The R function `apply` was designed to compute a given function on the rows or the columns of an array. So, the first command applies the function `mean` to each column (second dimension appearing as the second argument of the function `apply`). Similarly, one compute the vector `SDNORM` of the column standard deviations.

```
> SDMORN <- apply(MORN.mat,2,sd)
> SDMORN
   range  nbticks        vol  volratio  l2slope ticksep
3.288790 3.969787 8.634e-08 5.299e-02 10.11494 0.540103
```

We can use the function `apply` to check that we did exactly what we intended to do by computing the means and the variances of the columns of the standardized matrix.

```
> apply(NORMMORN.mat,2,mean)
     range   nbticks       vol volratio  l2slope   ticksep
 -1.05e-16 4.82e-16 2.59e-16 2.83e-17 3.46e-16  1.27e-15
> apply(NORMMORN.mat,2,var)
  range nbticks vol volratio l2slope ticksep
      1       1   1        1       1       1
```

This shows that we succeeded in turning the explanatory variables into variables with empirical mean zero and empirical variance one.

We now consider the problem of the prediction of the afternoon value of the volatility ratio at noon, i.e. when our information consists of the values of the six morning indicators. We shall see in Sect. 5.6 how to use the set of all 6 indicators as regressors, but because our sample size is only 59, we cannot hope to use kernel regression with more than 2 regressors (and even that may be a bit of a stretch). Based on the intuition developed with least squares linear regression, when it comes to the prediction of the values of `AFT.mat[,4]`, a sensible way to choose two explanatory variables out of the six morning indicators should be to find those morning indicators with the largest correlation with `AFT.mat[,4]`. However, basing our choice on this criterion alone could lead to disaster. Indeed, while trying to maximize the correlation with the response, we also need to make sure that the correlation between the two regressors which we choose is as small as possible. Indeed, if they carry the same information, they will be equally correlated with `AFT.mat[,4]`, but the two of them together will not be more predictive than any single one of them taken separately. We do not dwell on this problem here. We simply refer to Problem 5.19 for an attempt to use Principal Component Analysis to find uncorrelated variables summarizing efficiently the information in the set of six variables. For the purposes of the present discussion, we restrict ourselves to choosing `MORN.mat[,4]` and `MORN.mat[,5]`, hoping that it is a reasonable choice. Recall that `MORN.mat[,4]` is the morning value of the volatility ratio whose afternoon value we try to predict, and that `MORN.mat[,5]` is some form of measure of the average rate at which the transactions occur during the morning.

The left pane of Fig. 5.6 contains a three-dimensional scatterplot of the 59 observations of these three variables, while the right pane contains the two dimensional horizontal projection given by the 2-d scatterplot of the two explanatory variables. The three dimensional scatterplot is obtained by putting vertical bars over the points of the 2-d scatterplot of the explanatory variables, the heights of these bars being equal to the corresponding values of the response variable.

The two-dimensional scatterplot of the explanatory variables shows that the points are reasonably well spread throughout the rectangular region of the plot. Notice that the two plots of Fig. 5.8 would look exactly the same if we had used the normalized indicators instead of the raw ones, the only changes being in the axis labels. At this stage, it is a good idea to check for the presence of isolated points far from the bulk of the data. Indeed as we have seen in our discussion of linear regression, the latter are very influential, and they often distort the results of the regression. We do not have such distant points in the present situation. The three

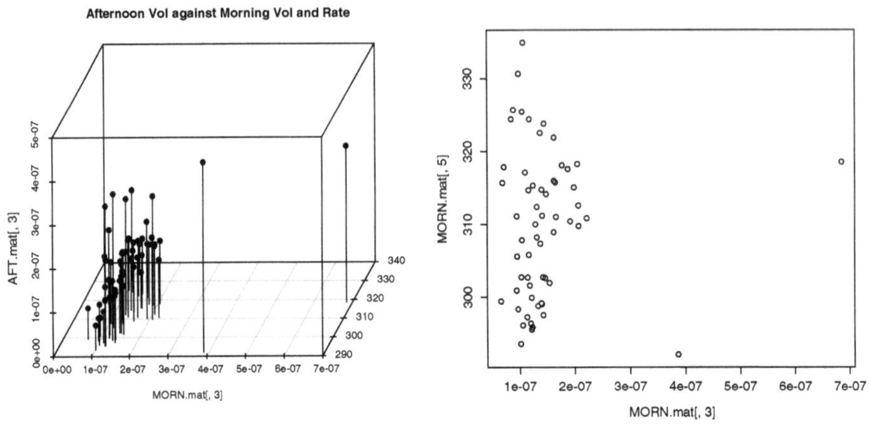

Fig. 5.6. Bar scatterplot of the afternoon volatility ratio `AFT.mat[,4]` against the morning volatility ratio `MORN.mat[,4]` and the arrival rate `MORN.mat[,5]` (*left*), and 2-d Scatterplot of `MORN.mat[,4]` and `MORN.mat[,5]` (*right*)

dimensional plot shows that, except perhaps for the observations on the far right, linear regression could be a possibility. Consequently, we perform a (bivariate) linear least squares regression. The results are shown in Fig. 5.7, and they do not confirm the early suspicion that a linear regression could do the trick.

As we are concerned about the effect of the two measurements with extreme values of MORN.mat[,4], we redid the analysis after removing measurements in rows 30 and 51 Clearly, it is very difficult to visualize why the regression plane would be pulled up or down simply due to the presence of these two observations. For observations to have influence only on the responses to neighboring observations, it is best to use kernel regression. So we perform a few kernel regressions to get a better sense of the response surface. The results are given in Fig. 5.9. Instead of using the function `kreg` described earlier in the text, we used a special purpose function designed for the computation and plot of two-dimensional kernel regression surfaces. This function is called `twoDkreg` and its main goal is to produce a surface plot, providing a graphical tool in the search for a reasonable bandwidth. For the sake of illustration, we used three different values of the bandwidth: the first one is presumably too small because the surface is too rough, the last one is presumably too large because the surface is too smooth, and presumably, as Goldilocks would say, the middle one is *just right*. The plots of Fig. 5.9 were produced with the commands:

```
> twoDkreg(cbind(NORMMORN.mat[,4],NORMMORN.mat[,5]),
      AFT.mat[,5],B=c(.3,.3),X1NAME="NormMornVolRatio",
      X2NAME="NormMornRate",YNAME="AftVolRatio")

> twoDkreg(cbind(NORMMORN.mat[,4],NORMMORN.mat[,5]),
      AFT.mat[,5],B=c(.5,.5),X1NAME="NormMornVolRatio",
      X2NAME="NormMornRate",YNAME="AftVolRatio")

> twoDkreg(cbind(NORMMORN.mat[,4],NORMMORN.mat[,5]),
```

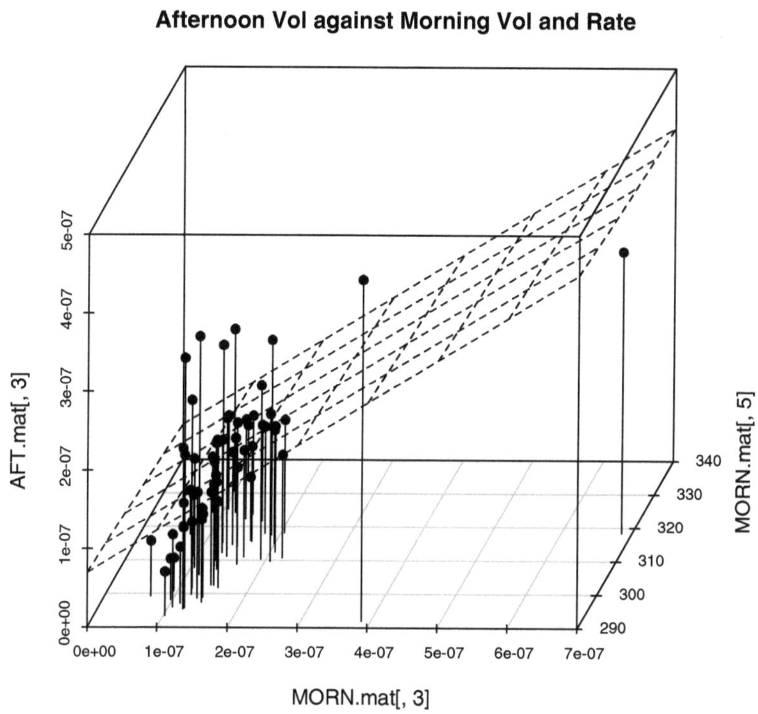

Fig. 5.7. Linear regression of the afternoon volatility ratio AFT.mat[,4] against the morning volatility ratio MORN.mat[,4] and the morning rate of arrival MORN.mat[,5] of the transactions

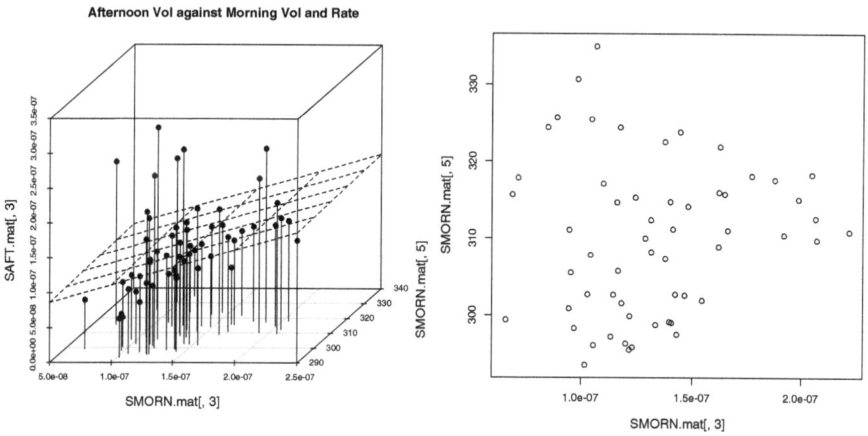

Fig. 5.8. Bar scatterplot of the afternoon volatility ratio AFT.mat[,4] against the morning volatility ratio MORN.mat[,4] and the arrival rate MORN.mat[,5] (*left*), and 2-D Scatterplot of MORN.mat[,4] and MORN.mat[,5] (*right*) after removing two extreme measurements

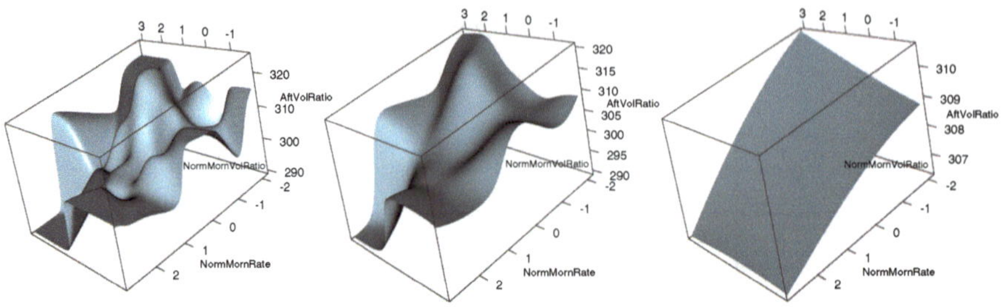

Fig. 5.9. Kernel regressions of the afternoon volatility ratio `AFT.mat[,4]` against the morning volatility ratio `MORN.mat[,4]` and the morning rate of arrival `MORN.mat[,5]` of the transactions in the original data matrix `MORN.mat`. We used a two dimensional Gaussian kernel with the bandwidths $b = 0.3$, $b = 1$ and $b = 2$ from *left* to *right* respectively

```
AFT.mat[,5],B=c(2,2),X1NAME="NormMornVolRatio",
X2NAME="NormMornRate",YNAME="AftVolRatio")
```

Given a value of the morning volatility ratio and a value of the morning rate, the prediction/estimate of the corresponding afternoon volatility ratio is obtained by first scaling these two values by subtracting the means (which can be found in the appropriate entries of `MEANMORN`) and dividing then by the standard deviations (which can be found in the appropriate entries of `SDMORN`) and reading off the value of the regression surface over the point of the plane given by the values of these scaled explanatory variables.

If we need to compute values of the regression function (i.e. predictions) for a large number of values of the explanatory variables, we pool these values in a matrix with the same number of columns as `MORN.mat` and we scale this matrix before we apply the computation of the kernel regression. However, we should now use the function `scale` with three parameters, the matrix of explanatory variables to be scaled, and two parameters, `center` and `scale` which are vectors with one entry for each explanatory variable. The value of `center` determines how column centering is performed: the j-th column of the matrix to be scaled has the j-th entry of the vector `center` subtracted from it. In our previous use of the function `scale`, the parameter `center` was not specified, and consequently, its default value `TRUE` was used. When the parameter `center` is missing or equal to `TRUE`, the centering is done by subtracting the column means (omitting NAs). Note that no centering is done if `center` is `FALSE`. The value of the parameter `scale` determines how column scaling is performed after centering. If `scale` is a numeric vector with length equal to the number of explanatory variables, then each column is divided by the corresponding entry of the vector `scale`. In our first use of the function `scale`, the parameter `scale` was not specified, and hence its default value was used. This default value is the boolean `TRUE`. Whenever this is the case, scaling is done by dividing the centered columns by their root-mean-square, and if the parameter `scale` is equal to the boolean `FALSE`, no scaling is done. We shall see more of the function `scale` in the forthcoming analysis of option pricing by non-parametric methods.

5.6 Projection Pursuit Regression

In their original proposal, the creators of the projection pursuit algorithm suggested writing the regression function $\varphi(\boldsymbol{x})$ of a model $y = \varphi(\boldsymbol{x}) + \epsilon$ in the form:

$$\varphi(\boldsymbol{x}) = \alpha + \sum_{j=1}^{m} \phi_j(\boldsymbol{a}_j \cdot \boldsymbol{x}). \tag{5.15}$$

Remember that we are working in the usual regression setting:

$$(\boldsymbol{x}_1, y_1), (\boldsymbol{x}_2, y_2), \ldots\ldots, (\boldsymbol{x}_n, y_n)$$

where the *explanatory variables* $\boldsymbol{x}_1, \boldsymbol{x}_2, \ldots, \boldsymbol{x}_n$ are p-dimensional. In other words, each $\boldsymbol{x}_i$ is a p-vector of the form $\boldsymbol{x}_i = (x_{i,1}, \ldots, x_{i,p})$ while the response variables $y_1, y_2, \ldots, y_n$ are univariate (i.e. scalar). As usual, we use bold face letters $\boldsymbol{x}$ to emphasize that we are dealing with a *multivariate* explanatory variable which we sometimes call an explanatory vector to emphasize that its dimension can be greater than one. The idea of the projection pursuit algorithm is to fight the *curse of dimensionality* inherent with large values of p, by replacing the p-dimensional explanatory vectors $\boldsymbol{x}_i$ by suitably chosen one-dimensional projections $\boldsymbol{a}_j \cdot \boldsymbol{x}_i$, hence the term *projection* in the name of the method. We now explain how the estimation of the quantities α, and $\phi_1(\boldsymbol{a}_1 \cdot \boldsymbol{x})$, ..., $\phi_m(\boldsymbol{a}_m \cdot \boldsymbol{x})$ appearing in formula (5.15) is performed. The projection pursuit algorithm is based on an inductive procedure in which residuals are recomputed and fitted at each iteration, and for the purposes of the present discussion, one should think of the observed values y_i as the starting residuals, i.e. the residuals of order zero. Each time one of the terms in the sum appearing in the right hand side of (5.15) is estimated, the actual estimates are subtracted from the current values of the residuals, providing in this way a new set of residuals from which we proceed to estimate the next term in (5.15). This recursive fitting of the residuals justifies the term *pursuit* in the name of the method.

The constant α is naturally estimated by the mean of the observations of the response:

$$\hat{\alpha} = \emptyset y = \frac{1}{n} \sum_{i=1}^{n} y_i.$$

This sample mean is subtracted from the observations (i.e. the residuals of order zero) to get the residuals of order one. Next we proceed to the estimation of the first direction $\boldsymbol{a}_1$ and the first function ϕ_1. Because of computational considerations, it is important to choose a specific procedure capable of selecting the best function ϕ for each choice of direction given by the unit vector $\boldsymbol{a}$. R does just that by choosing, for each candidate $\boldsymbol{a}$, the function $\phi_{\boldsymbol{a}}$ given by the super-smoother algorithm (implemented as the function supsmu) obtained from the scatterplot of the projections of the data points $\{\boldsymbol{x}_i\}_{i=1,\ldots,n}$ on the direction $\boldsymbol{a}$, i.e. the values of the scalar products $\boldsymbol{a} \cdot \boldsymbol{x}_i$, and the corresponding values y_i of the response y (or the current residuals if we are in the middle of the recursive fitting procedure).

The creators of the projection pursuit algorithm proposed to fit recursively the residuals (starting from the values of the response variable) with terms of the form $\phi_{\boldsymbol{a}_j}(\boldsymbol{a}_j \cdot \boldsymbol{x})$ and to associate to each such optimal term a figure of merit, for example the proportion of the total variance of the response actually explained by such a term. In this way, the whole procedure would depend upon only one parameter. One could choose this *tolerance* parameter first (typically a small number) and one would then fit the response and successive residuals until the figure of merit (i.e. the proportion of the variance explained by $\phi_{\boldsymbol{a}_j}(\boldsymbol{a}_j \cdot \boldsymbol{x})$) dropped below the tolerance parameter. This would automatically take care of the choice of the order m of the model (5.15).

Remark 2. Formula (5.15) is reminiscent of formula (5.2) for the basis (or feature) expansion least squares regression discussed earlier in the chapter. The main difference is the fact that, in the case of projection pursuit, the choice of the feature functions φ_j is driven by the data. This endogenous determination of the feature functions is more in the spirit of modern machine learning than traditional nonparametric regression.

5.6.1 The R Function ppr

Projection pursuit regression is implemented in R by the function ppr. A typical call to this function looks like:

```
data.ppr <- ppr(x,y,nterms,max.terms)
```

where x is the $n \times p$ matrix of the n observations of the p explanatory variables and y is the $n \times 1$ vector of corresponding responses. Equivalently, the function ppr can be called with a formula, say of the form y x1+x2+ .... +xp if x1, x2, $\cdots$, xp are the names of the p explanatory variables, in lieu of the parameters x and y. tt nterms and max.terms are integers whose values correspond to the number m of ridge functions in the decomposition (5.15) and the largest of the values of m tested by the program.

Had R implemented this algorithm in the way we described it, the proportion of the total variation in the response explained by the successive models obtained by increasing the number m of ridge functions would always be decreasing. Unfortunately, things are not that simple, and what we just described *is not exactly* the algorithm implemented in R.

When the arguments nterms and max.terms are set and the function ppr is run, the program does pretty much what we just described for m increasing from 1 to max.terms, and the sequential plot of the vector data.ppr$gofn would be decreasing if one could plot it at this stage. However, R does not stop after this *forward pass* over the data. It recomputes fitted models for $m =$ max.terms $- 1$,

$m =$ `max.terms` $- 2, \ldots, m =$ `nterms` in this reverse order. For each value of m in this range, the program runs an optimization procedure to find simultaneously all the $\boldsymbol{a}_j$ appearing in the model (5.15). Unfortunately, minimization procedures are not always reliable, especially when the dimension is large and the function is not convex. Their results depend strongly on the initializations. Here is what `R` is actually doing in order to initialize this minimization. For each value of m (varying from `max.terms` -1 down to `nterms`) the program considers the unit direction vectors $\hat{\boldsymbol{a}}_1, \ldots, \hat{\boldsymbol{a}}_{m+1}$ found to be optimal in the model fitted previously with $m+1$ terms, and it computes the variances of the values fitted by the super-smoother algorithm `supsmu`:

$$\sigma_j^2 = \frac{1}{n}\sum_{i=1}^{n} \phi_{\hat{\boldsymbol{a}}_j}(\hat{\boldsymbol{a}}_j \cdot \boldsymbol{x}_i)^2.$$

Notice that there is no need to subtract the mean in order to compute the variance because we already subtracted the mean $\o y$ of the response and after that the response (whether it is the actual response or the residual at a given stage of the recursion) is always centered. The minimization algorithm is then initialized with the m unit vectors $\hat{\boldsymbol{a}}_{j_1}, \ldots, \hat{\boldsymbol{a}}_{j_m}$ where the indices $j_1, \ldots, j_m$ are chosen in such a way that the variances $\sigma_{j_1}^2, \ldots, \sigma_{j_m}^2$ are the m largest among the $m+1$ variances $\sigma_1^2, \ldots, \sigma_{m+1}^2$. The rationale for this choice is simple. The relative size of the variance σ_j^2 is an indication of how important the contribution of $\hat{\boldsymbol{a}}_j$ (and of the corresponding function $\phi_{\hat{\boldsymbol{a}}_j}$) is. In this way, for each value of m, `R` finds a model of the form:

$$\varphi(\boldsymbol{x}) = \o y + \sum_{j=1}^{m} \phi_{\boldsymbol{a}_{m,j}}(\boldsymbol{a}_{m,j} \cdot \boldsymbol{x}). \tag{5.16}$$

However, the set $\{\boldsymbol{a}_{m,1}, \ldots, \boldsymbol{a}_{m,m}\}$ of unit vectors in the fitted model of order m, is not necessarily equal to the set of the first m vectors $\{\boldsymbol{a}_{m+1,1}, \ldots, \boldsymbol{a}_{m+1,m}\}$ of the set of unit vectors in the fitted model of order $m+1$. Even though it turns out that the plot of `data.ppr$gofn` is very often decreasing, it can increase at times, even though this is not the rule in general, as there is absolutely no reason why the proportion of the variation explained should be a decreasing function of m when computed in this way.

As a final remark, we mention the way in which `R` normalizes the functions ϕ_j appearing in the decomposition of the model (5.15). If one introduces the notation:

$$\beta_j = \sigma_j = \sqrt{\sum_{i=1}^{n} \phi_{\hat{\boldsymbol{a}}_j}(\hat{\boldsymbol{a}}_j \cdot \boldsymbol{x}_i)^2}, \qquad \varphi_j = \frac{1}{\sigma_j}\phi_{\hat{\boldsymbol{a}}_j}, \tag{5.17}$$

then obviously the model (5.15) can be rewritten in the form:

$$\varphi(x) = \o y + \sum_{j=1}^{m} \beta_j \varphi_j(\boldsymbol{a}_j \cdot \boldsymbol{x}) \tag{5.18}$$

and in this form, the contribution of each of the functions φ_j can be viewed as having mean 0 and variance 1.

5.6.1.1 **Running the** R **Function** ppr

As usual, if in doubt, get help on the function ppr by typing:

```
> help(ppr)
```

For a projection pursuit regression, the function ppr should be called at least twice. The first call to the projection pursuit function should be of the form:

```
> data.ppr <- ppr(x,y,nterms,max.terms)
```

where x is the $n \times p$ matrix of the n observations of the p explanatory variables and y is the $n \times 1$ vector of corresponding responses. It is recommended to choose a value of max.terms no larger than 10 (and significantly smaller if your data set is not large enough) and of nterms (which in any case should be smaller than max.terms) small for the first run, typically 1 or 2. Then, one should run the command

```
> plot(data.ppr$gofn)
```

to produce the plot of the goodness of fit figure of merit for the decompositions in n terms attempted by ppr for $n = \texttt{nterms}$, $n = \texttt{nterms} + 1, \cdots, n = \texttt{max.terms}$. The plotted value should be 0 for the values $n < \texttt{nterms}$. This plot is usually decreasing abruptly, before stabilizing. We should then choose for nterms as small a value of n as possible, while at the same time lowering the value of this figure of merit as significantly as possible. The next step is to rerun the function ppr with nterms equal to the value of n chosen in this way, making sure that we keep the same value for max.terms.

```
> data.ppr <- ppr(x,y,nterms=n,max.terms)
```

Prediction from a projection pursuit model fitted to data is done with the generic method predict in a standard way:

```
> data.ppr.pred <- predict(data.ppr,newdata=xpred)
```

where xpred is a matrix with the same number of columns as x, and containing one row for each of the vectors of explanatory variables for which a prediction (or regression estimate) is desired.

5.6.2 ppr **Prediction of the S&P Afternoon Indicators**

We revisit the analysis of the S&P indicators which was performed earlier in Sect. 5.5.2 with the help of kernel regression. This will lead to a first comparison of the performance of the two regression methods. We use the same notation, and since the introduction of the projection pursuit regression was motivated by problems with a large number of explanatory variables, we use for the prediction of any given afternoon indicator, the entire set of 6 morning indicators. Again, we choose to predict the volatility ratio. The projection pursuit regression is performed using the following R commands:

```
> PPR1 <- ppr(MORN.mat,AFT.mat[,4],nterms=1,max.terms=10)
> plot(PPR1$gofn, type="l",
            main="PPR of AftVolRatio against the 6 Morning Indicators")
```

The plot of the figure of merit `PPR1$gofn` is reproduced in Fig. 5.10.

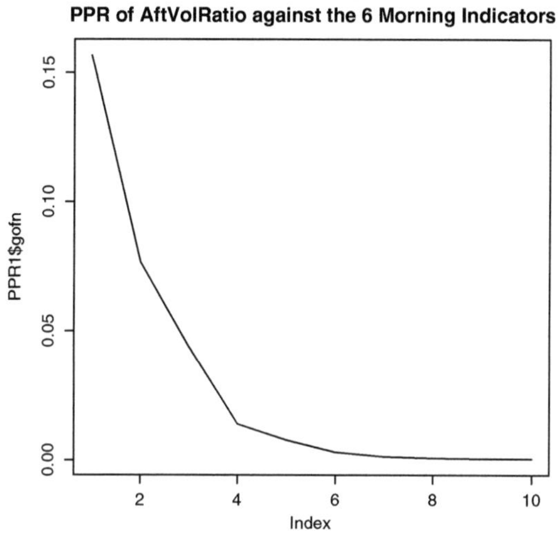

Fig. 5.10. Sequential plot of the figure of merit of the projection pursuit regression of the afternoon volatility ratio from the 6 S&P 500 morning indicators

From this plot, it seems that 8 should be the best choice, but given the small number of observations ($n = 59$) and the relatively high dimension of the explanatory variables ($p = 6$), a more reasonable choice is 4. So we re-run the projection pursuit algorithm with the command:

```
> PPR1 <- ppr(MORN.mat,AFT.mat[,4],nterms=4,max.terms=10)
```

Since the number of graphical tools is very limited when the number of explanatory variables is greater than 2 (see nevertheless the help file for a discussion of the tools provided to plot graphs of the functions ϕ_j) we decided to output the numerical figures of merit of this regression. We compute the relative sum of squares (analog of the $1 - R^2$ of the linear models) and the actual sum of the squares of the raw residuals.

```
> PPR1$gof
 [1] 0.01400715
> sum((PPR1$fitted.values - AFT.mat[,4])^2)
 [1] 0.01393516
```

These figures look very good. They should be compared to the corresponding figures which we would have obtained if we had applied other methods such as the kernel regression of the previous section or the neural network regression discussed in the

Notes & Complements at the end of the chapter. For the sake of comparison with the results of the kernel method which we used earlier, we perform projection pursuit regression using only the two morning indicators used earlier.

```
> PPR2 <- ppr(MORN.mat[,4:5],AFT.mat[,4],nterms=1,
              max.terms=10)
> plot(PPR2$gofn, type="l",
    main="PPR of AftVolRatio against 2 Morning Indicators")
```

The plot of the figure of merit PPR2$gofn is reproduced in Fig. 5.11.

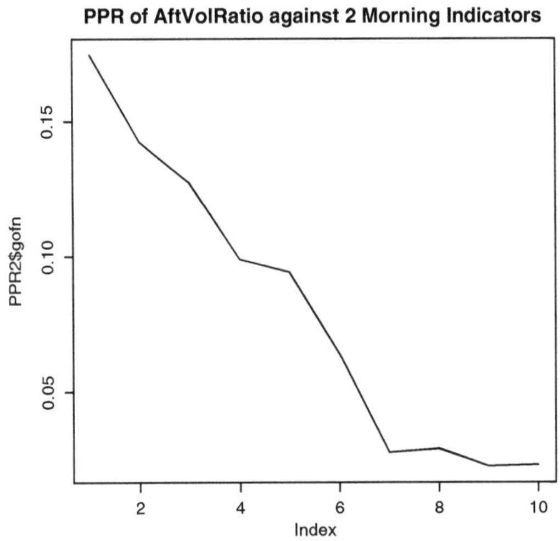

Fig. 5.11. Sequential plot of the figure of merit of the projection pursuit regression of the afternoon volatility ratio against 2 of the 6 S&P 500 morning indicators

From this plot, it seems that 7 should be the best choice. Notice that the choice of 7 for the number of ridge functions is more reasonable than earlier because of the number of regressors is only 2.

```
> PPR2 <- ppr(MORN.mat[,4:5],AFT.mat[,4],nterms=7,max.terms=10)
> PPR2$gof
[1] 0.02764523
```

This figure of merit is not as good as the number obtained earlier with the six explanatory variables, but it is much better than the one obtained with the kernel method for the value of the bandwidth we deemed reasonable. Indeed, computing the sum of square errors between the observations and the fitted values for the kernel regression with our choice of bandwidth would lead to much higher sums of square errors. In fact in order to get comparable values for the figure of merit of the regression, we would need to lower the bandwidth to values we rejected by fear that the result of the kernel regression would merely be the result of *fitting the noise*.

Despite the apparent superiority of projection pursuit, we would not recommend to give up the kernel regression solely on the basis of a comparison of the sums of square errors. Indeed, for pedagogical reasons, and for the sake of argument, we give in Fig. 5.12 the regression surface produce by this two-dimensional projection pursuit regression.

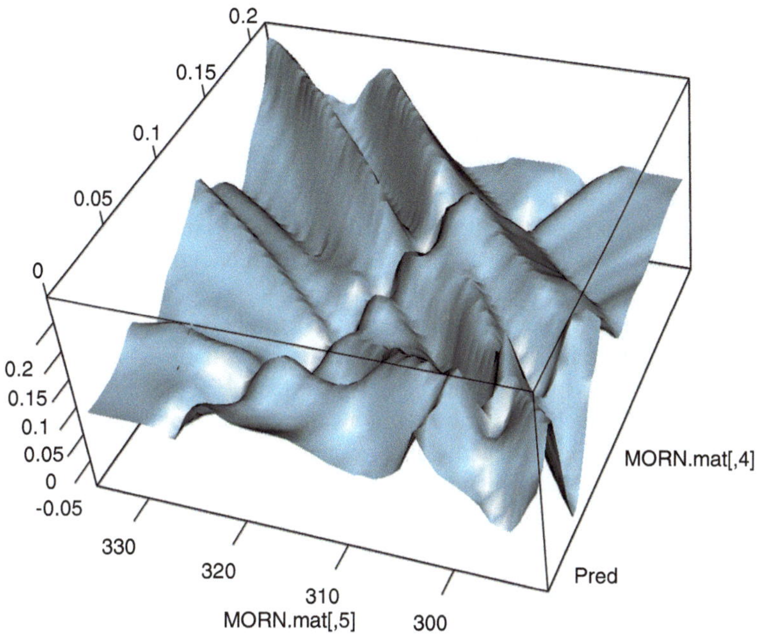

Fig. 5.12. Surface plot of the regression surface given by the two dimensional projection pursuit algorithm in the case of the morning and afternoon S&P data

The plot was produced with the R commands:

```
> n <- 50
> GRIDX <- seq(from=min(MORN.mat[,4]), to=max(MORN.mat[,4]), length=n)
> GRIDY <- seq(from=min(MORN.mat[,5]), to=max(MORN.mat[,5]), length=n)
> xmat <- rep(GRIDX, n )
> ymat <- rep(GRIDY, each = n)
> zmat <- predict(PPR2,cbind(xmat,  ymat))
> persp3d(GRIDX,GRIDY,zmat, aspect=c(1, 1, 0.5),
                col = "lightblue", xlab = "MORN.mat[,4]",
                ylab = "MORN.mat[,5]", zlab = "Pred")
```

We see from the plot reproduced in Fig. 5.12 that the regression surface is not as smooth as the one obtained in Fig. 5.9. Did we fit the noise or did the projection pursuit regression actually capture features that the kernel was oblivious to? The convoluted nature of the response surface evidenced by Fig. 5.12 is presumably an indication that we got a smaller sum of square errors at the cost of noise fitting, and that predictions from this regression could be unstable and unreliable. Moreover the special form of the response surface is worth noticing. Indeed, we chose a different perspective to evidence the structure of a projection pursuit regression surface as a superposition of surfaces which are constant in one direction (hence the name *ridge* functions). This special feature is clearly seen in Fig. 5.12.

Comparing the relative performance of the kernel and the projection pursuit regression methods is desirable. But one should not attempt to compare the results produced by following too closely the implementation prescriptions which we gave in this chapter. Indeed, the steps suggested for the implementation of the projection pursuit will very likely produce a good fit because the order of the model is based on an *in sample* criterion involving the sum of square errors between the actual observations and the fitted values. However, using the projection pursuit regression model obtained in this way for the prediction of the response to new values of the explanatory variables (which is often called *out of sample testing*) may not be a very good idea, since we are not sure that we did not over-fit the model to get a smaller sum of square errors.

On the other hand, the prescriptions we used to choose the bandwidth of the kernel regression were not driven by the desire to get a small sum of square errors. So, we should not be surprised if the regression surfaces obtained in this way do not fit the data as well. Much smaller bandwidth values would be needed for that. The rationale for not choosing the bandwidth too small comes from the desire to use the regression model for prediction purposes.

In short, we suggested implementing the projection pursuit algorithm to have a reasonable fit to the data, while we gave recommendations for the choice of kernel bandwidth with the prediction of future response values in mind. These goals may be incompatible, and fair comparisons may be difficult. Quantifying how well a regression model fits the data is relatively easy: just look at the sum of square errors. Quantifying the predictive value of a regression model is more complicated: one possible way to do it is to fit a regression model to a training sample and quantify the prediction power of the method by using such a model on a different data set. This strategy is explained in the next paragraph, and we will use it in the section on option pricing and in numerous problems.

5.6.3 More on the Comparison of the Two Methods

The question of the comparison of the performance of several nonparametric regression/prediction procedures is very delicate. Because they are based on statistical models with a solid theoretical foundation, the parametric regression methods (and especially the linear models) came with inferential tools for residual analysis. In the nonparametric world, the corresponding tools are only asymptotic (i.e. applicable when the sample size goes to ∞) and for this reason, they cannot be of much practical use for finite samples. As a consequence, practical *common sense recipes* are used instead. Since nonparametric methods can only be competitive when the sample size is relatively large, one often has the possibility of separating the data into two subsets, one used for *training* purposes, i.e. to fit a regression model, and the other one for *testing* purposes, i.e. to compare the figures of merit of several methods or several parameters, ... See the next section for an example. In financial applications, this is implemented in the form of *back testing*. In our specific example, after finding the kernel and the bandwidth, or the order and the parameters of the projection pursuit, using the June 1998 contract, we could use a different contract to test and compare the relative performances of the different methods or sets of parameter choices.

Confidence intervals offer a convenient way to quantify the accuracy of statistical estimates. In regression problems, confidence intervals are replaced by confidence bands, or sausages, containing the regression curves or surfaces. The computations needed to produce reliable confidence regions are usually very intensive. The most successful ones seem to have come out of *bootstrap*-like methodologies, but they are beyond the scope of this book.

5.7 Nonparametric Option Pricing

The goal of this section is to implement and compare several nonparametric methods of option pricing. We use real market data to compare the numerical performance of the various methods. After a couple of subsections devoted to introductory material on classical option pricing theory, the data are described in Sect. 5.7.2, and the details of the experiment are given in Sect. 5.7.3. The long Sect. 9.2.1 is not intended as a crash course on option pricing but merely as a convenient introduction to the mathematical theory of no-arbitrage option pricing, justifying the nonparametric approach used in this section.

5.7.1 Nonparametric Pricing Alternatives

Because of the shortcomings of the Black-Scholes formula illustrated above, we resort to nonparametric techniques to price liquid options. We choose options on the S&P 500 because on any trading day, there are a large number of options traded with various times to maturity and strikes. We shall consider two competing approaches. The second one will rely partly on the Black-Scholes formula, but the first one is fully nonparametric.

5.7.1.1 Fully Nonparametric Pricing

In this approach, we do not make any assumption on the functional relationship between the measurable variables S (current price of the underlying asset), K (strike price of the option), τ (time to maturity), and r (short interest rate), which we consider as explanatory variables, and the price C of the corresponding (European call) option which we regard as the response variable. So if we bind together the four explanatory variables into a row $\boldsymbol{X} = [S, K, \tau, r]$ and if we set $Y = C$ for the response variable, then we are in the classical setting of (multiple) regression, and we can use any of the techniques seen so far to explain and predict the price Y from the knowledge of $\boldsymbol{X}$. The goal of the first part of our experiment is to implement and compare the results of the kernel method and the projection pursuit regression in this setting.

5.7.1.2 *Semi-parametric Option Pricing*

By definition of the implied volatility, the knowledge of the value of an option is equivalent to the knowledge of its implied volatility. Indeed one can go from one to the other by evaluation of Black-Scholes formula (9.1). So instead of pricing the option by computing its values in US dollars via this formula, we first derive the implied volatility, and then we evaluate Black-Scholes formula. Next we argue that this implied volatility is a specific function of a smaller number of variables, and we attempt to derive this functional dependence by a nonparametric regression, typically a kernel regression since after reducing the number of variables, the dimension is presumably not an issue any longer. This approach is called semi-parametric because it is a combination of a purely nonparametric approach (using the kernel to predict the implied volatility) and a parametric strategy (characterizing the implied volatility by a small number of parameters and computing the Black-Scholes formula.

5.7.2 **Description of the Data**

The data are contained in the two R objects `TRGSP` and `TSTSP`. As you might guess from their names, these two data matrices are intended for training and testing purposes respectively. Each data file contains six columns, each row corresponding to an option. The meanings of these columns are as follows:

1. `SP` for the price of the index
2. `KK` for the strike price of the option
3. `TAU` for the time to expiration (in days)
4. `RR` for the spot interest rate
5. `ISIGMA` for the implied volatility
6. `CALL` for the price of the (European call) option

We chose to use repeated letters `KK` and `RR` instead of `K` and `R`, respectively, because R reserves some of the single letter names for protected objects. Also, you should not assume that the row numbers of the data frames have much to do with the dates at which these quotes were in force. The third column gives the *time to maturity* which is the length of time $\tau = T - t$ between the time of maturity T and the time t at which the quote was given. The actual dates t and T are not given in the data sets, only their differences is. For the sake of definiteness, all the quotes are from 1993, and some of the options expire in 1994. In particular, using sequential plots (like those given by the function `tsplot`) would not make much sense in the present situation.

Warning. The values of `TAU` found in the third column are given in days. In most applications, they need to be expressed in years before they can be used in the formulae as values of $\tau = T - t$. This change of units should be done by dividing all the entries of this column by 252. This is the average number of trading days in 1 year.

5.7.3 The Actual Experiment

For the purposes of this experiment, we wrote two simple `R` functions, `bscall` which computes the price of a European call option from the Black-Scholes formula (9.1), and `isig` which computes the implied volatility by inverting the same Black-Scholes formula. Notice that both functions require that the parameter `TAU` which stands for the variable τ be given in years, while the third columns of `TRGSP` and `TSTSP` give the numbers of days until maturity. We choose the convention of 252 trading days per year to convert the number of days in years. In order to make sure that this function does what it is supposed to do, one can use it with the arguments `SP`, `KK`, `TAU`, `RR`, and `ISIGMA`, and check that one does indeed recover the last columns of the training `TRGSP` and testing `TSTSP` data matrices respectively.

In order to check that the sixth column is equal to the result of the function `bscall` when applied to the first five columns, we compute the range (i.e. the two-dimensional vector whose entries are the minimum and maximum of the original vector) of the difference and we check that its entries are equal to 0, at least up to small rounding errors due to the fact that the Black Scholes formula cannot be inverted exactly, and to the fact that the computation of the implied volatility is merely the result of a numerical approximation.

```
> data(TRGSP)
> BSTRG <- bscall(TAU=TRGSP[,3]/252,K=TRGSP[,2],S=TRGSP[,1],
                              R=TRGSP[,4],SIG=TRGSP[,5])
> range(TRGSP[,6] - BSTRG)
[1] 0 0
> data(TSTSP)
> BSTST <- bscall(TAU=TSTSP[,3]/252,K=TSTSP[,2],S=TSTSP[,1],
                              R=TSTSP[,4],SIG=TSTSP[,5])
> range(TSTSP[,6] - BSTST)
 [1] 0 0
```

5.7.3.1 ***Fully Nonparametric Regression***

In this first part, we use the function `kreg` to predict the option prices in the testing sample `TSTSP` using as explanatory variables the current price of the underlying index, the strike price, the time to maturity, and the current value of the short interest rate. *Obviously, we do not use the implied volatility, that would be cheating!* In other words, we use the training data in `TRGSP` to determine the regression function which we then use to predict the prices of the options contained in `TSTSP`. We also compute the raw sum of square errors which we call SSE_1 and a per-option error. We shall use them later to compare the performance of this four dimensional kernel regression with the other methods explained above.

Fully Fledged Kernel. Since the function `kreg` uses only one bandwidth, we need to standardize the explanatory variables before running a multidimensional kernel regression. Also, we take this opportunity to emphasize the modicum of care needed to standardize a testing sample from the results of the normalization of the training sample. In other words, we need to standardize the regressor variables using the means and the standard deviations of the observations of the regressor variables contained in the training sample. This is a typical source of error when one first enters the business of nonparametric *prediction*. We follow the steps outlined in Sect. 5.5.2. Even though we use the function `scale` to standardize the regressors in the training sample in one single command, we also compute the means and the standard deviations of the explanatory variables because we will need them to standardize the testing sample.

```
X1 <- cbind(TRGSP[,1],TRGSP[,2],TRGSP[,3],TRGSP[,4])
MEANX1 <- apply(X1,2,mean)
MEANX1
[1] 441.91861365 439.84700000  50.33720000   0.03030124
SDX1 <- apply(X1,2,sd)
SDX1
[1] 6.538843e+00 2.248005e+01 2.532073e+01 8.098422e-04
SX1 <- scale(X1)
apply(SX1,2,mean)
[1] 5.049048e-16  8.992398e-16 -1.141206e-16 -1.491684e-15
apply(SX1,2,var)
[1] 1 1 1 1
```

We now normalize the entries of the testing sample by using the same normalization as in the normalization of the training sample. Notice that, for the sake of convenience, the second and third commands use the fact that the training sample is larger than the testing sample.

```
XPRED1 <- cbind(TSTSP[,1],TSTSP[,2],TSTSP[,3],TSTSP[,4])
SXPRED1 <- scale(XPRED1,center=MEANX1,scale=SDX1)
apply(SXPRED1,2,mean)
[1]  0.555335926  0.075311219 -0.002338005  0.639329493
apply(SXPRED1,2,var)
[1] 0.2279553 0.9240326 0.8392511 0.1225965
```

Notice that the columns of the standardized testing sample do not have mean zero and variance one. This is due to the fact that we have to use the means and standard deviations of the columns of the training sample if we want to pretend that we are going to predict the response for the individual elements of the testing sample, one by one, without assuming that we know the whole sample. Once this standardization is out of the way, we are ready to run the kernel regression. We use the Gaussian kernel function by setting the parameter `kernel` to gaussian in the command below. The results would be qualitatively the same if we use another kernel function. The choice of bandwidth is even more delicate than before. Indeed, it is not possible to eye-ball this choice by looking at plots of the regression surfaces obtained for different values of the bandwidth. A possible approach is to use the default value proposed by the program. This value is derived from mathematical results aiming at the identification of optimal choices for the bandwidth. Such mathematical results have been proved in the limit of sample sizes going to infinity. They give specific prescriptions for the choice of the optimal bandwidth in the univariate case, the multivariate case being handle by a scaling correction. Another possibility would be to implement a form of cross-validation. However, for the sake of simplicity, we use the default value. Using this bandwidth we get:

```
YPRED1.kreg <- kreg(SX1,TRGSP[,6],xpred=SXPRED1,
                                        kernel=gaussian,b=.05)
YPRED1 <- YPRED1.kreg$ypred
SSE1 <- sum((YPRED1 - TSTSP[,6])^2)
SSE1
[1] 635.652
sqrt(SSE1/length(YPRED1))
[1] 1.127521
```

Projection Pursuit. The implementation of projection pursuit is simpler because we do not have to standardize the explanatory variables. As before, we use the option data from TRGSP as a training sample to which we fit the regression model, which we then use to predict the prices of the options of the data set TSTSP. As before, we use as explanatory variables the current price of the underlying index, the strike price, the time to maturity, and the current value of the short interest rate. And as before, we compute the sum of square errors which we now call SSE_2, as a figure of merit.

```
YPRED2.ppr <- ppr(TRGSP[,1:4],TRGSP[,6],nterms=1,max.terms=10)
plot(YPRED2.ppr$gofn,type="l",
    main="PPR for the 4 Explanatory Variables")
```

The plot of YPRED2.ppr$gofn suggests using any integer between 2 and 9 for the number of terms to include in the pursuit. We choose 2 to be parsimonious, and we rerun the algorithm accordingly.

```
XPRED2 <- as.matrix.data.frame(TSTSP[,1:4])
YPRED2 <- predict(YPRED2.ppr,newdata=XPRED2)
SSE2 <- sum((YPRED2 - TSTSP[,6])*(YPRED2 - TSTSP[,6]))
SSE2
[1] 133.9494
sqrt(SSE2/length(YPRED2))
[1]  0.5175895
```

The above result shows a very significant improvement. It is a clear proof of the difficulties the kernel has when it has to handle high dimensions. Comparing the two fully nonparametric methods confirms the fact that the projection pursuit has a better control of the curse of dimensionality.

5.7.3.2 **Semi-parametric Estimation**

The prediction methods used above are based on a brute force approach, using only the raw data, and no particular knowledge of the specifics of the actual problem at hand. We now inject some *finance* into the statistical mix, and as we are about to see, this is going to take us a long way.

The price of the option is now derived from a prediction of its implied volatility, by computing the function `bscall`. Here we follow the market practice which favors implied volatility over price. However, the thrust of the financial input that we mentioned above is in the way we predict the implied volatility. Instead of blindly mining the four dimensional training sample, we use a two dimensional explanatory vector made of financially meaningful variables: the time to maturity τ, and the moneyness $M = e^{r(T-t)}S/K$. As before, r is the short interest rate, S is the current price of the underlying asset, and K is the strike price of the option. This ratio M is called the *moneyness* of the option because it compares the index futures price $e^{r(T-t)}S$ to the strike K. For this reason, the option is said to be *in the money* when $M > 1$, *out of the money* when $M < 1$ and *at the money* when $M = 1$. Note that, in the computation of the moneyness M, the time to maturity $\tau = T - t$ has to be expressed in years because the spot interest rate r is quoted annually.

At this stage, it is not important which nonparametric regression we use to predict the implied volatility from these two explanatory variables. Indeed, in two dimensions, similar figures of merit can be achieved if we choose the smoothing parameters appropriately. We decided to use the kernel method for the sake of definiteness.
In order to complete the program described above, we first compute the moneyness for each of the options of the training and testing samples.

```
TRGM <- exp(TRGSP[,4]*TRGSP[,3]/252)*TRGSP[,1]/TRGSP[,2]
TSTM <- exp(TSTSP[,4]*TSTSP[,3]/252)*TSTSP[,1]/TSTSP[,2]
```

Next, we play the same *standardization* game to prepare for the kernel regression/prediction. Since this is not the first time that we are going through this exercise, we limit ourselves to giving the R commands.

```
XX3 <- cbind(TRGSP[,3]/252,TRGM)
MEANXX3 <- apply(XX3,2,mean)
SDXX3 <- apply(XX3,2,sd)
SXX3 <- scale(XX3)
apply(SXX3,2,mean)
                       TRGM
-1.786882e-17 -1.978569e-15
apply(SXX3,2,sd)
     TRGM
   1    1
```

Now that we are done with the standardization of the training sample, we normalize the testing sample explanatory variables by using the means and standard deviations computed from the training sample.

```
XXPRED3 <- cbind(TSTSP[,3]/252,TSTM)
SXXPRED3 <- scale(XXPRED3,center=MEANXX3,scale=SDXX3)
apply(SXXPRED3,2,mean)
                        TSTM
-0.002338005  0.085645044
apply(SXXPRED3,2,sd)
                  TSTM
0.9161065 0.9754558
```

For the sake of illustration, we give the scatterplots of the normalized regressor variables in Fig. 5.13. As expected, the points of the training sample cover the region we expect a standardized bivariate normal sample would cover. On the other hand, it appears that the options of the testing sample come in three very distinct subgroups according to the values of the time to maturity. This should not be too bad of a problem given the fact that, each point in the testing sample seems to be well surrounded by a large number of points of the training sample. That will certainly help the kernel regression to do its job. Finally, we compute semi-parametric predictions of the option

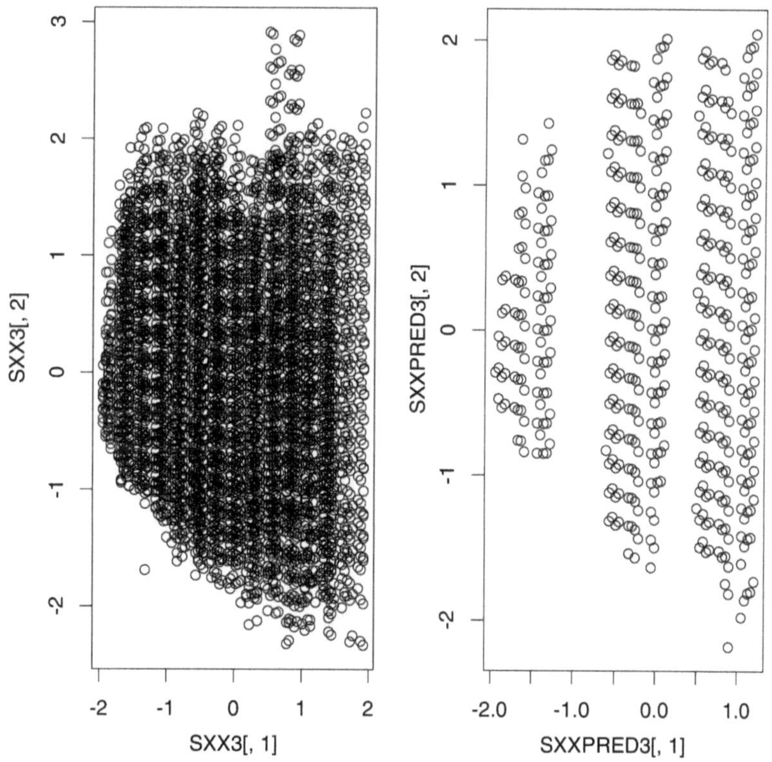

Fig. 5.13. Scatterplots of the standardized explanatory variables in the training sample (*left*) and the testing sample (*right*)

prices of the testing sample (i.e. the entries in the column CALL in TSTSP) by first computing the kernel prediction of the implied volatility, and then by plugging these predictions into the Black-Scholes formula. We compute the sum of square errors as before, and we call it SSE_3. As before, we choose to work with the Gaussian kernel function, hoping that this will prevent divisions by 0 in our tests, and as before, the choice of the bandwidth is the crucial difficulty to overcome. Fortunately, the fact that we are working with a two dimensional regressor vector makes it possible to visualize the properties of the regression surface. So for the purposes of this project, we limit ourselves to the simplest method of all: we compare the plots of the regression surfaces for several values of the bandwidth, and we pick the bandwidth leading to the most reasonable implied volatility surface. This procedure relies heavily on our past experiences: it is more of an art form than a quantitative approach, however, it will be good enough here. To illustrate some of the features leading to our choice, we reproduce in Fig. 5.14, the regression surfaces computed above a common grid of points in the (TAU, M)-plane, for three values of the bandwidth. The regression for the largest of the three bandwidths is too smooth and misses many of the features of the regression surface visible when the bandwidth is equal to 0.1, while the regression for the smallest of the two bandwidth is too rough, including effects which are presumably only due to noise. After extensive experimentation with the value of

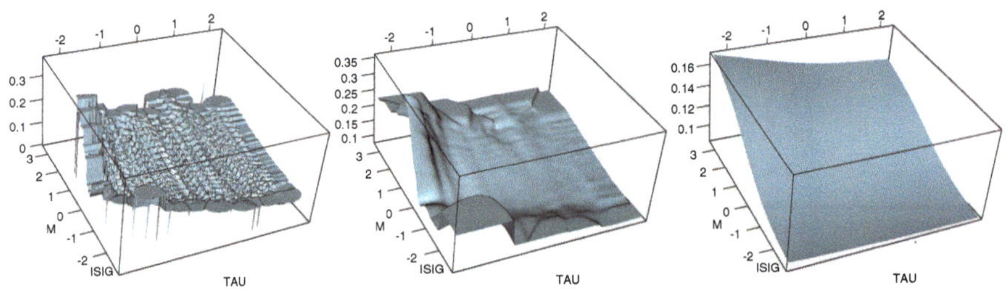

Fig. 5.14. Regression of the implied volatility on the time to expiration and the moneyness. From *left* to *right*, kernel regression surfaces for the bandwidths $b = 0.01$, $b = 0.1$ and $b = 1.0$

the bandwidth, we decided that $b = 0.033$ was a reasonable choice. According to the semi-parametric strategy outlined above, once the bandwidth has been chosen, we predict the implied volatilities of the testing sample by nonparametric regression, and we plug the predictions so-obtained into the Black-Scholes formula in order to obtain the semi-parametric predictions of the option prices. Finally, we compute the sum of square errors as before.

```
YPRED3.kreg <- kreg(SXX3,TRGSP[,5],xpred=SXXPRED3,
                                  kernel=gaussian,b=0.0032)
ISIGMAPRED <- YPRED3.kreg$ypred
YPRED3 <- bscall(TAU=TSTSP[,3]/252,K=TSTSP[,2],S=TSTSP[,1],
                                  R=TSTSP[,4], SIG=ISIGMAPRED)
```

```
SSE3 <- sum((YPRED3 - TSTSP[,6])^2)
SSE3
[1] 129.5386
sqrt(SSE3/length(YPRED3))
[1] 0.5089964
```

Next we compare the performance of the different methods used to predict the price of the (European call) options on the S&P 500, and we comment on the differences in the results of the various methods.

5.7.4 Numerical Results

We first compare the results of the three methods by comparing the sums of squares SSE1, SSE2, and SSE3 obtained with our implementations (i.e. for the particular choices of the smoothing parameters which we made) of the three methods. Obviously, the semi-parametric method gives much better results.

Fig. 5.15. Scatterplot of the predicted call prices against the actual prices. *Left*: 4- dimensional kernel regression. *Center*: projection pursuit regression. *Right*: semi-parametric regression based on the kernel prediction of the implied volatility and the Black-Scholes formula

As a last attempt to compare the predictions given by the three regression procedures, we plot the predicted values against the actual prices of the call options in the testing sample. The three plots are given in Fig. 5.15. They confirm the conclusions based on the numerical scores of merit compared earlier: the semi-parametric regression based on the 2-dimensional kernel gives the best results, and the 4-dimensional kernel is the worst because the points are scattered further away from the diagonal. Also, this plot shows that the kernel underestimates the highly priced options. It would be interesting to go back to the data and try to explain why.

As mentioned earlier, Problems 5.21 and 5.22 give examples of situations for which the conclusions are slightly different.

5.7.4.1 ***State Price Density***

According to the discussion leading to (9.3), we have:

$$C_{T,K}(t,S) = e^{-r(T-t)}\mathbb{E}\{f_K(S_T)|S_t = S\} = e^{-r(T-t)}\int f_K(s')p(s')ds'$$

for some density $p(s')$, where we use the notation $f_K(x) = (x-K)^+ = max\{x - K, 0\}$ for the pay-off function of the European call option. The density $p(s')$ plays such an important role that it is given a special name: the state-price density. Its importance comes from the fact that it can be used to price any kind of European contingent claim with the same maturity. Indeed, if an option pays the amount $f(S_T)$ at maturity T, simple arbitrage arguments tell us that its price at time t should be given by the risk neutral expectation:

$$e^{-r(T-t)}\mathbb{E}\{f(S_T)\} = e^{-r(T-t)}\int_0^\infty f(s')p(s')\,ds'.$$

Using the fact that:

$$\frac{\partial f_K(x)}{\partial K} = \begin{cases} -1 \text{ if } K < x \\ 0 \quad \text{if } x < K \end{cases} \qquad \text{and} \qquad \frac{\partial^2 f_K(x)}{\partial K^2} = \delta_x(K)$$

where δ_x denotes the Dirac delta function at x, and allowing ourselves to interchange the derivatives and the integration, we get:

$$\begin{aligned}\frac{\partial^2 C_{K,T}(t,s)}{\partial K^2} &= e^{-r(T-t)}\frac{\partial^2}{\partial K^2}\int f_K(x)p(x)\,dx \\ &= e^{-r(T-t)}\int \frac{\partial^2 f_K(x)}{\partial K^2}p(x)\,dx \\ &= e^{-r(T-t)}\int \delta_x(K)p(x)\,dx \\ &= e^{-r(T-t)}p(K).\end{aligned}$$

This shows that all other variables being fixed momentarily, the state price density is proportional to the second derivative of the call price when viewed as a function of the strike price, more precisely:

$$p(K) = e^{r(T-t)}\frac{\partial^2 C_{K,T}(t,s)}{\partial K^2}. \tag{5.19}$$

We propose to compute the state price density in each of the following cases:

1. $S = 445$, $r = 3\,\%$, $\tau = 21/252$,
2. $S = 445$, $r = 3\,\%$, $\tau = 45/252$.

In order to avoid to reproduce too large a number of R commands, we summarize some of the steps in words. First we choose a grid of values of the strike K and the

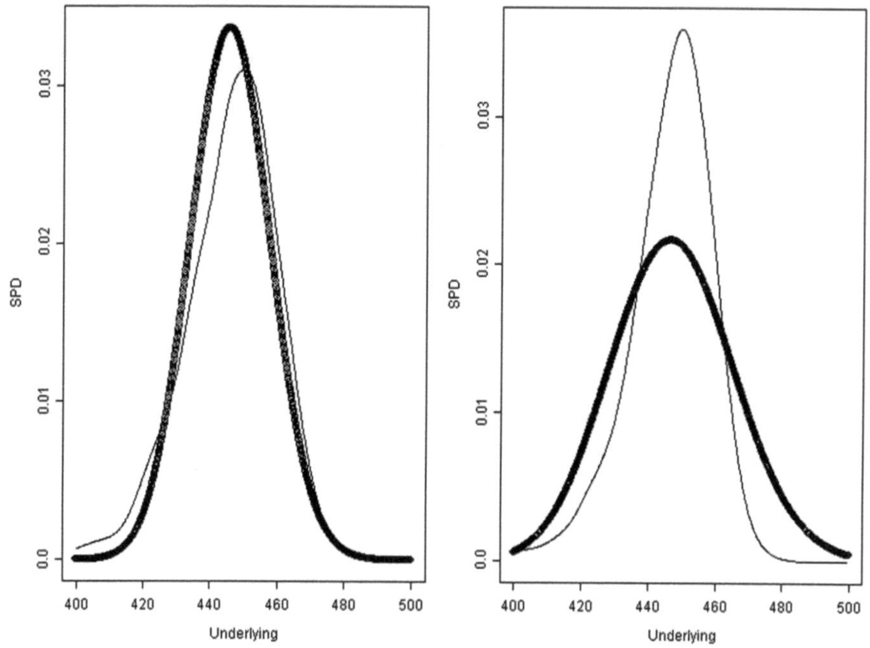

Fig. 5.16. Graphs of the estimates of the state price densities (*thin line*) together with the corresponding lognormal density of the Black-Scholes theory (*thick line*). In both cases, we assumed that the underlying was $S = 445$ and that the short rate was $r = 3\,\%$. The *left pane* is for options with 21 days to maturity, while the *right pane* is for options with 45 days to maturity

underlying S over which we compute the density. We choose a regular grid SEQK of 1,024 points between $K = 400$ and $K = 500$. In each case, we compute the moneyness with the prescribed values of S, r, and τ, and each of the values of K in the grid. We then standardize the constant τ and the vector of these values of the moneyness using the means and standard deviations of the training sample. We put the result in a matrix called SSPD. Once this is done, we compute the predictions of the implied volatilities and the corresponding call prices, and finally, we compute the second derivative, as approximated by the second difference (properly normalized by the size of the grid step which we called DELTAK).

```
SPD.kreg <- kreg(SXX1,OTRGSP[,5],xpred=SSPD,
                                    kernel=gaussian,b=.033)
SPDSIGMA <- SPD.kreg$ypred
SPDY <- bscall(TAU=rep(21/252,NP),K=seq(from=400,to=500,
          length=NP),S=rep(445,NP),R=rep(.03,NP),SIG=SPDSIGMA)
DENS2 <- diff(SPDY,differences=2)/(DELTAK^2)
```

The plots are given in Fig. 5.16. In each case, we plot of the state price density and we superimpose the theoretical state price density suggested by the Black-Scholes pricing paradigm (as used in our derivation of the Black-Scholes formula). Many interesting facts can be derived from these plots. We mention just a few to illustrate the interpretation of our results. Let us concentrate on the left plot of Fig. 5.16. For this maturity of 21 days, the fact that the Black-Scholes density is below the empirical estimate on the left and right most parts of the graph, implies that the options in the money (both puts and calls) are underpriced by the Black-Scholes formula. Notice that this underpricing becomes an overpricing for the longer maturity of 45 days.

APPENDIX: KERNEL DENSITY ESTIMATION & KERNEL REGRESSION

This appendix represents an excursion away from the practical bent of this chapter. We revisit the topic of density estimation, elucidating its strong connections with the nonparametric regression methods. We first start with an informal motivation.

The kernel regression is, as any other regression method, a way to estimate the value of the regression function:

$$\boldsymbol{x} \hookrightarrow \mathbb{E}\{Y|\boldsymbol{X} = \boldsymbol{x}\}.$$

Since the above right hand side is an expectation, it can be computed (at least in theory) as a sum when the distribution of Y is discrete, or as an integral when the distribution of Y has a density. In this case:

$$\mathbb{E}\{Y|\boldsymbol{X} = \boldsymbol{x}\} = \int y f_{\boldsymbol{X}=\boldsymbol{x}}(y)\, dy$$

if we denote by $f_{\boldsymbol{X}=\boldsymbol{x}}(y)$ the conditional density of Y given that $\boldsymbol{X} = \boldsymbol{x}$. By the definition of conditional probability, the conditional density appearing in this formula is given by the ratio:

$$f_{Y|\boldsymbol{X}=\boldsymbol{x}}(y) = \frac{f_{(\boldsymbol{X},Y)}(\boldsymbol{x}, y)}{f_{\boldsymbol{X}}(\boldsymbol{x})}$$

of the joint density $f_{(\boldsymbol{X},Y)}$ of $\boldsymbol{X}$ and Y to the (marginal) density $f_{\boldsymbol{X}}$ of $\boldsymbol{X}$. Consequently, the regression function can be rewritten as:

$$\mathbb{E}\{Y|\boldsymbol{X} = \boldsymbol{x}\} = \int y \frac{f_{(\boldsymbol{X},Y)}(\boldsymbol{x}, y)}{f_{\boldsymbol{X}}(\boldsymbol{x})} dy = \frac{1}{f_{\boldsymbol{X}}(\boldsymbol{x})} \int y f_{(\boldsymbol{X},Y)}(\boldsymbol{x}, y) dy \qquad (5.20)$$

since the denominator (which does not depend upon y) can be pulled out of the integral. From these formulae it seems reasonable to expect that being able to estimate densities should lead to regression estimation procedures.

As we stated earlier, the purpose of this appendix is to shed some light on the intimate relationship between the kernel density estimation method presented in Chaps. 1 and 3, and the kernel regression method presented in this chapter. The present discussion is an aside of conceptual interest, and it can be skipped in a first reading. To set the stage, we come back to the regression setting defined by a sample:

$$(x_1, y_1), \ldots\ldots, (x_n, y_n)$$

and for the sake of simplicity, we assume that the regressor variables are univariate (i.e. $p = 1$). It should be obvious that the following discussion also applies to the general case $p \geq 1$ without any change. We proceed to the estimation of the regression function by estimating the two densities entering the formula by the kernel method. First we estimate the density $f_X(x)$, and in order to do so we choose a kernel function K and a bandwidth $b > 0$. We get:

$$\hat{f_X}(x) = \frac{1}{nb}\sum_{j=1}^{n} K\left(\frac{x - x_j}{b}\right) \tag{5.21}$$

for the kernel estimate of the density $f_X(x)$. Next we work on the estimation of the joint density $f_{(X,Y)}(x, y)$ and we choose a tensor product kernel $\tilde{K}(x, y) = K(x)k(y)$ for some symmetric kernel k in the y variable, and a bandwidth b' which can be equal to or different from b, it will not matter in the end. In doing so we get:

$$\begin{aligned}\int y f_{(X,Y)}(x, y)dy &= \int y \frac{1}{nbb'}\sum_{j=1}^{n} K\left(\frac{x - x_j}{b}\right) k\left(\frac{y - y_j}{b'}\right) dy \\ &= \frac{1}{n^2bb'}\sum_{j=1}^{n} K\left(\frac{x - x_j}{b}\right) \int y\, k\left(\frac{y - y_j}{b'}\right) dy \quad (5.22)\end{aligned}$$

and if we compute the integral appearing in the right hand side of (5.22) using the substitution $z = (y - y_j)/b'$ we get:

$$\begin{aligned}\int yk\left(\frac{y - y_j}{b'}\right) dy &= \int (b'z + y_j)k(z)b'dz \\ &= b'^2 \int zk(z)dz + b'y_j \int k(z)dz \\ &= b'y_j \quad (5.23)\end{aligned}$$

if we use the facts that $\int k(z)dz = 1$, by definition of a kernel, and $\int zk(z)dz = 0$, because of our symmetry assumption. Plugging formula (5.23) into formula (5.22) we get:

$$\int y f_{(X,Y)}(x, y)dy = \frac{1}{nb}\sum_{j=1}^{n} y_j K\left(\frac{x - x_j}{b}\right) \tag{5.24}$$

Finally, plugging formulae (5.21) and (5.24) into the expression (5.20) defining the regression function, we obtain:

$$\mathbb{E}\{Y|X = x\} = \frac{\sum_{j=1}^{n} y_j K\left(\frac{x-x_j}{b}\right)}{\sum_{j=1}^{n} K\left(\frac{x-x_j}{b}\right)} \tag{5.25}$$

which is exactly the formula giving the kernel regression estimate. In other words, the kernel regression is consistent with the notion of kernel density estimation. In fact, if one agrees to use formula (5.21) as a density estimator, then the definition of the regression function forces on us formula (5.25) as the only reasonable regression estimate!!

PROBLEMS

Ⓣ **Problem 5.1** *The National Automobile Dealers Association (NADA) hired a statistician to develop a used car pricer to estimate the price a used car will sell at a dealer auction. NADA has an extensive data base of auction prices, and the statistician is expected to deliver a program which, given the Make, Model and Color of a given car, pulls out all the records of sales of cars of the same Make, Model and Color from the data base, and create a four dimensional regression model where the response variable is the sale price, and the four explanatory variables are (1) the model year of the car, (2) the mileage of the car, (3) a score (number between* 0 *and* 1*) reporting the general condition of the car, (4) the equipment level of the car (an integer between* 0 *and* 100 *computed from a proprietary algorithm) reporting the* extras *such as AC, Automatic Transmission, electric windows, navigation system, ...*

If the contract of the statistician stipulates that, given the Make, Model and Color, the price estimate should be computed from the above four explanatory variables by a kernel regression, what should the first step of the pricer algorithm be?

Ⓣ **Problem 5.2** *The central administration of a multi-campus state university purchased* 500 *identical cars for its campus operations. Three years later, these cars were sold in a series of auctions at prices* $p_1, p_2, \cdots, p_{500}$*. At the times of the sales, the mileages of these cars were* $m_1, m_2, \cdots, m_{500}$ *respectively. One month later, it is discovered that the car company from which the fleet of cars were purchased had offered for free a* 501*-th car at the time of the original sale, and that this car had been assigned to the ombudsman's office, and hardly ever used during the 3 years. The mileage* m_{501} *of this car, and all the other mileage figures are expressed in miles.*

1. *It is suggested to guess the price that this car will fetch at an auction by using a kernel regression with a triangular kernel. Given that*

$$m_{501} = 9,500 \qquad \text{and} \qquad \min_{i=1,\cdots,500} m_i = 41,000$$

what is the smallest bandwidth which can be used in order for the computation to be possible.

2. *Assuming that it is eventually decided to use the box kernel instead, and assuming that the mileages* $m_1, m_2, \cdots, m_{500}$ *form a sample uniformly distributed over the interval* $[41{,}000, 52{,}000]$*, what is the smallest value of the bandwidth for which the expected number of sold cars entering the computation of the estimate is at least* 100*?*

Ⓣ **Problem 5.3** *The mathematical results which you are asked to prove in this problem were evidenced numerically in the two problems above. Given a kernel function* K *and* n *couples* $(x_1, y_1), \ldots, (x_n, y_n)$ *of real numbers, compute the limits:*

$$\lim_{b \searrow 0} \varphi_{b,K}(x) \qquad \textit{and} \qquad \lim_{b \nearrow \infty} \varphi_{b,K}(x)$$

of the kernel smoother when the bandwidth goes to 0 or ∞.

You can choose one of the four kernel functions given in the text (pick your favorite) or you can try to give this proof for a general kernel function if you feel comfortable enough.

Ⓣ **Problem 5.4** *Given two functions f and g their convolution $f * g$ is the function defined by the formula:*

$$[f * g](x) = \int_{-\infty}^{+\infty} f(y)g(x-y)\,dy$$

1. *Use a simple substitution to check that $f * g = g * f$.*
2. *Prove that:*

$$K_{triangle} = K_{box} * K_{box}$$

and that:

$$K_{parzen} = K_{triangle} * K_{box}.$$

Recall the definitions of these kernels given in Table 5.1.

3. *Use* R *to plot the four kernel functions given in Table 5.1. Explain your work, and try to create the plots without using any mathematical formula.*

Ⓣ **Problem 5.5** 1. *Let K denote a density function in p dimensions, and for each positive number $b > 0$ let us set:*

$$K_b(x) = b^{-p} K(\frac{1}{b}x), \qquad x \in \mathbb{R}^p.$$

Now we assume that we have a sample $(x_1, y_1), \cdots, (x_n, y_n)$ where $x_i \in \mathbb{R}^p$ and $y_i \in \mathbb{R}$ for $i = 1, \cdots, n$. For any $x \in \mathbb{R}^p$, determine the real number m minimizing the function

$$m \hookrightarrow \sum_{i=1}^{n} K_b(x - x_i)|m - y_i|^2$$

and explain how the kernel regression can be viewed as a weighted least squares regression.

2. *We now study the behavior of the kernel estimator when one of the observations moves away from the data without bound. For the sake of simplicity we shall assume that $p = 1$. Fix $i \in \{1, \cdots, n\}$ and explain how the regression curve at any given point $x \in \mathbb{R}$ changes if*
   2.1 *We replace the measurement (x_i, y_i) by $(x_i, y_i \pm c)$ and let $c \to \infty$.*
   2.2 *We replace the measurement (x_i, y_i) by $(x_i \pm c, y_i)$ and let $c \to \infty$.*
3. *You were asked to develop a nonparametric pricer for call options using the (current) value S of the underlying asset, the strike κ, the time to maturity τ and the short interest rate r as explanatory variables. Assuming that you already implemented a nonparametric kernel regression method with kernel function K, bandwidth b, response variable P (the price of the option) and these four explanatory variables, how would you respond to the request for the estimation of the "Deltas" of the options? FYI: the Delta of an option is the derivative (sensitivity) of the price of the option P with respect to the value S of the underlying.*

Ⓣ **Problem 5.6** *We work with a sample $(x_1, y_1), \cdots, (x_n, y_n)$ where $x_i \in \mathbb{R}^p$ and $y_i \in \mathbb{R}$ for $i = 1, \cdots, n$ and we assume that k is an integer between 1 and n. We investigate the non-parametric regression method known as the k-nearest neighbors (k-NN) which, for any*

$x \in \mathbb{R}^p$, proposes the value $\varphi_k(x)$ equal to the average of the responses y_i for which the explanatory vector or variable x_i is one of the k nearest neighbors of x when how "near" x_i is of x is given by the Euclidean distance $\|x_i - x\| = \left(\sum_{j=1}^p |x_{i,j} - x_j|^2\right)^{1/2}$.

1. *Show that the k-NN regression is equal to a weighted sum the values of the response variable by showing that*

$$\varphi_k(x) = \sum_{i=1}^n w_i(x) y_i$$

for a specific set of weights $w_i(x)$ for which you are expected to give a formula. Compute these weights and the value of the regression function in the case $p = 1$, $n = 5$,

$$\{(x_i, y_i)\}_{1 \le i \le 5} \qquad \textit{is given by} \qquad \{(1,5), (7,12), (3,1), (2,0), (5,4)\}$$

for $x = 4$ and $k = 3$.

2. *Compute*

$$\lim_{k \searrow 1} \varphi(x) \qquad \textit{and} \qquad \lim_{k \nearrow n} \varphi(x)$$

and explain in words what happens to the regression curve/surface when k decreases to 1, and when it increases to n.

3. *Let us assume that $p = 1$ and that the values x_i form an equidistant regular grid $x_i = x_0 + i\Delta x$. Give values of k and $b > 0$ so that, if x is not too close to one of the grid points, the weights of the k-NN regression and of the kernel regression with the box kernel and bandwidth b are the same. Recall that the box kernel is the function*

$$K(x) = \begin{cases} 1/2 & \text{if } |x| \le 1 \\ 0 & \text{otherwise.} \end{cases}$$

4. *Let us now assume $n = 100$, that the values $x_1, \cdots, x_n$ of the explanatory variable are uniformly distributed over the interval $[41, 52]$, and that $x = 10.5$, and that we want to use a k-NN regression with $k = 10$. What is the probability that the k-NN regression gives the average of the values of the response variable corresponding to the values of x_i satisfying $x_i \le 45$.*

Ⓣ **Problem 5.7** *Figure 5.7 gives the* goodness of fit *plot of a projection pursuit. Remember that such a plot gives the relative sum of square errors as a function of the number of ridge functions included in the pursuit. If you had to run projection pursuit on the same data, how many ridge functions would you include in your projection pursuit? Explain your answer.*

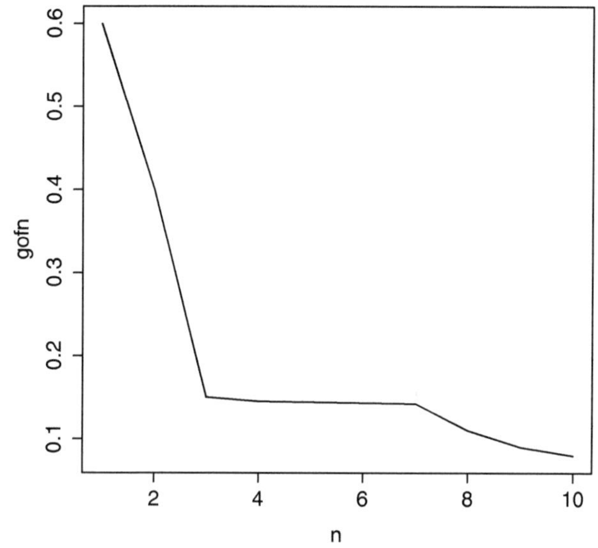

Ⓣ **Problem 5.8** *We consider the function φ of three real variables x_1, x_2 and x_3, defined by*

$$\mathbf{x} \hookrightarrow \varphi(\mathbf{x}) = 8x_1(x_2 + x_3).$$

for all $\mathbf{x} = (x_1, x_2, x_3) \in \mathbb{R}^3$

1. *Find two three dimensional vectors* $\mathbf{a}_1 = (a_{1,1}, a_{1,2}, a_{1,3})$, $\mathbf{a}_2 = (a_{2,1}, a_{2,2}, a_{2,3})$ *so that the equality*

$$\varphi(\mathbf{x}) = \varphi_1(\mathbf{a}_1 \cdot \mathbf{x}) + \varphi_2(\mathbf{a}_2 \cdot \mathbf{x})$$

holds true for all $\mathbf{x} = (x_1, x_2, x_3) \in \mathbb{R}^3$ *if the functions* φ_1 *and* φ_2 *are defined by*

$$\varphi_1(z) = 2z^2 \quad \text{and} \quad \varphi_2(z) = -2z^2 \qquad \text{for} \quad z \in \mathbb{R}.$$

2. *Let us assume that* $\mathbf{x}_1 = (x_{1,1}, x_{1,2}, x_{1,3}), \cdots, \mathbf{x}_n = (x_{n,1}, x_{n,2}, x_{n,3})$ *is a set of* $n = 512$ *points in* $\mathbb{R}^3$ *and that* $y_1, \cdots, y_n$ *are* n *real numbers satisfying*

$$y_i = 4x_{i,1}(x_{i,2} + x_{i,3}) + \epsilon_i, \qquad i = 1, \cdots, n$$

where the ϵ_i *are realizations of* $n = 512$ *independent identically distributed Gaussian random variables with mean* 0 *and variance* 0.04. *Assuming further that you are attempting a projection pursuit regression of the responses* y_i *against the explanatory variables* $x_{i,1}$, $x_{i,2}$ *and* $x_{i,3}$, *sketch the plot of the goodness of fit* (`gofn`) *you would get if you were to do a projection pursuit regression with possible numbers of ridge functions varying from* `nterms=1` *to* `max.terms=10`. *Give the number* `nterms` *of ridge functions you would choose, and explain what kind of ridge functions* φ_i *you expect the program will find.*

ⒺⓈⓉ **Problem 5.9** *The goal of this problem is to illustrate one of the most important feature of the projection pursuit regression: its ability to detect interactions between the regressor variables. We first show that the function* $f(x_1, x_2) = x_1 x_2$ *is easily written in the form used by the projection pursuit algorithm.*

1. *Determine μ, β_1 and β_2 so that the identity:*

$$x_1 x_2 = \mu + \sum_{j=1}^{2} \beta_j \phi_j(\boldsymbol{a}_j^t \boldsymbol{x})$$

holds for all $\boldsymbol{x}^t = [x_1, x_2]$ with $\boldsymbol{a}_1^t = [1, 1]$, $\boldsymbol{a}_2^t = [1, -1]$, $\phi_1(t) = t^2$ and $\phi_2(t) = -t^2$.
2. *Create data vectors $X1$ and $X2$ of length $n = 512$ and with entries forming independent samples from the uniform distribution $U(-1, +1)$ over the interval $[-1, 1]$. Create the vector Y according to the formula:*

$$Y = X1 * X2 + \epsilon,$$

where $\epsilon = \{\epsilon_j\}_{j=1,\ldots,512}$ is a Gaussian white noise (i.e. a sequence of independent identically distributed normal random variables with mean zero) with standard deviation $\sigma = 0.2$. Give a scatterplot of Y against $X1$, and of Y against $X2$.
3. *Run the projection pursuit algorithm to regress Y on $X1$ and $X2$ with a number of terms between* `min.term`= 2 *and* `max.term`= 3. *Produce scatterplots to visualize the graphs of the functions $\phi_j(t)$ found by the algorithm (if in doubt, read the on line help file for the function* `ppt`*). Does that fit with the results of the computations done in question 1 above? Finally, do a scatterplot of Y versus $\hat{Y}$ and another scatterplot of the residuals versus $\hat{Y}$ and comment.*
NB: recall that we use the notation ^{t} to denote the transpose of a vector or a matrix.

Ⓣ **Problem 5.10** 1. *Let us assume that $\mathbf{x}_1 = (x_{1,1}, x_{1,2}), \cdots, \mathbf{x}_n = (x_{n,1}, x_{n,2})$ is a set of $n = 512$ points in $\mathbb{R}^2$, and that $y_1, \cdots, y_n$ are observations of n real random variables $Y_1, \cdots, Y_n$ satisfying*

$$Y_i = 6x_{i,1}x_{i,2} - 3x_{i,1}^2 + \epsilon_i$$

where the ϵ_i are n independent identically distributed Gaussian random variables with mean 0 and variance 0.04. Let us also assume that an operator unaware of the relationship between the $\mathbf{x}_i$'s and the y_i's performs a projection pursuit regression of the y_i's against the $\mathbf{x}_i$'s. What do you expect the order of the model (i.e. the number of ridge functions used), the direction vectors $\mathbf{a}_j$ and the ridge functions φ_j will be? Explain your answers.

2. *For this second question, we assume that the relationship between the $\mathbf{x}_i$'s and the y_i's is of the form*

$$y_i = \frac{1}{\epsilon_i} e^{-3x_{i,1}^2} x_{i,2}^{6x_{i,1}}$$

where the ϵ_i's are realizations of n independent identically distributed random variables. Propose a transformation of the data and a distribution for the ϵ^i such that the transformed problem is equivalent to that of question 1.

Ⓣ **Problem 5.11** 1. *Let us assume that $\mathbf{x}_1 = (x_{1,1}, x_{1,2}), \cdots, \mathbf{x}_n = (x_{n,1}, x_{n,2})$ is a set of $n = 512$ points in $\mathbb{R}^2$, and that $y_1, \cdots, y_n$ are n real numbers satisfying*

$$y_i = 4x_{i,1}x_{i,2} + \epsilon_i, \qquad i = 1, \cdots, n$$

where the ϵ_i are n independent identically distributed Gaussian random variables with mean 0 and variance 0.04. Let us also assume that an operator unaware of the relationship between the $\mathbf{x}_i$'s and the y_i's performs a projection pursuit regression of the y_i's against the $\mathbf{x}_i$'s. What do you expect the order of the model (i.e. the number of ridge functions used), the direction vectors $\mathbf{a}_j$ and the ridge functions φ_j will be? Explain your answers.

2. *For this second question, we assume that the relationship between the* $\mathbf{x}_i$*'s and the* y_i*'s is of the form*

$$y_i = 4x_{i,1}x_{i,2}\epsilon_i, \qquad i = 1, \cdots, n$$

where the ϵ_i*'s are* n *independent identically distributed random variables. We now assume that the operator can only perform* linear regressions. *How should the explanatory and response variables be transformed in order to rewrite the problem as a linear regression, and what should the common distribution of the* ϵ_i*'s be for the least squares estimates of the parameters of the transformed problem to be maximum likelihood estimators?*

Ⓔ **Problem 5.12** *The data set* `GEYSER` *gives the characteristics of* 222 *eruptions of the "Old Faithful Geyser" during August* 1978 *and* 1979. *The first column gives the duration (in minutes) of the eruption and the second column gives the time until the following eruption (in minutes). Make sure that the* `R` *object you use has* 222 *rows and* 2 *columns.*

1. *Perform a least squares linear regression of the second column on the first and assess the significance (or lack thereof) of the regression. Give an interpretation to your results.*
2. *Perform several kernel regressions (say* 5*) using the* `normal` *kernel with* `ksmooth` *and varying values for the bandwidth in order to see the two extreme regimes (very small bandwidths and very large bandwidths) discussed in the text. Choose the value of the bandwidth which you find most reasonable, and compare the resulting regression curve to the result of part 1.*

Ⓔ **Problem 5.13** *Create a vector* `TIME` *containing the integers ranging from* 1 *to the number of entries in the vector* `SHIP`.

1. *Compute the least squares regression line and the least absolute deviations regression line of* `SHIP` *versus* `TIME` *and superimpose these two lines onto the scatterplot of* `TIME` *and* `SHIP`. *Compare the ways in which these lines account for the upward trend in the data and explain the differences.*
2. *The purpose of this question is to fit a more general polynomial (i.e. of degree possibly greater than* 1*) to the* `SHIP` *data. Perform polynomial regressions of degrees* 2, 4, 6, *and* 8 *successively, plot the results, and choose the value of the degree which seems the most reasonable.*
3. *The purpose of this question is to use natural splines to smooth the data* `SHIP`. *Vary the number of degrees of freedom (i.e. the parameter* `df`*). Use the values* 2, 6, 10, 14 *and* 18 *for the number of degrees of freedom and for each of them fit a natural spline to the data. Explain how the smoothed curve changes with the value of the number of degrees of freedom and choose the one which seems the most reasonable.*
4. *Finally we smooth the* `SHIP` *data using a kernel smoother* `ksmooth` *with a* `normal` *kernel function. Use the values* 1, 5, 20, 50 *and* 125 *for the bandwidth, and for each of these values, superimpose the graph of the kernel scatterplot smoother on the actual scatterplot of the original data. Explain how the smoothed curve changes with the bandwidth.*

Ⓔ **Problem 5.14** *This problem is concerned with the data set* `VINEYARD`.

1. *Superimpose (on the same plot) the scatterplot of* `NB` *and* `LUGS90`, *the graph of the polynomial regression of* `LUGS90` *on* `NB` *with a polynomial of degree* 3, *and the graph of the polynomial regression (still of degree* 3*) of the column of* 50 *measurements obtained by removing the first row and the last row from* `LUGS90` *on the column of* 50 *values obtained by removing the first row and the last row from* `NB`.

2. *Redo the same thing with a natural spline regression with degree of freedom* $df = 5$.
3. *Again, redo the same thing with smoothing spline regression.*
4. *Still redo the same thing, but with the kernel regression* `smooth` *with* `normal` *kernel function and a bandwidth equal to* 5.
5. *Compare the results obtained in questions 1 and 2 to the results obtained in questions 3 and 4. Explain which features of the regression algorithms are responsible for the differences.*
   Among the other regression methods seen in the text, which methods would give results of the type obtained in questions 1 and 2 and which methods would give results of the type obtained in question 3 and 4?

Ⓔ **Problem 5.15** *The purpose of this problem is to analyze the data set* `ROCK` *with the goal to illustrate the inner workings of the* `R` *implementation of the projection pursuit regression algorithm. The data consist of* 48 *measurements on* 4 *cross-sections of each of* 12 *oil-bearing rocks. Our goal is to predict the permeability* y *from the other* 3 *measurements (the total area, the total perimeter and a measure of roundness of the pores in the rock cross section).*

1. *Run the projection pursuit algorithm to perform such a prediction.*
2. *Assess the quality of the fit by computing the sum of square residuals.*
3. *Produce the plots of the first three additive terms in the projection pursuit regression formula and give the 2-D scatterplots of the response against the fitted values and the residuals respectively.*

Ⓔ **Problem 5.16** *The purpose of this problem is to analyze some of the features of the basketball data introduced in Problem 4.9. The first two questions are an attempt to evaluate the predictive value of a variable (*`age` *in the present situation).*

1. *Use projection pursuit to regress* `points` *against* `height`, `minutes` *and* `age`. *Redo the regression without using the variable* `age` *and compare with the previous result.*
2. *Similarly, use projection pursuit regression to explain the number of assists per game as given by the variable* `assists` *in terms of the explanatory variables* `height`, `minutes` *and* `age`. *Compare with the result obtained without using the predictor variable* `age`.
3. *Choose (at random if possible)* 10 *players, create a new data set, say* `BTST` *(for Basketball TeST sample), with these* 10 *rows and another data set, say* `BTRG` *(for Basketball TRaininG sample) with the remaining* 86 *players. Keep only the columns corresponding to the variables* `height`, `minutes` *and* `points`. *Use projection pursuit regression to fit a model to the training sample* `BTRG` *and use the coefficients of the model to predict the values of the variable* `points` *for the values of the variables* `height` *and* `minutes` *of the* 10 *players used to create the test sample* `BTST`. *Compute and note the residual sum of squares for these* 10 *predictions*

Ⓔ **Problem 5.17** *Given the names* `X1`, `X2`, `X3` *and* `Y` *of the four columns of the data set* `MIND`, *the goal is to regress the last variable* `Y` *against the first three columns using projection pursuit.*

1. *Determine the best order for the model.*

2. *Run a model of the order determined in question 1 and give the (directional) unit vectors found by the program. Comment.*

3. *Plot the functions* φ *found by the program. Can you make a guess at the functional dependence of* `Y` *upon* `X1`. `X2` *and* `X3`*?*

Ⓔ **Problem 5.18** *The goal of this problem is to mimic the volatility ratio prediction experiment described in the text in order to predict the afternoon value of the volatility indicator. The data needed for this problem are contained in the data matrices* `MORN.mat` *and* `AFT.mat` *used in the text.*

1. *Give the two dimensional scatterplot of the morning volatility and arrival rate indicators. Plot a three-dimensional scatter plot of these two explanatory variables together with the afternoon volatility indicator, plot the result of the least squares linear regression, and explain the results by identifying overly influential data points.*
2. *Plot the regression surfaces obtained by kernel regression (use the kernel function* `gaussian` *and three bandwidths which you will choose carefully after standardizing the explanatory variables), and confirm the explanations given in question 1 above.*
3. *Remove the excessively influential measurements identified earlier, recompute the kernel regression on the remaining data, plot the new regression surface, and compute the residual sum of squares.*
4. *Perform a projection pursuit analysis of the reduced data set, plot the projection pursuit regression surface, compute the new residual sum of squares and compare them both with the results obtained in question 3 above.*

Ⓔ **Problem 5.19** *The goal of this problem is to demonstrate the use of principal components analysis in the selection of a minimal informative set of explanatory variables. We use data sets* `SMORN.mat` *and* `SAFT.mat` *included in the library* `Rsafd`.

1. *Use the kernel method to regress the afternoon volatility ratio (third column of the data matrix* `SAFT.mat`*) against the six variables of* `SMORN.mat`*, and compute the sum of square errors. (NB: explain your bandwidth choice).*
2. *Perform a principal component analysis of the* `SMORN` *data set and compute the vector of the daily values of the two most important components* `PC1` *and* `PC2`*. Use the kernel method to regress the afternoon volatility ratio against* `PC1` *and* `PC2`*, and compute the sum of square errors. Again, you will need to justify your choice of bandwidth. Compare with the sum of square errors found in question 1, the sum of square errors found in the experiment with the kernel method described in the text. Comment.*

Ⓔ **Problem 5.20** *This problem is a sequel to Problem 5.19. It uses the data contained in the matrices* `SMORN.mat` *and* `SAFT.mat` *containing the six indicators computed for the September S&P 500 futures contract from the mornings and afternoons transactions of the period ranging from June 1, 1998 to September 5, 1998.*

1. *Use the regression function computed with the kernel regression of the afternoon volatility ratio of the modified June contract data (fourth column of* `SAFT.mat`*) against the morning volatility ratio and rate of arrival indicators of the same modified June contract data (fourth and fifth columns of* `SMORN.mat`*) to predict the afternoon volatility ratios given in the fourth column of* `SAFT.mat` *from the morning volatility ratio and rate of arrival indicators of* `SMORN.mat`*. Compute the sum of square errors.*
   In other words, use the June contract data as training sample and the September contract data as testing sample.
2. *Redo the same thing using as explanatory variables the first two (linear combinations) principal components of the six indicators found in the PCA analysis of the June matrix data done in Problem 5.19. Comment.*

Ⓔ **Problem 5.21** *The goal of this problem is to perform the option pricing analysis described in the text on a different data set. We consider a training sample of* 5,000 *trades of European call options on the S&P 500 index which took place in 1993, and we consider a testing sample of* 500 *different trades which took place the same year. The data are contained in the data sets* TRGSP *and* TSTSP *included in the library* Rsafd.

1. *Perform the analysis described in the text and compute the three figures of merit, especially the squared error per option.*
2. *Compare the performances of the three prediction procedures on the testing sample. Comment on the differences with the results reported in the text. Hint: when dealing with the semiparametric method, use the scatterplots of the explanatory variables as a guide.*

Ⓔ **Problem 5.22** *Like Problem 5.21 above, the present problem aims at the implementation on a different data set, of the option pricing analysis described in the text. We are still considering European call options on the S&P 500 index, but the data are more recent since we are using quotes from the year 2000. Also, the samples are larger and more importantly, selected at random. The data are contained in the data sets* TRGSP2000 *and* TSTSP2000 *included in the library* Rsafd.

1. *Perform the analysis described in the text and compute the three figures of merit, especially the squared error per option.*
2. *Compare the performances of the three prediction procedures on the testing sample. Comment on the differences with the results reported in the text, and the results of the experiment conducted in Problem 5.21 above. As before, use the scatterplots of the explanatory variables as a guide to understand and possibly explain the differences.*

Ⓔ **Problem 5.23** *This problem illustrates the use of temperature data, and especially the numbers of Heating Degree Days and the numbers of Cooling Degree Days (HDD and CDD for short), to predict the price of a commodity. The idea for this problem has its origin in a claim found in the introductory article written by Geoffrey Considine for the Weather Resource Center of the Chicago Mercantile Exchange website.*

We first describe the data. They are contained in the data set CORNTEMP *included in the library* Rsafd. *It is a* 642×3 *– data frame of numbers. Each row corresponds to a month, starting from January 1960 (the first row), and ending December 1998 (the last row). The first column gives the official price a farmer could get for a bushel of corn in Iowa that month. This variable is called* MCorn *for Monthly Corn price. The second column gives the monthly average heating degree days in Des Moines as measured at the meteorological station of the airport. Heating Degree Days (HDD's) and Cooling Degree Days (CDD's) are discussed in details in Chap. 6. For the purposes of the present problem it is enough to know that the higher the temperature, the larger the number of CDD's and the smaller the number of HDD's, and the cooler the temperature, the larger the number of HDD's and the smaller the number of CDD's. So we expect large numbers of CDD's and small numbers of HDD's in the summer months, and correspondingly, large numbers of HDD's and small numbers of CDD's in the winter months. For each month of the period under consideration, the entry of the second column of the data matrix was computed by summing up the HDD's of the month in question, and dividing the total by the number of days in the month. This variable is called* MDMHDD *for Monthly Des Moines HDD. The third column gives the monthly CDD averages for Des Moines. Its entries were computed in a similar way. This variable is called* MDMCDD.

The goal of the problem is to predict the price of corn offered to Iowa farmers in November (resp. December) each year, from the the knowledge of temperatures in Des Moines during the summer of that same year (or possibly previous years) as captured by the average numbers of CDD's and HDD's.

1. *For each year between* 1960 *and* 1998*,*
   - *Extract the November price of corn in Iowa;*
   - *Choose and compute regressor variables involving the values of* `MCorn`, `MDMHDD` *and* `MDMCDD` *up to (and possibly including) July of the same year but not later. You are allowed to use up to four variables (but remember that, as we repeated over and over, the smaller this number the better).*

   Once this is done, regress the November price of corn in Iowa on the regressor variables you chose according to the rules of the second item above. You are expected to perform three regressions, at least one of them being linear, and at least one of them being non-parametric. For each of these regressions, you need to give the list of the steps you take, and the parameters you use, so that your results can be reproduced. Also, in each case you will compute the proportion of the variance of the response variable explained by the regression.
2. *Same question as above, replacing the November price of corn by the December price, and the July limit by August.*

Ⓔ **Problem 5.24** *The data to be used for this problem are contained in the data frame* `PPRICE` *for which each row corresponds to a particular date.*

◇ *The first column contains the values of a variable named* `GasSpot` *which gives the spot price of natural gas on that date.*
◇ *The second column contains the values of a variable named* `SDTemp` *which gives the average temperature in San Diego over the* 31 *days preceding the date in question.*
◇ *The third column contains the values of a variable named* `PPower` *which gives the average over the* 5 *days preceding the date in question, of the spot price of firm on peak electric power at the Palo Verde station.*
◇ *Finally, the fourth column contains the values of a variable named* `FPower` *which gives the average over the 2 weeks following the date in question, of the spot price of firm on peak electric power at the Palo Verde station.*

Form a data frame `TRG` *with the first* 250 *rows of* `PPRICE`*. The entries of* `TRG` *correspond to the period from 2/4/1999 to 2/3/2000. You shall also need to form a data frame* `TST` *with the last* 80 *rows of* `PPRICE`*. The entries of* `TST` *correspond to the period from 7/13/2001 to 11/9/2001. We avoid the period in between because of the extreme volatility of the natural gas and power prices. This does not mean that we are not interested in studying periods of high volatility, quite the contrary. It is merely because the economic fundamentals were not the only driving force during this crisis period.*
The goal of the problem is to predict the values of the average price of (firm on peak) electric power over the next 2 weeks from past values of explanatory variables such as the weather (as quantified by the average temperature in San Diego), the price of natural gas, and possibly past values of the price of electricity at the same location. We use the data in `TRG` *as a training sample to fit a regression model, and we compute the predictions given by such a model for the data in the testing sample* `TST`.
Warning. *The variable* `PPower` *should not be used in the first four questions. Moreover, for all the predictions considered below, the figure of merit should be the square root of the mean squared error.*

1. *Fit a least squares linear regression model for* `FPower` *against* `GasSpot` *and* `SDTemp` *using the data in* `TRG`*, use this model to predict the values of* `FPower` *in* `TST` *from the corresponding values of the explanatory variables, and compute the figure of merit.*
2. *Same question with least absolute deviations linear regression instead.*
3. *Same question using projection pursuit. Explain your work.*
4. *Same question using kernel regression. Again make sure that you explain all the steps you take, and justify your choice of kernel function and bandwidth.*
5. *Fit a least squares linear regression model for* `FPower` *against* `GasSpot`, `SDTemp` *and* `PPower` *using the data in* `TRG`*, as before, use the fitted model to predict the values of* `FPower` *in* `TST` *from the corresponding values of the three explanatory variables, and compute the figure of merit.*
6. *Same question with least absolute deviations linear regression instead.*
7. *Same question using projection pursuit.*
8. *Compare the numerical results obtained with the various methods, and explain why they could have been expected.*

Ⓔ **Problem 5.25** *The data to be used in this problem are contained in the data frame* `CRUDE` *containing a numerical matrix with* 3,325 *rows and* 12 *columns, and the numeric vector* `COma` *of length* 3,325.

In both data structures, each row corresponds to a date, the first one being 4/18/1989, and the last one being 8/12/2002. Each row of the vector `COma` *contains the average of the crude oil spot price over the period of* 5 *days starting on (and including) the date indexing the row. Each row of the matrix* `CRUDE` *gives the prices of the* 12 *futures contracts of crude oil as traded the day before. Form a data frame* `TRGCRUDE` *and a vector* `TRGCOma` *with the first* 2,500 *rows of* `CRUDE` *and* `COma` *respectively. You shall also need to form a data frame* `TSTCRUDE` *and a vector* `TSTCOma` *with the last* 825 *rows of* `CRUDE` *and* `COma` *respectively.*

The goal is to predict the values of the average spot price over the next 5 days from the prices of the crude oil futures contracts traded the day before, by fitting a regression model to the training data contained in the data sets `TRGxxx`*, and using the model to compute predictions for the values of the response in the testing sample* `TSTCOma`.

Warning. *For all the predictions considered in this problem, the figure of merit should be the square root of the mean squared error. It is very important that you explain your work in detail. In particular, explain your choices for the order of the models, the kernel functions, and the bandwidths you use.*

1. *Fit a least squares linear regression model for* `COma` *against the* 12 *explanatory variables given by the prices of the futures contracts the day before using the data in* `TRGCRUDE` *and* `TRGCOma`*, use this model to predict the values of* `COma` *in* `TSTCOma` *from the corresponding values of the explanatory variables, and compute the figure of merit.*
2. *Fit a projection pursuit regression model for* `COma` *against the* 12 *explanatory variables given by the prices of the futures contracts the day before using the data in* `TRGCRUDE` *and* `TRGCOma`*, use this model to predict the values of* `COma` *in* `TSTCOma` *from the corresponding values of the explanatory variables, and compute the figure of merit.*
3. *Perform the PCA of the data in* `TRGCRUDE`*, plot the first four loadings, give the proportions of the variance they explain, and compute the first two principal components.*
4. *Fit a one dimensional kernel regression model for* `COma` *against the first principal component using the data in* `TRGCRUDE` *and* `TRGCOma`*, use this model to predict the values of* `COma` *in* `TSTCOma` *from the corresponding values of the explanatory variables, and compute the figure of merit.*

5. *Fit a two dimensional kernel regression model for* `COma` *against the first two principal components using the data in* `TRGCRUDE` *and* `TRGCOma`*, use this model to predict the values of* `COma` *in* `TSTCOma` *from the corresponding values of the explanatory variables, and compute the figure of merit.*
6. *Compare the numerical results obtained with the various methods, and explain why they could have been expected.*

Ⓣ **Problem 5.26** *Spread options are popular financial instruments. They are ubiquitous in the commodity markets. Some of the most popular spread options are traded as spark spread, crack spread, and calendar spread options. Typically, a spread option is written on the spread (the difference) between two underlying prices or indexes. For spark spread option, the underlyings are electricity and natural gas. Natural gas is widely used to generate electricity, so it is natural to expect that electricity and natural gas prices are highly correlated. For these reasons, the spark spread option is a useful instrument for hedging the risks of electricity generation.*

Assume T is the maturity time of the option. The electricity price at maturity is X_T, and the natural gas price at maturity is Y_T. Then the payoff of a zero-strike spread option is $\max(X_T - \alpha Y_T, 0)$ where α is a pre-determined factor (called Heat Rate in the case of spark spread options) specified in the contract. For the sake of definiteness, we will use $\alpha = 10$ in this problem. On any given day, say today, the factors affecting spread option prices are:

- *The current electricity price x;*
- *The current price of natural gas, say y;*
- *The time to maturity T.*

In this problem, we assume zero interest rate, i.e. $r = 0$. We use the data contained in the two R *objects* `TRGSS` *and* `TSTSS`*. These two data matrices are intended for training and testing purposes respectively. In the data files, each row corresponds to one trading day. Each data file has eight columns. The first two columns give the electricity and natural gas prices on that day. The next six columns give prices of zero-strike options with times to maturity 2 months, 4 months, 6 months, 8 months, 10 months, and 12 months respectively.*

1. *Fit a regression model to the option prices in the training sample* `TRGSS`*. Use as explanatory variables the current prices of the two underlying assets and the time to maturity. Use this regression model to predict the option prices in the test sample* `TSTSS` *and compute the sum of squared errors as figure of merit. Implement three different regression methods: linear regression, kernel regression and projection pursuit. For each of them explain in detail all the steps you take, compute the figure or merit, compare the results and comment.*
2. *We assume that at maturity, X_T and Y_T are log-normal random variables, more precisely, exponentials of jointly Gaussian random variables with standard deviations σ_1, σ_2 and Pearson correlation coefficient ρ. In such a situation, the option price is given by the following Margrabe formula*

$$S(x, y, T) = x\Phi(d_+) - \alpha y \Phi(d_-)$$

$$d_\pm = \frac{\ln(\frac{x}{\alpha y}) \pm \frac{\sigma^2}{2}T}{\sigma\sqrt{T}} \qquad \sigma = \sqrt{\sigma_1^2 + \sigma_2^2 - 2\rho\sigma_1\sigma_2} \tag{5.26}$$

where we use the notation Φ for the cumulative distribution function of the standard normal (Gaussian) distribution. In real life, electricity and natural gas prices do not

necessarily have the joint distribution of correlated log-normal random variables. We introduce the notion of implied correlation *to account for this fact. It is defined in analogy with the implied volatility of options on single-underlying interests. Specifically, for a given option quoted on the market, and assuming that all the other parameters x, y, T, σ_1, and σ_2 are known and fixed, it is defined as the value of the parameter ρ such that the price given by Margrabe formula equals the market price of the option. The implied correlation is a popular method to quote spread option prices.*
For the purpose of this problem, assume that $\sigma_1 = 0.6$ and $\sigma_2 = 0.3$. The implied correlation can be calculated in the following way:

(i) For each option price, assuming that the values of x, y, T and α are known, the unique value of the effective volatility σ appearing in Margrabe's formula can be calculated by inverting the Black-Scholes type formula in the same way we compute the implied volatility of a standard European call option. Compute these values for all the options in the traning and testing samples. You can use the `Rsafd` *function* isig *used in the text;*

(ii) Solve for ρ in the formula $\sigma = \sqrt{\sigma_1^2 + \sigma_2^2 - 2\rho\sigma_1\sigma_2}$ and use the solution to compute for each of the options in the training and testing samples the implied correlation of the option.

Fit a regression model to the implied correlation calculated from the data in the training sample `TRGSS`*. Use as explanatory variables the current prices of the two underlying interests and the time to maturity. Predict the implied correlation for each option in the test sample* `TSTSS`*, and from this implied correlation, compute the option price using Margrabe formular. As before, use successively a linear regression, a kernel regression and a projection pursuit regression explaining in detail all the steps you take. Compare the results to the results of the previous question and comment.*

Ⓔ **Problem 5.27** *The data set* `SUBSP` *to be used for this problem is a single column containing* 1,000 *rows giving the daily returns on a sub-index of the S&P 500 index (the Electronic Equipment subindex, to be specific) during the period beginning January 1993 and ending December 31, 1996. In other words, the first row gives the closing on January 4, 1993, the second row the closing on January 5, 1993 ...... and finally the last row contains the values of the subindex at the closing bell on Tuesday December 31, 1996. The markets are open approximately* 250 *days each calendar year so that* `SUBSP` *contains 4 years worth of daily quotes.*

The goal is to produce a forecasting for the future changes in the index. We shall use the first 3 years for training of our prediction system and we shall test them on the last year.

1. *Construct a kernel regression using the first* 750 *daily values of the sub-index to predict the* 5*-day move in the future (i.e. the variable which on day* `I` *has the value* `SUBSP[I+5]-SUBSP[I]`*) using only the past changes over the last day, the last week, the last 3 weeks and the last 12 weeks (i.e. the variables which on day* `I` *have the values* `SUBSP[I]-SUBSP[I-1]`, `SUBSP[I]-SUBSP[I-5]`, `SUBSP[I]-SUBSP[I-15]` *and* `SUBSP[I]-SUBSP[I-60]`*). Use this kernel regression model to predict the next* 245 *values of the 5 day increment in the index (i.e. the same response variable* `SUBSP [I+5]-SUBSP[I]` *for* `I=751,....,995`*) and compute the sum of the squares of the errors made.*
2. *Compute the same sum of square errors using now the latest value of the* 5*-day increment available (i.e.* `SUBSP[I]-SUBSP[I-5]`*) instead of the kernel prediction and compare to the result of question 1. Any comment?*

Ⓔ **Problem 5.28** *The data needed for this problem are contained in the data frames* `CO2TRG` *and* `CO2TST`. `CO2TRG` *contains information on the European call options traded in 2008 on EUA futures contracts (EUA stands for European Union Emission). Each row corresponds to one transaction. The first column gives the volume of the transaction (i.e. the number of options transacted at once), the second column gives the value of the underlying futures contract on which the option is written, the third column gives the time to maturity in years, the fourth column the strike of the option, and the last column gives the actual price at which the option was bought or sold. The data frame* `CO2TST` *contains the same information for the European call options transacted in 2009.*

For each of the following questions, you are expected to use the data contained in `CO2TRG` *as a* training sample *and the data contained in* `CO2TST` *as a testing sample, very much in the spirit of the discussion of nonparametric option pricing of Sect. 5.7 of the text.*

1. *Use* `CO2TRG` *to build a projection pursuit model (you will need to explain which explanatory variables you choose to use), and predict the prices of the options in* `CO2TST` *without using the fifth column of* `CO2TST`. *Compute the mean square error per option associated with your model.*
2. *Use* `CO2TRG` *to build a kernel regression model and predict the prices of the options in* `CO2TST` *without using the fifth column of* `CO2TST`. *You need to explain which explanatory variables you choose, how you compute them in order to include them in the model, and how you choose the bandwidth(s). Compute the mean square error per option associated with your model*
3. *Compare the numerical performances of the two models and explain why you think they are what they are.*

Ⓔ **Problem 5.29** *This problem uses the timeSeries* `Options.ts` *contained in the library* `Rsafd`. *For each day of the period ranging from January 4, 1996 to December 12, 1999, its last column gives the value of the VIX, while each of the* 26 *consecutive pairs of columns comprising the* 52 *first columns, gives the log-moneyness and the price on that day, of an option on the S&P500. These options have the same maturity which was chosen to match the maturity used to compute the VIX index. Negative log-moneyness corresponds to put options, positive values of the log-moneyness to call options. The goal of the problem is to* reverse engineer *by regression, the construction of the VIX index as a weighted average of call and put prices with weights depending upon the log-moneyness.*

1. *Construct a matrix of explanatory variables with* $N = 1{,}000$ *rows and* $p = 6$ *columns, say* `X` *such that the entries of* `X[,1]` *give the prices of the options whose log-moneyness are closest to* 1, `X[,2]` *the prices of the options whose log-moneyness are closest to* 2, `X[,3]` *to* 3, `X[,4]` *to* -1, `X[,5]` *to* -2 *and finally* `X[,6]` *to* -3.
2. *Regress the VIX index against the* 6 *explanatory variables comprising* `X`. *Use first a linear regression and then a projection pursuit regression. In each case, compute the mean squared error, compare the results of the two regressions and comment.*
3. *Run a kernel regression and compare the results.*

NOTES & COMPLEMENTS

The use of natural splines was proposed as a transition step from the set of fully parametric regression procedures presented in the previous chapter into a set of local regression methods leading to the nonparametric procedures discussed in this chapter. Besides natural splines, R also offers an implementation of B-splines . See the classic text by de Boor [26] for details on the zoology of the various families of splines. We recommend using these splines as an investigation tool, possibly as a coding device in additive models, but we encourage the user to be extremely careful using the results of spline regressions for prediction purposes, too many pitfalls are to be avoided. The rationale behind the definition of the smoothing splines is completely different from the typical search for a regression function in a parametric family. The fact that the argument of the minimization of the penalty function used to defined smoothing splines is indeed a spline is highly nontrivial. It is a result of Kimberdoff and Whaba that the solution of this optimization problem is in fact a spline of order $m + 1$.

The scatterplot smoothers described in the chapter have been chosen to illustrate the notion of locality in regression. Their main role is to prepare for the introduction of the multivariate nonparametric methods which occupy the rest of the chapter.

The so-called bootstrap method of estimation of the instantaneous forward-rate curve was introduced by Fama and Biss in [34]. Details on the implementations used by the Japanese and US Central Banks to produce their yield curves can be found in the comprehensive document [37] made available by the Bank of International Settlements.

Regression methods based on basis or feature function expansions are extremely powerful, especially when fast numerical algorithms are available for the computations of the decompositions onto the basis or feature functions. This is the case for Fourier expansions based on the famous Fast Fourier Transform (FFT) algorithm, wavelet expansions and wavelet packet expansions. Readers interested in applications of wavelets to statistics and finance are referred to the books of Wickerhaüser [97], Bruce and Gao [12], Carmona, Hwang, and Torresani [14], Gençay, F. Selçuk, and B. Whitcher [39], and Nason [71] for example. As explained in the text, while the statistical properties of regressions based on expansions do suffer from the curse of dimensionality, the complexity of their implementations is essentially immune, except possibly for the evaluation of the feature/basis functions φ_j, to increase in the dimension of the explanatory variable $\mathbf{x}$. This is one of the reasons why these methods were combined with Monte Carlo methods to compute prices of American options on large baskets of underlying instruments. The resulting pricing algorithms are widely used in the financial industry. They are known under the name of Longstaff Schwarz algorithm because of the seminal paper [62], which followed an attempt in the same spirit by Van Roy and Tsitsiklis [83].

Multi-dimensional kernel regression is very appealing because of the simplicity and clarity of the principle on which it is based: the algorithm mines the data to find records of the explanatory (vector) variable which are similar to the point at which the regression is being computed, and it returns a weighted average of the responses according the weights being proportional to the measures of similarities. Rigorous mathematical results can be proven on the desirable properties of this regression method. In particular optimality of the kernel function and of the bandwidth choices can be proven, but these results are unfortunately asymptotic in nature, i.e. valid only in the limit $n \to \infty$ of large sample sizes. Despite the simplicity of the rationale behind the kernel method, the mathematical proofs remain very technical. They have to deal with a subtle balance between the errors due to the bias and the variance of the estimator. Indeed, sampling from the conditional distribution of the response for neighboring values of the explanatory variable introduces a bias which, contrary to the case of the parametric methods seen in the previous chapter, cannot be avoided in the present situation. The reader interested in the mathematical analysis of the kernel regression as well as other non-parametric regression methods is referred to the book of Haerdle [43] and Hastie, Tibshirani, and Friedman [45].

The projection pursuit algorithm was originally proposed by Friedman and Stuetze in 1981 in a short article in the Journal of the American Statistical Association. The implementation of the general idea of projection pursuit is not limited to regression problems. It has seen applications to density estimation, classification and pattern recognition problems. More recently, a variation on the same idea was re-discovered, and it gained a lot of attention in the signal processing and image analysis communities: this new flavor of projection pursuit was proposed by Mallat and Zhang. It was called *matching pursuit*, but the idea remains the same. We limited our discussion to projection pursuit regression as it is implemented in R.

Neural networks were very popular in the late 1980s, especially in statistical pattern recognition circles. The R library `nnet` contributed by Venables and Ripley contains an implementation of the so-called feed-forward neural networks. See Sect. 11.4 of their book [94]. For the sake of simplicity they restrict their discussion to the case of one single hidden layer. Formally, such a regression/prediction procedure models the dependence of the response variable y upon the explanatory variables $\boldsymbol{x}$ by a function of the form:

$$y \approx \varphi_{weights}(\boldsymbol{x}) = \phi_{out}\left(\alpha + \sum_{j=1}^{p} w_{in,out,j}x_j + \sum_{j=1}^{nbunits} w_{2,j}\phi_{in}\left(\alpha_j + \sum_{\ell=1}^{p} w_{1,\ell}x_\ell\right)\right) \tag{5.27}$$

which is best understood in the light of the following comments:

- The numbers α_k play the roles played by the intercepts in linear regression. As in the case of linear regression, they can be subsumed by introducing a *dummy* input unit, which contains the number 1 irrespective of the observation, and adding weights for the outputs of this unit;
- The activation functions ϕ_{out} and ϕ_{in} are usually the same (typically the logistic function) but it happens quite often that the output activation ϕ_{out} is chosen to be linear;
- The p weights $w_{in,out,j}$ are used for the direct links between the input units and the output units;
- The p weights $w_{1,\ell}$ are used for the links between the input units and the units in the hidden layer while the weights $w_{2,j}$ are used for the links between the units in the hidden layer and the output unit.

A neural network of the type given by formula (5.27) is fit to the data $(\boldsymbol{x}_1, y_1), \ldots, (\boldsymbol{x}_n, y_n)$ by choosing the number of units in the hidden layer, the type of activation functions and possible direct links and most importantly by choosing the weights. This choice is usually made by solving the optimization problem:

$$\arg \min_{weights} \sum_{i=1}^{n} |y_i - \varphi_{weights}(\boldsymbol{x}_i)|^2. \tag{5.28}$$

In other words, for each possible choice of a set of weights we compute the fitted values $\varphi_{weights}(\boldsymbol{x}_i)$ of all the observations in the training sample, and the figure of merit given in formula (5.28) for this set of weights is the sum of the square discrepancies with the actual observed values y_i's. The optimal set of weights should minimize this sum of square errors.

Unfortunately, as experience shows, this minimization problem is very difficult because of the existence of many local minima and many of which are not satisfactory because they lead to poor predictions.

Because of the delicate optimization behind the fit of a neural net to data, implementations are difficult to come by. Venables and Ripley provide a function `nnet` to train a *feed-forward* one layer neural networks. A generic call to this function should look like:

```
> x.nnet <- nnet(X,Y,size=NB,Wts=W0,rang=R,linout=B,
                                    skip=BB,maxit=MAXIT)
```

where the parameters have the following meaning.

- `X` is the $n \times p$ design matrix, one row per observation in the training sample (whose size is denoted by n) and one column per explanatory variable;
- `Y` is the $n \times 1$ vector of the values of the response variable in the training sample;
- `size` gives the number of units in the hidden layer;
- `Wts` is the optional set of weights used to initialize the optimization;
- When the argument `Wts` is missing, the optimization procedure is initialized with random weights in the range $[-R, +R]$ where the number R is given as the parameter `rang`.
- `linout` is a boolean variable which, when set to `TRUE`, will force the output activation function ϕ_{out} to be linear or affine;
- `skip` is a boolean variable which when set to `TRUE` will force the existence of direct links from the input units to the output unit(s).

We found neural network regression difficult to use because of the difficulties in the choice of network and the search for parameters, and we chose not to discuss it in the text because of its poor performance compared to the other methods presented in the book. The advent of cheap fast computer with large memory prompted a new wave of interest in neural network among the machine learning community. The use of new forms of neural nets with an increased number of layers goes often under the name of deep learning.

The section on nonparametric option pricing and state price density estimation was inspired by Yacine Ait-Sahalia's Ph.D. thesis, and we use part of his data for illustration. The paper [51] is the first article we know of, where options are systematically priced from historical data using (modern) nonparametric regression procedures including neural networks. The later contribution [85] of Ait-Sahalia and Lo concentrated on the use of the kernel method. Their reason for choosing the kernel over the other nonparametric regression procedures comes from their desire to price more general contingent claims with (possibly) very complex payoff functions. Indeed, the kernel regression is more amenable to the estimation of the second derivative of the regression function with respect to the strike price. We decided to add the use projection pursuit for the sake of comparison.

Some very popular nonparametric regression methods have not been mentioned in the text. The *k-nearest neighbors* method is one of them. See Problem 5.6. It is very similar in spirit to the kernel regression when one uses the box kernel. Indeed in both cases the value of the regression function at a given point is given by the average of the observed responses for a set of neighboring explanatory vectors. The difference is only in the definition of these neighboring points: we choose all the points within a certain distance (given by the value of the bandwidth) in the case of the kernel regression independently of their number, while we choose the k-nearest points in the case of the k nearest neighbors regression. The smoothing parameter is now the number k of neighbors involved in the averaging. Numerical results are very similar, and like in the case of the kernel method, nonparametric density estimation can also be done by the k nearest neighbors method. However, adjusting implementations to allow for categorical explanatory variables appears to be easier with the k nearest neighbors method. The interested reader is referred to Silverman's book [88] for a detailed account of the k nearest neighbors method in the context of density estimation, and to Kohonen's book [57] for artificial intelligence and machine learning applications.

Classification and clustering can be viewed as regression problems for which the response variable can only take finitely many values. All the nonparametric methods discussed above can be adapted to solve classification problems. However, because they offer intuitive interpretations, tree based methods remain the most appealing classification procedures. R offers a set of methods to manipulate trees for regression and classification, based on the fundamental work of Breiman, Friedman, Olshen, and Stone [9]. As in the case of regression, classification suffers from the curse of dimensionality, and principal component analysis is viewed as a reasonable technique for reducing the dimension of the explanatory vectors. Alternatives based on coding and computing efficiency arguments have been proposed, for example wavelet packets. The interested reader is referred to Wickerhauser's book [97] for details. Finally, we close with a reference to a recent attempt to bring ideas and techniques for data mining and machine learning that were developed in the artificial intelligence community under the umbrella of mathematical statistics. These techniques go under the name of boosting and bagging and they are intended to combine several classification or regression algorithms into a single, better performing one. This might seem like an impossible dream, but it does happen in some cases, and theoretical arguments can be given to justify such seemingly unrealistic expectations. See, for example, the book of Witten and Frank [98], and the graduate text by Hastie, Tibshirani, and Friedman [45].

Part III

TIME SERIES & STATE SPACE MODELS

6

TIME SERIES MODELS: AR, MA, ARMA, & ALL THAT

Time series are ubiquitous in everyday manipulations of financial data. They are especially well suited to the nature of financial markets, and models and methods have been developed to capture time dependencies and produce forecasts. This is the main reason for their popularity. This chapter is devoted to a general introduction to the linear theory of time series, restricted to the univariate case. Later in the book, we will consider the multivariate case, and we will recast the analysis of time series data in the framework of state space models in order to consider and analyze nonlinear models.

6.1 Notation and First Definitions

The goal of time series analysis is to analyze data containing finite sequences of measurements, each coming with a time stamp, these time stamps being ordered in a natural fashion. The purpose of the analysis is to quantify the dependencies across time, and to take advantage of these correlations to explain the observations at hand, and to infer properties of the unobserved values of the series.

We have already encountered many instances of time series (recall, for example, the coffee futures data as plotted in Fig. 3.4). In most cases, we transformed the data to reduce the serial correlation to a minimum, and we used statistical techniques completely indifferent to the order of the data: in this way, we did not use any possible serial dependence in the data. It is now time to investigate the various ways one can model this dependence, and take advantage of the properties of these models. However, there was one instance where we had to deal with the effect of the dependencies over time of the data entries. This was the case of the utility indexes, whose data we reproduce now.

```
           ENRON.index DUKE.index UTILITY.index
01/04/1993    135.0000   104.2857      99.94170
01/05/1993    135.3714   103.5714      99.49463
01/06/1993    132.8571   104.2857      99.86034
```

R. Carmona, *Statistical Analysis of Financial Data in R*, Springer Texts in Statistics, DOI 10.1007/978-1-4614-8788-3_6, © Springer Science+Business Media New York 2014

01/07/1993	130.7143	103.5714	98.70023
......	......	......	......
12/28/1993	166.4571	125.0000	107.15202
12/29/1993	170.7429	123.9429	106.75023
12/30/1993	169.3143	124.2857	106.12351
12/31/1993	165.7143	121.0857	104.95227

The first column gives a set of dates, some form of daily time stamps, while the next three columns contain the numerical values of the three indexes on these dates. This is the typical structure of time series data which we consider in this chapter and the next.

6.1.1 Notation

Most statistical problems deal with data in the form

$$x_0, x_1, \ldots, x_n. \tag{6.1}$$

In the regression applications considered so far, the order in which the observations were collected did not play any role. We are now interested in applications for which the order of the x_i's plays a crucial role in the interpretation of the data, as well as in the definition of the inferential problems we consider.

In the applications we are considering now, the label n of the observation x_n corresponds to a time stamp, say t_n, giving the time at which the measurement was taken. As always, it is convenient to view the observations (6.1) as realizations of random variables $X_0, X_1, \ldots, X_n$ which we shall sometimes denote $X_{t_0}, X_{t_1}, \ldots, X_{t_n}$ when we want to emphasize the role of the time stamps. These $n+1$ random variables will most often be regarded as a subset of a (possibly infinite) sequence $\{X_t\}$ of random variables. The x_i's (and hence the X_i's) can be scalars as in this chapter, in which case we talk about univariate time series, or vectors as in the next chapter, in which case we talk about multivariate time series. As before, we try to use regular fonts for scalars and bold-face fonts for vectors.

Most of this chapter is devoted to the analysis of time series models. A *model* is a set of prescriptions for the joint distributions of the random variables (or random vectors in the multivariate case)

$$X_{i_1}, X_{i_2}, \ldots, X_{i_k}$$

for all possible choices of the finite ordered sequence $i_1 < i_2 < \cdots < i_k$ of time stamps. These joint distributions are completely determined by the model in some cases, while in other cases, only partial information is provided by the prescriptions of the model.

6.1.2 Regular Time Series and Signals

Regular time series are sets of measurements taken at regular time intervals. In other words, the time stamps $\{t_j\}_{j=0,1,\ldots,n}$ are of the form $t_j = t_0 + j\Delta t$ for

$j = 0, 1, \ldots, n$. Such a sequence of times is determined by its start t_0, its length $n + 1$, and the time interval Δt between two successive times. Note that, instead of giving the sampling interval Δt, one can equivalently give the sampling frequency, or the time of the final measurement. Once the time sequence has been defined, one can then give the sequence of corresponding measurements separately. This characterization of the sequence of time stamps of a regular time series by three characteristics is fundamental in the time series packages implemented in most of the statistical software computer packages.

Figure 6.1 gives an example of such a regular time series. It is a speech signal which we created by recording the short sentence "`how are you`", digitizing the sound file, and collecting the resulting numerical values in an `R` numerical vector which we called `HOWAREYOU`. Figure 6.1 was produced with the command:

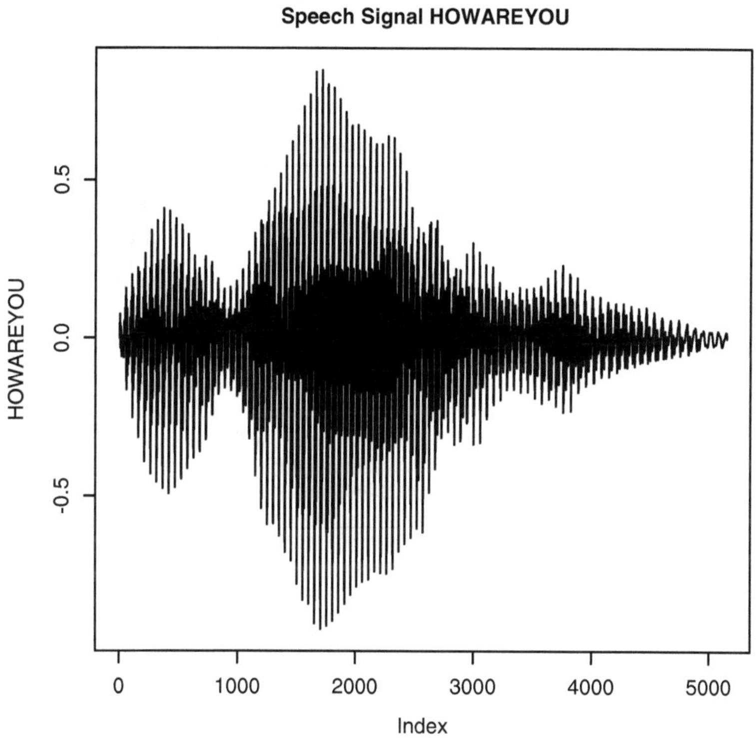

Fig. 6.1. Plot of the sound "How Are You" digitized at 8,000 Hz

```
plot(HOWAREYOU,type="l",main="Speech Signal HOWAREYOU")
```

In such a plot, the time stamps used to label the elements of the signal are simply successive integers starting from one. They are referred to as `INDEX` in the plot. This should be contrasted with what comes next.

In part because of their frequent occurrence in applications to signal analysis (as traditionally performed by electrical engineers), regular time series are often called signals. The library `stats` of `R` provides objects of class `ts` to manipulate signals, but we shall not use them.

6.1.3 Calendar and Irregular Time Series

Most of the financial time series do not have the good taste to be regular in the sense given above. They differ from the regular time series discussed above in several ways, and mostly by the fact that the time stamps are given by dates and times, thus their name calendar time series. Even though calendar time series are particular cases of a larger class of irregular time series, they will be the only ones considered here. Oftentimes, these data are daily, and gaps due to weekends and holidays create irregularities. Figure 6.2 gives the daily closing prices of the S&P 500 index on the New York Stock Exchange (NYSE for short) for the period ranging from January 3, 1950 to August 20, 2010.

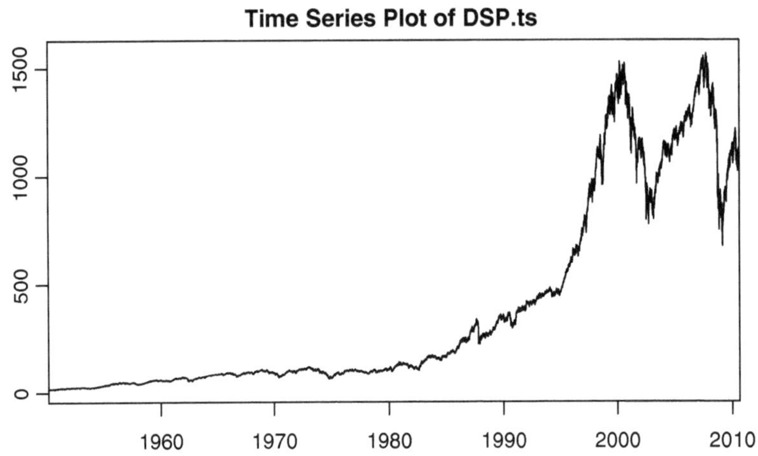

Fig. 6.2. `timeSeries` plot of the daily closing values of the S&P 500 index produced by the command `plot(DSP.ts)`

Here is a (very) small subset of the data used to produce the plot.

```
     Date     Open     High      Low    Close
    .....    .....    .....    .....    .....
17-Sep-01  1092.54  1092.54  1037.46  1038.77
10-Sep-01  1085.78  1096.94  1073.15  1092.54
 7-Sep-01  1106.40  1106.40  1082.12  1085.78
 6-Sep-01  1131.74  1131.74  1105.83  1106.40
 5-Sep-01  1132.94  1135.52  1114.86  1131.74
 4-Sep-01  1133.58  1155.40  1129.06  1132.94
```

```
31-Aug-01 1129.03 1141.83 1126.38 1133.58
30-Aug-01 1148.60 1151.75 1124.87 1129.03
29-Aug-01 1161.51 1166.97 1147.38 1148.56
28-Aug-01 1179.21 1179.66 1161.17 1161.51
  .....     .....     .....     .....     .....
```

As we can see, the time stamps can be very irregularly spaced at times. As illustrated in this snapshot, the regularity of the measurements can be affected by unexpected events. However, the scale of a typical plot of a multi-year data series would not allow us to see the gaps due to weekends and holidays and extraordinary market closure events.

6.1.4 **Creating and Plotting** `timeSeries` **Objects in** `R`

The manipulation of calendar time series in `R` in done via objects of class `timeSeries`. These objects contain a slot `positions` for the time stamps, and a slot `data` for the actual values of the numerical measurements. Typically, `positions` is a vector of dates or dates and times, while `data` is a numerical matrix with one row for each entry of the vector `positions`. One creates an object of class `timeSeries` with the constructor function `timeSeries` whose use is illustrated below. The numeric vector of the weekly values of the S&P 500 index was used earlier in Chap. 2 when we computed the weekly returns on the index. We make it into a `timeSeries` object by *attaching* the time stamps giving the weeks of the measurements. We first create the vector `SPWEEKS` of dates with the function `timeSequence`, and we bundle this vector of dates with the vector `WSP` of closing values into a `timeSeries` object `WSP.ts` which we then plot with the command `plot`. All this is done with the following commands:

```
SPWEEKS <- timeSequence(from = "1950-01-03",
                by = "week",length.out = 3163)
WSP.ts <- timeSeries(positions=SPWEEKS,data=WSP)
plot(WSP.ts)
```

The generic method `plot` can be used with `timeSeries` objects. The resulting plot is given in Fig. 6.3. At this scale, it is difficult to differentiate this plot from the plot of the daily closing values of the index.

6.1.5 **High Frequency Data**

The widespread availability of high frequency data changed the landscape of financial data analysis, and spurred for better or worse the development of high frequency trading. Our discussion of the morning and afternoon indicators in Chap. 5 on nonparametric regression was a first example involving high frequency data. Here, we consider another example more in line with the current discussion of time series.

High frequency data are the result of a different data collection process: a record is added to the data file each time a new transaction takes place. These data are also called transaction data, or tick data. They offer a unique insight into actual trading

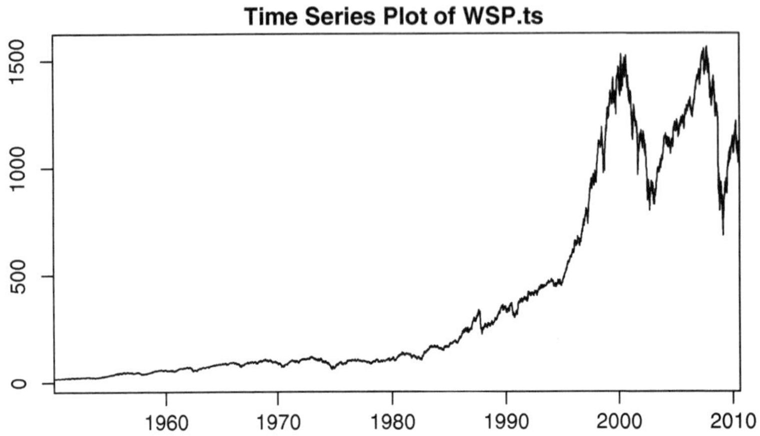

Fig. 6.3. `timeSeries` plot of the weekly closes of the S&P 500 index produced by the command `plot(WSP.ts)`

processes and market microstructure. In this subsection, we study the most important features of high-frequency time series data, identifying striking differences with lower-frequency data, and introducing new tools and new methods tailored to the new challenges presented by this new type of data.

Tick-by-tick data are available for liquid futures contracts, and in this section, we consider the example of a futures contracts on the S&P 500 index for the sake of illustration. Here is the way the data of the September 1998 contract look like.

"date"	"time"	"close"
19971008	14:53:38	1,013.20
19971017	10:59:16	986.00
19971027	10:02:13	960.00
19971103	10:28:51	968.00
19971105	09:08:44	975.00
19971106	10:59:21	969.00
19971124	12:52:52	986.90
19971209	10:58:18	1,015.00
19971210	09:22:05	1,005.70
19971224	09:27:21	968.00
...	...	...

We notice that the time stamps can be very sparse. This is due to the fact that these trades occurred on days very far from the maturity of the contract: speculators are actively trading contracts closer to delivery! However, the situation changes dramatically when we look at the data later in the life of the contract. Indeed, one sees that not only the transactions are more frequent, but also that a large number of transactions appears to happen simultaneously since they have the same time stamp. Given the fact that each row contains only one number besides the date and time, we shall assume that this number is the price at which the transaction was settled, not a bid or ask price. This information is not always given by the data provider, and the data analyst may be forced to make this sort of assumption. One of the unexpected surprises with high-frequency financial data is the fact that the notion of price is not clearly defined. Strangely enough, there are many reasons for that. The first one is clear from the data reproduced below: different values can be quoted with the same time stamp, so what is the price at that time?

"date"	"time"	"close"
...	...	...
19980804	11:05:00	1,103.50
19980804	11:05:00	1,103.00
19980804	11:06:00	1,102.80
19980804	11:06:00	1,102.60
19980804	11:06:00	1,102.50
19980804	11:06:00	1,102.40
19980804	11:06:00	1,102.20
19980804	11:06:00	1,102.00
...	...	...
19980804	11:06:00	1,102.50
19980804	11:06:00	1,102.40
19980804	11:06:00	1,102.20
19980804	11:06:00	1,102.00
19980804	11:07:00	1,101.70
...	...	...

Whether one looks at this particular portion of the data set or not, it happens very often that, many seconds do not appear because there is no transaction at these times. Another idiosyncrasy of high-frequency data is the fact that the bid and ask prices do not make sense all the time. Indeed, when the frequency is high enough, the time interval between two transactions is so small that the price cannot move out of the bid-ask spread, muddying both the definition of the notion of price and of bid-ask spread at the same time. We avoid this issue by considering that the tick value given by the data provider is the settlement price.

The data discussed above are included in the library `Rsafd`. They are contained in the data frame `SPsep98` whose top we reproduce below:

```
head(SPsep98)
      date       time   close
1 19971008 14:53:38 1013.2
2 19971017 10:59:16  986.0
3 19971027 10:02:13  960.0
4 19971103 10:28:51  968.0
5 19971105 09:08:44  975.0
6 19971106 10:59:21  969.0
```

This data frame has three columns. The first one, named `date`, gives the date of the quote. Notice the format! The second column, named `time`, gives the time of the day at which the quote was provided, and finally, the third column, named `close` gives the actual quote (so we assume).

In order to create an object of class `timeSeries` with these data, we first create a vector of positions with the days and times given in the first two columns of the data frame. We first use the function `makeDate` with the column `date`, specifying the format from which the date should be read in the parameter `in.format`. In this case, the upper case "Y" indicates that the first four characters should be understood as the year, the lower case "m" indicates that the next two characters should be understood as an integer giving the month of the year, and the lower case "d" indicates that the last two characters should be understood as an integer giving the day of the month. The command using the function `timeDate` gives another example of formatted reading, more in the `unix` style this time. In any case, its output is a vector which can be used as a vector of positions for an object of class `timeSeries`. The extra parameter `units` in the constructor `timeSeries` is used to specify a name for the numerical values of the data component of the time series.

```
SPsep98day <- makeDate(SPsep98[,1],in.format="Ymd")
SPsep98POS <- timeDate(paste(SPsep98day,SPsep98[,2]),
format = "%Y-%m-%d %H:%M:%S")
SPsep98.ts <- timeSeries(positions=SPsep98POS,
data=SPsep98[,3], units="Sep98")
head(SPsep98.ts)
                     Sep98
1997-10-08 14:53:38 1013.2
1997-10-17 10:59:16  986.0
1997-10-27 10:02:13  960.0
1997-11-03 10:28:51  968.0
1997-11-05 09:08:44  975.0
1997-11-06 10:59:21  969.0
plot(SPsep98.ts)
```

The corresponding `timeSeries` plot is reproduced in Fig. 6.4.

Fig. 6.4. `timeSeries` plot of the high frequency quotes of the S&P 500 futures contract maturing on September 1998 as produced by the command `plot(SPsep98.ts)`

This plot shows clearly the specific features of the data which we identified earlier. The left part of the plot contains only a few transactions, and because the plotting program interpolates linearly between points, we see an artificial piecewise linear pattern for the price of the futures contract. The right part of the plot is more typical of volatile financial time series.

Remark on Quantized Ticks. Price changes from one transaction to the next are quoted in multiples of tick size. This tick size varies from one exchange to another. Typical values are (or used to be) one 8th and one 16th of a dollar. This practice is obsolete on a certain number of exchanges. For example, all New York Stock Exchange (NYSE) and New York Mercantile Exchange (NYMEX) stocks are traded in decimals since January 29, 2001. Nevertheless, practitioners should be aware of the fact that high-frequency data, and especially historical high-frequency raw data which has not been pre-processed, quite often take only discrete values: this can introduce numerical artifacts, and in particular spurious correlation.

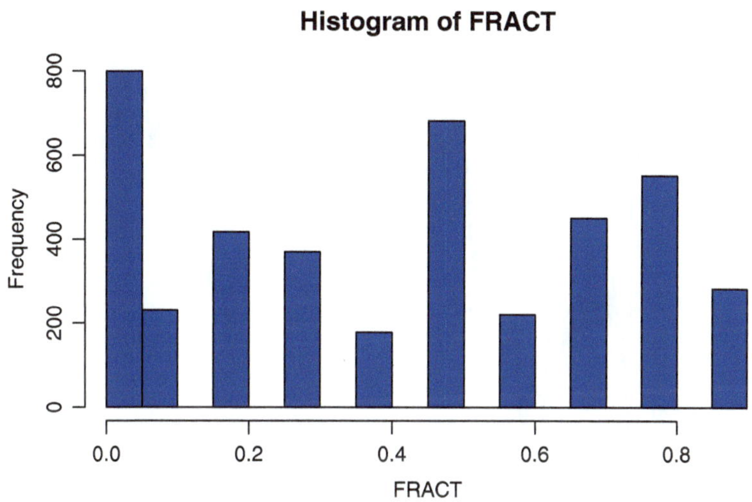

Fig. 6.5. Histogram of the fractional part of the June 15, 1998 quotes of the S&P 500 September 1998 futures contract. The discrete nature of the data shows clearly

We illustrate this fact with the S&P 500 data considered in this subsection. We computed the fractional parts of the transaction prices (obtained by removing the integer parts to the actual prices), and we plotted their histogram in Fig. 6.5. The quantification effect appears clearly.

6.1.6 `TimeDate` Manipulations

It is very easy to develop specific functions satisfying the needs of most time series data analysis. In particular, the library `Rsafd` contains a certain number of "home grown" functions which we wrote to make `timeDate` manipulations easy. Among them, the functions `begday` and `noon` extract the beginning of a given day, and noon of the same day. In other words, given a `timeDate`, the first function extracts the day, and returns a `timeDate` including hours, minutes and seconds of the beginning of that same day.

The following commands illustrate the algebraic manipulations on `timeDate` objects, and show how one can extract subsets of a time series.

```
DAY <- timeDate("08/12/1998")
SP081298 <- SPsep98.ts[seriesPositions(SPsep98.ts)
         >= DAY & seriesPositions(SPsep98.ts)<DAY+24*3600]
plot(SP081298)
```

and the resulting plot is reproduced in Fig. 6.6.

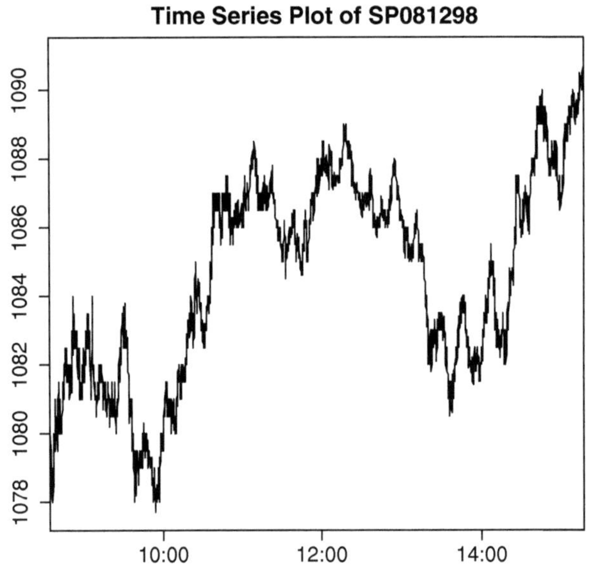

Fig. 6.6. Plot of 1 day extracted from the `timeSeries` object `SPsep98.ts`

One could as well extract three consecutive days instead of one.

```
SPDAYS <- SPsep98.ts[seriesPositions(SPsep98.ts)>=DAY
              & seriesPositions(SPsep98.ts)<DAY+3*24*3600]
plot(SPDAYS)
```

and the resulting plot is given in Fig. 6.7. As explained earlier, the plotting function performs a linear extrapolation to join the last quote of 1 day to the first quote of the following day. However, the scale of the horizontal axis is uniform throughout the time domain and no special shaded vertical bar is produced (like in `S-Plus`) to

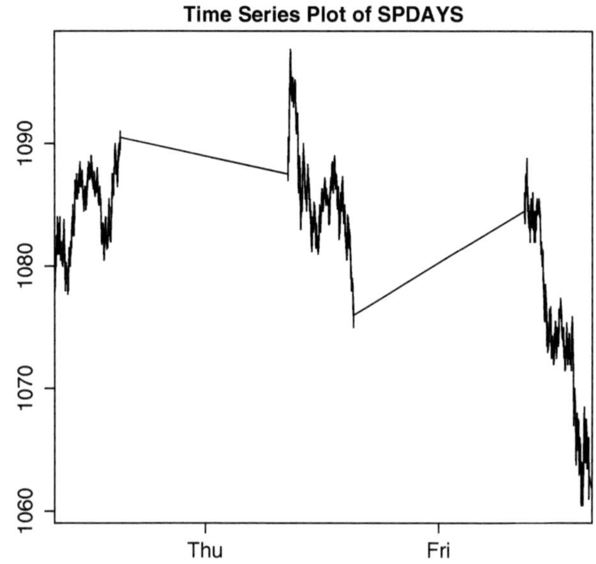

Fig. 6.7. Plot of the short `timeSeries` object obtained by extracting three consecutive days from the `timeSeries` object `SPsep98.ts`

warn the casual reader that the part of the plot in these bars does not correspond to actual quotes.

Working with regular time series is very convenient for many reasons, not the least being the fact that we can take advantage of all the tools we introduce in this book. For this reason, in order to avoid the technical difficulties inherent to the analysis of irregular time series, we consider only models for regular time series, and when we fit these models to real data, we implicitly assumed that the actual data has already been aligned, and act as if the time series data were regular. However, one should keep in mind that aligning a time series comes at a cost: the clear loss of information in time periods with high activity, and the introduction of undesired artifacts in periods of low trading activity.

6.2 Time Dependent Statistics and Stationarity

Now that we know the type of data amenable to practical time series analysis, we turn to the discussion of the first properties of the theoretical models. Throughout this section we assume that we are given such a theoretical model $\{X_t\}_t$ for a time series.

6.2.1 Statistical Moments

The mean function μ_X is defined as the (deterministic) function of time given by:

$$t \hookrightarrow \mu_X(t) = \mathbb{E}\{X_t\}.$$

Similarly, we define the variance function var_X (resp. the standard deviation function σ_X) as the (deterministic) function of time given by:

$$t \hookrightarrow \text{var}_X(t) = \text{var}\{X_t\} = \mathbb{E}\{(X_t - \mu_X(t))^2\},$$
$$(\text{resp. } t \hookrightarrow \sigma_X(t) = \sqrt{\text{var}_X(t)} = \mathbb{E}\{(X_t - \mu_X(t))^2\}^{1/2}).$$

Even though these statistics can capture some of the time dependent features of the series, they do not carry any information on the way individual random variables entering the series depend upon each other. With this in mind, we consider statistical moments involving several X_t's simultaneously. The auto-covariance function γ_X is defined as the (deterministic) function of two instants given by:

$$\gamma_X(s,t) = \text{cov}\{X_s, X_t\} = \mathbb{E}\{(X_s - \mu_X(s))(X_t - \mu_X(t))\}.$$

As one would expect, the auto-correlation function ρ_X is defined as the (deterministic) function of two instants given by:

$$\rho_X(s,t) = \text{cor}\{X_s, X_t\} = \frac{\mathbb{E}\{(X_s - \mu_X(s))(X_t - \mu_X(t))\}}{\sigma_X(s)\sigma_X(t)}.$$

We shall also use the notion of partial auto-correlation function. Its definition is best understood in the framework of stationary time series once the notion of linear prediction has been introduced. So stay tuned if you want to know more about partial auto-correlation functions.

The above concepts are limited to moments of orders one and two. They have been regarded as sufficient for quite some time, and many time series analysis tools have been designed to study models only on the basis of their first two moment functions. There are several reasons for that: first, these moments characterize entirely the Gaussian models. Moreover, their attractiveness is increased by the fact that the least squares methods only depends upon these moments. Unfortunately they are not sufficient to handle nonlinearities for which the use of higher order moments is required. Despite the fact that the importance of non-Gaussian and nonlinear time series is increasingly recognized, we shall not discuss these matters any further in this chapter. See the Notes & Complements at the end of the chapter for references.

6.2.2 The Notion of Stationarity

Stationarity is a crucial property of stochastic processes and time series models. It will be a *sine qua non* condition for the implementation of most estimation and prediction algorithms. The notion of stationarity can be described in the following manner.

6.2.2.1 Mathematical Definitions

A time series model for $\{X_t\}_t$ is said to be stationary if all its statistics remain unchanged after time shifts, i.e. if they are the same as the statistics of $\{X_{t_0+t}\}_t$ for all possible choices of t_0. This notion of stationarity is sometime called strong stationarity. The first obvious consequence of stationarity is that the mean function of a stationary time series is constant. The same holds for the variance and the standard deviation functions. Moreover, the auto-covariance function, and consequently the auto-correlation function as well, are functions of the difference between its arguments. This means that:

$$\mu_X(t) = \mu_X, \quad \text{var}_X(t) = \sigma_X^2, \quad \sigma_X(t) = \sigma_X, \quad \text{and} \quad \gamma_X(s,t) = \gamma_X(t-s), \tag{6.2}$$

for some constants μ_X and σ_X, and for some function of one variable for which we still use the notation γ_X.

There is a weaker notion of stationarity which will be useful in the sequel. A time series model is said to be weakly stationary if its mean function is constant, and its auto-covariance function is a function of the difference of its arguments. Obviously, a stationary series (in the strong sense given above) is necessarily weakly stationary as implied by formula (6.2). However, the converse is not true in general. In a nutshell, (strong) stationarity means that the moments of all orders are invariant under time shifts, while weak stationarity merely requires that the moments of order 1 and 2 are shift invariant.

Remark. Despite this fact, the two notions of stationarity coincide for Gaussian time series models, i.e. when all the finite dimensional marginal distributions are multivariate Gaussian. Indeed, a multivariate Gaussian distribution is entirely determined by its mean vector and its variance/covariance matrix, which in turn is determined by the covariances of the random variables taken two by two.

6.2.2.2 Linear Prediction and Partial Auto-Correlation Function (PACF)

Building a model for time series data is a required step in the prediction of future values. So, once a model for $\{X_t\}_t$ has been chosen, in many practical applications the goal is to produce, for a given time t (which we shall refer to as the present time), predictions for the (future) values of the outcomes x_{t+1}, x_{t+2}, $\ldots$ of the random variables X_{t+1}, X_{t+2}, $\ldots$. Obviously, these predictions should be non-anticipative in the sense that they should be based solely on the (present and past) observations x_t, x_{t-1}, $\ldots$. In other words, we are not allowed to use a crystal ball to look into the future when it comes to predictions!

Given the observations $X_0 = x_0$, $X_1 = x_1$, …, $X_t = x_t$ up to the present time t, and a mean zero random variable Z, we shall use the notation $E_t^{(m)}(Z)$ for the best prediction of Z by linear combinations of the m values $x_s - \mu_X$ for $t - m + 1 \leq s \leq t$. In other words, when viewed as a function of the random variables

$X_{t-m+1} - \mu_X$, $X_{t-m+2} - \mu_X$, ..., $X_t - \mu_X$, $E_t^{(m)}(Z)$ is the linear combination $\alpha_1(X_{t-m+1} - \mu_X) + \alpha_2(X_{t-m+2} - \mu_X) + \cdots + \alpha_m(X_t - \mu_X)$ which minimizes the quadratic error:

$$\mathbb{E}\{\|Z - \alpha_1(X_{t-m+1} - \mu_X) + \alpha_2(X_{t-m+2} - \mu_X) + \cdots + \alpha_m(X_t - \mu_X)\|^2\}. \quad (6.3)$$

Notice that, in the Gaussian case (i.e. in the case of (jointly) Gaussian time series models), this prediction operator is given by the conditional expectation:

$$E_t^{(m)}(Z) = \mathbb{E}\{Z|X_t, X_{t-1}, \ldots, X_{t-m+1}\}.$$

The best linear predictor $E_t^{(m)}(Z)$ is a linear combination of the $X_{t-m+1} - \mu_X$, $X_{t-m+2} - \mu_X$, ..., $X_t - \mu_X$, so it belongs to the linear space generated by these random variables. Moreover, since $E_t^{(m)}(Z)$ minimizes the quadratic error (6.3), it minimizes the distance between Z and this linear space. Consequently, $E_t^{(m)}(Z)$ can be viewed as the orthogonal projection of the random variable Z onto the linear space generated by the random variables $X_{t-m+1} - \mu_X$, $X_{t-m+2} - \mu_X$, ..., $X_t - \mu_X$. In fact, as we shall see in the sequel, this interpretation as an orthogonal projection is a great help when it comes to guessing and proving the properties of the best linear prediction operator $E_t^{(m)}$. The first instance is the following: because of the properties of orthogonal projections, $Z - E_t^{(m)}(Z)$ is orthogonal to all of the $X_j - \mu_X$ and consequently:

$$\mathbb{E}\{(Z - E_t^{(m)}(Z))(X_j - \mu_X)\} = 0, \qquad j = t-m+1, t-m+2, \ldots, t-1, t.$$

The random variable $Z - E_t^{(m)}(Z)$ is often referred to as an innovation because it represents, in a minimal way, the information needed to produce the outcome of Z, which cannot be given by linear combinations of the past values $X_{t-m+1} - \mu_X$, $X_{t-m+2} - \mu_X$, ..., $X_t - \mu_X$.

We shall use the notation E_t without the superscript $^{(m)}$ when $m = t$, in other words, when we use the entire available past to construct the prediction. This special prediction operator will be used extensively in Chap. 7 when we discuss filtering.

With the notion of best linear predictor at hand, it is now easy to define the partial auto-correlation coefficients. The k-th partial auto-correlation coefficient $\phi_{k,k}$ is defined as the last coefficient in the linear combination giving $E_t^{(k)}(X_{t+1})$. In other words, if:

$$E_t^{(k)}(X_{t+1} - \mu_X) = \alpha_k(X_t - \mu_X) + \alpha_1(X_{t-1} - \mu_X) + \cdots + \alpha_1(X_{t-k+1} - \mu_X)$$

then we set $\phi_{k,k} = \alpha_1$. In this way, one sees that the partial auto-correlation coefficient $\phi_{k,k}$ measures the correlation between X_{t+1} and X_{t-k+1} (or equivalently, between X_t and X_{t-k} because of stationarity) after adjustment for the intermediate lagged variables.

Because the best linear predictor is an orthogonal projection, our knowledge of Euclidean geometry tells us that it can be computed in terms of inner products. In this way one can prove that the partial auto-correlation coefficient $\phi_{k,k}$ is given by the formula:

$$\phi_{k,k} = \Gamma_{X,k}^{-1}\gamma_{X,k}, \tag{6.4}$$

where the $k \times k$ matrix $\Gamma_{X,k}$ and the k-dimensional vector $\gamma_{X,k}$ are defined by:

$$\Gamma_{X,k} = [\gamma_X(i-j)]_{i,j=1,\ldots,k} \qquad \gamma_{X,k} = [\gamma_X(1), \gamma_X(2), \ldots, \gamma_X(k)]^t. \tag{6.5}$$

This equivalent form of the definition of the partial auto-correlation function is preferable because it is immediately amenable to estimation. Indeed, if we assume momentarily that we know how to estimate the auto-covariance function γ_X from given data $x_0, \ldots, x_n$, see the discussion leading to (6.8) below, then the sequence $\{\phi_{k,k}\}_k$ of partial auto-correlations will be estimated by the sequence $\{\hat{\phi}_{k,k}\}_k$ given by:

$$\hat{\phi}_{0,0} = 1 \qquad \text{and} \qquad \hat{\phi}_{k,k} = \hat{\Gamma}_{X,k}^{-1}\hat{\gamma}_{X,k}, \tag{6.6}$$

where the $k \times k$ matrix $\hat{\Gamma}_{X,k}$ and the k-dimensional vector $\hat{\gamma}_{X,k}$ are computed from the estimate $\hat{\gamma}_X$ of the auto-covariance function via the formulae (6.5). The empirical estimate $\hat{\gamma}_X$ will be defined in the next subsection.

The notion of partial auto-correlation function may seem obscure at this stage, but please, bear with me for a short while: soon we shall give an enlightening interpretation of the partial auto-correlation function in terms of auto-regressive models of increasing orders fitted to the time series.

6.2.2.3 ***Time Averages as Statistical Estimates***

One of the main consequences of the stationarity of a time series is the fact that the *theoretical moments* introduced earlier can be computed (or at least estimated) by time averages. Indeed, when stationarity holds, the moment empirical estimates (see formulae (6.7) and (6.8) below) are time averages which converge when the number of terms used in the sum increases without bound. The existence of these limits is a good sign since it gives stability of the empirical estimates, but unfortunately the limits they converge to can still be random, and they can change with the particular realization of the time series. Fortunately, there are many situations in which these limits are not random. Time series with this property are called *ergodic*. Roughly speaking, ergodicity allows us to replace space averages such as expectations, co-variances, ..., by time averages. Ergodicity is a very subtle mathematical property and it is difficult to check that a given set of numbers $x_1, \ldots, x_n$ is a sample realization from an ergodic time series model. For this reason, we shall take ergodicity for granted in the sense that, whenever we find that a time series model is stationary, we shall implicitly assume that it is also ergodic. While we may not be able to justify this assumption in practice, it is fair to say that without it, we wouldn't be able to do much!

In practice, if $x_1, \ldots, x_n$ are observations from a time series model $\{X_t\}$ which we assume to be stationary, then we use the *time average*

$$\hat{\mu}_X = \bar{x} = \frac{1}{n}\sum_{i=1}^{n} x_i \tag{6.7}$$

as empirical estimate of the mean μ_X. As explained above, the stationarity of the series implies that:

$$\lim_{N\to\infty} \frac{1}{N}\sum_{i=1}^{N} X_i = \tilde{\mu}_X$$

almost surely, i.e. for all typical sequences of observations, where $\tilde{\mu}_X$ is a random variable. Moreover, ergodicity implies that the object $\tilde{\mu}_X$ (which was thought to be random) is in fact a deterministic number which is just the true value of the mean μ_X of the series. This is the justification for the use of the time average (6.7) as an estimate of the mean. Similarly, the auto-covariance function $\gamma_X(h)$ is estimated by time averages of the form:

$$\hat{\gamma}_X(h) = \frac{1}{n}\sum_{i=1}^{n-|h|} (x_i - \bar{x})(x_{i+|h|} - \bar{x}), \tag{6.8}$$

which are defined both for positive and negative lags h as long as $-n < h < n$. It is important to notice that the larger the lag absolute value $|h|$, the smaller the number of terms in the above sum. If we were to compute a confidence interval for the estimate $\hat{\gamma}_X(h)$ we would see that this interval grows very fast with $|h|$. As a consequence, it is wise to restrict the estimation of the auto-covariance function to lags which are small compared to the sample size n. In R, you can specify this maximum number of lags for which the auto-covariance function is estimated. If you don't, the program uses a multiple of the logarithm of n as a proxy. Indeed this number is usually much smaller than the sample size. Notice also that we divide by n while there are only $n - |h|$ terms in the summation. This departure from the standard definition of the empirical auto-covariance function becomes irrelevant for large samples, since when n is large, dividing by n or $n-|h|$ does not make much difference, especially with the limitation we imposed on the size of $|h|$. Moreover, dividing by n guarantees that the function $\widehat{\gamma_X}$ is nonnegative definite, a mathematical property of crucial importance for spectral theory. Since we shall not address spectral theory issues in these lectures, we do not go any further down this avenue.

As one could expect, the estimate of the sample auto-correlation function is defined by:

$$\hat{\rho}_X(h) = \frac{\hat{\gamma}_X(h)}{\hat{\gamma}_X(0)}, \qquad -n < h < n. \tag{6.9}$$

The distribution theory of the estimators $\hat{\mu}_X$, $\hat{\gamma}_X$ and $\hat{\rho}_X$ is well understood in the case of Gaussian time series, and confidence intervals and tests can be derived for these estimates. In the general case of possibly non-Gaussian series, only approximate tests are available. We discuss some of them in our discussion of the white noise series below.

6.2.3 The Search for Stationarity

The success of time series analysis depends strongly upon the satisfaction of two somewhat independent conditions. First one needs to massage the data into a stationary time series, and second, model the resulting stationary time series and estimate the parameters of the model.

In this subsection, we consider the first of these two objectives. Unfortunately, there is no universal recipe to turn a given time series into a stationary one, and experience will have to lead the analyst in this endeavor. For the record, we review several of the most commonly used procedures. We first discuss general strategies in an abstract setting, postponing the implementation of these ideas to the next subsection.

Statistical tests are required to quantify the extent to which a search for stationarity is successful. As a general rule, tests are not very powerful when the alternative hypothesis is too general. So rather restrictive alternative hypotheses are used in the commonly used tests for stationarity. We discuss the basic form of these tests at the end of Sect. 6.3.7. They are known under the name of Dickey-Fuller tests, or unit-root tests.

6.2.3.1 ***Removing Trends and Seasonal Components***

We first consider the issues connected with the analysis of the mean. We already mentioned that it is common practice to subtract the mean μ_X, and model the series $X_t - \mu_X$ instead of X_t. This is especially convenient when the mean function $t \hookrightarrow \mu_X(t)$ is constant, since in this case, μ_X can easily be estimated. However, we cannot expect to be able to do that in general. Fortunately, it happens quite often that the mean function depends upon time in a manner which can be identified. Let us assume for example that the random variables X_t, the observations of which produced the data x_t, can be reasonably well described, by an equation of the form:

$$X_t = T_t + S_t + R_t, \tag{6.10}$$

where both T_t and S_t are deterministic, and R_t is random and mean zero. So in other words, the mean $\mu_t = \mu_X(t)$ is decomposed into the sum of two components T_t and S_t to which we want to give specific interpretations. Typically, we assume that T_t is a deterministic monotone (increasing or decreasing) function of t, that S_t is a deterministic periodic function of t, and R_t is a mean-zero stationary time series. The function $t \hookrightarrow T_t$ is called trend, and $t \hookrightarrow S_t$ seasonal component while clearly, $\{R_t\}_t$ is called the remainder.

Several simple techniques can be used to identify the deterministic components T_t and S_t. If we momentarily rename the stationary time series R_t by ϵ_t, Eq. (6.10) becomes

$$X_t = \mu_t + \epsilon_t, \tag{6.11}$$

which is exactly the type of equation amenable to regression analysis. Indeed, regression techniques are the methods of choice for the identification of the deterministic components μ_t appearing in (6.11), and as we saw in our first examples dealing with energy indexes, if it is understood that the noise term ϵ_t can exhibit a significant dependent structure. Moreover, the fact that we expect the regression function μ_t to be the sum of a monotone function and a periodic function which we would like to identify separately will force us to use specific regression methods which we did not discuss yet. We did not discuss monotone regression, nor did we discuss Fourier methods of regression, so we will not elaborate on how these two components are usually identified. We shall merely refer to the help file of the `R` function `stl` which we use extensively for that purpose.

6.2.3.2 *Stabilization of the Variance*

Setting $\tilde{X}_t = X_t - \mu_X(t)$, it is always possible to write a time series as the sum of a deterministic function plus a mean-zero time series since $X_t = \mu_X(t) + \tilde{X}_t$. If the function $\mu_X(t)$ can be estimated from the data, (see examples below) it can be subtracted from the original data, and our modeling efforts should concentrate on the mean-zero time series $\tilde{X}_t$. It happens frequently that the standard deviation varies significantly with time, and since a local variance is more difficult to estimate than a local mean function, this may be a serious hurdle.

Variance stabilization transformations are ways to correct for these variations when the variance function is an explicit function of the mean in the sense that $\text{var}_{\tilde{X}}(t) = \varphi(\mu_{\tilde{X}}(t))^2\sigma^2$ for some constant $\sigma > 0$, and a known function φ. In such a case, it is easy to show that the variance of the time series $Y_t = \psi(\tilde{X}_t)$ is essentially constant and equal to σ^2 if the function ψ is such that $\psi'(\mu) = 1/\varphi(\mu)$. The time series $\{Y_t\}_t$ is more likely to be stationary, and in any case, it is more amenable to analysis than the original series. One will also demand that the function ψ be invertible in order to be able to return to $\tilde{X}_t$ (and X_t) after analysis of the time series $\{Y_t\}_t$.

Examples of variance stabilization transformations include the function $\psi(\mu) = \log(\mu)$ (which is often used in the analyses of time series of financial returns) and the function $\psi(\mu) = \sqrt{\mu}$. According to the discussion above, the transformation $\psi(\mu) = \log(\mu)$ is recommended when the variance varies like the square of the mean (i.e. $\varphi(\mu) \sim \mu$), while the transformation $\psi(\mu) = \sqrt{\mu}$ is recommended when the variance varies like the mean (i.e. $\varphi(\mu) \sim \sqrt{\mu}$) as we find in many situations involving the Poisson distribution.

6.2.3.3 *The Use of Differentiation*

If $\mu(t)$ is a constant function, a mere differentiation should make it disappear. This intuition from calculus can be implemented in the case of time series by introducing the analog of the differentiation operator. We shall use the suggestive notation ∇ for this operator. At the model level, ∇ is defined by:

$$\nabla X_t = X_t - X_{t-1} \tag{6.12}$$

and for a (finite) data set $x = (x_0, x_1, \ldots, x_N)$, the (first) difference $y = \nabla x$ is given by $y = (x_1 - x_0, x_2 - x_1, \ldots, x_N - x_{N-1})$. Notice that y_t is defined for $t = 1, 2, \ldots, N$ while x_t is defined for $t = 0, 1, 2, \ldots, N$! It is now time to give some respectability to a practice we have used several times already: taking differences to turn a non-stationary time series into a stationary one. Recall that the typical instance is the computation of the log-returns of a financial time series. So we elevate this example to the rank of definition: we shall say that a time series is integrated of order 1, or that it has one unit-root, if its difference is stationary. We shall use the notation $I(1)$ for the class of these time series.

In the same way a simple difference can *kill* a constant term, two successive differences will get rid of a linear trend (a function μ of the form $\mu(t) = at + b$). Iterating the definition (6.12) of the first difference operator, we find definition formulae for the higher order difference operators. For example, the second order difference operator is given by:

$$\begin{aligned}\nabla^2 X_t = \nabla[\nabla X]_t = [\nabla X]_t - [\nabla X]_{t-1} &= X_t - X_{t-1} - (X_{t-1} - X_{t-2}) \\ &= X_t - 2X_{t-1} + X_{t-2}.\end{aligned}$$

More generally, by successive applications of the difference operator one can remove any kind of polynomial trend! This remark makes it plain how useful the higher order difference operators can be. To keep up with the definition introduced above, we say that a time series is integrated of order p, or that it has p unit-roots, if its p-th order difference is stationary, and we denote the class of time series with this property by $I(p)$. In other words:

$$\{X_t\}_t \in I(p) \qquad \text{means that} \qquad \{\nabla^p X_t\}_t \quad \text{is stationary}$$

Obviously, the notation $I(0)$ will be used for stationary time series. We use the notation ∇ to conform with the conventions used in most of the textbooks on time series analysis, and to emphasize the analogy with the differentiation of functions of a (continuous) variable t.

6.2.4 The Example of the CO_2 Concentrations

We choose to illustrate the discussion of the search for stationarity with the classical example of the concentration of CO_2 above the Mauna Loa volcano in Hawaii. The data comprise monthly measurements. They range from March 1958 to December 2008. These data are contained in a data set `CO2` included in the `Rsafd` library. The values contained in the numeric vector `CO2` represent monthly concentrations adjusted to represent 2,400 h on the 15th day of each month. Units are parts per million by volume (ppmv) expressed in the 2003A SIO manometric mole fraction scale. We construct a `timeSeries` object `co2.ts` which we plot using the following commands.

```
POS <- timeSequence(from="1958-03-01",
                    to="2008-12-01", by="month")
co2.ts <- timeSeries(positions=POS,data=as.vector(CO2),
                                                 units="CO2")
plot(co2.ts)
```

The plot is reproduced in Fig. 6.8. Notice that the distribution of the R version you are using may already contain data sets with the name CO2. Installing the library Rsafd should mask these data sets. The numeric vector seriesData(co2.ts) is a matrix with only one column, but as a matrix, its rows and its columns can have *names*. Even though it is not a requirement, we will make a habit of choosing for the row names of the data matrix of a timeSeries object, strings of characters representing the actual dates of the measurements (i.e. the dates which appear as entries of the vector seriesPositions), and for the column names, the character strings used in the parameter units.

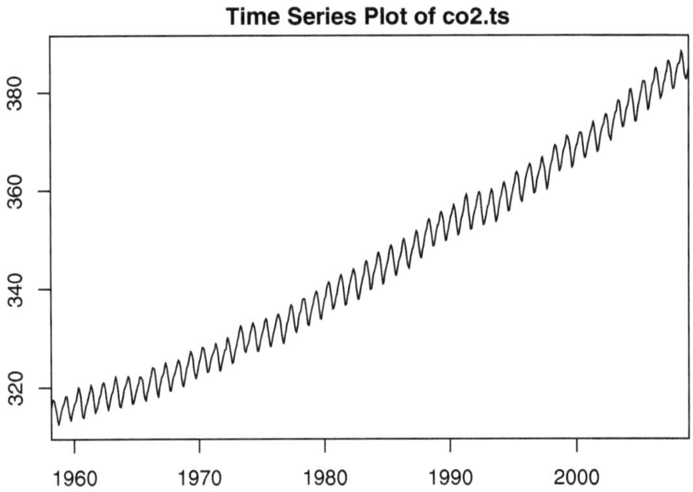

Fig. 6.8. Plot of the timeSeries object created for the analysis of CO_2 concentrations

This plot reveals an upward trend and a cyclic behavior whose period seems to be 12 months (none of these remarks should come as a surprise). The first step of the analysis is to identify and remove the trend and seasonal components. R provides a function to do just that. It is called stl. Unfortunately, it cannot be used with timeSeries objects, so we wrote a wrapper called sstl whose usefulness we proceed to illustrate. The following R commands were used to produce the plots contained in the following three figures.

```
co2.stl <- sstl(co2.ts,FREQ=12,TWIND=0.2,TDEGREE=1)
plot(co2.stl$trend)
plot(co2.stl$sea[1:120])
```

```
plot(co2.stl$rem)
```

The function `sstl` returns a list of three `timeSeries` objects named `trend`, `sea` and `rem` for the trend, seasonal and remainder components respectively. We set the parameter `FREQ` to 12 because our data are monthly and we expect yearly cycles in the data fluctuations. The parameters `TWIND` and `TDEGREE` give respectively, the size of the sliding window (as a proportion of the total length of the series), and the degree of the smoothing function used in `loess` involved in the regression procedure extracting the seasonal component. The plots produced by the three `plot` commands above are given in Figs. 6.9–6.11, respectively.

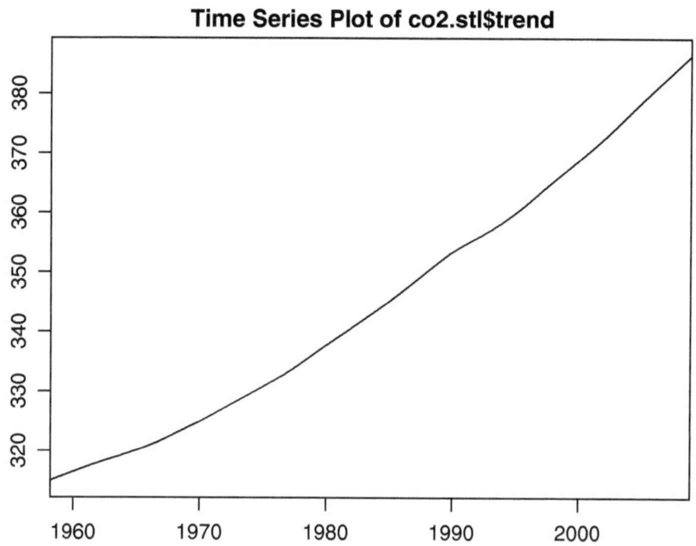

Fig. 6.9. Plot of the trend `timeSeries` object created by the function `sstl` for the CO_2 concentration data

The true seasonal component is possibly smoother than the periodic time series `co2.stl$sea`. However, for the purpose of the present analysis, we shall use this seasonal component despite its undesirable roughness.

Assuming that the remainder time series is stationary seems like a reasonable assumption, and despite the lack of quantitative evidence, we shall assume that this is indeed the case.

We shall come back to this example and continue its analysis after we introduce the auto-regressive and moving average models.

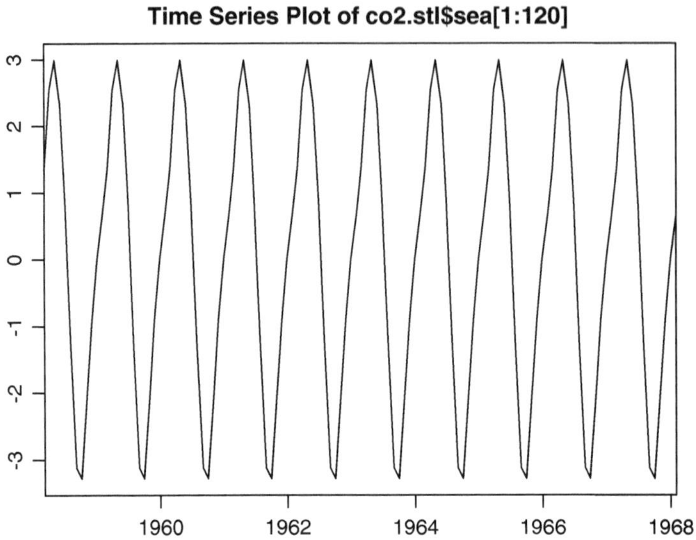

Fig. 6.10. Plot of the first 10 years (i.e. 120 first entries) of the seasonal `timeSeries` object created by the function `sstl` for the CO_2 concentration data. Notice that by construction, the mean of this component is zero

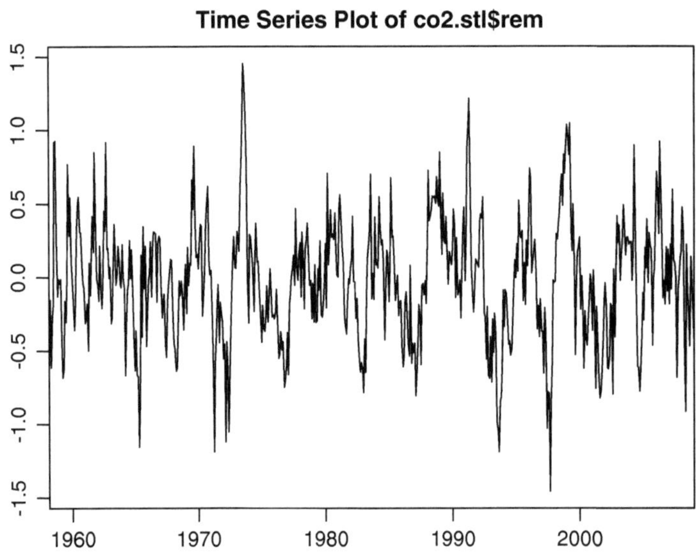

Fig. 6.11. Plot of the remainder `timeSeries` object created by the function `sstl` for the CO_2 concentration data

6.3 First Examples of Models

We review the most commonly used examples of time series models. We start with the fundamental example of a white noise which we encountered several times already, and which will serve as a crucial building block for the more sophisticated models studied in this chapter.

6.3.1 White Noise

A finite sequence

$$w_0, w_1, \ldots\ldots, w_n$$

of real numbers is said to form a white noise if they are observations from a (possibly infinite) sequence of independent and identically distributed (i.i.d. for short) mean zero random variables, say $\{W_t\}_t$. Such a sequence is obviously stationary and $\mu_W = \mathbb{E}\{W_t\} = 0$ by definition. Also, if $s \neq t$ we have:

$$\text{cov}\{W_s, W_t\} = \mathbb{E}\{W_s W_t\} = 0$$

because of the independence assumption. Consequently:

$$\gamma_W(h) = \begin{cases} \sigma^2 \text{ if } h = 0 \\ 0 \quad \text{otherwise} \end{cases} \tag{6.13}$$

where σ^2 denotes the common variance of the random variables W_t.

6.3.1.1 *Simulation*

We summarize the discussion presented in the introductory session to `R` reproduced in an appendix at the end of the book. We generate a numeric vector of length $N =$ 1,024 of realizations of independent Gaussian random variables with mean 0 and variance $\sigma^2 = 4$ with the `R`- command `WN <- rnorm(1024,0,4)`. We give the plot of this white noise sequence in Fig. 6.12 which also contain a separate plot of the first 70 entries of this white noise vector. The top pane shows the entire vector of length 1,024, while the bottom pane zooms into a sub-vector of length 70. The appearance of this second plot is of a much smoother looking curve. This shows that the wiggly nature of the plot, as well as its roughness, depends strongly upon the scale used to look at the series.

6.3.1.2 *Testing for White Noise*

The need to check that the residuals of a fitted model form a white noise is going to be a nagging problem recurring throughout the remainder of the book. For the purposes of the present discussion, we assume that these residuals form a stationary series. We tackle the problem of the serial correlation with a graphical tool first (see first bullet point below), and then, with a quantitative test in the third bullet point. We

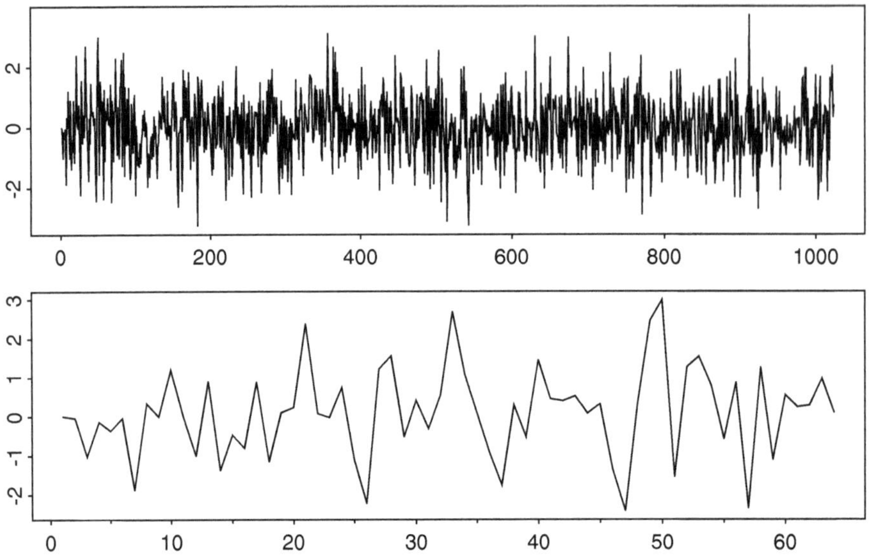

Fig. 6.12. Gaussian white noise time series with variance $\sigma^2 = 4$

also discuss the issue of identification of the marginal distribution of the individual terms in the second bullet point.

- The first available tool is graphical. In the same way Pavlov's dogs adjusted their behaviors, we should develop the reflex of plotting the auto-correlation function of any time series we bump into. The estimate which we can compute from the data makes sense when the time series is stationary, but its plot may be useful even when the time series is not. The following R commands can be used to compute and plot the auto-covariance function and the auto-correlation function of the white noise time series created earlier.

```
WNacov <- acf(WN,40,type="covariance")
WNacor <- acf(WN,40,type="correlation")
```

The results are given in Fig. 6.13. The looks of the two plots are identical. After all, the values of these two functions differ only by a scaling factor: the common variance of all the entries of the series. In other words, these two plots only differ through the scales on the vertical axis. Indeed the auto-correlation function is normalized to start from the value 1 for lag 0. Notice also the presence of a band around the horizontal axis of the auto-correlation function. It gives approximate 95 % confidence limits. These limits should not be taken too seriously. The confidence interval is only approximate: it is given as a tool to identify the values which are not significantly different from 0. Let us emphasize once more that, precisely because of the definition (6.8) of the estimated auto-covariance function (and auto-correlation function as well) it is not reasonable to compute the estimates for too large a value of the lag. Indeed, for the estimate to be useful, it needs to be based on a summation containing as many terms as possible. And since this number of terms is limited by the value of the lag, a compromise has to be reached. In this example we chose to compute and plot the acf coefficients for 40 lags. The maximum number of lags for which these functions are computed is usually chosen as a multiple of the logarithm of the length of the time series. Since the auto-correlation function of a white noise is non-zero only for lag zero, the confidence band should contain all the values except for the first one. This visual test is useful in detecting obvious correlations.

- Plotting a histogram (and a density estimate) of the entries of the series should be done to check for normality. In fact as we saw in the first chapter, a Q-Q plot is preferable. Such a plot could give an indication that the marginal distribution is Gaussian, and hence that the series entries could actually be independent (which is, as we know, much stronger than uncorrelated). Even though this graphical test could be complemented by a goodness of fit test (such as a χ^2 or a Kolmogorov-Smirnoff test), it will only give information on the marginal distribution: it is very difficult to test for *joint normality* of all the marginal distributions! If the normality of the marginal distributions were to be rejected, the apparatus introduced in the first chapter would be needed to fit GPD's.
- There are several powerful tests for the white noise hypothesis. The most common ones come under the name of *portmanteau* tests. They are based on the fact that, under the null hypothesis of a Gaussian white noise, appropriately weighted sums of the squares of the estimates of the auto-correlation function should follow a χ^2 distribution. `R` has one such test, but unfortunately, it is buried in the residual analysis of the fit to an ARIMA model. See Sect. 6.4.2 below for details.

*6.3.1.3 **Warning: The Different Forms of White Noise***

There are many instances in which the terminology white noise is used even though the i.i.d. assumption is violated. In these instances, one merely demands that (6.13) holds in order for the series to be called a white noise. The reason for this confusing practice is that this weaker assumption is enough to justify most of the *least squares procedures.* In order to better understand what is at stake here, let us introduce clear definitions for the various forms of white noise. A time series $\{W_t\}_t$ is said to be a white noise if it satisfies the properties given in the two bullets which follow:

- The W_t are mean-zero i.e. $\mathbb{E}\{W_t\} = 0$ for all t;
- They have the same variance and they are uncorrelated:
  - In the strong sense, i.e. when all the W_t's are independent;
  - In the weak sense, i.e. when $\mathbb{E}\{W_t W_s\} = 0$ whenever $s \neq t$.

In this chapter, and in most of the remainder of the book, all the white noise time series will be assumed to be white noise series in the strong sense, i.e. with independent terms, even if in practice, we only check for the weak form of the white noise definition. We shall revisit this convention later in the last chapter when we discuss the so-called ARCH models. Hopefully, at that time, we shall be able to shed enough light on this problem to clear up some of the remaining ambiguities.

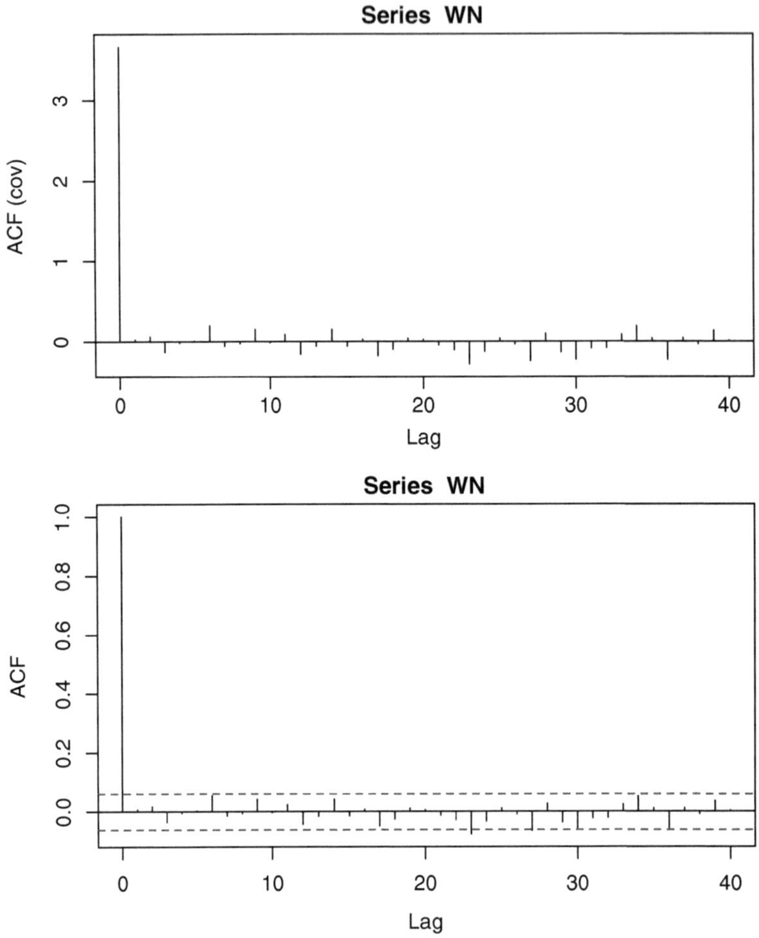

Fig. 6.13. Autocovariance (*top*) and autocorrelation (*bottom*) functions of a white noise time series

For a statistical analysis, reaching a white noise (in the strong sense) is the end of the *modeling road.* Indeed, the statistician massages the data to extract significant component, after significant component, ... until she gets residuals forming a white noise. Then she is done, unless of course these residuals still contain some structure which can be identified and extracted. This is only possible if this residual white noise is a white noise in the weak sense, not in the strong sense. But of course, one should not expect this task to be easy: after all, for all of its tools, linear analysis cannot go beyond a weak white noise. Extracting substance from its *guts* will require skill and finely-sharpened tools. We will give examples in our analysis of the ARCH/GARCH and stochastic volatility models later in Chap. 8.

6.3.2 **Random Walk**

In our introductory session to R, a random walk was defined as the integral of a white noise, and we used the R function cumsum to construct a sample of a random walk from a sample of a white noise. At the level of the models, we say that $\{X_n\}_n$ is a random walk if there exists a white noise $\{W_t\}_t$ such that:

$$X_n = X_0 + W_1 + W_2 + \cdots + W_n \ , \tag{6.14}$$

in other words, if the whole sequence $\{X_n\}_{n \geq 0}$ is determined by X_0 and the induction formula $X_{n+1} = X_n + W_{n+1}$. Usually, X_0 is assumed to be independent of the entire white noise sequence $\{W_t\}_t$. Notice that $\mu_X(n) = \mathbb{E}\{X_n\} = \mathbb{E}\{X_0\}$ for all n. However, even though the mean function is constant, the random walk is not stationary. Indeed:

$$\begin{aligned} \text{var}\{X_n\} &= \text{var}\{X_0\} + \text{var}\{W_1\} + \cdots + \text{var}\{W_n\} \\ &\quad + 2\text{cov}\{X_0, W_1\} + \cdots + 2\text{cov}\{X_0, W_n\} + 2\text{cov}\{W_1, W_2\} + \cdots \\ &\quad + 2\text{cov}\{W_1, W_n\} + \cdots\cdots \\ &= \text{var}\{X_0\} + n\sigma^2, \end{aligned}$$

which is obviously changing with n. Notice that the independence of the terms appearing in (6.14) guaranteed the fact that the variance of the sum is equal to the sum of the variances. Keep in mind that this is not true in general. Even though the random walk is not stationary, its first difference is. Indeed $X_t - X_{t-1} = W_t$ is a white noise (and hence is stationary). So $\{X_t\}_t \sim I(1)$. As we already mentioned, these processes are also called *root one* processes for reasons we shall discuss below.

6.3.2.1 ***Random Walk with Drift***

It happens quite often that the log-returns of market indexes have a small positive mean over significantly long periods of time. This remark is consistent with our discussion of the Samuelson's model for stock prices and stock indexes given in Sect. 5.7, and it justifies the introduction of the model

$$X_{n+1} = \mu + X_n + W_{n+1} \tag{6.15}$$

dubbed random walk with drift, the mean $\mu = \mathbb{E}\{X_{n+1} - X_n\}$ being called the drift. Summing both sides of (6.15) for different values of n, we obtain the analog of (6.14)

$$X_n = X_0 + n\mu + W_1 + W_2 + \cdots + W_n \ , \tag{6.16}$$

which shows that a random walk with drift is equal to a pure random walk (i.e. without a drift) plus a linear function with slope equal to the drift μ.

Random walk and random walk with drift models are not stationary. Their acf's decay very slowly. This behavior is very different from the fast decay of the acf's of the stationary models which we study in this book.

6.3.3 **Auto Regressive Time Series**

The random walk model satisfies the induction equation:

$$X_t = X_{t-1} + W_t$$

giving the value of the series at time t in terms of the preceding value at time $t-1$ and a noise term. This shows that X_t is a good candidate for a (least squares) regression on its past value X_{t-1}. We now define a large class of time series with this property.

A mean-zero time series $X = \{X_t\}_t$ is said to be auto-regressive of order p (with respect to a white noise $W = \{W_t\}_t$) if:

$$X_t = \phi_1 X_{t-1} + \phi_2 X_{t-2} + \cdots + \phi_p X_{t-p} + W_t \tag{6.17}$$

for some set of real numbers ϕ_1, ϕ_2, …, ϕ_p. More generally, we say that $X = \{X_t\}_t$ is auto-regressive of order p if there exists a number μ_X (which will necessarily be the common mean of the random variables X_t) such that the series $\{(X_t - \mu_X)\}_t$ is auto-regressive of order p in the sense given above. In any case, we use the notation $X \sim AR(p)$. AR models are very important because of their simplicity, and because of the efficient fitting algorithms which have been developed. We shall give several example of their usefulness.

6.3.3.1 ***Identification of the Coefficients***

We explain how to estimate the coefficients of a stationary autoregressive model when the order of the model is known. See the next subsection for a discussion of several possible ways to determine the order of the model. So we momentarily assume that we know the order of the autoregressive model, and we present a general strategy on a specific example: for the sake of definiteness, we assume that the order of the autoregressive series is equal to 2, and we try to estimate the coefficients of the model. Since the model is of the form:

$$X_t = \phi_1 X_{t-1} + \phi_2 X_{t-2} + W_t, \tag{6.18}$$

our goal is to estimate the values of ϕ_1 and ϕ_2 and possibly of the variance of the noise W_t. Multiplying both sides of this definition by X_t and taking expectations we get:

$$\begin{aligned}\mathbb{E}\{X_t^2\} &= \phi_1\mathbb{E}\{X_tX_{t-1}\} + \phi_2\mathbb{E}\{X_tX_{t-2}\} + \mathbb{E}\{X_tW_t\}\\ &= \phi_1\mathbb{E}\{X_tX_{t-1}\} + \phi_2\mathbb{E}\{X_tX_{t-2}\} + \phi_1\mathbb{E}\{W_tX_{t-1}\}\\ &\qquad + \phi_2\mathbb{E}\{W_tX_{t-2}\} + \mathbb{E}\{W_tW_t\}\end{aligned}$$

after re-injecting formula (6.18) into the last expectation of the first equality. First, we notice that $\mathbb{E}\{W_tX_{t-1}\} = 0$ and $\mathbb{E}\{W_tX_{t-2}\} = 0$ because X_{t-1} and X_{t-2} depend only on the past values W_{t-1}, W_{t-2}, W_{t-3}, … which are independent of W_t. Next we rewrite the remaining expectations in terms of the auto-covariance function γ_X. We get:

$$\gamma_X(0) = \phi_1\gamma_X(1) + \phi_2\gamma_X(2) + \sigma^2 \tag{6.19}$$

if we denote by σ^2 the variance of the white noise. Next, multiplying both sides of (6.18) by X_{t-1} and taking expectations we get:

$$\mathbb{E}\{X_tX_{t-1}\} = \phi_1\mathbb{E}\{X_{t-1}X_{t-1}\} + \phi_2\mathbb{E}\{X_{t-2}X_{t-1}\} + \mathbb{E}\{W_tX_{t-1}\}$$

or, in terms of the auto-covariance function γ_X:

$$\gamma_X(1) = \phi_1\gamma_X(0) + \phi_2\gamma_X(1), \tag{6.20}$$

where, as before, we used the fact that $\mathbb{E}\{W_tX_{t-1}\} = 0$ which holds because W_t and X_{t-1} are uncorrelated (remember that X_{t-1} is a function of the W_s for $s \le t-1$). Finally, multiplying both sides of (6.18) by X_{t-2} and taking expectations as before, we get:

$$\mathbb{E}\{X_tX_{t-2}\} = \phi_1\mathbb{E}\{X_{t-1}X_{t-2}\} + \phi_2\mathbb{E}\{X_{t-2}X_{t-2}\} + \mathbb{E}\{W_tX_{t-2}\}$$

or, equivalently:

$$\gamma_X(2) = \phi_1\gamma_X(1) + \phi_2\gamma_X(0) \tag{6.21}$$

since $\mathbb{E}\{W_tX_{t-2}\} = 0$ as well. Summarizing what we just did, assuming that we know the first few values of the auto-covariance function, the numbers $\gamma_X(0)$, $\gamma_X(1)$ and $\gamma_X(2)$ to be specific, we derived a system of three equations, (6.19)–(6.21), from which we can solve for ϕ_1, ϕ_2 and σ^2. These equations are called the *Yule-Walker equations* of the model. It is straightforward to solve them, even in the general case of a model of order p, since these equations form a linear system of $(p+1)$ equations with $(p+1)$ unknowns. The solution of this system provides a way to compute the coefficients ϕ_i and σ^2 of the models in terms of the first $p+1$ values $\gamma_X(0)$, $\gamma_X(1)$, $\cdots$, $\gamma_X(p)$ of the auto-covariance function γ_X. Replacing these theoretical values by estimates $\hat{\gamma}_X(k)$ computed as explained earlier, one obtains estimates $\hat{\phi}_i$ and $\hat{\sigma}^2$ of the parameters of the model. This method is very easy to implement, and in fact, it is the standard way to fit an auto-regressive model to sample data once the order has been determined.

6.3.3.2 *Finding the Order*

In some sense, estimating the order of an AR is a special case of the estimation of the dimension of a linear model. So no one will be surprised if the method of choice for such an estimation is based on a parsimonious balance between a quantitative measure of fit to the data, and the dimension of the proposed model (typically the number of parameters needed). When fitting general models, as in the case of linear models, information criteria are most commonly used. We shall abide by the rule and use a form of the AIC criterion. However, despite its rather universal character, the AIC criterion is subsumed by more powerful tools in specific cases. As we are about to explain, theoretical properties of the partial auto-correlation function will identify features to look for in order to make sure that an autoregressive model is appropriate, while at the same time, giving us a sharp estimate of the order of the model. Moreover, when we try to fit a moving average model in next subsection, we will see that a vanishing auto-correlation function will be an indication for a moving average model, and the lag at which it vanishes will provide a sharp estimate of the order.

As we are about to see, the best tool for identifying the order of an AR series is the partial auto-correlation function.

6.3.3.3 *Using the Partial Auto-Correlation Function to Find the Order*

The concept of auto-regressive process can be used to enlighten the definition of partial auto-correlation function. Indeed, the definition (6.17) of an auto-regressive process and the definition (6.3) of the best linear prediction operator say that the partial auto-correlation coefficient $\phi_{k,k}$ is the last coefficient obtained when one tries to force an $AR(k)$ model on $\{X_t\}_t$ (whether or not $\{X_t\}_t$ is an auto-regressive process). For this reason, if $X \sim AR(p)$, we should have:

$$\phi_{k,k} = 0$$

whenever $k > p$. This property has very important practical implications, for, if one tries to find out if a time series $x_0, \ldots, x_n$ is a sample from an $AR(p)$, and if one can estimate the partial auto-correlation coefficients $\phi_{k,k}$'s from the data, these estimates should be zero (or essentially zero), whenever k is greater than the order p. This property will be used to suggest auto-regressive models with a specific order. Remember that, according to (6.4)–(6.6), γ_X and $\hat{\gamma}_X$ completely determine the partial acf's $\{\phi_{kk}\}_k$ and $\{\hat{\phi}_{kk}\}_k$ respectively.

In `R`, the partial auto-correlation function is computed by including the option `"partial"` in the command `acf`. This option instructs the program to evaluate formula (6.6).

6.3.3.4 *Prediction*

Let us assume that the time series $\{X_t\}_t$ is an $AR(p)$ process for which we know the order p, the coefficients $\phi_1, \ldots, \phi_p$ and the variance σ^2 of the noise. Let us also assume that we have observed the outcomes $X_s = x_s$ of the series up to now, i.e. for $s \leq t$. Our goal is to predict the future values $X_{t+1}, X_{t+2}, \ldots$ of the series. We shall denote by $\hat{X}_{t+1|t} = E_t(X_{t+1})$, $\hat{X}_{t+2|t} = E_t(X_{t+2})$, ... these predictions. Recall the notation $E_t(\,\cdot\,)$ for the prediction operator introduced earlier. As we already mentioned, it could be interpreted as the conditional expectation given the information of all the past observations up to and including time t, or as the orthogonal projection onto the linear space generated by all the random variables known at that time. In any case, rewriting the definition of the $AR(p)$ model as:

$$X_{t+1} = \phi_1 X_t + \cdots + \phi_p X_{t-p+1} + W_{t+1}$$

and applying the prediction operator to both sides of the definition we get:

$$\hat{X}_{t+1|t} = \phi_1 X_t + \cdots + \phi_p X_{t-p+1}$$

since $E_t(W_{t+1}) = 0$ by the very definition of a white noise, and since $E_t(X_s) = X_s$ if $s \le t$. Similarly, rewriting the definition as:

$$X_{t+2} = \phi_1 X_{t+1} + \phi_2 X_t + \cdots + \phi_p X_{t-p+2} + W_{t+2}$$

and applying once more the prediction operator to both sides, we get:

$$\begin{aligned}\hat{X}_{t+2|t} &= \phi_1 \hat{X}_{t+1|t} + \phi_2 X_t + \cdots + \phi_p X_{t-p+2} \\ &= \phi_1(\phi_1 X_t + \cdots + \phi_p X_{t-p+1}) + \phi_2 X_t + \cdots + \phi_p X_{t-p+2} \\ &= (\phi_1^2 + \phi_2) X_t + (\phi_1\phi_2 + \phi_3) X_{t-1} + \cdots + \phi_1\phi_p X_{t-p+1}.\end{aligned}$$

This procedure can be repeated at will, and we can compute the prediction $\hat{X}_{t+k|t}$ for any prediction horizon k. Such a prediction appears as a linear combination of the p most-recently observed values of the series with coefficients computed inductively from the coefficients of the model. Also, because these predictions are given by explicit formulae, one can compute a confidence interval for each of them. Unfortunately, these predictions converge very fast toward the mean of the series, zero in the present situation. So predictions will be uninformative (i.e. plainly equal to the mean) for long prediction horizons.

The recursive formulae giving the predictions for all the finite horizons are easy to program. However, we will not need to write such a program as they are implemented in `R` by the generic function `predict`. We illustrate its use, together with the anti-climatic convergence of the prediction toward the mean of the signal, in Sect. 6.5.3 when we discuss options on the temperature.

6.3.3.5 *Monte Carlo Simulations & Scenarios Generation*

Random simulation should not be confused with prediction. For example, it would be unreasonable to add a white noise to a series of predicted values to create a Monte Carlo sample from the series. Proceeding in this way may seem silly, but it is a common error with novices.

For the sake of the present discussion, we assume as before that we know the order and the parameters of the model, as well as the values of X_t, X_{t-1}, ..., X_{t-p+1}, and that we would like to generate N Monte Carlo sample scenarios for the values of X_{t+1}, X_{t+2}, ..., X_{t+M}.

The correct procedure is to generate N samples of a white noise time series of length M, say $\{W_s\}_{s=t+1,\ldots,t+M}$, with variance σ^2, and then to use the parameters ϕ_1, ϕ_2, ..., ϕ_p and the definition of the AR(p) model to generate samples of the AR model from these N samples of the white noise. In other words, for each of the N

given samples $W_{t+1}^{(j)}, \ldots, W_{t+M}^{(j)}$ of the white noise, we generate the corresponding Monte Carlo scenarios of the series (which we denote with a tilde) by computing recursively the values $\tilde{X}_{t+1}^{(j)}, \ldots, \tilde{X}_{t+M}^{(j)}$ from the formula:

$$\tilde{X}_{t+k}^{(j)} = \phi_1 \tilde{X}_{t+k-1}^{(j)} + \phi_2 \tilde{X}_{t+k-2}^{(j)} + \cdots + \phi_p \tilde{X}_{t+k-p}^{(j)} + W_{t+k}^{(j)}, \qquad k = 1, 2, \ldots, M,$$

for $j = 1, 2, \ldots, N$, given the fact that the "tilde"s over the X's (i.e. the simulations) are not needed when the true values are available, i.e. $\tilde{X}_{t+k-p}^{(j)} = X_{t+k-p}$ whenever $k \leq p$.

This simulation procedure is very easy to implement. `R` does it as part of a more general procedure called `arima.sim` which can be used for more general ARIMA models. We give the details of its use later in Sect. 6.4.2.

6.3.4 Moving Average Time Series

A time series $X = \{X_t\}_t$ is said to be a moving average time series of order q (with respect to a white noise $W = \{W_t\}_t$) if:

$$X_t = W_t + \theta_1 W_{t-1} + \theta_2 W_{t-2} + \cdots + \theta_q W_{t-q} \tag{6.22}$$

for some real numbers $\theta_1, \theta_2, \ldots, \theta_q$. In such a case we use the notation $X \sim MA(q)$.

While the definition formula of an AR process was recursive, the definition formula (6.22) is explicit in terms of the white noise, and as a consequence, some properties of the MA time series can be derived easily from the very form of this formula.

- MA *series are stationary.* This is an immediate consequence of the stationarity of the white noise and the fact that X_t bears to $(W_t, W_{t-1}, \ldots, W_{t-q})$ the same relation as X_{t+t_0} to $(W_{t+t_0}, W_{t+t_0-1}, \ldots, W_{t+t_0-q})$;
- *The auto-covariance function and the auto-correlation function of an MA(q) series vanish for lags greater than q.* Indeed, the definition formula (6.22) shows that X_{t+s} only depends upon W_{t+s}, W_{t+s-1}, …, W_{t+s-q}. Consequently, if $t+s-q > t$ (i.e. if $s > q$), this set of white noise terms is disjoint from the set of terms on which X_t depends. This shows that in this case, the random variables X_{t+s} and X_t are independent and $\gamma_X(s) = \mathrm{cov}\{X_{t+s}, X_t\} = 0$.

As we stated earlier, this last property will be instrumental in the identification of the order of the model.

6.3.4.1 *Prediction and Simulation*

Simulation of an MA process is trivial. Indeed it is straightforward to follow the definition of the process to generate the values of the desired samples of the MA from the samples of the white noise. The prediction problem is more delicate, and we shall not dwell on it here because of its technical nature. Nevertheless, anticipating a little bit what comes next, we would simply say that, if it were possible to invert an MA model and rewrite it as an AR model (quite likely with different coefficients) then one could use the procedure reviewed above in the case of AR processes. This off-the-wall idea, is not too far-fetched, and it is worth keeping it in mind when one reads the section on invertibility.

6.3.4.2 A Simulation Example

Even though the auto-correlation function is a powerful tool in itself, it is often very instructive to visualize the serial correlations in a graphical way. We illustrate this fact with the simple example of a simulated $MA(2)$ series. The following `R` – commands create a sample of size 1,024 from a normally distributed moving average time series with coefficients $\theta_1 = \theta_2 = 1$ and unit noise variance $\sigma^2 = 1$. This simulation is a naive implementation of the definition of a moving average process. We shall see later in this chapter, that `R` offers more powerful simulation methods for general ARIMA processes.

```
SNOISE <- rnorm(1026)
MA2 <- SNOISE[1:1024] + SNOISE[2:1025] + SNOISE[3:1026]
plot(SNOISE[1:1024], type="l")
plot(MA2,type="l")
```

The plots are reproduced in Fig. 6.14. The moving average series appears to be *smoother* than the white noise. As we pointed out in the Simulation part of Sect. 6.3.1, a comparison based on the roughness or smoothness of the series can be very misleading when series are plotted with different time scales. Fortunately, this is not the case here, and this impression is real. The plot of the auto-correlation function of the `MA2` series is given by the `R`-command:

```
acf(MA2)
```

The result is given in Fig. 6.15. It confirms what we already learnt: the auto-correlation function of an $MA(2)$ time series vanishes after lag 2. Figure 6.16 shows four scatterplots. It was produced with the command:

```
lag.plot(MA2,lags=4,layout=c(2,2))
```

From left to right and top to bottom, they show the scatterplots of all the possible values of the couples (X_t, X_{t-1}), the couples (X_t, X_{t-2}), the couples (X_t, X_{t-3}), and finally the couples (X_t, X_{t-4}) which can be formed from the data. Obviously, the point pattern in the first scatterplot indicates that any two successive entries in the series are correlated. The second scatterplot shows that a significant dependence still exists between entries two time lags apart. However, this dependence does not seem to be as strong. On the other hand, the circular patterns characteristic of the last two scatterplots suggest that there should be no correlation between entries three and four time units apart. This confirms what we learned from the serial correlations in a moving average process.

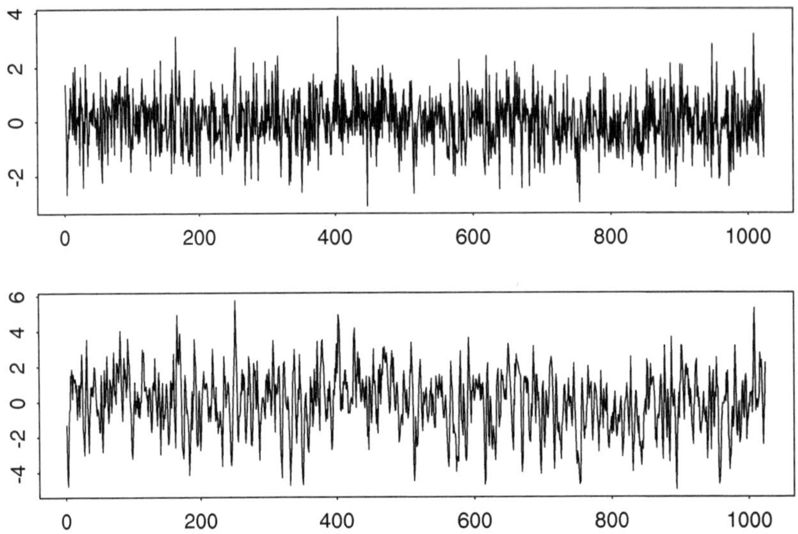

Fig. 6.14. White noise SNOISE (*top*) and a serially correlated MA(2) series (*bottom*) constructed from SNOISE

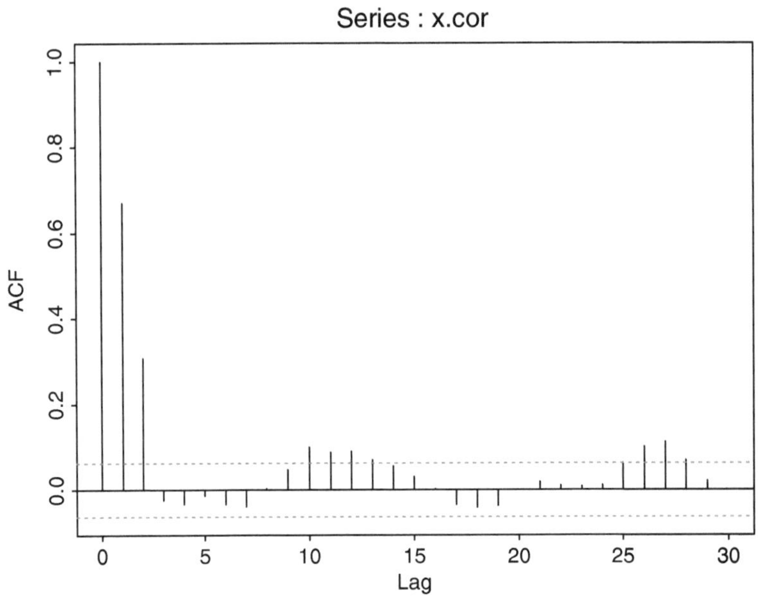

Fig. 6.15. Auto-correlation function of the serially correlated series MA2

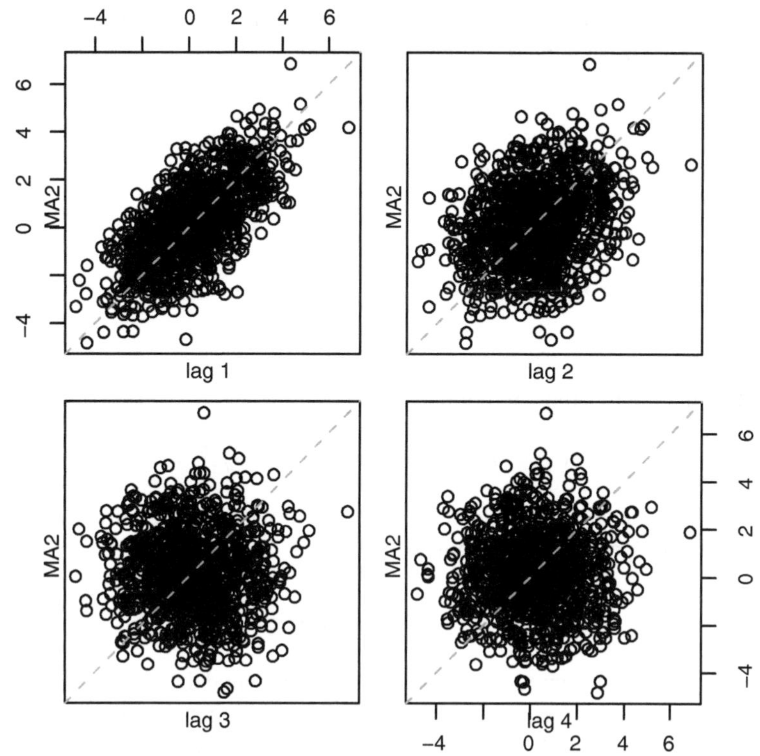

Fig. 6.16. Lag-plot of the series MA2 showing the serial dependence

6.3.5 Using the Backward Shift Operator B

The definitions given above (as well as the computations to come) provide another justification for the introduction of the backward shift operator B. If X is a time series, then BX is the time series which is equal to X_{t-1} at time t. In other words,

$$BX_t = X_{t-1}.$$

The operator B so defined is called the backward shift operator. For the record we notice that the differentiation operator ∇ can be rewritten in terms of the backward shift operator. Indeed, we have:

$$\nabla = I - B,$$

which follows from the fact that $\nabla X_t = X_t - X_{t-1} = IX_t - BX_t = (I - B)X_t$, if we denote by I the identity operator which leaves the time series unchanged. Applying the operator B twice to the time series X, we get:

$$B^2 X_t = BBX_t = BX_{t-1} = X_{t-2}.$$

By induction, we get that $B^k X_t = X_{t-k}$. The definition (6.17) of an auto-regressive time series of order p can be rewritten in the form:

$$\phi(B)X_t = W_t \tag{6.23}$$

where the function ϕ is the polynomial defined by:

$$\phi(z) = 1 - \phi_1 z - \phi_2 z^2 - \cdots - \phi_p z^p. \tag{6.24}$$

Similarly, the definition (6.22) of a moving average time series of order q can be rewritten in the form:

$$X_t = \theta(B)W_t \tag{6.25}$$

where the function θ is the polynomial defined by:

$$\theta(z) = 1 + \theta_1 z + \theta_2 z^2 + \cdots + \theta_q z^q. \tag{6.26}$$

The forthcoming analysis of auto-regressive and moving average time series relies heavily on the properties of these polynomials.

6.3.6 Linear Processes

The definitions of autoregressive and moving average models seem to indicate that these models are very different. This first impression is somewhat misleading, for in fact they are both members of the same family of linear models which we are about to define. Again, we restrict ourselves to the case of mean-zero time series for the sake of a simpler presentation. A mean-zero time series $X = \{X_t\}_t$ is said to be a linear process if there exists a white noise $W = \{W_t\}_t$ such that the representation:

$$X_t = \sum_{j=-\infty}^{\infty} \psi_j W_{t-j} \tag{6.27}$$

holds for some (possibly doubly infinite) sequence $\{\psi_j\}_j$ of real numbers satisfying $\sum_j |\psi_j| < \infty$. Restated in a developed form, the definition of a linear process says that:

$$X_t = \cdots\cdots + \psi_{-2}W_{t+2} + \psi_{-1}W_{t+1} + \psi_0 W_t + \psi_1 W_{t-1} + \psi_2 W_{t-2} + \cdots\cdots$$

and the condition imposed on the $|\psi_j|$'s is only here to guarantee that this doubly infinite sum has a meaning. Notice that the fact that this doubly infinite series converges is not a trivial fact, for the sequence of white noise terms W_t is not bounded in general, and consequently, multiplying it by a summable sequence may not be enough to give a meaning to the sum. It is only because of the independence of the terms of the white noise series, that the above definition of X_t is meaningful.

Linear processes are stationary by construction. Indeed, shifting the time indices for X amounts to shifting the time indices for the white noise and, since a time shifted white noise is still a white noise, the shifted version of X bears to the shifted white noise the same relationship as X bears to the original white noise W, so its distribution is the same. A particularly interesting class of linear processes is given by the processes for which $\psi_j = 0$ whenever $j < 0$. These processes are called *causal*. They are characterized by the fact that they are functions of the past of the white noise. Indeed, they are of the form:

$$X_t = \psi_0 W_t + \psi_1 W_{t-1} + \psi_2 W_{t-2} + \cdots\cdots .$$

In particular, we see that any moving average time series is causal since it is of this form by definition, the sum being in fact a finite sum.

6.3.6.1 *Auto Covariance Function (ACF)*

The computation of the auto-covariance function of a linear process goes as follows:

$$\begin{aligned}
\gamma_X(k) = \mathbb{E}\{X_t X_{t-k}\} &= \mathbb{E}\left\{\left(\sum_{i=-\infty}^{+\infty} \psi_i W_{t-i}\right)\left(\sum_{j=-\infty}^{+\infty} \psi_j W_{t+k-j}\right)\right\} \\
&= \sum_{i=-\infty}^{+\infty} \sum_{j=-\infty}^{+\infty} \psi_i \psi_j \mathbb{E}\{W_{t-i} W_{t+k-j}\} \\
&= \sigma^2 \sum_{\substack{-\infty < i,j < +\infty \\ t-i=t+k-j}} \psi_i \psi_j \\
&= \sigma^2 \sum_{i=-\infty}^{+\infty} \psi_i \psi_{k+i} \qquad (6.28)
\end{aligned}$$

Despite its apparent complexity, formula (6.28) can be extremely useful when it comes to actually computing the auto-covariance functions of the AR, MA and ARMA models.

Obviously, one can use formula (6.28) to compute the auto-covariance function of a *MA*(q) process just by inspection. Indeed one sees immediately that $\gamma_X(k)$ will be zero whenever $|k| > q$. Otherwise the value of $\gamma_X(k)$ is given by formula (6.28) where the summation is only taken from $i = 0$ to $i = q - |k|$.

It is also possible to use formula (6.28) to obtain a closed-form formula for the auto-covariance function of AR(p) and ARMA(p,q) models when p is small.

6.3.7 **Causality, Stationarity and Invertibility**

Recall that in this section, we concern ourselves with time series $X = \{X_t\}_t$ whose definition relies on a white noise series $W = \{W_t\}_t$.

We noticed that *MA*(q) time series were stationary and causal by definition. What about the $AR(p)$ series? In order to answer this question we consider first the particular case $p = 1$ for the sake of illustration. Iterating the definition we get:

$$\begin{aligned} X_t &= W_t + \phi_1 X_{t-1} \\ &= W_t + \phi_1 (W_{t-1} + \phi_1 X_{t-2}) \\ &= W_t + \phi_1 W_{t-1} + \phi_1^2 X_{t-2}. \end{aligned}$$

We can continue the procedure, using the definition to replace X_{t-2}. We get:

$$\begin{aligned} X_t &= W_t + \phi_1 W_{t-1} + \phi_1^2 (W_{t-2} + \phi_1 X_{t-3}) \\ &= W_t + \phi_1 W_{t-1} + \phi_1^2 W_{t-2} + \phi_1^3 X_{t-3} \end{aligned}$$

and reiterating the substitution n times we get:

$$X_t = W_t + \phi_1 W_{t-1} + \phi_1^2 W_{t-2} + \cdots + \phi_1^n W_{t-n} + \phi_1^{n+1} X_{t-n-1}.$$

At this stage we would like to take the limit $n \to \infty$, and get a representation of the form:

$$X_t = W_t + \phi_1 W_{t-1} + \phi_1^2 W_{t-2} + \cdots + \phi_1^n W_{t-n} + \cdots \tag{6.29}$$

which would guarantee, if the above infinite sum converges, that the time series X is linear (thus stationary) with a causal representation. The above sum is convergent when $|\phi_1| < 1$, while it diverges in the case $|\phi_1| > 1$. This result is due to the geometric nature of the sequence of coefficients of the shifted white noise. In the case $\phi_1 = 1$, X_t is a random walk, and we know that we cannot have stationarity. The case $\phi_1 = -1$ is the same because the minus sign can be absorbed in the white noise term.

The above result is quite general when restated appropriately. In full generality it says that, an $AR(p)$ time series is stationary and causal when the infinite series (6.29) converges. In fact, it can be proved mathematically that this is the case when the (complex) roots of the polynomial $\phi(z) = 0$ lie outside the unit disk of the complex plane (i.e. are of modulii greater than 1).

To check that this is not different from what we just saw in the case $p = 1$, notice that in this case the polynomial $\phi(z)$ is given by $\phi(z) = 1 - \phi_1 z$, and consequently there is only one root, since there is only one root of the equation $1 - \phi_1 z = 0$, namely the number $z = 1/\phi_1$. We saw earlier that the series was stationary and causal if and only if $|\phi_1| < 1$. This is the same thing as $|1/\phi_1| > 1$, which is the condition given in terms of the roots of the polynomial $\phi(z)$.

When it makes sense, formula (6.29) is called the moving average representation (or MA representation) of the autoregressive process X. The existence of this moving average representation can be understood in the following light. If we use the definition of an $AR(p)$ time series in the form $\phi(B)X_t = W_t$, we can formally write:

$$X_t = \frac{1}{\phi(B)} W_t$$

and this would be the desired representation if the rational fraction $1/\phi(z)$ (which is just the inverse of a polynomial) could be written as an infinite power series (i.e. a series in powers of z). Indeed, if we revisit one more time the case $p = 1$, we see that $\phi(z) = 1 - \phi_1 z$ and so:

$$\frac{1}{\phi(z)} = \frac{1}{1 - \phi_1 z} = 1 + \phi_1 z + \phi_1^2 z^2 + \cdots$$

and, even though we shall not try to justify it, the root condition introduced earlier guarantees that the series converges.

The definition of causality states that X_t can be expressed as a function of W_t and its past values $W_{t-1}, W_{t-2}, \ldots$. Trying to give symmetric roles to the two series $X = \{X_t\}_t$ and $W = \{W_t\}_t$, one may wonder when W_t can be expressed as a function of X_t and its past values $X_{t-1}, X_{t-2}, \ldots$. If this is the case, we say that the time series X is *invertible*.

Recall that, according to its definition, a time series X is an $AR(p)$ if it has the representation:

$$X_t - \phi_1 X_{t-1} - \phi_2 X_{t-2} - \cdots - \phi_p X_{t-p} = W_t.$$

This form of the definition says that an $AR(p)$ series is always invertible in the sense of the above definition. What about moving average series? You have probably already noticed a form of duality between auto-regressive and moving average properties. This duality is real as we are about to see one more time. The situation of moving average processes with respect to invertibility is the same as the situation of auto-regressive processes with respect to causality. In order to see that clearly, we consider the simple case of $q = 1$. The definition of an $MA(1)$ series is $X_t = W_t + \theta_1 W_{t-1}$. It can be rewritten as:

$$W_t = X_t - \theta_1 W_{t-1}.$$

Iterating this definition, we replace W_{t-1} and we get:

$$W_t = X_t - \theta_1(X_{t-1} - \theta_1 W_{t-2}) = X_t - \theta_1 X_{t-1} - \theta_1^2 W_{t-2}$$

and if we play the same substitution game over and over we get:

$$W_t = X_t - \theta_1(X_{t-1} - \theta_1 W_{t-2}) = X_t - \theta_1 X_{t-1} - \theta_1^2 X_{t-2} - \cdots \tag{6.30}$$

the sum of this infinite series making sense only when $|\theta_1| < 1$. When this is the case, formula (6.30) is called the auto-regressive representation (or the AR representation) of X. As in the case of moving average representations, this result is quite general. It applies to all the moving average processes as long as the coefficients are such that the above series converges, and this is indeed the case when all the (complex) roots of the characteristic polynomial $\theta(z)$ are outside the unit disk (i.e. are of modulus greater than 1).

Let us show the duality in action one more time. If we rewrite the definition of a $MA(q)$ time series as $X_t = \theta(B) W_t$, then the method used above (in the particular

	AR(p)	MA(q)
Stationarity causality	$\lvert z\rvert > 1$ if $\phi(z) = 0$	Always
Invertibility	Always	$\lvert z\rvert > 1$ if $\theta(z) = 0$

Table 6.1. Table of the *invertibility, causality and stationarity* properties illustrating the duality between the auto-regressive and moving average processes

case of a $MA(1)$ model) to show invertibility of the model, can be viewed as a way to rewrite the left hand side of:

$$\frac{1}{\theta(B)} X_t = W_t$$

as an infinite series in positive powers of B. When this infinite series can be derived as a convergent infinite expansion, it provides an AR-representation on which the invertibility condition appears clearly (Table 6.1).

Remarks.

1. The need for the statistical estimates provided by time averages was convenient for justifying the importance of stationarity. At this stage of the analysis, it is practically impossible to justify the fuss about causality and invertibility: why would we care so much about the fact that X_t is a function of the past values of the noise, and why would we need to know that the noise W_t is in turn a function of the past values of the observed series. The importance of these issues will become clear in Chap. 7 when we consider partially observed systems, and filtering issues. For the time being, the reader will have to take our word for it: these matters are of crucial importance in practical applications.
2. Notice that adding a constant to a model can have very different effects depending on the model in question. For example, adding a constant term to a regression amounts to adding an intercept. Adding a constant to the equation of a moving average model merely changes the constant mean of the series. On the other hand, adding a constant term to the equation of an auto-regressive model adds a linear drift, i.e. a linear function of time to the series.

6.3.7.1 ***Unit Root Test***

Let us assume that the logarithms of a stock price follow either one of the two models

$$X_t = \phi_1 X_{t-1} + W_t$$
$$X_t = \phi_0 + \phi_1 X_{t-1} + W_t$$

for some strong white noise $\{W_t\}$. Testing whether the log-price is a random walk is testing the null hypothesis $H_0 : \ \phi_1 = 1$ against the alternative $H_1 : \ \phi_1 < 1$.

Notice that when, $\phi_0 = 0$, this alternative guarantees a stationary time series. Such a test is called a unit-root test. Given observations $x_1, \ldots, x_n$ of the log-prices, the least squares estimates of ϕ_1 and of the variance σ^2 of the white noise are given in either model by:

$$\hat{\phi}_1 = \frac{\sum_{i=1}^n x_i x_{i-1}}{\sum_{i=1}^n x_{i-1}^2}, \quad \text{and} \quad \hat{\sigma}^2 = \frac{1}{n-1}\sum_{i=1}^n (x_i - \hat{\phi}_1 x_{i-1})^2 \tag{6.31}$$

where we set $x_0 = 0$ by convention. The quantity:

$$DF = \frac{\hat{\phi}_1 - 1}{\sqrt{\hat{\sigma}^2}} \tag{6.32}$$

is called the Dickey-Fuller test statistics. Its distribution converges when the sample size n grows indefinitely, and the limit can be identified both in the case $\phi_0 = 0$ and in the case $\phi_0 \neq 0$, leading to tests of the hypothesis H_0 against H_1. Problem 6.11 is devoted to a direct Monte Carlo computation of the critical values of the test. The library `Rsafd` contains a function `DF.test` performing unit root tests on numeric vectors and univariate `timeSeries` objects. Its use is illustrated in the next subsection below.

6.3.8 **ARMA Time Series**

A time series $X = \{X_t\}_t$ is said to be an auto-regressive moving average time series of order p and q if there exists a white noise $W = \{W_t\}_t$ such that:

$$X_t - \phi_1 X_{t-1} - \cdots - \phi_p X_{t-p} = W_t + \theta_1 W_{t-1} + \cdots + \theta_q W_{t-q} \tag{6.33}$$

for some real numbers $\phi_1, \ldots, \phi_p$, and $\theta_1, \ldots, \theta_q$. In such a case we use the notation $X \sim ARMA(p, q)$. Using the shift operator B this definition can be rewritten in the form:

$$\phi(B)X_t = \theta(B)W_t$$

for the polynomials $\phi(z)$ and $\theta(z)$ defined above in formulae (6.24) and (6.26) respectively. Since one can formally write:

$$X_t = \frac{\theta(B)}{\phi(B)}W_t \qquad \text{and} \qquad \frac{\phi(B)}{\theta(B)}X_t = W_t$$

one sees why the stationarity and causality properties only depend upon its auto-regressive part, and its invertibility only depends upon its moving average part. More precisely:

- The $ARMA(p, q)$ series is stationary and causal if all the (complex) roots of $\phi(z)$ have modulii greater than 1;
- The $ARMA(p, q)$ series is invertible if all the (complex) roots of $\theta(z)$ have modulii greater than 1,

since the use of the polynomials $\phi(z)$ and $\theta(z)$ shows that we have the stationarity and the causality of the $ARMA(p,q)$ series if we can prove the stationarity and causality of its $AR(p)$ part, while we have the invertibility of the $ARMA(p,q)$ series if we can prove the invertibility of its $MA(q)$ part.

Remark. By definition, the unit-root test introduced earlier was appropriate for AR models. An extension to ARMA models was proposed. It goes under the name of augmented Dickey-Fuller test. The function `DF.test` of the library `Rsafd` provides an implementation whose use we now illustrate.

```
TST <- DF.test(rnorm(1024))
TST$p.value
[1] 0.01
```

In fact, the series is *so stationary* that the function `DF.test` gives a warning stating that the p-value is in fact smaller than the output!

```
TST <- DF.test(co2.ts)
TST$p.value
[1] 0.7766632
```

which is certainly not a surprise given the rapidly increasing nature of the series.

6.3.9 ARIMA Models

A time series $\{X_t\}_t$ is said to be an ARIMA process if, when differentiated finitely many times, it becomes an ARMA time series. More precisely, one says that $\{X_t\}_t$ is an $ARIMA(p,d,q)$ if its becomes an $ARMA(p,q)$ after d differences. So using the notation introduced earlier:

$$X \sim ARIMA(p,d,q) \Longleftrightarrow \nabla^d X \sim ARMA(p,q).$$

Recall that the operator $\nabla = I - B$ is the first difference operator. Equivalently, using the definition of ARMA processes in terms of polynomials in the shift operator B we see that:

$$X \sim ARIMA(p,d,q) \Longleftrightarrow \phi(B)(I-B)^d X = \theta(B)W$$

for a white noise W and polynomials ϕ and θ.

6.4 Fitting Models to Data

The purpose of this section is to take advantage of the theoretical properties derived in the first part of the chapter in order to identify models appropriate for given time series data, and to fit these models to data.

6.4.1 **Practical Steps**

We now summarize the steps recommended for fitting the models we discussed in this chapter.

6.4.1.1 ***Searching for Stationarity***

The first step of a time series analysis is the identification and the removal of the trend and seasonal components. We should

1. Transform the data into a stationary time series if needed:
   - Remove trends (regression);
   - Remove seasonal components;
   - Differentiate successively;
2. Check if the data form a white noise, in which case model fitting is over.

We shall see in the last chapter that some forms of white noise in the weak sense are still amenable to model fitting, but for the purposes of this chapter, reaching a white noise is our stop criterion.

Once we transformed the original data into what we believe to be a stationary time series, one visualizes the serial correlation by computing and plotting the auto-correlation function. If its values vanish after a finite lag, then we fit a moving average model of order given by the last lag with a non zero correlation. If not we proceed to fitting an auto-regressive model.

6.4.1.2 ***Fitting an AR***

In order to fit an AR model we follow the steps given below:

1. Computation of the AIC criterion and the partial auto-correlation function to determine the order of the model;
2. Estimation of the auto-regression coefficients and of the variance of the noise by solving the Yule-Walker equations;
3. Computation of the residuals and testing for white noise to decide if the model fitting is complete or if it needs to be pursued further.

Even though this method is used by default in `R`, solving the Yule-Walker equations is not the only way to estimate the coefficients of an AR model. For example, `R` offers the option to use Burg's method as an alternative. This method is based on spectral and information theoretic arguments. We shall not discuss it here.

6.4.1.3 ***Fitting an MA***

In order to fit an MA model we follow the steps given below:

1. Computation of the auto-correlation function and checking that it vanishes after some lag;

2. Confirmation of the order of the model with the AIC criterion;
3. Estimation of the moving average coefficients by maximum likelihood;
4. As always, computation of the residuals and testing for white noise to decide if the model fitting is complete or if it needs to be pursued further.

If the order of the model has already been determined, `R` has a method for determining the coefficient of the model. We shall see how to use this function below.

6.4.1.4 *Fitting an ARMA*

In order to fit an ARMA model it is recommended to follow the steps:

1. Attempt to fit an AR model to the data and computation of the residuals;
2. Attempt to fit an MA model to the residuals of the AR model fitted first, or to the original data if the AR fit was not deemed satisfactory;
3. Using the AR-order p and the MA-order q determined by steps 1 and 2 above, fit of an ARMA(p,q) model by maximum likelihood;
4. Analysis of the residuals and testing for white noise.

6.4.1.5 *Fitting an ARIMA*

If the search for stationarity is done by successive differentiations, and if an ARMA model is fitted to the result, then we have fitted an ARIMA model, the order of which is the triplet (p, d, q) of integers, p standing for the order of the AR component, d for the number of differences computed to get to stationarity, and q for the order of the MA component.

So like Mister Jourdain was producing prose without knowing it, we have been fitting ARIMA models by first differentiating a time series until we reached a stationary series, and then fitting an ARMA model to the result.

6.4.2 `R` Implementation

We illustrate the prescriptions given above with the analysis of a couple of simple examples.

6.4.2.1 *Example of an AR-Analysis*

In order to illustrate how one fits an AR model, we revisit the example of the CO_2 concentration data. Remember that, after extraction of a trend and a seasonal components, we were left with a stationary looking remainder time series `co2.stl$rem` which we now try to model. As shown in Fig. 6.17, the plot of the auto-correlation function of `co2.stl$rem` produced by the function `acf` decays at first, but does not want to vanish for large lags. Fortunately, it does not show periodic serial dependence between successive entries of the series, typically with period 12, which would have been a strong indication that the seasonal component was not removed entirely.

Since this acf-plot does not seem to vanish after a specific lag, it does not point toward a moving average model, and it is reasonable to attempt to fit an auto-regressive time series first. A first attempt to fit an auto-regressive model should look like:

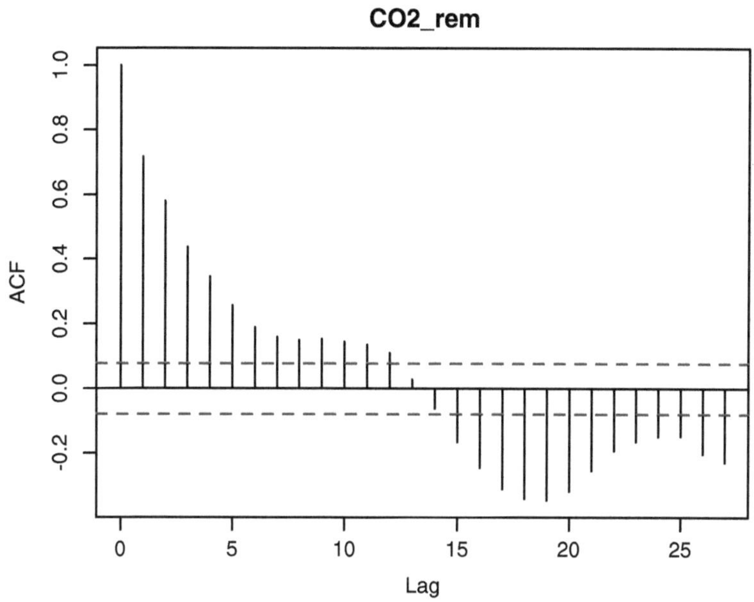

Fig. 6.17. Plot of the auto-correlation function of the timeSeries `co2.stl$rem` as produced by the command `acf(co2.stl$rem)`

```
co2.ar <- ar(co2.stl$rem)
co2.ar
Call:
ar(x = co2.stl$rem)
Coefficients:
       1        2        3        4        5        6
  0.5914   0.1571  -0.0277   0.0514  -0.0018  -0.0301
       7        8        9       10       11       12
  0.0147   0.0226   0.0436   0.0000   0.0406   0.0848
      13       14       15       16       17
 -0.0387  -0.0277  -0.0605  -0.0405  -0.0691

Order selected 17  sigma^2 estimated as  0.0812
```

R provides a powerful function `ar` to fit auto-regressive models. Using the command `ar` with the time series as only parameter gives the program total *carte blanche* for the choice of the order of the model. From a look at the output `co2.ar` of the program we see that the order 17 was chosen. In order to understand why, we plot the partial auto-correlation function with the command `acf(co2.stl$rem, type="partial")`. The result is shown in Fig. 6.18.

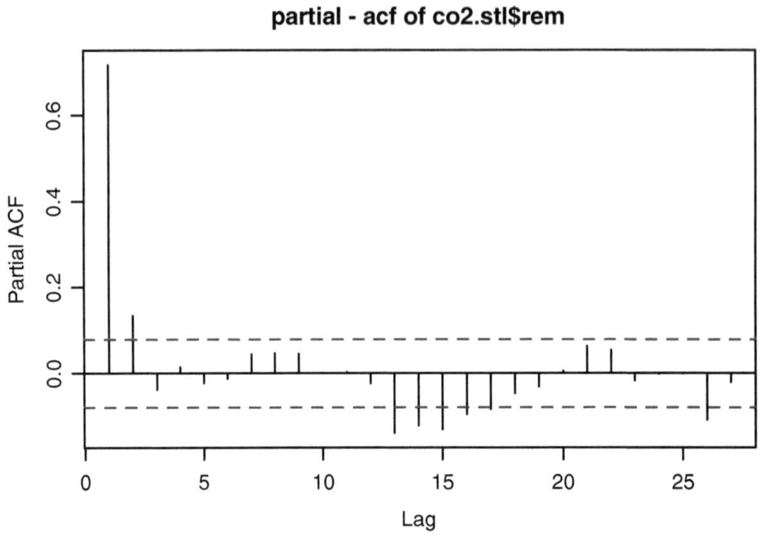

Fig. 6.18. Plot of the partial auto-correlation function of the timeSeries `co2.stl$rem` as produced by the command `acf(co2.stl$rem,type="partial")`

First, one clearly sees that the plot of a partial auto-correlation function produced by the function `acf` with the option `type="partial"` is very different from the plot of the mere auto-correlation function produced by the function `acf` without specifying the option `type="partial"`. Moreover, one sees that, except for one bar barely sticking out of the confidence band, all of the partial auto-correlation values in the confidence band after lag 17, explaining why the program chose such a value for the order of the model. However, it appears that not many partial auto-correlation values are significantly different from zero after lag 2, and it may be more parsimonious to work with an $AR(2)$ model, especially given the relatively small length of the data. Despite a poor readability due to the scale of the first values, this is confirmed by the plot of the AIC reproduced in Fig. 6.19. Remember that the AIC is a form of *universal* criterion used to determine the order of a model. Again, we understand why the program chose the order 17, but we also see why working with order 2 could be better. So we decide to fit an $AR(2)$ model to the remainder term.

```
myco2.ar <- ar(co2.stl$rem,order=2)
myco2.ar
Call:
ar(x = co2.stl$rem, order.max = 2)
Coefficients:
     1        2
0.6407   0.1517

Order selected 2  sigma^2 estimated as  0.08483
```

To convince ourselves that this was the right thing to do, we produce a sequential plot of the residuals of the model and their auto-correlation function Notice that we

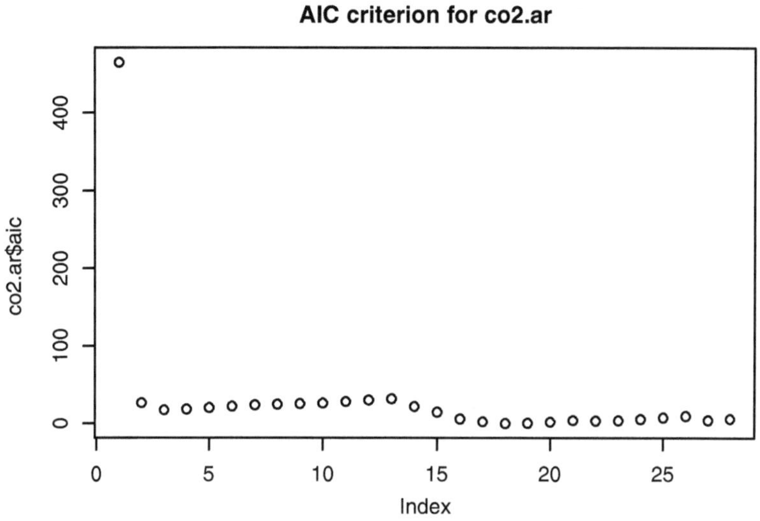

Fig. 6.19. Sequential plot of the AIC (Akaike Information Criterion) for the AR model for the remainder series `co2.stl$rem`

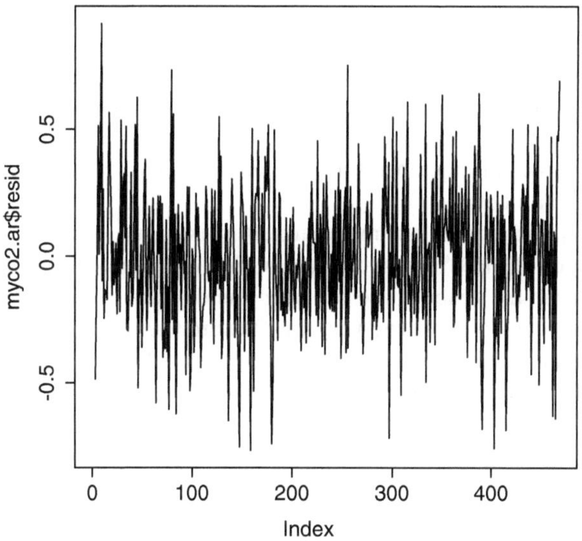

Fig. 6.20. Sequential plot of residuals of the $AR(2)$ model for the remainder series `co2.stl$rem` as given by the command `plot(myco2.ar$resid)`

compute the auto-correlation function of the vector of residuals starting with index 3. Indeed, because we work with an AR(2) model, the first two residuals are NAs, and their presence would crash the function `acf` (see title of Fig. 6.21).

We learn from the discussion of the statistical properties of the residuals of a linear model that raw residuals are heteroskedastic and that they are serially correlated.

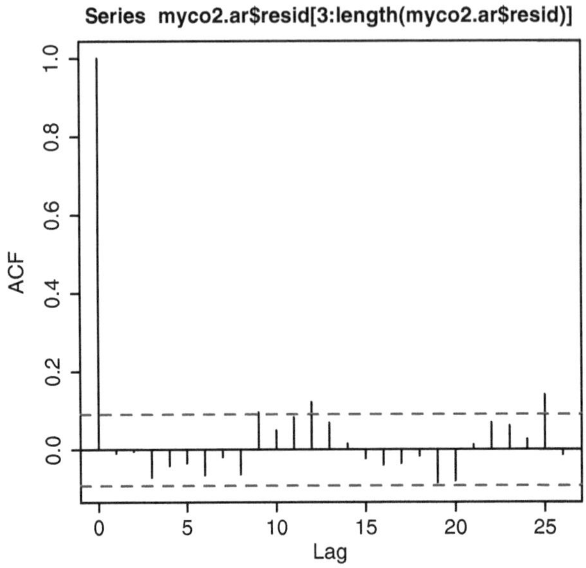

Fig. 6.21. Auto-correlation function of the residuals of the $AR(2)$ model given by the command `acf(myco2.ar$resid[3:length(myco2.ar$resid)])`

However, a look at the above plot seems to indicate that the auto-correlation function of these raw residuals is not much different from the auto-correlation of a white noise, and for the time being, we will consider that this is a sign that our statistical modeling is complete. Remember, this will not be the case when we stop restricting ourselves to linear models.

A natural application of our modeling effort is to use our fitted model to predict future values of the CO_2 concentration. We discussed earlier how to compute such predictions in the general case of $AR(p)$ models and we explained that, *unfortunately*, predictions do **converge very fast** toward the mean of the series (zero in the present situation of a remainder time series) when they try to look too far ahead in the future. Let us see a practical illustration of these ideas in the specific case at hand. In R, we use the generic function `predict` to compute predictions from a model. In particular, if we want to predict one year worth of monthly values of the CO_2 remainder time series we do:

```
myco2.pred12 <- predict(myco2.ar,
                newdata=seriesData(co2.stl$rem),n.ahead=12)
myco2.pred12
$pred
Time Series:
Start = 611
End = 622
Frequency = 1
  CO2_rem    CO2_rem    CO2_rem    CO2_rem    CO2_rem    CO2_rem
```

```
-0.427262 -0.371504 -0.313034 -0.267114 -0.228821 -0.197321
  CO2_rem    CO2_rem    CO2_rem    CO2_rem    CO2_rem    CO2_rem
-0.171329 -0.149896 -0.132221 -0.117645 -0.105624 -0.095711

$se
Time Series:
Start = 611
End = 622
Frequency = 1
[1] 0.29126 0.34592 0.38272 0.40524 0.41996 0.42967 0.43615
[8] 0.44050 0.44344 0.44543 0.44677 0.44768
```

The output of the function `predict` is a list with a component `$pred` for the values of the desired predictions, and `$se` for the estimations of the prediction errors. We may want to plot the predictions so-obtained with the command `plot(myco2.pred12$pred)`. The results are reproduced in Fig. 6.22. Note the need for the parameter `newdata` which has to be a vector or a matrix. Note also that the result is not a `timeSeries` object (hence the labels on the x-axis). To confirm what we mentioned several time on the anti-climatic lack of content of longer horizon predictions, we can compute the predictions for 10 years worth of remainder values (using the value `n.ahead=120`). The result is plotted in Fig. 6.23. The plot was produced with the commands

```
plot(myco2.pred120$pred)
abline(h=mean(seriesData(co2.stl$rem)))
```

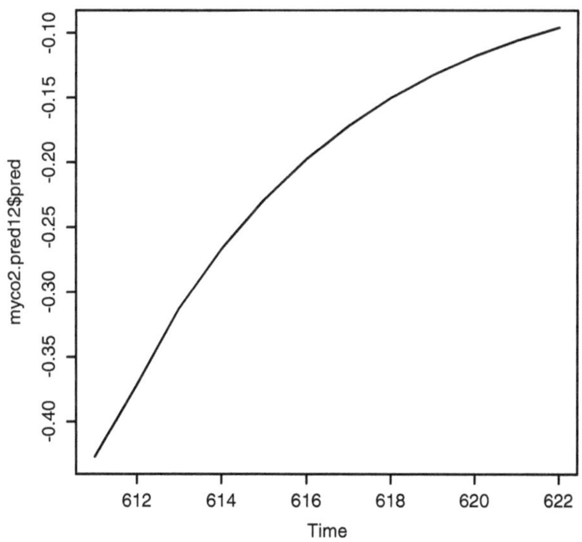

Fig. 6.22. Prediction of the values of the next 12 months for the $AR(2)$ model fitted to the CO_2 remainder component

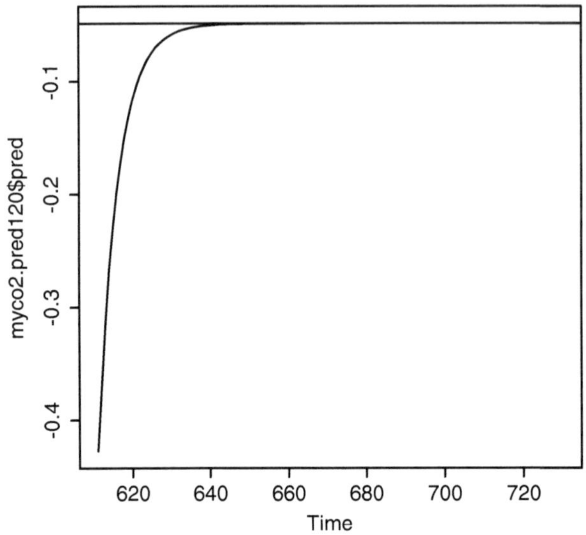

Fig. 6.23. Prediction of the values of the next 120 months for the $AR(2)$ model fitted to the CO_2 remainder component

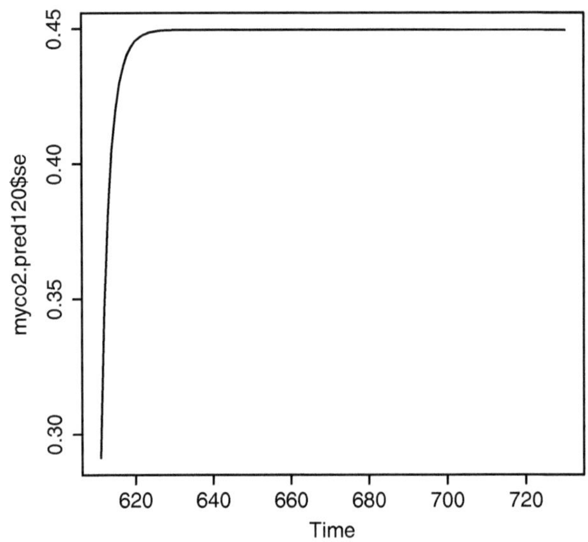

Fig. 6.24. Prediction of the errors on the predictions of the next 12 months for the $AR(2)$ model fitted to the CO_2 remainder component

Note the rapid convergence toward the mean. We can also plot the prediction errors using the command `plot(myco2.pred120$se)`. The results are reproduced in Fig. 6.24. We see that the estimates of the prediction errors are increasing fast before stabilizing near a steady asymptotic level.

6.4.2.2 Fitting a Moving Average Model

Instead of using a real life time series for which we never know whether or not a model is appropriate, we illustrate the fitting procedure with a simulated time series. In this case, we know in advance what to expect. We could use the series MA2 constructed earlier by bare hands, but in order to illustrate the simulation capabilities of R we choose to construct a brand new series. We use the function arima.sim to simulate a sample from an $MA(4)$ model and we immediately compute the empirical estimate of the auto-correlation function of the time series generated in this way.

```
MA4 <- arima.sim(1024,model=list(ma=c(-.5,-.25,.1,.6)))
plot(MA4,type="l")
acf(MA4)
```

In R, ARIMA models are given by specifying a parameter called model. The latter is a list. Its ar component gives the coefficients ϕ_i of the AR-polynomial $\phi(z)$ while the component ma gives the values of the coefficients θ_j of the MA-polynomial $\theta(z)$. We give further details below. The two plots are given in Fig. 6.25. The values of the auto-correlation function do not seem to be significantly different from zero after lag 4. This is consistent with a moving average model of order 4. Naturally, we then try to fit an $MA(4)$ model. If we assume that we only know the order of the moving average (and not the actual coefficients), we fit a univariate $MA(4)$ model

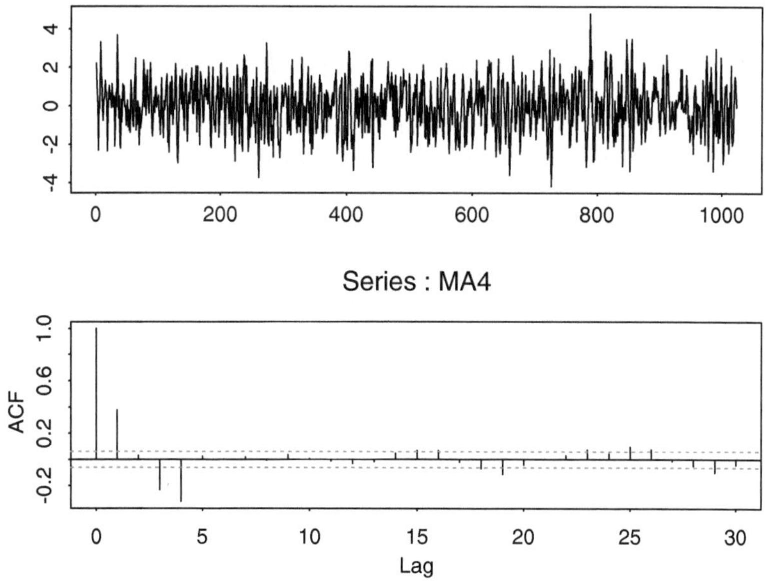

Fig. 6.25. *Top*: Moving Average (of order 4) simulated series. *Bottom*: Corresponding auto-correlation function. Notice that it is vanishing for lags larger than 4

with the R function arima which will be discussed in the next subsection. Printing the results gives the following output:

```
MA4FIT <- arima(MA4,order=c(0,0,4))
MA4FIT
 Call:
arima(x = MA4, order = c(0, 0, 4))

Coefficients:
          ma1      ma2     ma3     ma4  intercept
      -0.4941  -0.2300  0.0542  0.6328    -0.0138
s.e.   0.0247   0.0292  0.0284  0.0249     0.0300

sigma^2 estimated as 0.9975:
log likelihood = -1454.41,   aic = 2920.82
```

The object returned by the function arima contains the information about the model, together with the value of the AIC criterion, the maximum value of the log-likelihood as computed, the estimates of the coefficients (those at which the maximum of the likelihood was attained) and the corresponding standard errors. The experiment is satisfactory in the sense that as expected, the estimated coefficients are reasonably close to the actual coefficients used to generate the MA sample.

6.4.2.3 ***Fitting an ARIMA Model***

If we already know the order of an ARIMA model, in other words, if we know the three integers (d for the order of differentiation, p for the auto-regressive order and q for the moving average order), then it is easy to fit an ARIMA model by maximum likelihood. The R function arima will do that for us if we give the argument order=c(p,d,q). The output is an arima model object. Besides the input parameters, it has an attribute called model. The latter is a list with ar and ma components which give the ϕ and θ coefficients of the ARMA part of the model. See the appendix at the end of this chapter for a discussion of the sign conventions used by R.

Remark. Clearly, the function arima requires the knowledge of the order of the model. It cannot be used to fit an ARIMA model if one does not already know the order (p, d, q), it can only be used to fit an ARIMA model of a specific order.

6.4.2.4 ***Diagnostics***

One of the most insidious mischiefs of powerful computer statistical packages is the fact that it is always possible to fit a model to data, even when the model in question is not appropriate. So it is a good practice, once a model is fitted, to run diagnostics in search of confirmation. For example, after fitting an $ARMA(2,1)$ model to a time series TS, the resulting model can be diagnosed by the R function tsdiag.

```
> TS.fit <- arima(co2,order=c(2,0,1))
> tsdiag(TS.fit)
```

Unfortunately, diagnostics for time series fits are not as developed in `R` as they are in `S-Plus`. In particular, the generic function `tsdiag` does not return numerical statics. It merely plots the standardized residuals, the autocorrelation function of the residuals, and the p-values of the LjungBox version of the portmanteau test for all lags up to the value of the parameter `gof.lag`.

6.4.2.5 ***Simulation***

`R` offers a function to effortlessly produce samples of ARIMA time series. This function is called `arima.sim`. The parameters of the model need to be provided each time the function is called. These parameters should be in the form of the output of the function `arima`.

```
SIM <- arima.sim(model, n=100, innov=NULL, n.start=100)
```

The parameter `n` gives the length of the series to simulate, while `n.start` gives the number of generated values to discard. The use of `n.start` is desirable if we want to make sure that the simulation is representative of an equilibrium situation. By default, the function `arima.sim` uses a normal random sequence created with the function `rnorm` as a sample for the white noise from which the sample of the ARIMA model is created. When the `innov` parameter is provided, the simulation is not based on such a normal random sequence. Instead, the sequence `innov` is used for that purpose. This is especially useful if we want to repeat simulations with different features while keeping the same source of randomness, but it is also convenient if we want to simulate series with a non-Gaussian white noise with heavy tails. We can use the simulation techniques developed in Chap. 2 to simulate samples from a heavy tail distribution, and use such a sample as the parameter `innov`.

Other parameters can be used. The interested reader is invited to check the help file of the function `arima.sim`. See our analysis of the Charlotte temperature below for an illustration of the use of simulation for Monte Carlo computations of probabilities and expectations.

6.4.2.6 ***Prediction***

In the general case of an ARIMA model, like in the particular case of AR models, predictions of the future values are computed with the generic function `predict`. A typical call to this function has the form:

```
> PRED <- predict(fit, n.ahead = 6)
```

where the object `fit` is of class `arima`, i.e. has been created by the function `arima`, and where `n.ahead` is the number of consecutive future values of the series to be predicted, starting from the end of the series. The object `PRED` is a list containing a vector `PRED$pred` of length `n.head` for the actual predictions, and a vector `PRED$se` for the estimates of the standard errors of the predictions. See the analysis of the temperature at Charlotte NC for an example.

6.5 PUTTING A PRICE ON TEMPERATURE

There is a renewal of interest in temperature time series in many economic sectors. We believe that this current wave is mostly due to the great excitement caused by the growing use of weather derivatives for risk management, by the prospect of large speculative profits, and presumably by the mystery surrounding the difficult pricing issues associated with these instruments. A better understanding of the statistics of these series will help dissipate this mystery. This is what we attempt to do in this section.

Our goal is to develop an understanding of the statistics of these time series which would be appropriate for the pricing of meteorological derivatives. Options on heating and cooling degree days are actively traded on the *over the counter* (OTC for short) market. Moreover, as of September 22nd 1999, the Chicago Mercantile Exchange (CME, for short) started offering futures contracts on monthly indexes computed from the cumulative HDD's and CDD's in three cities, as well as options on these futures contracts. Contrary to our use of non-parametric regression to price options on the S&P 500 index, we shall not attempt to price temperature options in the same direct way. Price discovery is too much of a challenge in these markets, price data are too scarce, and illiquidity problems are *muddying the water* too much for us to take a chance at this touchy business. But before we actually start addressing the pricing issues, we devote the following two subsections to the introduction of the terminology and notation needed for the analysis of these financial instruments.

6.5.1 **Generalities on Degree Days**

Temperature data are readily available, especially on the internet. For a very large number of locations, usually US meteorological stations (near or at airports), daily temperatures are provided free of charge. The information is given in the form of a high and a low for each day. We use the standard notation $MinT$ and $MaxT$ for the minimum and the maximum temperatures on a given day. These temperatures are expressed in Farenheit degrees. From these, one usually computes the so-called average temperature $AvgT$ defined as:

$$AvgT = \frac{MinT + MaxT}{2}. \tag{6.34}$$

This definition may not correspond exactly to the intuitive notion we have of the average temperature on a given day. Nevertheless, it is the quantity which is used to define the important concepts of heating degree day (HDD for short) and cooling degree day (CDD for short). On each given day, say t, the heating degree day HDD_t is defined as the number:

$$HDD_t = (65 - AvgT_t)^+ = \max\{65 - AvgT_t, 0\}. \tag{6.35}$$

The intuitive notion of heating degree day is simple: this number tells us by how many degrees one should heat on day t in order to keep the temperature at a minimum level. The threshold at which one does not need to heat any longer is (arbitrarily) set at the level 65. Apparently, this is the temperature at which the furnaces used to be turned on. But let's face it, the notion of temperature at which one should start the heater is very subjective: residents of Southern California and Minnesota may not agree on the choice of this threshold. In any case, the value of 65 is generally accepted, and it is used in all degree days computations. We shall accept it as is. Formula (6.35) says that $HDD_t = 0$ on days where the average temperature (as given by $AvgT_t$) is greater than 65, and that on the other days, HDD_t is equal to the amount by which the average temperature falls under 65. Similarly, the cooling degree day CDD_t is defined as the number:

$$CDD_t = (AvgT_t - 65)^+ = \max\{AvgT_t - 65, 0\}, \tag{6.36}$$

and its intuitive interpretation is similar. Some daily temperature services provide daily values for $AvgT$, HDD and CDD along with the values of $MinT$ and $MaxT$.

Given a time period P, which can be a month, or a summer, or any other kind of pre-specified time period, the cumulative numbers of degree days for this period are defined as:

$$HDD^{(P)} = \sum_{t \in P} HDD_t \qquad \text{and} \qquad CDD^{(P)} = \sum_{t \in P} CDD_t.$$

As we shall see shortly, these quantities are the *underlying indexes* on which the temperature futures and option contracts are written.

6.5.2 **Temperature Options**

There are many kinds of weather related financial instruments. We already mentioned catastrophic bonds when we discussed the PCS index in Chap. 2. In this section, we concentrate on options on the temperature. We motivate their introduction with a few simple examples chosen only for the sake of illustration. Gas retailers are bound to lose money during mild winters because of low demand. Notice also that extremely cold winter can also be the source of losses since, tied by delivery contracts, the gas retailer may have to buy gas at a prohibitive price on the spot market to satisfy an unexpectedly high demand. Similarly, electric power utility companies suffer from cool summers because of lower demand, but they also suffer from too hot a summer when they are forced to purchase electricity on the spot market to satisfy an unexpectedly high demand. The highly speculative and volatile nature of the electricity spot market, especially in the de-regulated markets, has been a source of great concern both for users and producers. So it seems natural to expect that gas producers and retailers will try to protect themselves against warm winters, and electricity producers will try to protect themselves against cool summers. They are the *naturals* for a market which brings together the banking and the insurance sides of the financial industry.

The instruments we discuss below have been the most popular derivatives since the inception of this market in the late 1990s. Even though we mentioned only energy providers as naturals for this market, the astute reader will reckon that any business exposed to weather risk (and according to some estimates this may represent up to 70 % of the economy) should benefit from the existence of a liquid market for such instruments.

Futures contracts on monthly cumulative degree days have been introduced by the CME, but because of the small volume, we shall restrict our attention to the OTC market. There, the buyer of an option will receive the amount $\xi = f(DD)$ at the end of the period P, the pay-off function f being computed on the cumulative index $DD = HDD^{(P)}$ for the HDD seasons or $DD = CDD^{(P)}$ for the CDD seasons. A CDD season typically starts May 15 and ends September 15. It can also start June 15. The HDD season starts November 15 (December 15 in some cases) and ends usually March 15. The remaining months are called the shoulder months, and the lack of underlying instruments during these months is a peculiarity of the OTC weather market.

6.5.2.1 *Examples of Pay-Off Functions*

Here are some examples of pay-off functions which have been used in the over the counter markets. Figure 6.26 gives the plots of two of the most popular pay-off functions.

Call with a Cap. For these options, the pay-off $\xi = f(DD)$ is of the form:

$$\xi = \min\{\alpha(DD - K)^{+}, C\}. \tag{6.37}$$

The graph of the pay-off function f is given on the left pane of Fig. 6.26. The buyer of such an option pays the up-front premium at the signature of the contract, and if $DD > K$ at the end of the strike period, the writer (seller) of the option pays the buyer the nominal pay-off rate α times the amount by which DD exceeds the strike, the overall payment being limited by the cap C. The values of US$ 2,500.00 and US$ 5,000.00 have been quite commonly used for the pay-off rate α, while typical

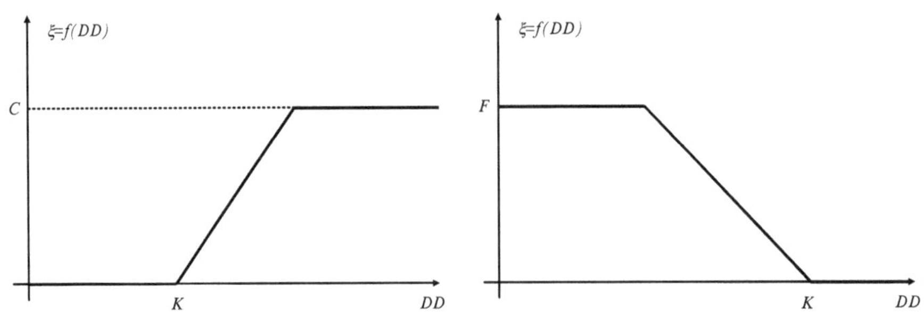

Fig. 6.26. Pay-off functions of a call with a cap (*left*) and a put with a floor (*right*)

values of the cap C are US$ 500,000.00 or US$ 1,000,000.00. See real life examples later in the text and in the problems at the end of the chapter.

The option is said to be *at the money* when the strike K is near the historical average of DD, and *out of the money* when K is far from this average. As far as we know, most of the options written so far have been deep out of the money: buyers are willing to give up some upside for some downside protection.

There is no agreement on what is a reasonable period over which to compute the so-called *historical average.* Where to strike an option? Some recommend to use no more than 10–12 years in order to capture recent warming trends (urbanization, global warming, ...), while others suggest to use longer periods to increase the chances of having a stable historical average.

Example. The marketing department of my favorite cruise line has known for quite some time that the summer sales suffer when the Spring is warm in the North East of the US. People do not suffer from cabin fever which usually follows cold winters, and they tend not to rush to warmer horizons for their summer vacations. So Royal Caribbean will want to buy an out of the money call on Spring CDD's, possibly with a cap since after all, it will sell a minimum number of cruises no matter what.

Put with a Floor. For these options, the pay-off $\xi = f(DD)$ is given by:

$$\xi = \min\{\alpha(K - DD)^+, F\}. \tag{6.38}$$

The graph of the pay-off function f is given on the right pane of Fig. 6.26. In exchange for the premium, the buyer of such an option receives, at the end of the strike period, if $DD < K$, α times the amount by which the strike exceeds DD, the overall payment being limited by the floor F.

Example. In order to hedge the risk that a warm winter hampers its sale numbers, a heating oil provider will buy a put on HDD's over the winter season.

Collar. Buying a collar contract is essentially equivalent to being long a call with a cap and short a put with a floor. More precisely, the pay-off is given by:

$$\xi = \min\{\alpha(DD - K_c)^+, C\} - \min\{\beta(K_p - DD)^+, F\}. \tag{6.39}$$

The graph of the pay-off function f is given in Fig. 6.27. It is possible to adjust all the parameters $\alpha, \beta, C, F, K_c, K_p$ for the premium to be zero. Since there is no up-front cost to get into one of these contracts, this form of collar is quite popular.

Example. In order to protect its revenues against the possible losses of a mild Winter, a local Gas Company will get into a collar contract with no up-front payment. It will pay the counter party if the Winter is cold (which is not a problem since it will enjoy the income from gas sales) and the counter party will pay the Gas Company if the Winter is warm (offsetting the losses due to slow gas sales).

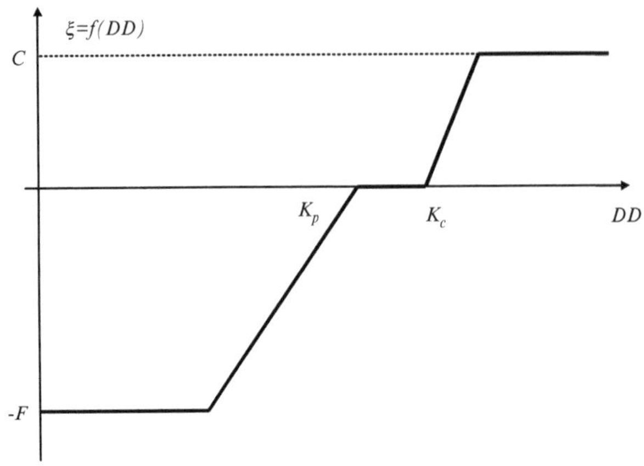

Fig. 6.27. Pay-off function of a collar

The weather derivatives introduced above can be viewed as European *vanilla* options, if we view them as written on an underlying of the form $HDD^{(P)}$ or $CDD^{(P)}$. But it might be easier to view them as written on the temperature $AvgT_t$ as underlying. In this case, they are more of the Asian type than of the European type since the payoff is computed on the average of the underlying number of daily degree days over a period ending at maturity. For this reason, pricing may be more difficult in this case.

It is clear that a good understanding of the underlying is necessary before we attempt to price some of these derivatives. In any case, the time evolution of the temperature underlying indexes is very different from the time evolution of the indexes and the stock prices considered so far. This is a strong indication that Samuelson's geometric Brownian motion model is not appropriate, and that we need a better grasp of the statistics of the underlying index before we can proceed to the analysis of these derivatives.

6.5.3 **Statistical Analysis of Temperature Historical Data**

Temperature options are written on a non-financial underlier, so the classical models of option pricing need to be revisited in order to understand the dynamics of these new derivatives. We propose to illustrate the major issues by the detailed analysis of the statistical properties of the temperature at a given meteorological station. We chose the location of Charlotte, NC for the sake of illustration and for reasons to be uncovered later.

Because of the numerous examples of weather derivatives discussed in the text and the problem sets at the end of this chapter and the following one, the library `Rsafd` contains several temperature data sets. They are stored as objects of class `timeSeries` containing a slot for the time stamps and a slot for the actual temperature data. These `timeSeries` objects are usually named after the location where the temperature measurements were taken, with an extension .ts. Recall that creating a timeSeries object can be done with the constructor function timeSeries as illustrated earlier in the chapter. Note that the object Charlotte.ts contains the daily minimum, maximum and average temperatures from January 1, 1961 to March 31, 1999. So the timeSeries object Charlotte.ts is multivariate (trivariate to be specific).

```
head(Charlotte.ts)
              MinT MaxT AvgT
1961-01-01    35   55   45
1961-01-02    28   52   40
1961-01-03    29   43   36
1961-01-04    22   52   37
1961-01-05    24   58   41
1961-01-06    25   59   42
plot(Charlotte.ts[1:1800,])
```

The purpose of the command plot(Charlotte.ts[1:1800,) is to plot the first 1,800 entries of the series. This amount of data is enough to highlight the periodicity in the data, and not too much to clutter the plot. The result is reproduced in Fig. 6.28. When applied to a multivariate timeSeries object, the generic function plot tries to plot all the components of the time series (i.e. all the columns of the data matrix seriesData).

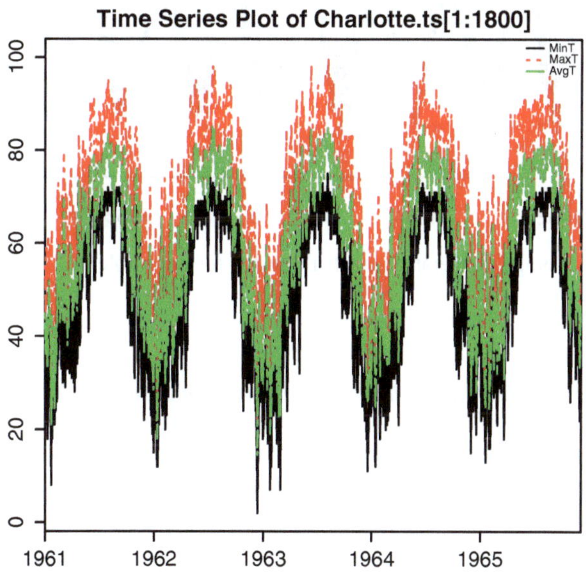

Fig. 6.28. Daily average temperature at Charlotte NC

We concentrate on the daily average temperature (i.e. the third column of `Charlotte.ts`) and to visualize the seasonality of the data more clearly, we plot only the first 10 years of the data.

```
par(mfrow=c(2,1))
plot(Charlotte.ts[,3])
plot(Charlotte.ts[1:3650,3])
par(mfrow=c(1,1))
```

The results of these plots are reproduced in Fig. 6.29.

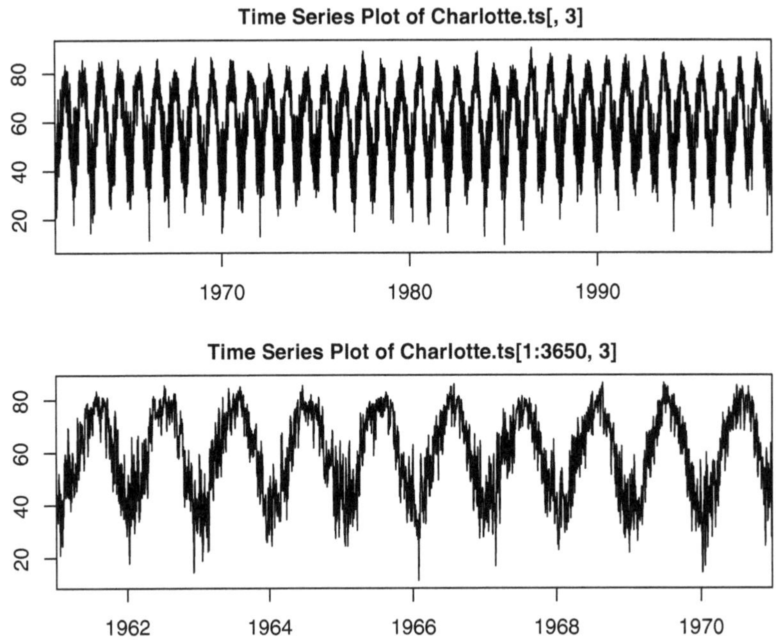

Fig. 6.29. Daily average temperature at Charlotte NC for the whole data set (*top*) and for the first 10 years (*bottom*)

6.5.3.1 *Identifying and Removing the Seasonal Component*

The first order of business is to identify and remove the trend and seasonal components in the data. As in the case of the analysis of the CO_2 concentration data, we use the function `sstl` to identify and extract these components.

```
Charlotte.stl <- sstl(Charlotte.ts[,3],TWIND=0.75)
```

Note that we used `TWIND=0.75` instead of `TWIND=0.05` as before in the case of the CO_2 data. The sliding window used to compute the trend and the seasonal component by successive smoothings has length `TWIND` times the length of the series.

But since the period is now 365 instead of 12, the window has to be much larger in order to cover enough *periods* hence our choice of the value for `TWIND`. We plot separately the resulting three component produced by the function `sstl` (Fig. 6.30).

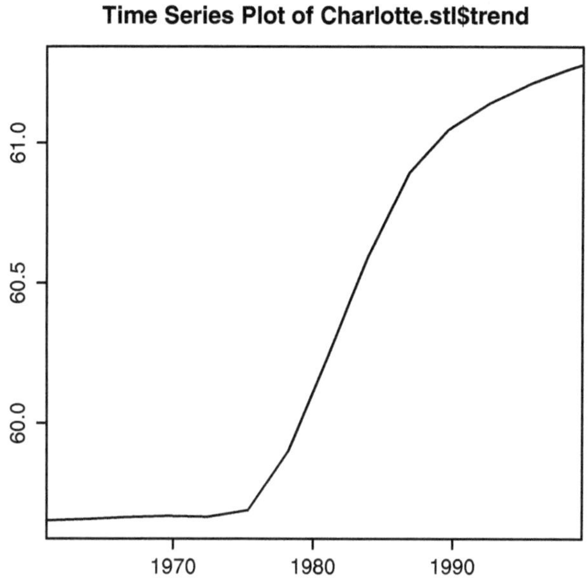

Fig. 6.30. Trend component of the daily average temperature at Charlotte NC, as produced by the command `plot(Charlotte.stl$trend)`

The trend component identified by the function `sstl` is upward (almost two degrees over a period of 40 years), fact which could be linked to global warming, but possibly to other reasons. The increase in the baseline level of the temperature seems to start in the mid-1970s as the curve is flat on the left part of the plot. There are many reasons for such a break: it could have been produced by a change in location of the measuring station, or a change in urbanization around the station, or .... We know that such a case has been made for Charlotte, and this is why we chose this example. Some traders claim that they made significant profits taking advantage of this knowledge. It is important to keep in mind that this sort of uncertainty about information concerning the data collection often remained private, and suspicions about the nature of the data hindered transparency and altogether hurt the market of temperature derivatives. Looking at the second component of `Charlotte.stl` partially plotted in Fig. 6.31, we notice that the seasonal component identified by the function `sstl` is not very smooth.

One could wish for a smoother seasonal component as it is very likely that many of the irregularities are noise artifacts, or features that should be modeled as part of the stationary remainder component. For the sake of definiteness, we shall accept the

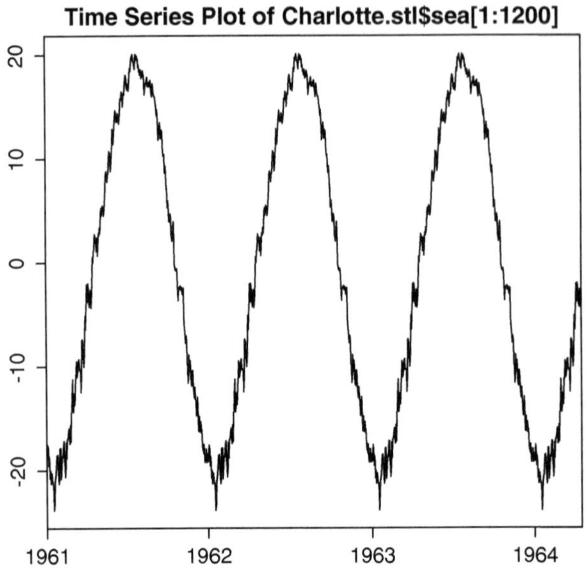

Fig. 6.31. Part of the seasonal component of the daily average temperature at Charlotte NC, as produced by the command `plot(Charlotte.stl$sea[1:1200])`

seasonal component identified by the function `sstl` and refrain from constructing a smoother seasonal component (Fig. 6.32).
Finally we plot in Fig. 6.32 the remainder term which we treat as a stationary time series and try to model as an auto-regressive process.

6.5.3.2 *Analysis of the Serial Correlations*

As explained in the text above, in order to detect and identify the possible serial correlations present in the data, the first step is to compute and plot the auto-correlation function of the series. This plot of the auto-correlation function is given in Fig. 6.33.

It shows significant serial correlations. But since this auto-correlation function does not seem to vanish after a finite number of lags, we should not try to fit a moving average model, so we first try to fit an auto-regressive model.

6.5.3.3 *Could an AR Model Do the Trick?*

We fit an auto-regressive model to the data (the remainder term, to be specific), and we check that the order chosen by the program is reasonable using the commands:

```
ChAR <- ar(Charlotte.stl$rem)
ChAR$order
[1] 8
```

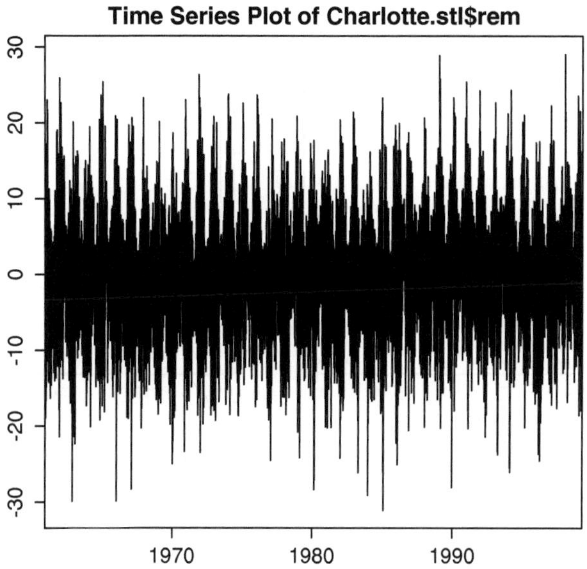

Fig. 6.32. Remainder component of the daily average temperature at Charlotte NC, as produced by the command `plot(Charlotte.stl$rem)`

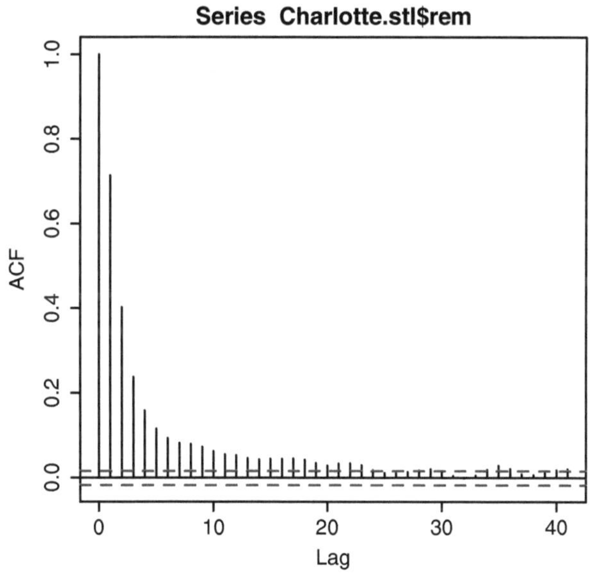

Fig. 6.33. Auto-correlation function of the remainder component of the daily average temperature at Charlotte NC, as produced by the command `acf(Charlotte.stl$rem)`

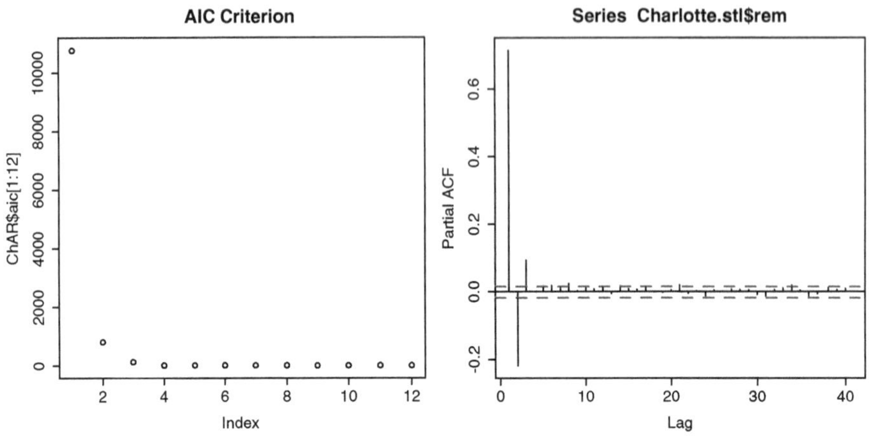

Fig. 6.34. AIC criterion for the remainder component of the daily average temperature at Charlotte NC (*left*) and partial auto-correlation function (*right*)

```
plot(ChAR$aic, main="AIC Criterion")
acf(Charlotte.stl$rem,type="partial")
```

Given the relatively small number of years used in the fit, the order 8 chosen by the function `ar` looks suspiciously large. In order to avoid over-fitting, we double check for reasons leading to the choice of such a large order by plotting the AIC and the partial auto-correlation function. The plots of the AIC criterion and of the partial auto-correlation function reproduced in Fig. 6.34 are consistent with the choice of 8 for the order of the model. However, despite the fact that a careful look at the plot of the AIC confirms that the minimum is attained for the order 8 (fact which is rather difficult to see on the plot given in the left pane of Fig. 6.34 because of the large scale of the first few values) the plot also shows that the major relative drop occurs before the value 4 of the argument. This suggests that an $AR(3)$ model could be appropriate. In order to confirm this guess, we checked once more the partial auto-correlation function in the right part of Fig. 6.34. Except for a minor blip for lag 8, the partial auto-correlation function vanishes for lags greater than or equal to 4, so 3 is the right order, and in an effort to use a more parsimonious model, on the basis of these two plots, we choose to fit an $AR(3)$

```
MyChAR <- ar(Charlotte.stl$rem, order =3)
```

imposing the order to the function `ar` by setting the parameter `order` to `3`. We assess the goodness of the fit by looking at the raw residuals shown in Fig. 6.35. Even though consistent with a white noise time series, looking at it is not sufficient. Using Q-Q plots and tools reviewed in the first chapter of the book, one easily convince ourselves that these residuals look pretty much Gaussian. But most importantly, the plot of the auto-correlation function reproduced in Fig. 6.36 looks very much like

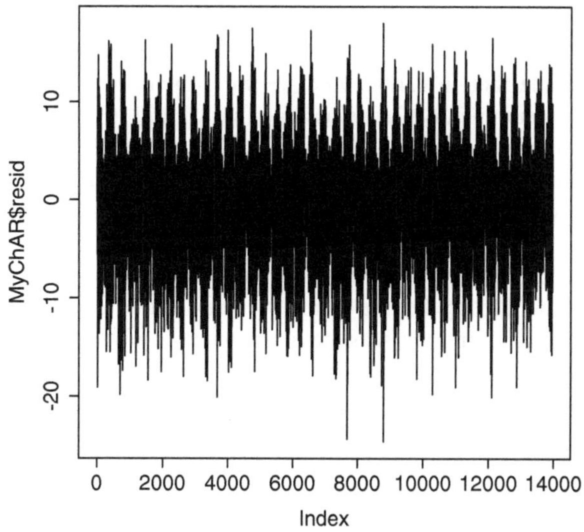

Fig. 6.35. Sequential plot of the raw residuals of the $AR(3)$ model fitted to the remainder component of the daily average temperature at Charlotte NC

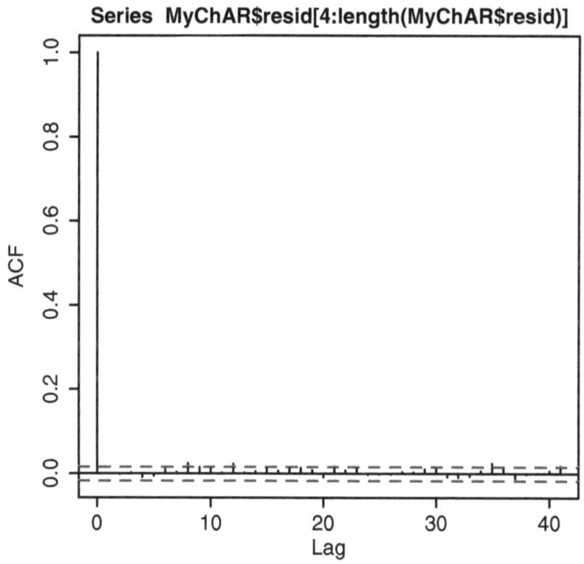

Fig. 6.36. Auto-correlation function of the raw residuals of the $AR(3)$ model fitted to the remainder component of the daily average temperature at Charlotte NC

the plot of the auto-correlation function of a white noise: it seems that all the serial correlation in the data was captured by the $AR(3)$ model. It looks like with are done with the modeling part!

6.5.3.4 Application to Temperature Options

Let us now imagine that in early 1999, we were interested in buying protection against a hot 1999 summer. So let us assume that we did just that by buying a call option on CDD's for the period covering July and August 1999 with a given strike K. How could we have used the analysis done so far to quantify the risk/reward profile of such a purchase? This analysis requires that we look $n = 153$ days ahead in the future.

Our first reaction should be to take the prediction formulae developed for AR models, and use them to predict the summer values of the remainder term, add the predictions of the trend and the seasonal component to obtain a prediction for the actual daily temperatures, from which we could finally compute a prediction for the number of cooling degree days and the pay-off of the option. Unfortunately, this approach is non-sensical. This strategy has at least two fatal flaws. First, the prediction horizon is too long for the actual predictions of the remainder term to be significantly different from zero. Indeed, as we saw earlier, the predictions of the future values of an AR model converge exponentially fast toward the mean. But most importantly, we need predictions for nonlinear functions of the daily temperatures, and computing a nonlinear function of a prediction can lead to unexpected results.

We choose to illustrate this last point with a revisit of the Black-Sholes formula. As explained in Chap. 5, the price $C = C_{T.K}(t, S)$ at time t when the value of the underlying stock is S, of call option with strike K and maturity T is given by the expectation

$$C_{T.K}(t, S) = e^{-r(T-t)}\mathbb{E}\{(S_T - K)^+\} \tag{6.40}$$

where r denotes the short interest rate (assumed to be deterministic and constant) and where S_T is a log-normal random variable. To be specific, S_T is the exponential of a Gaussian random variable with mean $\tilde{\mu} = \log S + (r - \sigma^2/2)(T - t)$ and variance $\tilde{\sigma}^2 = \sigma^2(T - t)$ where σ is the (implied) volatility. Under these assumptions the prediction of S_T (as computed at time t) is

$$\hat{S}_T = e^{\tilde{\mu}+\tilde{\sigma}^2/2} = Se^{(r-\sigma^2/2)(T-t)}e^{\sigma^2(T-t)/2} = Se^{r(T-t)}$$

which is often called the forward price. Now, since the present value of the pay-off of the option is $e^{-r(T-t)}(S_T - K)^+$, replacing the unknown (future) value S_T of the stock at maturity by its prediction $\hat{S}_T = \mathbb{E}\{S_T\}$ would give the value $e^{-r(T-t)}(\hat{S}_T - K)^+ = e^{-r(T-t)}(\mathbb{E}\{S_T\} - K)^+$ which amounts to interchanging the expectation and the positive part function $x \hookrightarrow x^+$, and this gives the **wrong** value!

However, and despite the fact that we may not be able to use these predictions of pricing and risk management purposes, we still go over the steps leading to predictions for the future values of the temperature at Charlotte NC. Our goal is to extend the last year of average daily temperature data by the values of the predictions of the next 153 daily average temperatures given by the model.

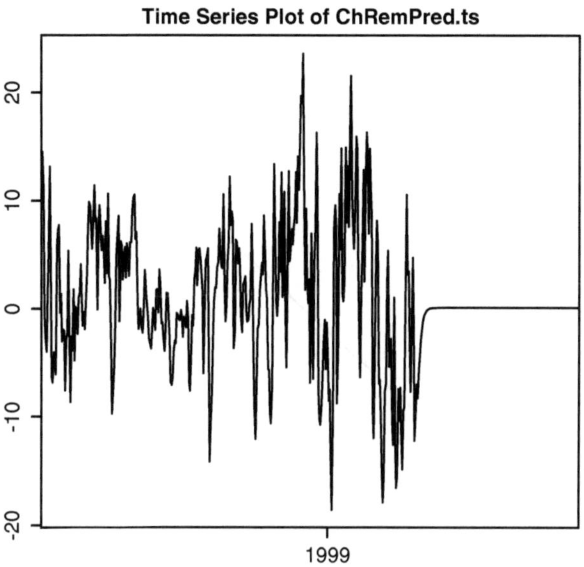

Fig. 6.37. Predictions for the $AR(3)$ model fitted to the remainder component of the daily average temperature at Charlotte NC

As defined earlier in the chapter, prediction is a linear operation, so we can predict separately the future values of the trend, seasonal and remainder components, and add them up to obtain predictions for the future values of the daily average temperature.

We first produce predictions from our $AR(3)$ model for the remainder term.

```
ChPred <- predict(object=MyChAR, newdata=
                seriesData(Charlotte.stl$rem), n.ahead = 153)
LL <- length(seriesData(Charlotte.stl$rem))
RemPred <- c(seriesData(Charlotte.stl$rem)[(LL-364):LL],
                              ChPred$pred)
TODAY <- seriesPositions(Charlotte.stl$rem)
                [length(seriesPositions(Charlotte.stl$rem))]
TODAY
   GMT
   [1] [1999-03-31]
TOMORROW <- timeDate("1999-04-01",format="%Y-%m-%d")
POS <- c(seriesPositions(Charlotte.stl$rem)[(LL-364):LL],
              timeSequence(from=TOMORROW,length=153,by="day"))
ChPred.ts <- timeSeries(positions=POS,data=RemPred)
plot(ChPred.ts)
```

The first command is the most important. It produces the desired predictions in the form of a numeric vector. The role of the other commands is to produce an object of class `timeSeries` with the right time stamps. The plot of these remainder predictionsl is given in Fig. 6.37. Next, we compute predictions of the trend component. The trend being a deterministic function, the predictions of future values are not given by expectations computed in a statistical model fitted to data. They are ob-

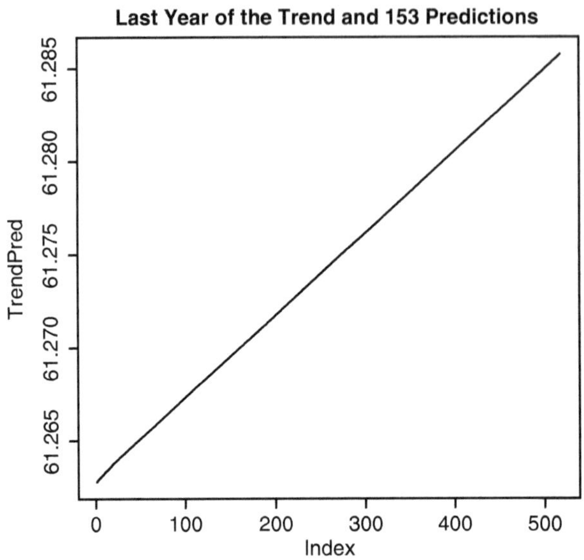

Fig. 6.38. Predictions for the trend component of the daily average temperature at Charlotte NC

tained by *extrapolation*, extending the function 153 points in the future. But since we do not really know the true values of the trend (what we have is only an estimate provided by the function `sstl`), we proceed by smoothing the current estimate by natural splines, and then using the function `predict` for the linear model providing the natural splines smoothing.

```
TREND <- seriesData(Charlotte.stl$trend)
LL <- length(TREND)
XX <- (LL-364):LL
TREND.ns <- lm(TREND[XX] ~ ns(XX,df=6))
TrendPred <- c(TREND[XX],predict(TREND.ns, newdata=
                          data.frame(XX=(LL+1): (LL+153))))
plot(TrendPred,type="l")
title("Last Year of the Trend and 153 Predictions")
```

The plot of the trend predictions is given in Fig. 6.38.
Finally we compute the predictions of the seasonal component. This is clearly the easiest of the three steps. Indeed, because of the periodicity of the seasonal component, the extrapolation is given by the values one period earlier.

```
SEA <- seriesData(Charlotte.stl$sea)
SeaPred <- c(SEA[(LL-364):LL],SEA[(LL-364):(LL-364+152)])
plot(SeaPred,type="l")
title("Last Year of the Seasonal Component and 153 Predictions")
```

The plot of the predictions of the seasonal component is given in Fig. 6.39.

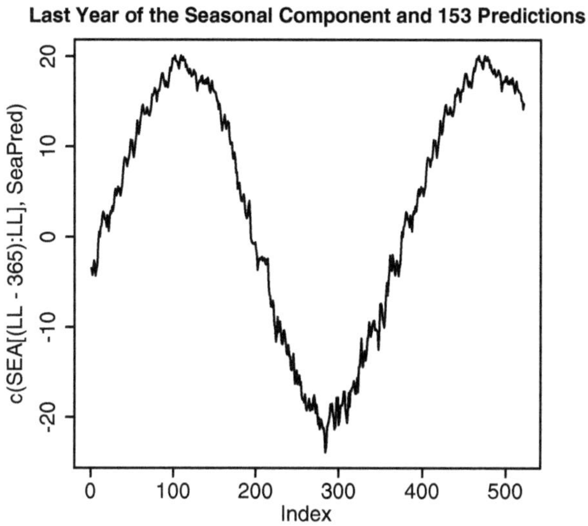

Fig. 6.39. Predictions for the seasonal component of the daily average temperature

As explained earlier, we need to add the predictions of the trend and the seasonal components to the predictions of the remainder component to obtain the temperature predictions.

```
TempPred <- TrendPred + SeaPred + RemPred
TempPred.ts <- timeSeries(positions=POS,data=TempPred)
plot(TempPred.ts)
```

The result is given in Fig. 6.40.

6.5.3.5 *Monte Carlo Computations Within the Fitted Model*

If one is interested in computing the probability that a given option will actually be exercised, or computing the present value of an expected payoff, because one cannot interchange prediction (an expectation) and a nonlinear function of the underlying, we need to resort to Monte Carlo computations of approximations of these probabilities and expectations.

Let us first review the main steps in the Monte Carlo estimation of the probability that the option will end up being exercised. This is the probability that the number $CDD^{(P)}$ is greater than or equal to the strike K. A Monte Carlo estimate of this probability can be computed in the following way:

- Choose a large number of scenarios, say $N = 10{,}000$;
- Use the AR model fitted to the remainder component to generate N samples $x_t^{(j)}$, $x_{t+1}^{(j)}$, ..., $x_T^{(j)}$ where t corresponds to March 31st, 1999, and T to August 31st, 1999, and $j = 1, 2, \ldots, N$;

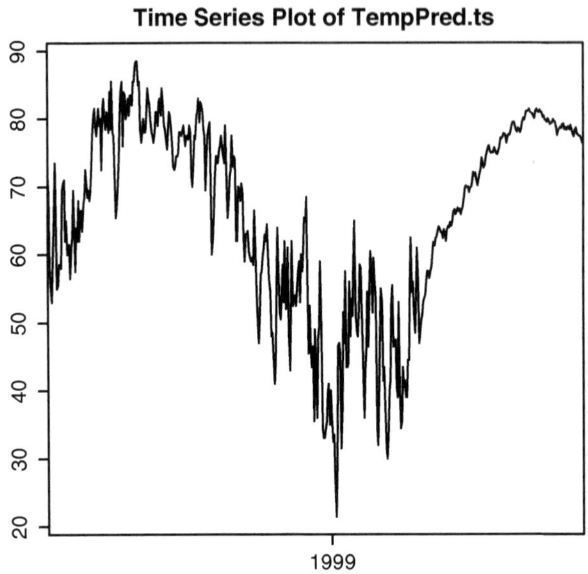

Fig. 6.40. Predictions for the $AR(3)$-based model fitted to the daily average temperature in Charlotte NC

- Add to each of these simulation scenarios, the predictions/extrapolations of the trend and the seasonal component to obtain N scenarios of the temperature in Charlotte starting from March 31st, 1999, and ending August 31st, 1999;
- For each single one of these N scenarios, compute the number of CDD's between July 1st and August 31st;
- Compute the number of times this total number of CDD's is greater that the strike K, and divide by N.

The above procedure is very versatile. Not only can it be used to estimate probabilities, but it can also be used to estimate expected values. For example, if a tick rate is agreed upon (say $\alpha = \$5{,}000$ per degree day) one can estimate the expected amount the writer of the option will have to pay to the buyer of the option. As before, we can generate a large number of Monte Carlo scenarios for the temperature, compute the pay-off of the option for each of these scenarios, and compute the average (over the scenarios) of the cost to the writer. Similarly, one can compute the expected amount the buyer of the option will receive. In fact, given the choice of utility functions for the buyer and the seller of the options, one can compute the expected terminal utilities for both of them, and these numbers should give a good indication of what the writer would be willing to sell the option for, and of what kind of premium the buyer would be willing to pay in order to get the protection provided by the option.

PROBLEMS

Ⓣ **Problem 6.1** *Let us assume that* $\{W_t\}_t$ *is a variance-one white noise, and let us consider the time series* $\{X_t\}_t$ *defined by:*

$$X_t = W_t + (-1)^{t-1} W_{t-1}.$$

Compute the mean and auto-covariance functions of the time series $\{X_t\}_t$. *Is this time series stationary? Say why.*

Ⓣ **Problem 6.2** *1. Let us assume that the random variable* W *is* $N(0,1)$. *Compute the values of the moments:*

$$\mathbb{E}\{W\}, \qquad \mathbb{E}\{W^2\}, \qquad \mathbb{E}\{W^3\}, \qquad \mathbb{E}\{W^4\}, \qquad \mathbb{E}\{W^5\}, \qquad \mathbb{E}\{W^6\}.$$

2. *Let us now assume that* W *is* $N(0,\sigma^2)$. *Determine (as functions of* σ*) the values of the same moments.*
3. *Let us assume that* $\{W_t\}_{t=0,\dots,N}$ *is a Gaussian white noise with variance* σ^2, *and for each* $\alpha \in \mathbb{R}$, *let us define the time series* $\{X_t\}_{t=0,\dots,N}$ *by the formula:*
$$X_t = \alpha W_t + W_t^3, \qquad t = 0, \dots, N. \tag{6.41}$$
Compute the mean, variance and auto-covariance functions of the time series $\{X_t\}_{t=0,\dots,N}$. *Is it stationary?*
4. *Obviously,* X_t *strongly depends upon* W_t, *however, find a value of* α *for which* W_t *and* X_t *are uncorrelated?*

Ⓣ **Problem 6.3** *Let us assume that* $\theta \in (-1,+1)$ *is known, that* $\{W_t\}$ *is a Gaussian white noise with variance one, and that* $\{W'_t\}$ *is a Gaussian white noise with variance* θ^2. *Show that the MA(1) time series* $\{X_t\}_t$ *defined by* $X_t = W_t + \theta W_{t-1}$ *and the time series* $\{Y_t\}_t$ *defined by* $Y_t = W'_t + \frac{1}{\theta} W'_{t-1}$ *have the same auto-covariance functions. Do they have the same auto-correlation functions?*

Ⓣ **Problem 6.4** *1. Find the AR representation of the MA(1) time series*

$$X_t = W_t - 0.4 W_{t-1}$$

where $\{W_t\}_t$ *is a Gaussian white noise with variance* σ^2.

2. *Find the MA representation of the AR(1) time series*
$$X_t - 0.2 X_{t-1} = W_t$$
where, as before, $\{W_t\}_t$ *is a Gaussian white noise with variance* σ^2.

Ⓣ **Problem 6.5** *Let us consider the ARMA time series* $\{X_t\}_t$ *defined by:*

$$X_t - 0.6 X_{t-1} = W_t - 0.9 W_{t-1}$$

where $\{W_t\}_t$ *is a white noise with variance one.*

1. *Rewrite the model using the shift operator* B.
2. *Is the model stationary? Say why.*
3. *Is the model invertible? Say why.*

4. *Express the model in an MA representation if it exists.*
5. *Express the model in an AR representation if it exists.*

Ⓣ **Problem 6.6** *For each of the following ARMA(1,1) models:*

(i) $X_t - X_{t-1} = W_t - 1.5W_{t-1}$
(ii) $X_t - 0.8X_{t-1} = W_t - 0.5W_{t-1}$
for which we assume that $\{W_t\}_t$ *is a* $N(0, \sigma^2)$ *white noise,*

1. *Rewrite the model using the backward shift operator* B *and determine the polynomials* $\phi(B)$ *and* $\theta(B)$.
2. *Check if the model is stationary and/or invertible. Explain your answers.*

Ⓣ **Problem 6.7** *Let us assume that* $\{W_t\}_t$ *is a white noise process with mean* 0 *and variance* $\sigma^2 = 1$. *Consider the time series* $\{X_t\}_t$ *defined by:*

$$X_t + 0.4X_{t-1} = W_t - 2.5W_{t-1} + W_{t-2}.$$

1. *Rewrite the model using the backward shift operator* B.
2. *Is the time series invertible? Say why.*
3. *Does an auto-regressive representation exist? If yes, give it.*
4. *Is the time series stationary (causal)? Say why.*
5. *Does a moving average representation exist? If yes, give it.*
6. *Is the time series* $\{Y_t\}_t$ *defined by* $Y_t = (-1)^t X_t$ *stationary? Explain your answer.*

Ⓣ **Problem 6.8** *Let us assume that the time series* $\{X_t\}_t$ *is defined by:*

$$X_t - 2X_{t-1} + X_{t-2} = W_t - 0.3W_{t-1} - 0.5W_{t-2}$$

where $\{W_t\}_t$ *is a* $N(0, \sigma^2)$ *white noise.*

1. *Rewrite the model using the shift operator* B.
2. *Is the time series stationary? Say why.*
3. *Is the second difference* $D_t = (1 - B)^2 X_t$ *stationary? Say why.*
4. *Compute the auto-covariance function of the second difference* D_t.

Ⓣ **Problem 6.9** *We assume that the data* $x_0, x_1, \ldots, x_n$ *satisfy*

$$x_t = \phi_0 + x_{t-1}, \quad t = 1, 2, \ldots, n, \tag{6.42}$$

where $\phi_0 \neq 0$ *is a real constant. Note that these data are purely deterministic.*

1. *Find an expression for* $x_1, x_2, \ldots, x_n$ *in terms of* x_0 *and* ϕ_0.
2. *Compute the sample mean* $\bar{x} := \frac{1}{n}\sum_{i=1}^{n} x_i$, *the sample second moment* $\frac{1}{n}\sum_{i=1}^{n} x_i^2$, *and the sample variance* $\frac{1}{n}\sum_{i=1}^{n}(x_i - \bar{x})^2$, *again as a function of* x_0, ϕ_0 *and* n.
3. *For each fixed lag* $h = 1, 2, \ldots, n-1$, *find an expression for the sample autocorrelation* $\hat{\rho}(h)$ *(as before as a function of* x_0, ϕ_0 *and* n*).*
4. *Show that* $\hat{\rho}(h) \to 1$ *as* $n \to \infty$ *for each fixed* h, x_0 *and* ϕ_0.

Ⓣ **Problem 6.10** *The goal of this problem is to show that the difference operator* ∇ *preserves stationarity. We assume that the time series* $\{X_t\}_t$ *is stationary.*

1. *Compute the mean function* $m_{\nabla X}(t)$ *and the variance function* $var_{\nabla X}(t)$ *in terms of the mean* μ_X *and variance* σ_X^2 *of the original series* $\{X_t\}_t$ *and its covariance function* $\gamma_X(h)$.

2. *For each lag* $h \geq 0$ *compute the covariance* $cov\{\nabla X_{t+h}, \nabla X_t\}$ *in terms of the statistics of the original series used in question 1.*
3. *Explain why the sequence* $\{\nabla X_t\}_t$ *is also stationary. As always, we use the notation* ∇ *for the difference operator* $\nabla S_t = S_t - S_{t-1} = (1 - B)S_t$*, and the notation* B *for the back-shift operator* $BS_t = S_{t-1}$ *for any function or series* $\{S_t\}_t$*.*
4. *Prove by induction that for each integer* $p \geq 1$*, the sequence* $\{\nabla^p X_t\}_t$ *is stationary.*

Ⓔ Ⓢ **Problem 6.11** *The goal of this problem is to use Monte Carlo simulations to compute the critical values of the Dickey-Fuller unit-root test described in the text, and to use the results to test real log-price data for stationarity. We assume that the time series* $\{X_t\}_t$ *is of the form* $X_t = \phi_1 X_{t-1} + W_t$ *for some strong white noise* $\{W_t\}_t$ *with unknown variance* σ^2*.*

1. *Generate a sample of size* $N = 5{,}000$ *from the standard normal distribution and use it to compute a sample of the same size for the time series* $\{X_t\}_t$ *with* $\sigma^2 = 1$ *and* $\phi_1 = 1$*.*
2. *For each* $n = 1, 2, \ldots, N$ *compute the values of the estimates* $\hat{\phi}_1$ *and* $\hat{\sigma}^2$*, and of the Dickey-Fuller statistic* DF *(given by formulae (6.31) and (6.32) in the text) from the sample of size* n *formed by the first* n *entries of the sample of size* N *generated in question 1 for* $\{X_t\}_t$*. Produce a sequential plot of the estimates of* ϕ_1 *and* σ^2 *and check that they do converge toward their true values (1 for both of them.)*
3. *We now fix* $N = 1{,}000$ *and* $NS = 500$*. Repeat the steps above* NS *times to produce* NS *independent samples of size* N*. Collect the results in a numerical matrix with* N *rows and* NS *columns which you call* `XX`*. For each* $n = 100, 110, 120, \ldots, 1{,}000$ *compute the values of the quantiles of the sample of size* NS *given by the* n*-th row of the matrix* `XX` *for the values* $q = 0.01, 0.02, 0.03, 0.04, 0.05, 0.1, 0.25, 0.5, 0.75, 0.9, 0.95, 0.96, 0.97, 0.98, 0.99$*, and organize the results in a numerical matrix with* 91 *rows and* 15 *columns which you call* `QQ`*.*
4. *Produce a sequential plot of each of the columns of* `QQ` *and check that each quantile converges as* n *grows indefinitely. It is in this sense that the distribution of the Dickey-Fuller statistic converges.*
   From now on we use the quantiles given by the last row of the matrix `QQ` *as a proxy for the critical values of the unit-root test. The goal of the last question is to investigate the stationarity of the daily prices of the Calpine stock contained in the data set* `CPN`*, as well as its log returns* `CPNLRet`*.*
5. *In each case, compute the Dickey-Fuller statistic* DF*, and give the p-values of the unit-root tests. Is the random walk assumption rejected at the level 1 %? Compute the p-values given by the implementation of the augmented Dickey-Fuller test provided by the function* `DF.test` *of the library* `Rsafd` *and compare the results.*

Ⓔ **Problem 6.12** *1. Compute and plot the auto-correlation function of the numeric vector* `RW` *contained in the library* `Rsafd`*, and discuss the dependence (or lack thereof) between its successive entries.*

2. *Compute the first difference (i.e. with lag* $k = 1$*), call it* `WN`*, and address the same question of the dependence of the successive entries. Compare the results with those obtained for the series* `RW`*. Are your observations consistent with the claim: "the series* `RW` *was created as the cumulative sum of the entries of a white noise series"?*
3. *This question deals with the* `timeSeries` *object* `HS.ts` *included in the library* `Rsafd`*. The data give the values of the daily closing values of the Hang Seng Index of Hong Kong Stock Prices from January 2, 1987 to June 25, 2003. Go through the steps of the analysis of questions 1. and 2. for the series* `HS.ts` *and compare the results. What does that tell you about the Hang Seng index?*

Ⓣ **Problem 6.13** *The goal of this problem is to complete the analysis of the* Random Walk with Drift *model introduced in Sect. 5.4.2 of the text. Changing slightly the notation (to replace the time index n by t) we assume that $\{X_t\}_{t\geq 0}$ satisfies*

$$X_t = \mu + X_{t-1} + W_t, \quad t = 1, 2, \dots,$$

where the constant μ is called the drift, where $X_0 = x_0$ is known (non-random) constant, and where the random variables $W_1, W_2, \dots$ are independent and identically distributed with mean 0 and variance σ^2.

1. *Show that $X_t = x_0 + \mu t + \sum_{i=1}^{t} W_i$ for $t \geq 1$.*
2. *Compute the mean, variance, and auto-covariance functions of $\{X_t\}_t$.*
3. *Is $\{X_t\}_t$ stationary? Say why.*
4. *Derive a simple expression for the process $\{\nabla X_t\}$ in terms of $\{W_t\}_t$, and compute its auto-covariance function. As usual, ∇ denotes the (first) difference operator satisfying $\nabla S_t = S_t - S_{t-1} = (1 - B)S_t$ and B is the back-shift operator such that $BS_t = S_{t-1}$ for any function or series S_t in t.*
5. *Is the process $\{\nabla X_t\}$ stationary? Say why.*
6. *Choose $x_0 = 1.5$, $\mu = 0.5$, and $n = 128$, and assume that the white noise is Gaussian. Generate a sample $w_1, \dots, w_n$ of length n of $\{W_t\}_t$, compute and plot the corresponding sample $x_0, x_1, \dots, x_n$ of $\{X_t\}_t$, and explain how you can read the values of x_0 and μ off the graph. Give the scatterplot of the vector of the values $x_1, \dots, x_n$ against the vector of the values $x_0, \dots, x_{n-1}$ and explain what you see.*

ⒺⓈ **Problem 6.14**

1. *Set the seed of the random generator to 14, generate a realization of length 1,024 of a $N(0,1)$ white noise $\{W_t\}_{t=1,\dots,1{,}024}$, and generate the corresponding realization of the AR(3) time series $\{X_t\}_{t=1,\dots,1{,}024}$ satisfying $X_0 = X_{-1} = X_{-2} = 0$ and:*
$$(1 - 0.07B - 0.02B^2 - 0.3B^3)X_t = W_t, \qquad t = 1, 2, \dots, 1{,}024.$$
2. *Fit autoregressive models of orders up to 9 and produce the corresponding AIC. Choose the best model according to this criterion, determine the coefficients and forecast the next 16 values of the time series. Produce a plot of the predictions together with an approximate 95 % confidence interval for these predictions.*
3. *With the same white noise series as before (which you can regenerate by resetting the seed to 14), generate a realization of length 1,000 of the ARMA(3,4) process X_t defined by:*
$$(1 - 0.07B - 0.02B^2 - 0.3B^3)X_t = (1 - 0.4B - 0.3B^2 - 0.2B^3 - 0.05B^4)W_t.$$
4. *Fit autoregressive models of orders up to 9 to the data generated in question 3, and produce the corresponding AIC values. What is the best model suggested by this criterion? Comment. Fit such a model, and as before, forecast the next 16 values of the time series and produce a plot of the predictions together with an approximate 95 % confidence interval for these predictions.*
5. *Ignoring the suggestion of AIC, fit an AR(3) model and compute the estimated residuals. Fit moving averages models (of orders up to 5) to the time series of estimated residuals and choose the best one. Use the ARMA model so-obtained to forecast the next 16 values of the original time series and produce a plot of the predictions together with an approximate 95 % confidence interval for these predictions. Compare the results with those obtained in part 4.*

Problems 6.15 and 6.16 use the data contained in the data set `Reno.ts` included in the library `Rsafd` as an object of class `timeSeries`. It gives the daily average temperature in Reno from March 1st 1937 to November 8, 2001.

Ⓔ **Problem 6.15** *The purpose of this problem is to perform a simple regression analysis of the Reno CDD's option discussed in the text.*

1. *Compute for each year in the period from 1961 to 2001 inclusive, the yearly cumulative number of CDD's in Reno from June 1st to September 30th. Compute the average of these yearly indexes over the last 15 years of this period, and call this average K.*
   From now on we consider a European call option with strike K computed above (i.e. at the money), with rate α equal to US$ 5,000 per degree day, and cap US$ 1,000,000, written on the cumulative number of CDD's in Reno from June 1, 2002 to September 30, 2002.
2. *Use the data at hand and a simple linear regression to explain the dependence of this yearly CDD index upon the year of the computation, and use this model to give an estimate of the value of the yearly CDD index in Reno for the period covered by the option. Give a 95 % prediction interval for this prediction.*
3. *Use this model to estimate the probability that the option described in the text is exercised.*
4. *Estimate the expected loss to the writer of the option (i.e. the party selling the option), and estimate the amount of reserve (in US$) the writer of the option should have in order to cover her losses 95 % of the time.*

Ⓔ **Problem 6.16** *Follow the steps of the analysis of the daily temperature in Charlotte given in Sect. 6.5 of the text to fit a model to the daily temperature in Reno, use this fitted model and follow the prescriptions given in the text to provide new answers to the questions 2 and 3 of Problem 6.15 above.*

Ⓔ **Problem 6.17** *In the last few years* correlation swaps *have been touted as the right tool to hedge or gain exposure to the* correlation *between stocks. Note that this form of correlation trading is different from what traders on a credit desk do when they* trade correlation *by taking positions in CDO tranches.*

The data to be used for this problem are contained in the data set `CorrSwap.ts`, *an object of class* `timeSeries` *whose* `data` *slot is a matrix whose first column gives the closing values of BNP Paribas stock and the second column gives the closing values of Deutsche Bank stock. We denote by* LL *the length of this time series, i.e.*
`LL=length(seriesPositions(CorrSwap.ts))`.

1. *Compute the log returns of both stocks.*
   We choose $L = 252$ for the length of a sliding time window, denote by $\mathcal{T}$ the set of time stamps in `seriesPositions(CorrSwap.ts)[(L+1):LL]`, *and for each time $t \in \mathcal{T}$, denote by $\mathcal{W}(t)$ the window of length L ending at t, namely the set of the L most recent (including date t) entries in* `CorrSwap.ts$pos` *relative to t.*
   1.1. *For each time $t \in \mathcal{T}$, compute the Pearson correlation coefficient of the vectors of log-returns in $\mathcal{W}(t)$, and create with these data, a* `timeSeries` *object which you call* `Pearson.ts`.
   1.2. *For each time $t \in \mathcal{T}$, compute the Kendall correlation coefficient of the vectors of log-returns in $\mathcal{W}(t)$, and create with these data, a* `timeSeries` *object which you call* `Kendall.ts`.
   1.3. *For each time $t \in \mathcal{T}$, compute the Spearman correlation coefficient of the vectors of log-returns in $\mathcal{W}(t)$, and create with these data, a* `timeSeries` *object which you call* `Spearman.ts`.
   1.4. *Plot the three correlation coefficient time series on the same graph and comment.*

2. *For each $t \in \mathcal{T}$, we consider a fourth correlation index based on the notion of copula.*
   - *Fit a GPD distribution to the BNP log-returns in the window $\mathcal{W}(t)$*
   - *Fit a GPD distribution to the DBK log-returns in the window $\mathcal{W}(t)$*
   - *Use the fitted marginals and estimate the parameter of a Gumbel copula for the joint distribution of the log-returns*

   and create with these parameter estimates, a `timeSeries` *object which you call* `Delta.ts`.
3. *For the purpose of this question we define a correlation swap through its cash flows as follows: the owner at time t of a correlation swap with time to maturity L and strike K will receive* $\max\{CC[t+L]-K,0\}$ *from, and pay* $\max\{K-CC[t+L],0\}$ *to the counter-party of the contract, at the time* $t+L$ *of maturity of the contract. Here,* $CC[s]$ *is a* correlation index *computed from the log-returns of BNP and DBK in a window of length L ending at time* s*. For the sake of simplicity, we assume that interest rate is 0, in other words, we ignore discounting. We will consider four different cases for the* correlation index *underlying the contract:*
   (i) *The* correlation index *is computed as the Pearson correlation coefficient;*
   (ii) *The* correlation index *is computed as the Kendall correlation coefficient;*
   (iii) *The* correlation index *is computed as the Spearman correlation coefficient;*
   (iv) *The* correlation index *is computed as the maximal likelihood estimate of the coefficient of the Gumbel copula fitted with the procedure described above.*

   For each day t starting with the 50*-th element of* $\mathcal{T}$*, and ending with* $\mathrm{LL}-\mathrm{L}$*, let us assume that you own a correlation swap with a time to maturity L and strike* $K=K_t$ *given by the average*
   $$K_t = \frac{1}{50}\sum_{t'\in\mathcal{W}'(t)} CC[t']$$
   where $\mathcal{W}'(t)$ *is the window of the last* 50 *days* t' *before and including t, for which the correlation index* $CC[t']$ *was computed.*

   Explain how this position would have done over time, i.e. compute the cumulative profits and losses for all these swaps, and compare the results for the four cases described above.

NOTES & COMPLEMENTS

The random walk and its continuous time analog were introduced at the very beginning of the twentieth century by Bachelier as a model for the stock market. Indeed, a time series plot of a sample from a random walk looks very much like a typical stock chart, and the fact that the increments of the random walks are independent is a good proxy for the expected efficiency of markets. The classical theory of linear processes with the AR, MA, ARIMA techniques presented in this chapter, is often associated with the names of Box and Jenkins who made the theory and the practice of these models popular through a series of books. See for example [8]. We refer the reader interested in the extensions of this theory to Priestley's book [76] for an introduction to nonlinear time series analysis, and to Rosentblatt's book [81] for an extensive discussion of the identification problems which can be handled with the use of higher order statistics.

Unit root tests and cointegration are important fixtures of econometric theory and practice, and our sketchy presentation does not do justice to their importance. The reader is referred to the financial econometrics books [13] by Campbell, Lo and McKinlay, [42] by Gourieroux and

Jasiak, and to the book [99] of Zivot and Wang for examples of `Splus` experiments which can easily be adapted to the `Rsafd` library. The `R` packages `tseries` and `urca` offer specialized functions for the unit root Dickey-Fuller test and co-integration tests. The function `DF.test` implementing the unit root Dickey-Fuller test in the library `Rsafd` is a mere wrapper around the function `add.test` of the library `tseries`.

An important component of stationary time series and signal analysis was purposely ignored in this chapter: spectral analysis. The latter is based on the mathematical theory of the Fourier transform, and it offers a dual perspective on the properties of signals and time series. One of its cornerstones is the Shannon sampling theorem for band-limited signals. For details on the sampling theorem and for the spectral analysis of the linear processes studied in this chapter, we refer the interested reader to standard textbooks such as [10] for example. We chose to stay in the time domain, and consequently ignore the frequency domain, not only to avoid the idiosyncrasies of complex analysis, but also to keep the treatment of causality and nonanticipativeness at a simple and intuitive level. `R` has an extensive set of powerful tools to analyze time series and signal series objects from this point of view, including the package `Rwave` implementation in `R` of the `Splus` library `Swave` developed as a companion to the book [14].

Many *charting* time series analysis tools have been developed with trading systems in mind. They usually go under the name of *technical analysis,* and they are based on indicators giving buy and sell signals from the behavior of various moving averages. These last two words are the only commonality with our presentation of the linear time series models.

Despite growing recognition of its importance, the weather market is still not a part of mainstream finance. We proposed the analysis of temperature options as a striking illustration of the theory presented in this chapter because we believe that temperature options offer an example of financial engineering at its best. The interplay between statistical modeling, mathematical analysis, financial risk management, and Monte Carlo computations makes it the epitome of financial engineering case studies. While proofreading the manuscript of this book, I discovered the collective monograph [5], and I highly recommend it to anyone trying to understand the weather market. Also, I noticed at the same time that a chapter on weather derivatives was added to a recent edition [50] of Hull's book. These two publications are a clear indication of the growing respectability of the weather market among the business and academic communities. The reader interested in statistical analysis for climate research is referred to the book [95] by von Storch and Zwiers.

7

MULTIVARIATE TIME SERIES, LINEAR SYSTEMS AND KALMAN FILTERING

This chapter is devoted to the analysis of the time evolution of random vectors. The first section presents the generalization to the multivariate case of the univariate time series models studied in the previous chapter. Modern accounts of time series analysis increasingly rely on the formalism and the techniques developed for the analysis of general stochastic systems. Even though financial applications have remained mostly immune to this evolution, because of its increased popularity and its tremendous potential, we decided to include this alternative approach in this chapter. The tone of the chapter will have to change slightly as we discuss concepts and theories which were introduced and developed in engineering fields far remote from financial applications. The practical algorithms were developed mostly for military applications. They led to many civil and technological breakthroughs. Here, we present the main features of the filtering algorithms, and we use financial examples as illustrations, restricting ourselves to linear systems.

7.1 Multivariate Time Series

Multivariate time series need to be introduced and used when significant dependencies between individual time series cannot be ignored. As an illustration, we discuss further the weather derivative market, and some of the natural issues faced by its participants. An active market maker in these markets will want to hold options written on HDD's and/or CDD's in different locations. One good reason for that may be the hope of taking advantage of the possible correlations between the weather (and hence the temperature) in different locations. Also large businesses with several units spread out geographically are likely to want deals structured to fit their diverse weather exposures, and this usually involves dealing with several locations simultaneously. Think for example of a chain of amusement parks, or a large retailer such as Wal-Mart or Home Depot: their weather exposures are tied to the geographic locations of the business units, and an analysis of the aggregate weather exposure cannot be done accurately by considering the locations separately and ignoring the correlations. As we are about to show, the simultaneous tracking of the weather

R. Carmona, *Statistical Analysis of Financial Data in R*, Springer Texts in Statistics, DOI 10.1007/978-1-4614-8788-3_7, © Springer Science+Business Media New York 2014

in several locations can be best achieved by considering *multivariate* time series. We discuss a specific example in Sect. 7.1.4 below.

The goal of this section is to review the theory of the most common models of multivariate time series, and to emphasize the practical steps to take in order to fit these models to real data. Most of the Box-Jenkins theory of the ARIMA models presented earlier can be extended to the multivariate setting. However, as we shall see, the practical tools become quite intricate when we venture beyond the auto-regressive models. Indeed, even though the definition of moving average processes can be generalized without change, fitting this class of models becomes even more cumbersome than in the univariate case. As a side remark we mention that R does not provide any tool to fit multivariate moving average models.

In terms of notation, we keep using the convention followed so far in the book: as a general rule, we use bold face characters for multivariate quantities (i.e. vectors for which $d > 1$) and regular fonts for univariate quantities (i.e. scalars for which $d = 1$). In particular, throughout the rest of this section, we shall use the notation $\mathbf{X} = \{\mathbf{X}_t\}_t$ to denote a multivariate time series. In other words, for each time stamp t, $\mathbf{X}_t$ is a d-dimensional random vector.

7.1.1 Stationarity and Auto-Covariance Functions

Most of what was said concerning stationarity extends without any change to the multivariate case. This includes the definition and the properties of the shift operator B, and the derivative (time differentiation) operator ∇. Remember that stationarity is crucial for the justification of the estimation of the statistics of the series by time averages. The structure of the auto-covariance/correlation function is slightly different in the case of multivariate time series. Indeed, for each value k of the time lag, the lag-k auto-covariance is given by a matrix $\gamma(k) = [\gamma_{ij}(k)]_{i,j}$ defined by:

$$\gamma_{ij}(k) = \mathbb{E}\{X_t^{(i)} X_{t+k}^{(j)}\} - \mathbb{E}\{X_t^{(i)}\}\mathbb{E}\{X_{t+k}^{(j)}\}.$$

In other words, the quantity $\gamma_{ij}(k)$ gives the lag-k cross-correlation between the i-th and j-th components of $\mathbf{X}$, i.e. the covariance of the scalar random variables $X_t^{(i)}$ and $X_{t+k}^{(j)}$.

7.1.2 Multivariate White Noise

As in the case of univariate series, the white noise series are the building blocks of the time series edifice. A multivariate stationary time series $\mathbf{W} = \{\mathbf{W}_t\}_t$ is said to be a white noise if it is:

- Mean-zero (i.e. $\mathbb{E}\{\mathbf{W}_t\} = \mathbf{0}$ for all t);
- Serially uncorrelated
  - Either in the strong sense, i.e. if all the $\mathbf{W}_t$'s are independent of each other;
  - Or in the weak sense, i.e. if $\mathbb{E}\{\mathbf{W}_t \mathbf{W}_s^t\} = 0$ whenever $s \neq t$.

The equality specifying the fact that a white noise needs to be mean zero is an equality between vectors. If d is the dimension of the vectors $\mathbf{W}_t$, then this equality is equivalent to a set of d equalities between numbers, i.e. $\mathbb{E}\{W_t^{(i)}\} = 0$ for $i = 1, \ldots, d$. As usual, we use the notation $W_t^{(i)}$ for the i-th component of the vector $\mathbf{W}_t$. On the other hand, the last equality is an equality between $d \times d$ matrices, and whenever $s \neq t$, it has to be understood as a set of $d \times d$ equalities $\mathbb{E}\{W_t^{(i)} W_s^{(j)}\} = 0$ for $i, j = 1, \ldots, d$.

This is exactly the same definition as in the scalar case in the sense that there is complete de-correlation in the time variable. But since the noise terms are vectors, there is also the possibility of correlation between the various components. In other words,

at each time t, the components of $\mathbf{W}_t$ *can be "correlated".*

Hence, in the case of a white noise, we have

$$\gamma_{i,j}(k) = \mathbb{E}\{W_t^{(i)} W_{t+k}^{(j)}\} = \gamma_{i,j}\delta_0(k)$$

where $\delta_0(k)$ is the usual *delta* function which equals 1 when $k = 0$ and 0 when $k \neq 0$, and $\gamma = [\gamma_{i,j}]_{i,j=1,\ldots,k}$ is a time independent variance/covariance matrix for a d-dimensional random vector. Using the jargon of electrical engineers we would say that the components are *white in time and possibly colored in space.*

7.1.3 Multivariate AR Models

A d-dimensional time series $\mathbf{X} = \{\mathbf{X}_t\}_t$ is said to be an auto-regressive series of order p if there exist $d \times d$ matrices A_1, A_2, ..., A_p and a d-variate white noise $\{\mathbf{W}_t\}_t$ such that:

$$\mathbf{X}_t = A_1\mathbf{X}_{t-1} + A_2\mathbf{X}_{t-2} + \cdots + A_p\mathbf{X}_{t-p} + \mathbf{W}_t. \tag{7.1}$$

As before we ignored the mean term (which would be a d-dimensional vector in the present situation) by assuming that all the components have already been centered around their respective means. Notice that the number of parameters is now $pd^2 + d(d+1)/2$ since we need d^2 parameters for each of the p matrices A and we need $d(d+1)/2$ parameters for the variance/covariance matrix Σ_W of the white noise (remember that this matrix is symmetric so that we need only $d(d+1)/2$ coefficients instead of d^2!). Except for the fact that the product of matrices is not commutative, AR models can be fitted in the same way as they are fitted in the univariate case, for example by solving the system of Yule-Walker linear equations obtained from the empirical estimates of the auto-covariance function and the consistency relations with the definition (7.1). In fact as we are about to see, fitting auto-regressive models with the `R` function `ar` can be done with multivariate time series in exactly the same way it is done with univariate series.

7.1.3.1 Fitting Multivariate AR Models in R

Before fitting a multivariate AR model we plot the auto-correlation function with the same command `acf` as in the univariate case. In order to give an illustration, we use the example of multivariate `timeSeries` object `TEMPS.ts` included in the library `Rsafd`. We shall show how to create such a tri-variate time series from univariate `timeSeries` objects in Sect. 7.1.4 below.

```
> acf(TEMPS.ts)
```

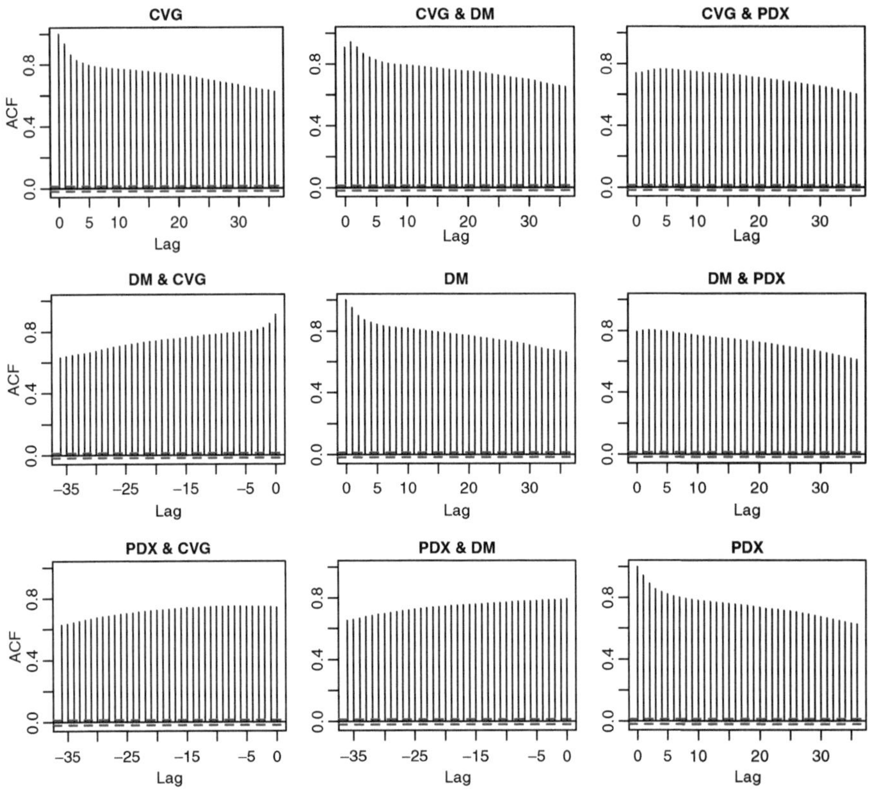

Fig. 7.1. Auto-correlation functions of the 3-variate time series of original temperature data

The acf plot is reproduced in Fig. 7.1. The output is a 3×3 matrix of 9 plots (the "acf" of a d-variate time series produces a $d \times d$ matrix of plots) of the quantities $\gamma_{i,j}(k)$ for the various lags k when the subscripts i and j stand for the indices of the components of the multivariate time series. These plots are full in the sense that the correlations are very high and decay very slowly. This is not due to the existence of strong statistical dependencies between the individual time series or the different lags, but instead to the existence of strong deterministic seasonal components which overwhelmed the computations of the estimates. We shall show in Sect. 7.1.4 how to remove the seasonal components and create a stationary 3-variate `timeSeries` object `TEMPS` with the remainder terms, time series to which we can reasonably try to fit an AR model. Before doing so, we produce the "acf" plot with the command:

```
> acf(TEMPS)
```

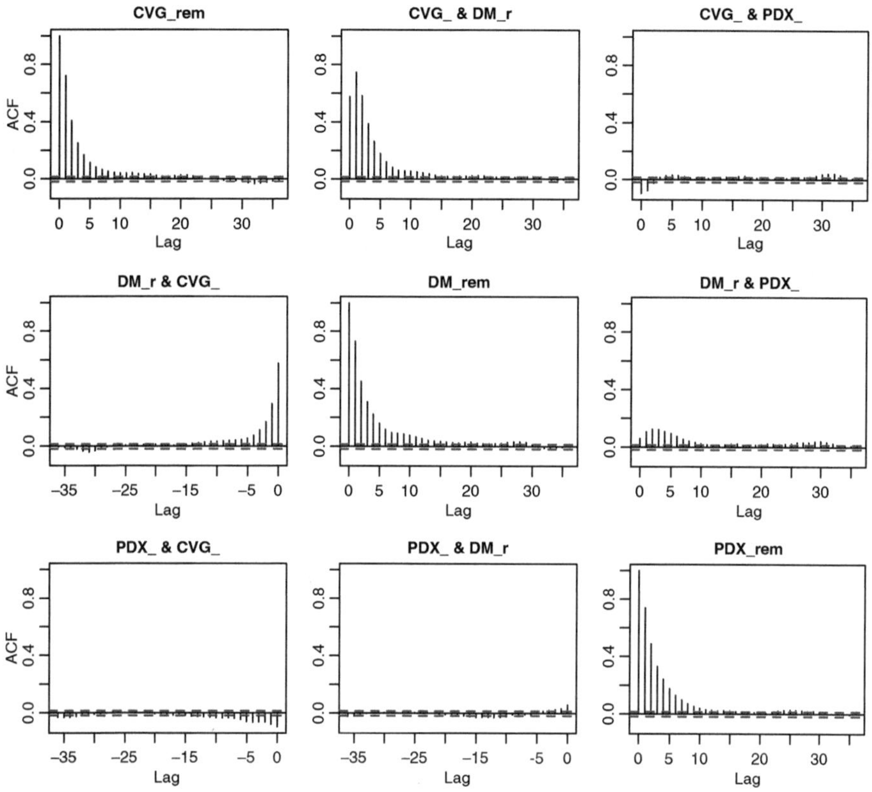

Fig. 7.2. Auto-correlation functions of the 3-variate time series of temperature remainders

The result is given in Fig. 7.2. We can now fit an AR model with the command:

```
> TEMPS.ar <- ar(TEMPS)
```

The order chosen by the fitting algorithm can be printed in the same way. The order chosen for the three city temperature remainders analyzed below is 5. The order was chosen because of the properties of the AIC, but as in the univariate case, the computation of the partial auto-correlation functions should be used to check that the choice of the order is reasonable.

```
> TEMPS.ar$order
[1] 5
> pacf(TEMPS)
```

The partial auto-correlation function can be obtained with the `R` function `pacf` or with the generic function `acf` provided we add the parameter `type` to `"partial"` as in `acf(TEMPS,type="partial")`. Figure 7.3 shows that this value is indeed reasonable.

7.1.3.2 *Prediction*

Except for the technical complexity of the computations which now involve matrices instead of plain numbers, our discussion of the univariate case applies to the multivariate case considered in this section.

For example, if $\hat{A}_1$, $\hat{A}_2$, ..., $\hat{A}_p$ and $\hat{\Sigma}_W$ are the estimates of a d-dimensional AR(p) model (i.e. the parameters of a model fitted to a d-variate data set), then at any instant t (by which we mean at the end of the t-th time interval, when we know the exact values of the outcome $\mathbf{X}_s$ for $s \leq t$), the prediction of the next value of the series is given by:

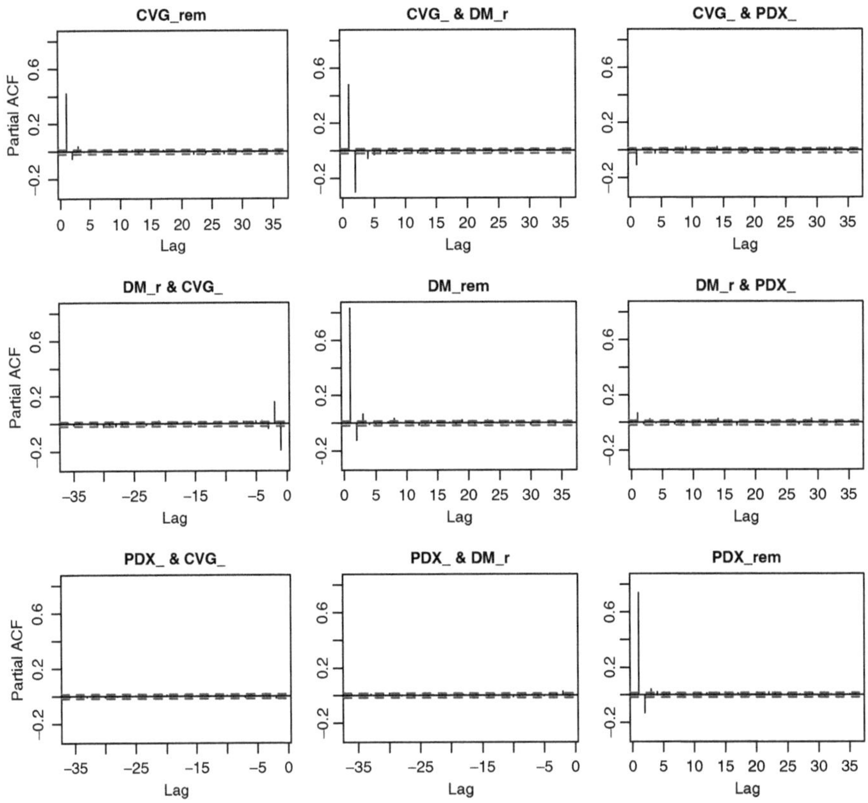

Fig. 7.3. Partial auto-correlation functions of the 3-variate time series of temperature remainders

$$\hat{\mathbf{X}}_{t+1} = \hat{A}_1 \mathbf{X}_t + \hat{A}_2 \mathbf{X}_{t-1} + \cdots + \hat{A}_p \mathbf{X}_{t-p+1}. \tag{7.2}$$

One can similarly derive the prediction of the value of the series two steps ahead, or for any finite number of time steps in the future.

$$\hat{\mathbf{X}}_{t+k} = \hat{A}_1 \hat{\mathbf{X}}_{t+k-1} + \hat{A}_2 \hat{\mathbf{X}}_{t+k-2} + \cdots + \hat{A}_p \hat{\mathbf{X}}_{t+k-p} \tag{7.3}$$

where we ought to replace the prediction $\hat{\mathbf{X}}_s$ by the observed value $\mathbf{X}_s$ whenever available, i.e. when $s \leq t$. As in the univariate case, when the time series is stable, the longer the horizon of the prediction, the closer it will be to the *long term average* of the series. We shall illustrate this fact on the numerical example we consider below.

For implementation purposes, the `R` generic function `predict` can be used in the multivariate case as well. The only difference is that for multivariate predictions, the generic function `predict` does not compute estimates of the prediction errors, returning NAs for these error estimates.

7.1.3.3 ***Monte Carlo Simulations & Scenarios Generation***

As in the univariate case, random simulation should not be confused with prediction. For the sake of the present discussion, let us assume that we have the values of the vectors $\mathbf{X}_t$, $\mathbf{X}_{t-1}$, …, $\mathbf{X}_{t-p+1}$ at hand, and that we would like to generate a specific number, say N, of Monte Carlo scenarios for the values of $\mathbf{X}_{t+1}$, $\mathbf{X}_{t+2}$, …, $\mathbf{X}_{t+M}$.

The correct procedure is to generate N samples of a d-dimensional white noise time series of length M, say $\{\mathbf{W}_s\}_{s=t+1,\ldots,t+M}$, with the variance/covariance matrix $\hat{\Sigma}_W$, and then to use the parameter estimates $\hat{A}_1$, $\hat{A}_2$, …, $\hat{A}_p$ and the recursive formula (7.1) giving the definition of the AR(p) model to generate samples of the AR model from these N samples of the white noise. In other words, for each of the N given samples $\mathbf{W}^{(j)}_{t+1}$, …, $\mathbf{W}^{(j)}_{t+M}$ of the white noise, we generate the corresponding Monte Carlo scenarios of the series by computing recursively the values $\hat{\mathbf{X}}^{(j)}_{t+1}$, …, $\hat{\mathbf{X}}^{(j)}_{t+M}$ from the formula:

$$\hat{\mathbf{X}}^{(j)}_{t+k} = \hat{A}_1 \hat{\mathbf{X}}^{(j)}_{t+k-1} + \hat{A}_2 \hat{\mathbf{X}}^{(j)}_{t+k-2} + \cdots + \hat{A}_p \hat{\mathbf{X}}^{(j)}_{t+k-p} + \mathbf{W}^{(j)}_{t+k}, \quad k = 1, 2, \ldots, M \tag{7.4}$$

for $j = 1, 2, \ldots, N$, given the fact that, like in the prediction case, the "hats" over the $\mathbf{X}$'s (i.e. the simulations) are not needed when the true values are available.

7.1.4 **Back to Temperature Options**

We come back to temperature data and temperature options to give a concrete example of the practical use of multivariate time series fitting.

7.1.4.1 ***Hedging Several Weather Exposures***

Let us assume for the sake of illustration, that most of the summer revenues of a company selling thirst quenchers come from sales in Des Moines, Portland and Cincinnati. Clearly, an unusually cool summer in these cities would hurt the company's revenues, so the chief risk officer decided to approach a financial institution looking for a solution to mitigate the possible losses linked to such a weather exposure. One can imagine that this financial institution offers to sell the three call options whose characteristics are given in Table 7.1.

	Option #1	**Option #2**	**Option #3**
Underlying series	CDD's in Des Moines	CDD's in Portland	CDD's in Cincinnati
Period	June, July & August	June, July & August	June, July & August
Strike	830	420	980
Rate	US$5,000	US$5,000	US$5,000
Cap	US$1,000,000	US$1,000,000	US$1,000,000

Table 7.1. Characteristics of the three call options offered in lieu of risk mitigation

A naive approach to the valuation of this protection would be treating the three options separately, performing three times the analysis outlined in our discussion of temperature options at a single location. This is definitely not a good idea, for it ignores the obvious dependencies between the three temperature series. In order to avoid this shortcoming, we propose to analyze the temperatures in these three different locations *simultaneously*. We read the temperature data from these three cities: Cincinnati (more precisely the meteorological station at Covington airport), Des Moines, and Portland into three time series. These time series are included in the library `Rsafd` as `timeSeries` objects `Covington.ts`, `DesMoines.ts` and `Portland.ts`. As explained in the previous chapter for univariate time series, and earlier in this chapter in the case of multivariate time series (recall Fig. 7.1 and the ensuing discussion), the auto-correlation function should not be computed on the original time series because of the spurious effects produced by deterministic trends and seasonal components. So we apply the function `sstl` separately to remove the three trend and seasonal components of each of the three time series. Once this is done, we bind the three remainder series into a 3-dimensional series which we proceed to model. Notice that the function `merge` can only bind two `timeSeries` objects at a time, so we need to use it twice.

```
> CVG.stl <- sstl(Covington.ts)
> DM.stl <- sstl(DesMoines.ts)
> PDX.stl <- sstl(Portland.ts)
> TEMPS <- merge(CVG.stl$rem, DM.stl$rem)
> TEMPS <- merge(TEMPS,PDX.stl$rem)
```

As in the case of univariate series, we first look for serial correlations by computing and plotting the auto-correlation function. We already explained how to fit a multivariate auto-regressive model to this tri-variate time series. We gave the R commands, we computed the order of the fitted model, and we plotted the partial auto-correlation function as a check. We now show how to use this model in order to estimate the probabilities and expectations we need to compute to understand the risks associated with the purchase of the three options described in Table 7.1. As in the univariate case, we use Monte Carlo techniques to estimate, for example, the probability that at least one of the three options will be exercised. We choose this simple problem for the sake of illustration. Indeed, this probability is not of much practical interest. It would be more interesting to associate a cost/penalty/utility to each possible temperature scenario, and to compute the expected utility over all the possible scenarios. In any case, this probability can be computed in the following way:

- Choose a large number of scenarios, say $N = 10{,}000$;
- Use the fitted model to generate N samples $\boldsymbol{x}_t^{(j)}, \boldsymbol{x}_{t+1}^{(j)}, \ldots, \boldsymbol{x}_T^{(j)}$ where t corresponds to March 31st, 1999, and T to August 31st, 1999, $j = 1, 2, \ldots, N$, and each $\boldsymbol{x}_{t+s}^{(j)}$ is a three dimensional vector representing the temperature remainders in the three cities, on day $t+s$ for the j-th scenario. Remember that according to the explanations given above, this is done by generating the white noise first, and then computing the $\boldsymbol{x}_{t+s}^{(j)}$ inductively from formula (7.4).
- Add the mean temperatures and the seasonal components to each of the three components to obtain N scenarios of the temperatures in Covington, Des Moines, and Portland starting from March 31st, 1999, and ending August 31st, 1999;
- For each single one of these N tri-variate scenarios, compute the number of CDD's between June 1st and August 31st in each of the three cities;
- Compute the number of times at least one of the three total numbers of CDD's is greater that the corresponding strike, and divide by N.

7.1.4.2 The Case of a Basket Option

Instead of purchasing three separate options, similar risk mitigation can be achieved by the purchase of a single option written on an index involving the temperatures in the three locations. The simplest instruments available over the counter are written on an underlying index obtained by computing the plain average, or the sum, of the indexes used above for the three cities. In other words, we would compute the total number of CDD's in Des Moines over the summer, the total number of CDD's in Portland over the same period, and finally the total number of CDD's in Cincinnati in the same way, add these three numbers, and compare this aggregate index to the strike of the option. In other words, instead of being the sum of the three separate pay-offs, the pay-off of the option is a single function of an aggregate underlying index. Such options are called *basket options*. Obviously, the averages could be weighted averages instead of plain averages, the analysis would be the same. Modeling the separate underlying indexes simultaneously is a must in dealing with basket options. It would be unreasonable, borderline suicidal, not to include the dependencies between these indexes in the model. Fortunately, the Monte Carlo analysis presented above can be performed in exactly the same way. For each of the N tri-variate scenarios generated above, one compute the numbers of CDD's in each of the three locations, and add them up together: this is the underlying index. It is then plain to decide if the option is exercised, or to compute the loss/utility associated to such a scenario for the three locations, ... and to compute probabilities of events of interest or desired expectations, by averaging over the scenarios.

7.1.5 **Multivariate MA & ARIMA Models**

This subsection is included for the sake of completeness only. None of the concepts and results presented in this subsection will be used in the sequel (including the problems).

A d-dimensional time series $\mathbf{X} = \{\mathbf{X}_t\}_t$ is said to be a moving average series of order q if there exist $d \times d$ matrices B_1, B_2, ..., B_q and a d-variate white noise $\{\mathbf{W}_t\}_t$ such that:

$$\mathbf{X}_t = \mathbf{W}_t + B_1\mathbf{W}_{t-1} + B_2\mathbf{W}_{t-2} + \cdots + B_q\mathbf{W}_{t-q}$$

for all times t. As before we ignored the mean term (which would be a d-dimensional vector in the present situation) by assuming that all the components have already been centered around their respective means. As in the case of a multivariate AR model, the number of parameters can be computed easily. This number of parameters is now $qd^2 + d(d+1)/2$. As in the univariate case, fitting moving average models is more difficult than for auto-regressive processes. It can be done by maximum likelihood, but the computations require sophisticated recursive procedures and they are extremely involved. This is the reason why `R` does not provide a function (similar to `ar`) to fit multivariate moving average time series.

A d-dimensional time series $\mathbf{X} = \{\mathbf{X}_t\}_t$ is said to be an ARMA(p,q) time series if there exist $d \times d$ matrices A_1, A_2, ..., A_p, B_1, B_2, ..., B_q, and a d-variate white noise $\{\mathbf{W}_t\}_t$ such that:

$$\mathbf{X}_t - A_1\mathbf{X}_{t-1} - A_2\mathbf{X}_{t-2} - \cdots - A_p\mathbf{X}_{t-p} = \mathbf{W}_t + B_1\mathbf{W}_{t-1} + \cdots + B_q\mathbf{W}_{t-q}$$

Finally, as in the univariate case, an ARIMA(p,r,q) time series is defined as a time series which becomes an ARMA(p,q) after r differentiations.

7.1.5.1 ***Prediction and Simulation in*** `R`

Contrary to the univariate case, `R` does not provide any function to produce random samples of multivariate ARIMA time series. It does not have a function to compute predictions of future values from a general multivariate ARIMA model either. However, the generic function `predict` can still be used with objects of class `ar` even when the latter are multivariate.

7.1.6 **Cointegration**

The concept of cointegration is of crucial importance in the analysis of financial time series. We define it in the simplest possible context: two $I(1)$ time series. See the Notes & Complements at the end of the chapter for details and references.

Two unit-root time series are said to be cointegrated if there exists a linear combination of its entries which is stationary. The coefficients of such a linear combination are said to form a cointegration vector. For example, it has been argued that yields of different maturities are cointegrated.

Example. We first discuss an academic example for the purpose of illustration. Let us consider two time series $\{X_n^{(1)}\}_{n=1,\dots,N}$ and $\{X_n^{(2)}\}_{n=1,\dots,N}$ defined by:

$$X_n^{(1)} = S_n + \epsilon_n^{(1)} \quad \text{and} \quad X_n^{(2)} = S_n + \epsilon_n^{(2)}, \qquad n = 1, \dots, N$$

where $S_n = S_0 + W_1 + \cdots + W_n$ is a random walk starting from $S_0 = 0$ constructed form the $N(0,1)$ white noise $\{W_n\}_{n=1,\dots,N}$, and where $\{\epsilon_n^{(1)}\}_{n=1,\dots,N}$ and $\{\epsilon_n^{(2)}\}_{n=1,\dots,N}$ are two independent $N(0, 0.16)$ white noise sequences which are assumed to be independent of $\{W_n\}_{n=1,\dots,N}$. Since the time series $\{X_n^{(1)}\}_{n=1,\dots,N}$ and $\{X_n^{(2)}\}_{n=1,\dots,N}$ are of the type *random walk plus noise,* they are both integrated of order 1, and in particular, non-stationary. However, the linear combination:

$$X_n^{(1)} - X_n^{(2)} = \epsilon_n^{(1)} - \epsilon_n^{(2)}, \qquad n = 1, \dots, N$$

is stationary since it is a white noise with distribution $N(0, 0.32)$. The random walk $\{S_n\}_n$ is a common trend (though stochastic) for the series $\{X_n^{(1)}\}_n$ and $\{X_n^{(2)}\}_n$, and it disappears when we compute the difference. Cointegration is the state of several time series sharing a common (non-stationary) stochastic trend, the remainder terms being stationary.

To emphasize this fact, we consider the case of two models of the same type *random walk plus noise,* but we now assume that the random walks are independent. More precisely, we assume that $\{X_n^{(1)}\}_{n=1,\dots,N}$ and $\{X_n^{(2)}\}_{n=1,\dots,N}$ are defined by:

$$X_n^{(1)} = S_n^{(1)} + \epsilon_n^{(1)} \quad \text{and} \quad X_n^{(2)} = S_n^{(2)} + \epsilon_n^{(2)}.$$

for two random walks $S_n^{(1)} = S_0^{(1)} + W_1^{(1)} + \cdots + W_n^{(1)}$ and $S_n^{(2)} = S_0^{(2)} + W_1^{(2)} + \cdots + W_n^{(2)}$ starting from $S_0^{(1)} = 0$ and $S_0^{(2)} = 0$ respectively, constructed from two independent $N(0,1)$ white noise series $\{W_n^{(1)}\}_{n=1,\dots,N}$ and $\{W_n^{(2)}\}_{n=1,\dots,N}$ of size $N = 1{,}024$, and where as before the two independent white noise sequences $\{\epsilon_n^{(1)}\}_{n=1,\dots,N}$ and $\{\epsilon_n^{(2)}\}_{n=1,\dots,N}$ have the $N(0, 0.16)$ distribution, and are independent of $\{W_n^{(1)}\}_{n=1,\dots,N}$ and $\{W_n^{(2)}\}_{n=1,\dots,N}$. As before, the time series $\{X_n^{(1)}\}_n$ and $\{X_n^{(2)}\}_n$ are both of the type *random walk plus noise,* but their trends are independent, and no linear combination can cancel them. Indeed it is easy to see that all the linear combinations of $\{X_n^{(1)}\}_n$ and $\{X_n^{(2)}\}_n$ are of the type *random walk plus noise*. Indeed:

$$\begin{aligned} a_1 X_n^{(1)} + a_2 X_n^{(2)} &= a_1 S_n^{(1)} + a_1 \epsilon_n^{(1)} + a_2 S_n^{(2)} + a_2 \epsilon_n^{(2)} \\ &= S_n + \epsilon \end{aligned}$$

where $\{S_n\}_n$ is the random walk starting from $S_0 = a_1 S_0^{(1)} + a_2 S_0^{(2)}$ constructed from the $N(0, a_1^2 + a_2^2)$ white noise $W_n = a_1 W_n^{(1)} + a_2 W_n^{(2)}$ and $\epsilon_n = a_1 \epsilon_n^{(1)} + a_2 \epsilon_n^{(2)}$ is a $N(0, 0.16 a_1^2 + 0.16 a_2^2)$ white noise. So for the *random walk plus noise* series $\{X_n^{(1)}\}_n$ and $\{X_n^{(2)}\}_n$ to be cointegrated, the random walk components have to be the same. Problem 7.7 is devoted to the illustration of these facts on simulated samples.

7.1.6.1 Testing for Cointegration

Two times series $\{X_n^{(1)}\}_n$ and $\{X_n^{(2)}\}_n$, are cointegrated if neither of these series is stationary, and if there exists a vector $\boldsymbol{a} = (a_1, a_2)$ such that the time series $\{a_1 X_n^{(1)} + a_2 X_n^{(2)}\}_n$ is stationary. As we already mentioned, such a vector is called a cointegration vector.

- Testing for cointegration is easy if the cointegration vector $\boldsymbol{a} = (a_1, a_2)$ is known in advance. Indeed, this amounts to testing for the stationarity of the appropriate time series $\{a_1 X_n^{(1)} + a_2 X_n^{(2)}\}_n$. This can be done with a unit root test. *à la Dickey-Fuller*.
- In general, the cointegration vector cannot be zero, so at least one of its components is non-zero. Assuming that for example that $a_2 \neq 0$ for the sake of definiteness, and dividing by a_2, we see that the series $\{(a_1/a_2) X_n^{(1)} + X_n^{(2)}\}_n$ has to be stationary, or equivalently (if we isolate the mean of this stationary series):
$$X_n^{(2)} = \beta_0 + \beta_1 X_n^{(1)} + \epsilon_n \tag{7.5}$$
for some mean-zero stationary time series $\{\epsilon_n\}_n$. Formula (7.5) says that in order for two time series to be cointegrated, a form of linear model with stationary errors should hold. But as we have seen at the beginning of Chap. 4, linear regression for time series is a *very touchy business*! Indeed, the residuals are likely to have a strong serial auto-correlation, an obvious sign of lack of independence, but also a sign of lack of stationarity. In such a case, one cannot use statistical inference and standard diagnostics to assess the significance of the regression. In the particular example treated in Chap. 4, we overcame these difficulties by considering the log-returns instead of the raw indexes. In general, when the residuals exhibit strong serial correlations, it is recommended to replace the original time series by their first differences, and to perform the regression with these new time series.

This informal discussion is intended to stress the fact that testing for cointegration is not a simple matter. The interested reader is encouraged to consult the Notes & Complements at the end of the chapter for references.

7.2 State Space Models: Mathematical Set Up

A state-space model is determined by two equations (or to be more precise, two systems of equations). The first equation:

$$\mathbf{X}_{n+1} = F_n(\mathbf{X}_n, \boldsymbol{V}_{n+1}) \tag{7.6}$$

describes the time evolution of the state $\mathbf{X}_n$ of a system. Since most of the physical systems of interest are complex, it is to be expected that the state will be described by a vector $\mathbf{X}_n$ whose dimension $d = d_X$ may be large. The integer n labels time and for this reason, we shall use the labels n and t interchangeably. Models are often defined and analyzed in continuous time, and in this case we use the variable t for time. Most practical time series come from sampling of continuous time systems. In these situations we use the notation t_n for these sampling times. Quite often t_n is of the form $t_n = t_0 + n\Delta t$ for a sampling interval Δt. But even when the sample times are not regularly spaced, we shall still use the label n for the quantities measured, estimated, computed, ... at time t_n.

For each $n \geq 0$, F_n is a vector-valued function (i.e. a function taking values in the space $\mathbb{R}^d$ of d-dimensional vectors) which depends upon the (present) time n, the (current) state of the system as given by the state vector $\mathbf{X}_n$, and a system noise $\boldsymbol{V}_{n+1}$ which is a (possibly multivariate) random quantity. Throughout this chapter we shall assume that the state equation (7.6) is linear in the sense that it is of the form:

$$\mathbf{X}_{n+1} = F_n\mathbf{X}_n + \boldsymbol{V}_{n+1} \tag{7.7}$$

for some $d \times d$ matrix F_n and a $d \times 1$ random vector $\boldsymbol{V}_{n+1}$. The system matrix F_n will be independent of n in most applications. The random vectors $\boldsymbol{V}_n$ are assumed to be mean zero and independent, and we shall denote by Σ_V their common variance/covariance matrix. Moreover, the noise term $\boldsymbol{V}_{n+1}$ appearing in (7.6) and (7.7) is assumed to be independent of all the $\mathbf{X}_k$ for $k \leq n$. This is emphasized by our use of the index $n+1$ for this noise term.

Notice that, except for a change in notation, such a linear state space system with state matrix constant over time, is nothing but a multivariate AR(1) model! Nothing very new so far. Except for the fact which we are about to emphasize below: such an AR(1) process may not be observed in full. This will dramatically enlarge the realm of applications for which these models can be used, and this will enhance their usefulness.

The second equation (again, to be precise, one should say the second system of equations) is the so-called observation equation. It is of the form:

$$\mathbf{Y}_n = G_n(\mathbf{X}_n, \mathbf{W}_n). \tag{7.8}$$

It describes how, at each time n, the actual observation vector $\mathbf{Y}_n$ depends upon the state $\mathbf{X}_n$ of the system. Notice that in general, we make several simultaneous measurements, and that, as a consequence, the observation should be modeled as a vector $\mathbf{Y}_n$ whose dimension d_Y will typically be smaller than the dimension $d = d_X$ of the state $\mathbf{X}_n$. The observation functions G_n are $\mathbb{R}^{d_Y}$-valued functions which depend upon the (present) time n, the (current) state of the system $\mathbf{X}_n$, and an observation noise $\mathbf{W}_n$ which is a (possibly multivariate) random quantity. Throughout this chapter we shall also assume that the observation equation (7.8) is linear in the sense that it is of the form:

$$\mathbf{Y}_n = G_n \mathbf{X}_n + \mathbf{W}_n \tag{7.9}$$

for some $d_Y \times d_X$ matrix G_n and a $d_Y \times 1$ random vector $\mathbf{W}_n$. As for the system matrix, in most of the applications considered in this chapter, the observation matrices G_n will not change with n, and we will denote by G their common value. The random vectors $\mathbf{W}_n$ modeling the observation noise are assumed to be mean zero, independent (of each other) identically distributed, and also independent of the system noise terms $\boldsymbol{V}_n$. We shall denote by Σ_W their common variance/covariance matrix.

The challenges of the analysis of state-space models are best stated in the following set of bullet points:

> At any given time n, and for any values $\boldsymbol{y}_1, \ldots, \boldsymbol{y}_n$ of the observations (which we view as realizations of the random vectors $\mathbf{Y}_1, \ldots, \mathbf{Y}_n$), find estimates for:

- The state vector $\mathbf{X}_n$ at the same time. This is the so-called *filtering* problem;
- The future states of the system $\mathbf{X}_{n+m}$ for $m > 0$. This is the so-called *prediction* problem;
- A past occurrence $\mathbf{X}_m$ of the state vector (for some time $m < n$). This is the so-called *smoothing* problem.

Note that as usual, when we say estimate, we always worry that an estimate does not have much value if it does not come with a companion estimate of the associated error, and consequently of the confidence we should have in these estimates.

Remark. Linearization of Nonlinear Systems In many applications of great practical interest, the observation equation is of the form

$$\mathbf{Y}_n = \Phi(\mathbf{X}_n) + \mathbf{W}_n \tag{7.10}$$

for some vector-valued function Φ whose i-th component $\Phi_i(\boldsymbol{x})$ is a nonlinear function of the components of the state $\mathbf{X}$. This type of observation equation (7.10) is in principle not amenable to the theory presented below because the function Φ is not linear, i.e. it is not given by the product of a matrix with the vector $\mathbf{X}$. The remedy is to replace this nonlinear observation equation by the approximation provided by the first order Taylor expansion of the function Φ:

$$\Phi(\mathbf{X}_n) = \Phi(\mathbf{X}_{n-1}) + \nabla\Phi(\mathbf{X}_{n-1})[\mathbf{X}_n - \mathbf{X}_{n-1}] + HOT's,$$

where we encapsulated all the higher order terms of the expansion into the generic notation $HOT's$ which stands for "Higher Order Terms". If we are willing to ignore the $HOT's$, which is quite reasonable when $\|\mathbf{X}_n - \mathbf{X}_{n-1}\|$ is not too large, or if we include them in the observation noise term, then once in this form, the observation equation can be regarded as a linear equation given by the observation matrix $\nabla\Phi(\mathbf{X}_{n-1})$. This linearization idea is at the basis of what is called the extended Kalman filter.

7.3 Factor Models as Hidden Markov Processes

Before we consider the filtering and prediction problems, we pause to explain the important role played by the analysis of state space systems in financial econometrics. This section provides a general discussion somewhat of an abstract nature, and it can safely be skipped by the reader only interested in the mechanics of filtering and prediction.

7.3.1 **Factor Models**

Attempts at giving a general definition of factor models are usually hiding their intuitive simplicity. So instead of shooting for a formal introduction, we shall consider specific examples found in the financial arena. Let us assume that we are interested in tracking the performance of d_Y financial assets whose returns over the n-th time period we denote by $Y_{1,n}$, $Y_{2,n}$, ..., $Y_{d_Y,n}$. The assumption of a d-factor model is that the dynamics of these returns are driven by d economic factors X_1, X_2, ..., X_d which are also changing over time, possibly in a random manner. So we denote by $X_{1,n}$, $X_{2,n}$, ..., $X_{d,n}$ the values of these factors at time n. The main feature of a linear factor model is to assume that the individual returns $Y_{i,n}$ are related to the factors $X_{j,n}$ by a linear formula:

$$Y_{i,n} = g_{i,1,n}X_{1,n} + g_{i,2,n}X_{2,n} + \cdots + g_{i,d,n}X_{d,n} + w_{i,n} \tag{7.11}$$

for a given set of (deterministic) parameters $g_{i,j,n}$ and random residual terms $w_{i,n}$. Grouping the d_Y returns $Y_{i,n}$ in a d_Y-dimensional return vector $\mathbf{Y}_n$, grouping the d factors $X_{j,n}$ into a d-dimensional factor vector $\mathbf{X}_n$, grouping the noise terms $w_{i,n}$ in a d_Y-dimensional *observation* vector $\mathbf{W}_n$, and finally grouping all the parameters $g_{i,j,n}$ in a $d_Y \times d$ matrix G_n, we rewrite the system of equations (7.11) in the vector/matrix form:

$$\mathbf{Y}_n = G_n\mathbf{X}_n + \mathbf{W}_n$$

which is exactly the form of the observation equation (7.9) of a linear state-space model introduced in the previous section. In analogy with the CAPM terminology, the columns of the observation matrix G_n are called the *betas* of the underlying factors.

7.3.2 Assumptions of the Model

Most econometric analyses rely heavily on the notion of information: we need to keep track at each time n of the information available at that time. For the purposes of this section, we shall denote by $\mathcal{I}_n$ the information available at time n. Typically, this information will be determined by the knowledge of the values of the return vectors $\mathbf{Y}_n$, $\mathbf{Y}_{n-1}$, $\dots$ and the factor vectors $\mathbf{X}_n$, $\mathbf{X}_{n-1}$, $\dots$ available up to and including that time. In order to take advantage of this information, the properties of the models are formulated in terms of conditional probabilities and conditional expectations given this information. To be more specific, we assume that, conditionally on the past information $\mathcal{I}_{n-1}$, the quantities entering the factor model (7.11) satisfy:

- The residual terms $\mathbf{W}_n$ have mean zero, i.e. $\mathbb{E}\{\mathbf{W}_n|\mathcal{I}_{n-1}\} = 0$
- The variance/covariance matrix Σ_W of the residual terms does not change with time, i.e. $\text{var}\{\mathbf{W}_n|\mathcal{I}_{n-1}\} = \Sigma_W$
- The residual terms $\mathbf{W}_n$ and the factors $\mathbf{X}_n$ are uncorrelated:

$$\mathbb{E}\{\mathbf{X}_n\mathbf{W}_n^t|\mathcal{I}_{n-1}\} = 0.$$

It is clear that the residual terms behave like a white noise, and for this reason, it is customary to assume directly that they form a white noise independent of the random sequence of the factors.

Remarks.

1. There is no uniqueness in the representation (7.11). In particular, one can change the definition of the factors and still preserve the form of the representation. Indeed, if U is any invertible $k \times k$ matrix, one sees that:

$$\mathbf{Y}_n = G_n\mathbf{X}_n + \mathbf{W}_n = G_nU^{-1}U\mathbf{X}_n + \mathbf{W}_n = \tilde{G}_n\tilde{\mathbf{X}}_n + \mathbf{W}_n$$

   provided we set $\tilde{G}_n = G_nU^{-1}$ and $\tilde{\mathbf{X}}_n = U\mathbf{X}_n$. So the returns can be explained by the new factors given by the components of the vector $\tilde{\mathbf{X}}_n$. Not only does this argument show that there is no hope for any kind of uniqueness in the representation (7.11), but it also shows that one can replace the original factors by appropriate linear combinations, and this makes it possible to make sure that the rank of the matrix G_n can be equal to the number of factors. Indeed, if that is not the case, one can always replace the original factors by a minimal set of linear combinations with the same rank.
2. The aggregation of factors by linear combinations does have advantages from the mathematical point of view, but it also has serious drawbacks. Indeed, some of the factors may be observable (this is, for example, the case if one uses interest rates, or other published economic indicators in the analysis of portfolio returns) and bundling them together with generic factors, which may not be observable, may dilute this desirable (practical) property of some of the factors.

7.3.3 Dynamics of the Factors

One way to prescribe the time evolution of the factors is by giving the conditional distribution of the factor vector $\mathbf{X}_n$ given its past values $\mathbf{X}_{n-1}, \mathbf{X}_{n-2}, \ldots$.

A typical example is given by the ARCH prescription which we will analyze in Chap. 8. This model is especially simple in the case of a one-factor model, i.e. when $d = 1$. In this case, the dynamics of the factor are prescribed by the conditional distribution

$$X_n|X_{n-1}, X_{n-2}, \ldots \sim N(0, \sigma_n^2) \qquad \text{with} \qquad \sigma_n^2 = \alpha_0 + \sum_{i=1}^{p} \alpha_i X_{n-i}^2. \tag{7.12}$$

In the present chapter, we restrict ourselves to linear factor models for which the dynamics of the factors are given by an equation of the form:

$$\mathbf{X}_n = F_n \mathbf{X}_{n-1} + \boldsymbol{V}_n \tag{7.13}$$

for some $d \times d$ matrix F_n (possibly changing with n) and a d-variate white noise $\boldsymbol{V}$ which is assumed to be independent of the returns $\mathbf{Y}$. In other words, the conditional distribution of $\mathbf{X}_n$ depends only upon its last value $\mathbf{X}_{n-1}$, the latter providing the conditional mean $F_n \mathbf{X}_{n-1}$, while the fluctuations around this mean are determined by the distribution of the noise $\boldsymbol{V}_n$. Given all that, we are now in the framework of the linear state-space models introduced in the previous section.

Remarks.

1. As we shall see in Sect. 7.6 on the state-space representation of time series, it is always possible to accommodate more general dependence involving more of the past values $\mathbf{X}_{n-2}, \mathbf{X}_{n-3}, \ldots$ in Eq. (7.13). Indeed, increasing the dimension of the factor vector, one can always include these past values in the current factor vector to reduce the dependence of $\mathbf{X}_n$ upon $\mathbf{X}_{n-1}$ only.
2. Factors are called exogenous when their values are dependent upon quantities (indexes, instruments, ...) external to the system formed by the returns included in the state vector. They are called endogenous when they can be expressed as functions of the individual returns entering the state vector. There is a mathematical result that states that any *linear* factor model can be rewritten in such a way that all the factors become endogenous. This anti-climatic result is limited to the linear case, and because of its counter-intuitive nature, we shall not use it.

7.4 Kalman Filtering of Linear Systems

This section is devoted to the derivation of the recursive equations giving the optimal filter for linear (Gaussian) systems. They were discovered simultaneously and independently by Kalman and Bucy in the late 1950s, but strangely enough, they are most of the time referred to as the Kalman filtering equations.

7.4.1 One-Step-Ahead Prediction

To refresh our memory, we restate the definition of the linear prediction operator originally introduced in Chap. 6 when we discussed partial auto-correlation functions. Given observations $\mathbf{Y}_1, \ldots, \mathbf{Y}_n$ up to and including n (which we should think of as the present time), and given a random vector $\boldsymbol{Z}$ with the same dimension as $\mathbf{Y}$, we denote by $E_n(\boldsymbol{Z})$ the best prediction of $\boldsymbol{Z}$ by linear combinations of the $\mathbf{Y}_m$ for $0 \leq m \leq n$. Since "*best prediction*" is implicitly understood in the least squares sense, $E_n(\boldsymbol{Z})$ is the linear combination $\alpha_1 \mathbf{Y}_1 + \cdots + \alpha_n \mathbf{Y}_n$ which minimizes the quadratic error:

$$\mathbb{E}\{\|\boldsymbol{Z} - (\alpha_1 \mathbf{Y}_1 + \cdots + \alpha_n \mathbf{Y}_n)\|^2\}.$$

The best linear prediction $E_n(\boldsymbol{Z})$ should be interpreted as the orthogonal projection of $\boldsymbol{Z}$ onto the span of $\mathbf{Y}_1, \ldots, \mathbf{Y}_n$. Unfortunately, in some applications, it may not be natural to restrict the approximation to linear combinations of the $\mathbf{Y}_j$'s. If we lift this restriction and allow all possible functions of the $\mathbf{Y}_j$'s in the approximation, then $E_n(\boldsymbol{Z})$ happens to be the conditional expectation of $\boldsymbol{Z}$ given the knowledge of the $\mathbf{Y}_m$ for $1 \leq m \leq n$. Notice that all the derivations below are true in both cases, i.e. whether $E_n(\boldsymbol{Z})$ is interpreted as the best linear prediction or the conditional expectation. In fact, when the random vectors $\boldsymbol{Z}$ and $\mathbf{Y}_m$ for $m \leq n$ are jointly Gaussian (or more generally jointly elliptically distributed), then the two definitions of $E_n(\boldsymbol{Z})$ coincide. For these reasons, we shall not insist on which actual definition of $E_n(\boldsymbol{Z})$ we use. For the sake of definiteness, we shall assume that the observations and the state vectors are jointly Gaussian and we shall understand the notation E_n as a conditional expectation.

The strength of the theory developed in this section is based on the fact that it is possible to compute the best one-step-ahead predictions recursively. More specifically, if the best prediction $\hat{\mathbf{X}}_n$ of the unobserved state $\mathbf{X}_n$ (computed on the basis of the observations $\mathbf{Y}_m$ for $1 \leq m \leq n-1$) is known, the computation of the next best guess $\hat{\mathbf{X}}_{n+1}$ requires only the knowledge of $\hat{\mathbf{X}}_n$ and the new observation $\mathbf{Y}_n$. Well, this is almost true. The only *little lie* comes from the fact that, not only should we know the current one-step-ahead prediction, but also its prediction quadratic error as defined by the matrix:

$$\Omega_n = \mathbb{E}\{(\mathbf{X}_n - \hat{\mathbf{X}}_n)(\mathbf{X}_n - \hat{\mathbf{X}}_n)^t\}.$$

7.4.2 Derivation of the Recursive Filtering Equations

We now proceed to the rigorous derivation of this claim, and to the derivation of the formulae which we shall use in the practical filtering applications that we consider in this text. This derivation is based on the important concept of innovation. The innovation series is given by its entries I_n defined as:

$$I_n = \mathbf{Y}_n - E_{n-1}(\mathbf{Y}_n).$$

The innovation I_n gives the information contained in the latest observation $\mathbf{Y}_n$ which was not already contained in the previous observations. The discussion which follows is rather technical and the reader interested in practical applications more than theoretical derivations can skip the next two pages of computations, and jump directly to the boxed recursive update formulae.

First we remark that the innovations I_n are mean zero and uncorrelated (and consequently independent in the Gaussian case). Indeed:

$$\mathbb{E}\{I_n\} = \mathbb{E}\{\mathbf{Y}_n\} - \mathbb{E}\{E_{n-1}(\mathbf{Y}_n)\} = \mathbb{E}\{\mathbf{Y}_n\} - \mathbb{E}\{\mathbf{Y}_n\} = 0$$

and a geometric argument (based on the properties of orthogonal projections), gives the fact that $I_n = \mathbf{Y}_n - E_{n-1}(\mathbf{Y}_n)$ is orthogonal to all linear combinations of the $\mathbf{Y}_1, \ldots, \mathbf{Y}_{n-1}$, and consequently orthogonal to $I_{n-1}, I_{n-2}, \ldots$ which are particular linear combinations. In other words, except possibly for the fact that they may not have the same variance/covariance matrix, the I_n form an independent (or at least uncorrelated) sequence, and they can be viewed as a white noise of their own.

We now proceed to the computation of the variance/covariance matrices of the innovation vectors I_n. Notice that $E_{n-1}(\mathbf{W}_n) = \mathbf{0}$ since $\mathbf{W}_n$ is independent of the past observations $\mathbf{Y}_1, \ldots, \mathbf{Y}_{n-1}$. Consequently, applying the prediction operator E_{n-1} to both sides of the observation equation we get:

$$E_{n-1}(\mathbf{Y}_n) = E_{n-1}(G\mathbf{X}_n) + E_{n-1}(\mathbf{W}_n) = GE_{n-1}(\mathbf{X}_n) = G\hat{\mathbf{X}}_n$$

and consequently:

$$\begin{aligned} I_n &= G\mathbf{X}_n + \mathbf{W}_n - G\hat{\mathbf{X}}_n \\ &= G(\mathbf{X}_n - \hat{\mathbf{X}}_n) + \mathbf{W}_n. \end{aligned}$$

$\mathbf{W}_n$ is independent of $\mathbf{X}_n$ by definition, moreover, $\mathbf{W}_n$ is also independent of $\hat{\mathbf{X}}_n$ because the latter is a linear combination of $\mathbf{Y}_1, \ldots, \mathbf{Y}_{n-1}$ which are all independent of $\mathbf{W}_n$. Consequently, the two terms appearing in the above right hand side are independent, and the variance/covariance matrix of I_n is equal to the sum of the variance/covariance matrices of these two terms. Consequently:

$$\begin{aligned} \Sigma_{I_n} &= G\mathbb{E}\{(\mathbf{X}_n - \hat{\mathbf{X}}_n)(\mathbf{X}_n - \hat{\mathbf{X}}_n)^t\}G^t + \mathbb{E}\{\mathbf{W}_n\mathbf{W}_n^t\} \\ &= G\Omega_n G^t + \Sigma_W. \end{aligned} \tag{7.14}$$

This very geometric argument also implies that the operator E_n which gives the best linear predictor as a function of $\mathbf{Y}_1, \ldots, \mathbf{Y}_{n-1}$, and $\mathbf{Y}_n$ can also be viewed as the best linear predictor as a function of $\mathbf{Y}_1, \ldots, \mathbf{Y}_{n-1}$ and I_n. In particular this implies that:

$$\begin{aligned} \hat{\mathbf{X}}_{n+1} &= E_{n-1}(\mathbf{X}_{n+1}) + \mathbb{E}\{\mathbf{X}_{n+1}|I_n\} \\ &= E_{n-1}(F\mathbf{X}_n + \boldsymbol{V}_{n+1}) + \mathbb{E}\{\mathbf{X}_{n+1}I_n^t\}\Sigma_{I_n}^{-1}I_n \\ &= F\hat{\mathbf{X}}_n + \mathbb{E}\{\mathbf{X}_{n+1}I_n^t\}\Sigma_{I_n}^{-1}I_n, \end{aligned} \tag{7.15}$$

where we computed the conditional expectation as an orthogonal projection, and where we used the fact that $E_{n-1}(\boldsymbol{V}_{n+1}) = \mathbf{0}$ since $\boldsymbol{V}_{n+1}$ is independent of $\mathbf{Y}_1$,

..., $\mathbf{Y}_{n-1}$, and $\mathbf{Y}_n$. Note that we assume that the variance/covariance matrix of the innovation is invertible. Also, note that:

$$\begin{aligned}\mathbb{E}\{\mathbf{X}_{n+1}I_n^t\} &= \mathbb{E}\{(F\mathbf{X}_n + \boldsymbol{V}_{n+1})[(\mathbf{X}_n - \hat{\mathbf{X}}_n)^t G^t + \mathbf{W}_n^t]\}\\ &= \mathbb{E}\{F\mathbf{X}_n(\mathbf{X}_n - \hat{\mathbf{X}}_n)^t G^t\}\\ &= F\mathbb{E}\{(\mathbf{X}_n - \hat{\mathbf{X}}_n)(\mathbf{X}_n - \hat{\mathbf{X}}_n)^t\}G^t\\ &= F\Omega_t G^t, \end{aligned} \tag{7.16}$$

where we used the facts that:

1. $\boldsymbol{V}_{n+1}$ is independent of $\mathbf{X}_n$, $\hat{\mathbf{X}}_n$ and $\mathbf{W}_n$;
2. $\mathbf{W}_n$ is independent of $\mathbf{X}_n$; and finally
3. $\mathbb{E}\{\hat{\mathbf{X}}_n(\mathbf{X}_n - \hat{\mathbf{X}}_n)^t\} = 0$.

At this stage, it is useful to introduce the following notation.

$$\begin{cases} \Delta_n = G\Omega_n G^t + \Sigma_W \\ \Theta_n = F\Omega_n G^t \end{cases} \tag{7.17}$$

Notice that (7.14) shows that the matrix Δ_n is just the variance/covariance matrix of the innovation random vector I_n. The matrix $\Theta_n \Delta_n^{-1}$ which appears in several of the important formulae derived below is sometimes called the "Kalman gain" matrix in the technical filtering literature. This terminology finds its origin in the fact that using (7.16) and (7.17), we can rewrite (7.15) as:

$$\hat{\mathbf{X}}_{n+1} = F\hat{\mathbf{X}}_n + \Theta_n \Delta_n^{-1} I_n. \tag{7.18}$$

This formula gives a (recursive) update equation for the one-step-ahead predictions, but hidden in the correction term $\Theta_n \Delta_n^{-1} I_t$, is the error matrix Ω_n. So this update equation for the one-step-ahead prediction of the state cannot be implemented without being complemented with a practical procedure to compute the error Ω_n. We do that now.

$$\begin{aligned}\Omega_{n+1} &= \mathbb{E}\{(\mathbf{X}_{n+1} - \hat{\mathbf{X}}_{n+1})(\mathbf{X}_{n+1} - \hat{\mathbf{X}}_{n+1})^t\}\\ &= \mathbb{E}\{\mathbf{X}_{n+1}\mathbf{X}_{n+1}^t\} - \mathbb{E}\{\mathbf{X}_{n+1}\hat{\mathbf{X}}_{n+1}^t\} - \mathbb{E}\{\hat{\mathbf{X}}_{n+1}\mathbf{X}_{n+1}^t\} + \mathbb{E}\{\hat{\mathbf{X}}_{n+1}\hat{\mathbf{X}}_{n+1}^t\}\\ &= \mathbb{E}\{\mathbf{X}_{n+1}\mathbf{X}_{n+1}^t\} - \mathbb{E}\{\hat{\mathbf{X}}_{n+1}\hat{\mathbf{X}}_{n+1}^t\}\end{aligned}$$

because:

$$\mathbb{E}\{\mathbf{X}_{n+1}\hat{\mathbf{X}}_{n+1}^t\} = \mathbb{E}\{\hat{\mathbf{X}}_{n+1}\mathbf{X}_{n+1}^t\} = \mathbb{E}\{\hat{\mathbf{X}}_{n+1}\hat{\mathbf{X}}_{n+1}^t\}.$$

Consequently:

$$\begin{aligned}\Omega_{n+1} &= \mathbb{E}\{(F\mathbf{X}_n + \boldsymbol{V}_{n+1})(F\mathbf{X}_n + \boldsymbol{V}_{n+1})^t\}\\ &\qquad - \mathbb{E}\{(F\hat{\mathbf{X}}_n + \Theta_n\Delta_n^{-1}I_n)(F\hat{\mathbf{X}}_n + \Theta_n\Delta_n^{-1}I_n)^t\}\\ &= F\mathbb{E}\{\mathbf{X}_n\mathbf{X}_n^t\}F^t + \mathbb{E}\{\boldsymbol{V}_{n+1}\boldsymbol{V}_{n+1}^t\}\\ &\qquad - F\mathbb{E}\{\hat{\mathbf{X}}_n\hat{\mathbf{X}}_n^t\}F^t + \Theta_n\Delta_n^{-1}\mathbb{E}\{I_nI_n^t\}\Delta_n^{-1}\Theta_n^t\\ &= F(\mathbb{E}\{\mathbf{X}_n\mathbf{X}_n^t\} - \mathbb{E}\{\hat{\mathbf{X}}_n\hat{\mathbf{X}}_n^t\})F^t + \Sigma_V + \Theta_n\Delta_n^{-1}\Theta_n^t\\ &= F\Omega_nF^t + \Sigma_V + \Theta_n\Delta_n^{-1}\Theta_n^t,\end{aligned}$$

where we used the facts that:

1. V_{n+1} is independent of $\mathbf{X}_n$;
2. I_n is independent of (or at least orthogonal to) $\widehat{\mathbf{X}_n}$;
3. The matrix Δ_n is symmetric since it is a variance/covariance matrix.

For later reference, we summarize the recursion formulae in a box:

$$\hat{\mathbf{X}}_{n+1} = F\hat{\mathbf{X}}_n + \Theta_n \Delta_n^{-1}(\mathbf{Y}_n - G\hat{\mathbf{X}}_n)$$
$$\Omega_{n+1} = F\Omega_n F^t + \Sigma_V - \Theta_n \Delta_n^{-1}\Theta_n^t$$

Recall the definitions (7.17) for the meanings of the matrices Δ_t and Θ_t.

7.4.2.1 ***Writing an*** R ***Function to Do Just That***

The R function kalman from the Rsafd library computes the one-step-ahead prediction using the above recursive formulas. Its code reads:

```
kalman <- function(FF,SigV,GG,SigW,Xhat,Omega,Y)
{
  Delta <- GG %*% Omega %*% t(GG) + sigW
   Theta <- FF %*% Omega %*% t(GG)
   X <- FF%*%Xhat + Theta%*%solve(Delta)%*%(Y-GG%*%Xhat)
   Om <- FF%*%Omega%*%t(FF) + SigV
                       - Theta%*%solve(Delta)%*%t(Theta)
   Ret <- list(xpred = X, error=Om)
   Ret
}
```

The following remarks should help understand the features of this R function.

1. As we already mentioned, the transpose of a matrix is obtained by the function t, so that t(A) stands for the transpose of the matrix A.
2. The inverse of a matrix A is given by solve(A) (see the online help for the explanation of this terminology).
3. R has a certain number of *reserved* symbols which cannot be used as object names if one does not want to mask the actual R objects. This is the case for the symbols t, c, ... which we already encountered, but also for F which means FALSE. For this reason we used the notation FF for the system matrix. Similarly we used the notation GG for the observation matrix.
4. A call to the function kalman must be of the form:

   ```
   > PRED  <- kalman(FF,SigV,GG,SigW,Xhat,Omega,Y)
   ```

   and the R object PRED returned by the function is what is defined on the last line of the body of the function. In the present situation, it is a list with two elements, the one-step-ahead prediction for the next state of the system, and the estimate of the quadratic error. These elements can be extracted from the list by PRED$xpred and PRED$error respectively.

5. The above code was written with pedagogy in mind, not efficiency. It is not optimized. In particular, the inversion of the matrix `Delta`, and the product of the inverse by `Theta` are computed twice. This is a waste of computer resources. This code can easily be streamlined, and made more efficient, but we refrained from doing so for the sake of clarity.

7.4.3 **Filtering**

We now show how to derive a recursive update for the filtering problem by deriving the zero-step-ahead prediction (i.e. the simultaneous estimation of the unobserved state) from the recursive equations giving the one-step-ahead prediction of the unobserved state. From this point on, we use the notation:

$$\hat{\boldsymbol{Z}}_{n|m} = E_m(\boldsymbol{Z}_n)$$

for the $(n - m)$ step(s) ahead prediction of $\boldsymbol{Z}_n$ given the information contained in $\mathbf{Y}_1, \ldots, \mathbf{Y}_m$. With this notation, the one step ahead prediction analyzed in the previous subsection can be rewritten as:

$$\hat{\mathbf{X}}_{n+1} = \hat{\mathbf{X}}_{n+1|n}.$$

For the filtering problem, the quantity of interest is the best prediction:

$$\hat{\mathbf{X}}_{n|n} = E_n(\mathbf{X}_n)$$

of $\mathbf{X}_n$ by a linear function of the observations $\mathbf{Y}_n, \mathbf{Y}_{n-1}, \ldots, \mathbf{Y}_1$. As in the case of the one-step prediction, we cannot find recursive update formulae without involving the update of the error covariance matrix:

$$\Omega_{n|n} = \mathbb{E}\{(\mathbf{X}_n - \hat{\mathbf{X}}_{n|n})(\mathbf{X}_n - \hat{\mathbf{X}}_{n|n})^t\}.$$

The innovation argument used earlier gives:

$$\begin{aligned} E_n(\mathbf{X}_n) &= E_{n-1}(\mathbf{X}_n) + \mathbb{E}\{\mathbf{X}_n I_n^t\}\mathbb{E}\{I_n I_n^t\}^{-1} I_n \\ &= \hat{\mathbf{X}}_n + \mathbb{E}\{\mathbf{X}_n (G(\mathbf{X}_n - \hat{\mathbf{X}}_n) + \mathbf{W}_n)^t\}\Delta_n^{-1} I_n \\ &= \hat{\mathbf{X}}_n + \Omega_n G^t \Delta_n^{-1} I_n \end{aligned}$$

and computing Ω_n as a function of $\Omega_{n|n}$ we get:

$$\Omega_n = \Omega_{n|n} + \Omega_n G^t \Delta_n^{-1} G \Omega_n^t.$$

We summarize these update formulae in a box for later references:

$$\hat{\mathbf{X}}_{n|n} = \hat{\mathbf{X}}_n + \Omega_n G^t \Delta_n^{-1}(\mathbf{Y}_n - G\hat{\mathbf{X}}_n)$$
$$\Omega_{n|n} = \Omega_n - \Omega_n G \Delta_n^{-1} G \Omega_n^t.$$

Remark. Later in the text, we will illustrate the use of the filtering recursions on the example of the "time varying beta's" version of the CAPM model for which the observation matrix G changes with n. A careful look at the above derivations shows that the recursive formulae still hold as long as we replace the matrix G by its value at time n. The above `R` function needs to be modified, either by passing the whole sequence $\{G_n\}_n$ of matrices as parameter to the function, or by adding the code necessary to compute the matrices G_n on the fly whenever possible. This will appear natural in the application that we give later in the chapter.

7.4.4 More Predictions

Using the same arguments as above we can construct all sorts of predictions.

7.4.4.1 k-Steps-Ahead Prediction of the Unobserved State of the System

Let us first compute the *two steps ahead prediction.*

$$\begin{aligned}\hat{\mathbf{X}}_{n+2|n} = E_n(\mathbf{X}_{n+2}) &= E_n(F\mathbf{X}_{n+1} + \boldsymbol{V}_{n+2}) \\ &= FE_n(\mathbf{X}_{n+1}) \\ &= F\hat{\mathbf{X}}_{n+1|n}\end{aligned}$$

with the notation of the previous subsection for the one-step-ahead prediction. We used the fact that $\boldsymbol{V}_{n+2}$ is independent of the observations $\mathbf{Y}_m$ with $m \leq n$. The above result shows how to compute the two-steps-ahead prediction in terms of the one-step-ahead prediction. One computes the *three steps ahead prediction* similarly. Indeed:

$$\begin{aligned}\hat{\mathbf{X}}_{n+3|n} = E_n(\mathbf{X}_{n+3}) &= E_n(F\mathbf{X}_{n+2} + \boldsymbol{V}_{n+3}) \\ &= FE_n(\mathbf{X}_{n+2}) \\ &= F\hat{\mathbf{X}}_{n+2|n} = F^2\hat{\mathbf{X}}_{n+1|n}\end{aligned}$$

Obviously, the k-steps-ahead prediction is given by the formula:

$$\hat{\mathbf{X}}_{n+k|n} = F^{k-1}\hat{\mathbf{X}}_{n+1|n} \tag{7.19}$$

and the corresponding error

$$\Omega_{n+k|n} = \mathbb{E}\{(\mathbf{X}_{n+k} - \hat{\mathbf{X}}_{n+k|n})(\mathbf{X}_{n+k} - \hat{\mathbf{X}}_{n+k|n})^t\}$$

is given by the formula:

$$\Omega_{n+k|n} = F^{k-1}\Omega_{n+1|n}(F^{k-1})^t.$$

Notice that $\Omega_{n+1|n}$ was denoted by Ω_{n+1} earlier in our original derivation of the one-step ahead prediction. We summarize these results in a box for easier reference.

$$\begin{aligned}\hat{\mathbf{X}}_{n+k|n} &= F^{k-1}\hat{\mathbf{X}}_{n+1|n}\\ \Omega_{n+k|n} &= F^{k-1}\Omega_{n+1|n}(F^{k-1})^t.\end{aligned}$$

7.4.4.2 *k* ***Steps Ahead Predition of the Observations***

Finally, we remark that it is also possible to give formulae for the k-steps-ahead predictions of the future observations. Indeed:

$$\begin{aligned}\hat{\mathbf{Y}}_{n+1|n} = E_n(\mathbf{Y}_{n+1}) &= E_n(G\mathbf{X}_{n+1} + \mathbf{W}_{n+1})\\ &= GE_n(\mathbf{X}_{n+1})\\ &= G\hat{\mathbf{X}}_{n+1|n}\end{aligned}$$

because $\mathbf{W}_{n+1}$ is independent of all the $\mathbf{Y}_m$ for $m \leq n$. More generally:

$$\hat{\mathbf{Y}}_{n+k|n} = GF^{k-1}\hat{\mathbf{X}}_{n+1|n} \tag{7.20}$$

and if we use the notation

$$\Delta_n^{(k)} = \mathbb{E}\{(Y_{n+k} - \hat{\mathbf{Y}}_{n+k|n})(Y_{n+k} - \hat{\mathbf{Y}}_{n+k|n})^t\}$$

for its prediction quadratic error, it follows that:

$$\Delta_n^{(k)} = GF^{k-1}\Omega_{n+1|n}(GF^{k-1})^t.$$

As before we highlight the final formulae in a box.

$$\begin{aligned}\hat{\mathbf{Y}}_{n+k|n} &= GF^{k-1}\hat{\mathbf{X}}_{n+1|n}\\ \Delta_n^{(k)} &= GF^{k-1}\Omega_{n+1|n}(GF^{k-1})^t.\end{aligned}$$

7.4.5 Estimation of the Parameters

The recursive filtering equations derived in this section provide optimal estimators, and as such, they can be regarded as some of the most powerful tools of data analysis. Unfortunately, their implementation requires the knowledge of the parameters of the model. Because of physical considerations, or because we are often in control of how the model is set up, the observation matrix G is very often known, and we shall focus the discussion on the remaining parameters, namely, the state transition matrix F, the variance/covariance matrices Σ_V and Σ_W of the system and observation noises, and the initial estimates of the state and its error matrix. This parameter estimation problem is extremely difficult, and it can be viewed as filtering's Achilles heel.

Several tricks have been proposed to get around this difficulty, including the parameters in the state vector being one of them. The present theory implies that this procedure should work very well when the parameters enter linearly (or almost linearly) in the state and observation equations. However, despite the fact that the theory is not completely developed in the nonlinear case, practitioners are using this trick in many practical implementations, whether or not the system is linear.

For the sake of completeness we describe the main steps of the maximum likelihood approach to parameter estimation in this setup. This approach requires the computation and optimization of the likelihood function. The latter can be derived when all the random quantities of the model are jointly Gaussian. This is the case when the white noise series $\{\boldsymbol{V}_n\}_n$ and $\{\mathbf{W}_n\}_n$ are Gaussian, and when the initial value of the state is also Gaussian (and independent of both noise series). So instead of looking for exact values of the initial state vector and its error matrix, one assumes that this initial state vector $\mathbf{X}_0$ is Gaussian, and we add its mean vector μ_0 and its variance/covariance matrix Σ_0 to the list of parameters to estimate. The computations of the previous subsection showed that the innovations satisfy:

$$I_n = \mathbf{Y}_n - G\hat{\mathbf{X}}_n$$

and we also argued the fact that they are independent (recall that we are now restricting ourselves to the Gaussian case), and we derived recursive formulae to compute their variance/covariance matrices Δ_n. The above equation, together with the fact that one can compute the one-step-ahead predictions $\hat{\mathbf{X}}_n$ recursively from the data, allows us to compute the innovations from the values of the observations, and reformulate the likelihood problem in terms of the innovations only. This simple remark simplifies the computations. Given all this, one can write down the joint density of the I_n's in terms of the observations $\mathbf{Y}_n = \boldsymbol{y}_n$ and the successive values of the matrices Δ_n and of the one-step-ahead predictions $\hat{\mathbf{X}}_n$ which one computes inductively from the recursive filtering equations. Because of the special form of the multivariate normal density, the new log-likelihood function is of the form:

$$-2\log L_{I_1,\ldots,I_n}(\Theta) = \text{cst} + \sum_{j=1}^{n} \log\det(\Delta_j(\Theta)) + \sum_{j=1}^{n} I_j(\Theta)\Delta_j(\Theta)^{-1}I_j(\Theta)^t$$

where cst stands for a constant of no consequence, and where we emphasized the dependence of the innovations I_j and their variance/covariance matrices Δ_j upon the vector Θ of unknown parameters, $\Theta = (\mu_0, \Sigma_0, F, \Sigma_V, \Sigma_W)$. Notice that the dimension of this parameter vector may be large. Indeed its components are matrices and vectors whose dimensions can be large. In any case, this log-likelihood function is a non-convex function of the multi-dimensional parameter Θ and its maximization is quite difficult. Practical implementations of optimization algorithms have been used by the practitioners in the field: Newton-Raphson, EM algorithm, ... are among those, but no single method seems to be simple enough and reliable enough for us to discuss this issue further at the level of this text.

7.5 Applications to Linear Models

Linear models were introduced in Chap. 4 as a convenient framework for linear regression. Their versatility made it possible to apply their theory to several specific classes of nonlinear regression problems such as polynomial and natural spline regression. We now recast these linear models in the framework of partially observed state space systems, and we take advantage of the filtering tools developed in this chapter to introduce generalized forms of linear models, and to tackle several new problems which could not have been addressed with the tools of Chap. 4.

7.5.1 State Space Representation of Linear Models

Combining the original notation of Sect. 4.5 with the notation introduced in this chapter for state space models, we give a new interpretation to the multiple linear regression setup:

$$y_n = \mathbf{X}_n \boldsymbol{\beta} + \epsilon_n \tag{7.21}$$

where $\{\epsilon_n\}_n$ is a white noise in the strong sense, where $\mathbf{X}_n = (x_{1,n}, x_{2,n}, \ldots, x_{d,n})$ is a $d = p+1$ dimensional vector of explanatory variables, and $\boldsymbol{\beta} = (\beta_1, \beta_2, \ldots, \beta_d)$ is a d-dimensional vector of unknown parameters. Notice that we are now using the lower case n to label the observations, while we were using the lower case i when we introduced the linear models in Sect. 4.5. The novelty of the present approach is to interpret equation (7.21) as the observation equation of a state space system whose dynamics are very simple since they do not change over time. To be more specific, we consider a state vector $\mathbf{X}_n$ always equal to the vector $\boldsymbol{\beta}$ of parameters. Hence, the dynamical equation reads:

$$\mathbf{X}_{n+1} = \mathbf{X}_n. \tag{7.22}$$

In other words, the state matrix F_n does not change with n, and is always equal to the $d \times d$ identity matrix, while the state noise is identically zero. Setting $G_n = \mathbf{X}_n$, Eq. (7.21) coincides with the observation equation (7.9) if we set $\mathbf{Y}_n = y_n$ and $\mathbf{W}_n = \epsilon_n$. This rewriting of a linear model as a state space model can appear artificial at times, but it makes it possible to apply the powerful tools of filtering theory to these models. We proceed to demonstrate by example some of the fringe benefits of this reformulation.

7.5.1.1 ***Recursive Estimation***

Recasting linear models as state space models is especially useful when we need to recompute the least squares estimates of the parameters after an observation is added. Indeed, if we denote by $\hat{\boldsymbol{\beta}}_n$ the least squares estimate of $\boldsymbol{\beta}$ computed from the data $(\boldsymbol{x}_1, y_1)$, ..., $(\mathbf{x}_n, y_n)$, then the Kalman recursive filtering equations give a very convenient way to compute the least squares estimate $\hat{\boldsymbol{\beta}}_{n+1}$ as an update of the previous estimate $\hat{\boldsymbol{\beta}}_n$ using the current observation $(\boldsymbol{x}_{n+1}, y_{n+1})$. If we use the notation K_n for the Kalman gain matrix introduced earlier, we get:

$$\hat{\boldsymbol{\beta}}_{n+1} = \hat{\boldsymbol{\beta}}_n + K_{n+1}(y_{n+1} - \boldsymbol{x}_{n+1}\hat{\boldsymbol{\beta}}_n), \tag{7.23}$$

with the corresponding update for the estimated error variance:

$$\Omega_{n+1} = [I - K_{n+1}\mathbf{X}_{n+1}]\Omega_n.$$

7.5.1.2 ***Recursive Residuals***

Some of the most efficient tests for change in a model are based on the analysis of the residuals $r_n = y_n - \mathbf{X}_n\hat{\boldsymbol{\beta}}_n$. Again, the recursive filtering equations derived in the previous section make it possible to update these residuals recursively without having to recompute them from scratch each time a new observation is made available. The recursive standardized residuals w_n are usually defined by the formula:

$$w_n = \frac{y_n - \mathbf{X}_n\hat{\boldsymbol{\beta}}_{n-1}}{\sqrt{1 + \mathbf{X}_n(\mathbf{X}_{n-1}^t\mathbf{X}_{n-1})^{-1}\boldsymbol{x}_n^t}}.$$

Unexpectedly, the recursive residuals defined in this manner form a sequence of independent $N(0, \sigma^2)$ random variables. This makes their distribution theory very easy. These remarkable properties were first identified and used by Brown, Durbin and Evans who derived a series of useful tests, called CUSUM tests, for the adaptive detection of changes in a linear model. It appears that `R` does not have any implementation of the CUSUM test. Its function `cusum` serves another purpose.

7.5.2 **Linear Models with Time Varying Coefficients**

This subsection discusses a generalization of linear models based on the idea introduced in the previous subsection. There, the state equation was trivial because the state did not change with time. Since filtering theory deals with state vectors varying with time, it is natural in this context to consider linear models where the parameter vector $\boldsymbol{\beta}$ can change with time. Indeed, the theory presented in this chapter allows to consider time-varying parameters $\boldsymbol{\beta}_n$ satisfying:

$$\boldsymbol{\beta}_{n+1} = F\boldsymbol{\beta}_n + \boldsymbol{V}_{n+1} \tag{7.24}$$

for some white noise $\{\boldsymbol{V}_n\}_n$ and some deterministic matrix F. Writing the corresponding linear model one component at a time as before, we get the same observation equation as (7.21):

$$y_n = \boldsymbol{x}_n\boldsymbol{\beta}_n + \epsilon_n \tag{7.25}$$

Notice that the update equation for $\hat{\boldsymbol{\beta}}_n$ is slightly more involved than (7.23), since we do not have $F = I$ and $\Sigma_V = 0$ as before.

We give a detailed implementation of this idea in the application to the CAPM discussed in the next subsection.

7.5.3 CAPM with Time Varying β's

The Capital Asset Pricing Model (CAPM for short) of Lintner and Sharpe was introduced in Sect. 4.5.5. There we argued that, even when empirical evidence was not significant enough to reject the zero-intercept assumption of the model, instability over time of the betas did not always support the model despite its economic soundedness and its popular appeal. We now try to reconciliate CAPM with empirical data by allowing the betas to vary with time. Recall that according to CAPM, we assume that, at each time t, the excess return $\tilde{R}_{j,t}$ of the j-th asset over the risk-free rate r of lending and borrowing is given, up to an additive noise term, by a multiple of the excess return $\tilde{R}_t^{(m)}$ of the market portfolio at the same time t. In other words, the CAPM model states that:

$$\tilde{R}_{j,t} = \beta_j\tilde{R}_t^{(m)} + \epsilon_{j,t} \tag{7.26}$$

for some white noise $\{\epsilon_{j,t}\}_t$ specific to the j-th asset.

7.5.3.1 Time Varying Reformulation

As we mentioned in Chap. 4, there is empirical evidence that the hypotheses underlying CAPM theory do not always hold. And even if one is willing to accept them for their appealing economic rationale, the estimates of the β_j appear to be quite unstable, varying with economic factors. Given that fact, we propose to generalize the CAPM model by allowing the betas to vary with time. For the sake of illustration, we choose the simplest possible model, and we assume for example that our time-varying betas follows a random walk. So switching to the notation n for time instead of t, for each index j, we assume that:

$$\beta_{j,n+1} = \beta_{j,n} + v_{j,n+1} \tag{7.27}$$

for some white noise $\{v_{j,n}\}_n$ with unknown variance σ_j^2. Choosing for state at time n the vector of betas under consideration, i.e. setting $\mathbf{X}_n = \boldsymbol{\beta}_n$, we can rewrite the various equations (7.27) in a vector form:

$$\mathbf{X}_{n+1} = \mathbf{X}_n + \boldsymbol{V}_{n+1}, \tag{7.28}$$

giving the dynamics of the state. Here the state matrix F is the $d_X \times d_X$ identity matrix where d_X is the number of stocks. For the sake of simplicity, we shall only consider this model one stock at a time, in which case $d_X = 1$, $\mathbf{X}_n = \beta_{j,n}$ and $\boldsymbol{V}_n = v_{j,n}$ since there is only one value for the index j, and the state matrix F is the number one. Since the excess returns can be observed, we use (7.26) as our observation equation. We can do that provided we set $\mathbf{Y}_n = \tilde{R}_{j,n}$, in which case $d_Y = 1$, we choose $\mathbf{W}_n = \epsilon_{j,t}$ for the observation noise, and the 1×1 matrix $\tilde{R}_t^{(m)}$ for the observation matrix G_n. Notice that the observation matrix changes with n. As we pointed out earlier, this is not a major problem. Indeed, even though the derivation of the recursive filtering equations was done with a time independent observation matrix G, the same formulae hold when it varies with n, we just have to use the right matrix at each time step.

So, given an initial estimate for the value of β_0, and given an initial estimate for its prediction error, we can implement the recursive filtering equations derived in this chapter to track the changes over time of the beta of each stock. Notice that this assumes that we know the state and observation variances σ_v^2 and σ_ϵ^2. Further analysis is proposed in Problem 7.13.

It is important to notice that the estimates of the betas are non-anticipative in the sense that the estimate $\hat{\boldsymbol{\beta}}_n$ at time n depends only upon the *past* values of the observed excess returns, i.e. the values of $\tilde{R}_{j,m}$ for $m \leq n$ and not on the values of $\tilde{R}_{j,m}$ for $m > n$.

7.5.3.2 ***Filtering Experiment***

For the sake of illustration, we revisit the example of energy companies before and after the crisis that followed Enron's collapse. We already discussed in Sect. 4.5.5, the case of American Electric Power (AEP). We now consider the weekly excess returns of another major electric utility, DUKE Electric Power, over the period starting 01/01/1995, and ending 12/10/2010. Our analysis of a time-dependent CAPM will shed some light on the effects of the energy crisis, and clearly exhibit facts which are impossible to uncover in the classical time-independent setup. The data for AEP weekly excess returns are included in the library `Rsafd` and the reader is asked to perform this "time-varying beta" analysis in Problem 7.13.

Figure 7.4 shows the result of the non-anticipative estimation of the time varying betas by the Kalman filter. It shows that after a couple of years, the filter estimate of beta starts hovering below the level one, consistent with the fact that the stock was not considered risky before 2002. But despite the fact that DUKE was one of the very few energy companies to weather the crisis with little damage, its beta became greater than one in 2002 implying (according to the commonly admitted interpretation of the size of the beta) that this stock became risky at that time. This phenomenon is present for all the companies that were affected by the energy crisis. This is illustrated for example in Problem 7.13. For the sake of completeness, we give and comment the `R`

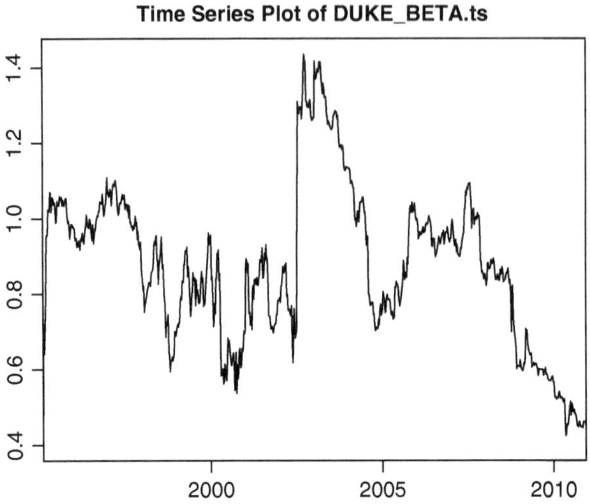

Fig. 7.4. Non-anticipative estimates of the DUKE time varying betas given by Kalman filtering over the period starting 01/01/1995 and ending 12/10/2010

code used to produce the results reproduced in Fig. 7.4. We first set up the parameters of the state space model with partial observations as explained above.

```
GG <- seriesData(MARKET.wer.ts)
NG <- length(GG)
SIGMAV <- .0001; SIGMAW <- .0001
YY <- seriesData(DUKE.wer.ts)
NY <- length(YY)
FF <- matrix(1,ncol=1,byrow=T)
SigV <- matrix(SIGMAV,ncol=1)
SigW <- matrix(SIGMAW,ncol=1)
```

The computations of the one-step-ahead predictions are stored in a matrix `Xhat` as follows:

```
Xhat <- matrix(rep(0,N),ncol=1)
Xhat[1] <- .4
Omega <- matrix(rep(0,N),ncol=1)
Omega[1] <- .035
for (n in 1:(N-1))
{
    KALMAN <- kalman(FF=FF, SigV=SigV, GG[n], SigW=SigW,
                      Xhat=Xhat[n], Omega=Omega[n], Y=YY[n])
    Xhat[n+1] <- KALMAN$xpred
    Omega[n+1] <- KALMAN$error
}
```

Then the (simultaneous) values of the filter estimate of the unobserved "beta" are computed, put in a `timeSeries` object and plotted.

```
DELTAn <- GG*Omega*GG + SIGMAW
Xhatnn <- Xhat + Omega*GG*(YY-GG*Xhat)/DELTAn
Omegann <- Omega -Omega*GG*GG*Omega/DELTAn
DUKE_BETA.ts <- timeSeries(positions=
                 seriesPositions(DUKE.wer.ts),data=Xhat)
plot.timeSeries(DUKE_BETA.ts)
```

7.6 State Space Representation of Time Series

As we already noticed, we are only considering state-space models whose dynamics are given by a multivariate AR(1) series. Since some of the components of the state vector may remain unobserved, we can always add new components to the state vector without changing the observations. This feature of the partially observed systems makes it possible to rewrite the equation defining an AR(p) model as an AR(1)-like equation, simply by adding the components $\mathbf{X}_{t-1}, \ldots, \mathbf{X}_{t-p}$ to the value of a (new and extended) state at time t. Adding the past values to the current value of the state enlarges the dimension, and this did not make sense when we were assuming that the entire state vector was observed. However, now that the observations are not necessarily complete, this transformation makes sense. This section takes advantage of this remark, but not without creating new problems. Indeed, a given set of observations vectors may correspond to many state space vectors, and consequently, as in the case of the factor models, we should be prepared to face a *lack of uniqueness* in the representation of a given stochastic system as a state space system.

7.6.1 The Case of AR Series

Let us first consider the simple example of an AR(1) model. Let us assume for example that $X \sim AR(1)$, and more precisely that:

$$X_t = .5X_{t-1} + W_t.$$

Switching to the index n, one can write:

$$\begin{cases} X_{n+1} = 0.5X_n + W_{n+1} \\ Y_n = X_n \end{cases}$$

This means that an AR(1) model is equivalent to a state space system described by the one dimensional (i.e. $d_X = 1$) state vector $\mathbf{X}_n = X_n$, its dynamics being given by the first of the equations above, the second equation giving the observation equation with $\mathbf{Y}_n = X_n$. Notice that the observations are perfect since there is no noise term in the observation equation.

Since this example is too simple to be indicative of what is really going on, we consider the case of an AR(2) model. Let us assume for example that:

$$X_t = 0.5X_{t-1} + 0.2X_{t-2} + W_t \tag{7.29}$$

for some white noise $\{W_t\}$. Switching once more to the notation n for the time stamp, we rewrite this definition in the form:

$$\begin{bmatrix} X_{n+1} \\ X_n \end{bmatrix} = \begin{bmatrix} 0.5 & 0.2 \\ 1 & 0 \end{bmatrix} \begin{bmatrix} X_n \\ X_{n-1} \end{bmatrix} + \begin{bmatrix} W_{n+1} \\ 0 \end{bmatrix} \tag{7.30}$$

Indeed, if we look at this equality between two vectors, the equality of the first component of the left hand side with the first component of the right hand side gives back the definition (7.29) of X_n, while the equality between the second components is merely a consistency relation giving no extra information. Now, if for each time index n we define the two-dimensional (i.e. $d_X = 2$) random vector $\mathbf{X}_n$ by:

$$\mathbf{X}_n = \begin{bmatrix} X_n \\ X_{n-1} \end{bmatrix} \tag{7.31}$$

the deterministic matrix F and the random vector $\boldsymbol{V}_n$ by:

$$F = \begin{bmatrix} 0.5 & 0.2 \\ 1 & 0 \end{bmatrix} \qquad \text{and} \qquad \boldsymbol{V}_n = \begin{bmatrix} W_n \\ 0 \end{bmatrix},$$

then Eq. (7.30) becomes:

$$\mathbf{X}_{n+1} = F\mathbf{X}_n + \boldsymbol{V}_{n+1}$$

which can be viewed as the equation giving the dynamics of the state $\mathbf{X}$. Using perfect observation, i.e. setting $\mathbf{Y}_n = X_n$, which corresponds to $G = [1, 0]$ and $\mathbf{W}_n \equiv 0$, we see that we can represent our AR(2) series as a state-space model with perfect observation. We now show how this procedure can be generalized to all the AR models.

Let us now assume that $\{X_n\}_n$ is a mean-zero time series of the most general AR(p) type given by the standard auto-regressive formula:

$$X_n = \phi_1 X_{n-1} + \cdots + \phi_p X_{n-p} + W_n \tag{7.32}$$

for some set of coefficients $\phi_1, \ldots, \phi_p$, and a white noise $\{W_n\}_n$ of unknown variance. For each time stamp n we define the p-dimensional column vector $\mathbf{X}_n$ by:

$$\mathbf{X}_n = [X_n, X_{n-1}, \ldots, X_{n-p+1}]^t.$$

There is no special reason to define $\mathbf{X}_n$ from its transpose, we are merely trying to save space and make typesetting easier! Using this state vector, we see that we can derive an observation equation of the desired form by setting $\mathbf{Y}_n = X_n$, $G = [1, 0, \ldots, 0]$, and $\mathbf{W}_n \equiv 0$. This is again a case of perfect observation. Finally we write the dynamics of the state vector $\mathbf{X}_n$ in the form of definition (7.32). Adding the consistency relations for the remaining components, this equation can be written in the form:

$$\mathbf{X}_{n+1} = \begin{bmatrix} X_{n+1} \\ X_n \\ \vdots \\ X_{n+2-p} \end{bmatrix} = \begin{bmatrix} \phi_1 & \phi_2 & \cdots & \phi_{p-1} & \phi_p \\ 1 & 0 & \cdots & 0 & 0 \\ \vdots & \vdots & \vdots & \vdots & \vdots \\ 0 & 0 & \cdots & 1 & 0 \end{bmatrix} \begin{bmatrix} X_n \\ X_{n-1} \\ \vdots \\ X_{n+1-p} \end{bmatrix} + \begin{bmatrix} W_{n+1} \\ 0 \\ \vdots \\ 0 \end{bmatrix}$$

which can be viewed as the state equation giving the dynamics of the state vector $\mathbf{X}_n$ if we define the state matrix F and the state white noise $\boldsymbol{V}_n$ by:

$$F = \begin{bmatrix} \phi_1 & \phi_2 & \cdots & \phi_{p-1} & \phi_p \\ 1 & 0 & \cdots & 0 & 0 \\ \vdots & \vdots & \vdots & \vdots & \vdots \\ 0 & 0 & \cdots & 1 & 0 \end{bmatrix} \quad \text{and} \quad \boldsymbol{V}_n = \begin{bmatrix} W_n \\ 0 \\ \vdots \\ 0 \end{bmatrix}.$$

7.6.2 The General Case of ARMA Series

We now assume that the time series $\{X_n\}_n$ satisfies:

$$X_n - \phi_1 X_{n-1} - \cdots - \phi_p X_{n-p} = W_n + \theta_1 W_{n-1} + \cdots + \theta_q W_{n-q}$$

for some white noise $\{W_n\}_n$ with unknown variance. Adding zero coefficient terms if necessary, we can assume without any loss of generality that $p = q + 1$. Using the notation introduced in the previous chapter this can be written in a condensed manner in the form $\phi(B)X = \theta(B)W$. Let U be an AR(p) time series with the same coefficients as the AR – part of X, i.e. satisfying $\phi(B)U = W$. If for example the moving average representation exists (but we shall not really need this assumption here) we saw that:

$$\phi(B)U = W \qquad \text{is equivalent to} \qquad U = \frac{1}{\phi(B)}W \tag{7.33}$$

and consequently:

$$X = \frac{\theta(B)}{\phi(B)}W = \theta(B)\frac{1}{\phi(B)}W = \theta(B)U$$

which implies that:

$$\begin{aligned} X_n &= U_n + \theta_1 U_{n-1} + \cdots + \theta_q U_{n-q} \\ &= [1, \theta_1, \ldots, \theta_q] \begin{bmatrix} U_n \\ U_{n-1} \\ \vdots \\ U_{n-q} \end{bmatrix} \end{aligned}$$

which can be viewed as an observation equation if we set:

$$\mathbf{Y}_n = X_n, \qquad G = [1, \theta_1, \ldots, \theta_q], \qquad \text{and} \qquad \mathbf{W}_n \equiv 0$$

and if we define the state vector $\mathbf{X}_n$ as the column vector with $p = q + 1$ row entries U_n, U_{n-1}, ..., U_{n-p+1}. So we have $d_Y = 1$ and $d_X = p = q + 1$. Now that we have the observation equation, we derive the state equation. Because of its definition (7.33), the time series U is an AR(p) series and $\phi(B)U = W$ can be rewritten in the standard form:

$$U_n - \phi_1 U_{n-1} - \cdots - \phi_p U_{n-p} = W_n$$

which is exactly the form we used in the previous subsection to rewrite the AR series in a state-space form. So using the same procedure (recall that we arranged for $p = q + 1$)

$$\mathbf{X}_{n+1} = \begin{bmatrix} U_{n+1} \\ U_n \\ \vdots \\ U_{n+2-p} \end{bmatrix} = \begin{bmatrix} \phi_1 & \phi_2 & \cdots & \phi_p \\ 1 & 0 & \cdots & 0 \\ \vdots & \vdots & & \vdots \\ 0 & 0 & \cdots & 1 \end{bmatrix} \begin{bmatrix} U_n \\ U_{n-1} \\ \vdots \\ U_{n+1-p} \end{bmatrix} + \begin{bmatrix} W_{n+1} \\ 0 \\ \vdots \\ 0 \end{bmatrix}$$

which can be viewed as the desired state equation giving the dynamics of the state vector $\mathbf{X}_n$, if we define the matrix F as the $p \times p$ matrix appearing in the above equation, and if we define the noise vector $\boldsymbol{V}_{n+1}$ as the p-dimensional column vector with components W_{n+1}, 0,. . ., 0.

Remarks.

1. As we already pointed out, the above representation is not unique.We gave an algorithmic procedure which works in all cases, but one should keep in mind that there are many ways to define a state vector and its dynamics, while keeping the same observations.
2. The above procedure can be implemented (modulo minor technical changes to accommodate the dimensions) in the case of multivariate time series, providing a state space representation of multivariate ARMA series.

7.6.3 Fitting ARMA Models by Maximum Likelihood

When explaining the limitations of `R` in fitting ARIMA models, we said that the standard way to fit an MA(q) was to use the maximum likelihood method, but that the computation of the likelihood function was difficult, and that the only practical way to do so was to use a recursive computation reminiscent of the filtering paradigm. Moreover, we also said at that time, that we suspected that this was the main reason why fitting MA models was not implemented in `R` in the multivariate case. Now that we have seen the recursive Kalman filters, and that we know how to rewrite ARMA models in a state space form, it is time to revisit this fitting issue and shed some light on the matter.

The only reason for not implementing the maximum likelihood estimation of an ARMA model is the lack of filtering tools.

Indeed, given the order of an ARMA model, one can

- Use the algorithmic procedure presented in the previous subsection to rewrite the model as a state space model;
- Follow the step outlined in Sect. 7.4.5 to compute the likelihood of any set of observations;
- Run our favorite optimization program to solve for the maximum of this likelihood function.

7.7 Example: Prediction of Quarterly Earnings

This section is devoted to a detailed discussion of a specific application of the recursive filters to the analysis of a financial problem. We show how to use the state space representations and the recursive filtering equations to analyze a scalar time series which appears naturally as the sum of trend, seasonal and irregular components. Contrary to the strategy adopted in the case of temperature data, we do not assume that the trend and seasonal components are deterministic, and we do not start by removing them to reduce the problem to the analysis of a stationary time series. Instead, we include the trend and seasonal components in a structural model which we represent as a partially observed state-space model, which we predict using the recursive filtering equations.

We choose to illustrate this approach with the example of the prediction of the quarterly earnings of a public company. Figure 7.5 gives the plot of the time series of

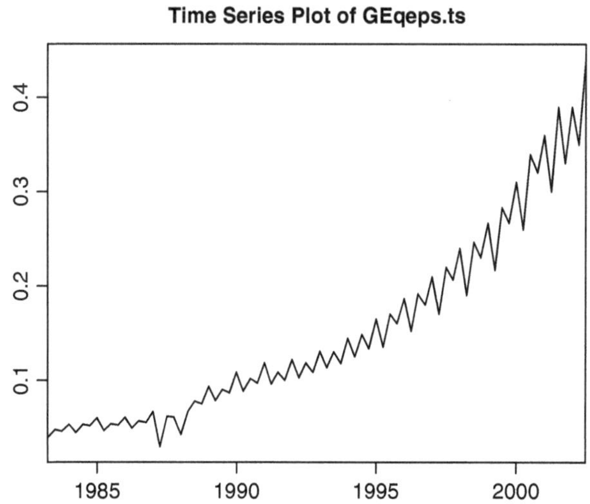

Fig. 7.5. Quarterly earnings per share of General Electric from March 31, 1983 to June 30, 2002

the quarterly earnings of General Electric starting March 31, 1983 and ending June 30, 2002. Earnings per share used to be published yearly in 1982 and before. This series is obviously non-stationary. In particular, it shows a significant upward trend and a seasonal component that cycles every four quarters, or once per year. Moreover, the scale of this seasonal component seems to fluctuate erratically, possibly getting larger over time. Since square root and logarithmic transformations do not seem to resolve the non-stationarity problem, we express the series directly as the sum of a trend component, a seasonal component and a white noise:

$$Y_n = T_n + S_n + \epsilon_n \tag{7.34}$$

where the series $\{T_n\}_n$ models the trend, $\{S_n\}_n$ models the seasonal component, and $\{\epsilon_n\}_n$ is a white noise. Because the trend could be suspected to be exponential, we choose to model the trend as a time series satisfying an evolution equation of the form:

$$T_{n+1} = \phi T_n + \epsilon_{n+1}^{(T)} \tag{7.35}$$

where $\{\epsilon_n^{(T)}\}_n$ is a white noise independent of $\{\epsilon_n\}_n$, and where the real coefficient ϕ is chosen to satisfy $\phi > 1$. The evolution equation of T_n is the equation of an AR(1), but the assumption we make on the coefficient, namely $\phi > 1$, guarantees that this AR series is not stationary. Instead of assuming that the seasonal component is a deterministic periodic function of period 4, we assume that it is a series satisfying a dynamical equation of the form:

$$S_{n+1} = -S_n - S_{n-1} - S_{n-2} + \epsilon_{n+1}^{(S)} \tag{7.36}$$

where $\{\epsilon_n^{(S)}\}_n$ is a white noise independent of the two previous ones. The idea behind this choice is that at least in expectation, S should sum up to zero over a complete period of four quarters. All these requirements can be encapsulated inside a partially observed state-space model in the following way. First we define the state vector:

$$\mathbf{X}_n = [T_n, S_n, S_{n-1}, S_{n-2}]^t$$

so that $d_X = 4$, and together with the obvious consistency conditions, Eqs. (7.35) and (7.36) can be used to give the dynamics of the state via the equation:

$$\mathbf{X}_{n+1} = \begin{bmatrix} T_{n+1} \\ S_{n+1} \\ S_n \\ S_{n-1} \end{bmatrix} = \begin{bmatrix} \phi & 0 & 0 & 0 \\ 0 & -1 & -1 & -1 \\ 0 & 1 & 0 & 0 \\ 0 & 0 & 1 & 0 \end{bmatrix} \begin{bmatrix} T_n \\ S_n \\ S_{n-1} \\ S_{n-2} \end{bmatrix} + \begin{bmatrix} \epsilon_{n+1}^{(T)} \\ \epsilon_{n+1}^{(S)} \\ 0 \\ 0 \end{bmatrix}$$

which can be viewed as the state equation giving the dynamics of the state vector $\mathbf{X}_n$, if we define the matrix F as the 4×4 matrix appearing in the above equation, and if we define the noise vector $\boldsymbol{V}_{n+1}$ as the 4-dimensional column vector with components $\epsilon_{n+1}^{(T)}$, $\epsilon_{n+1}^{(S)}$,0,0. Formula (7.34) used to define the model can now be used to give the observation equation for the one-dimensional observation vector $\mathbf{Y}_n = y_n$:

$$\mathbf{Y}_n = [1, 1, 0, 0] \begin{bmatrix} T_n \\ S_n \\ S_{n-1} \\ S_{n-2} \end{bmatrix} + \epsilon_n$$

which is of the desired form if we set $G = [1, 1, 0, 0]$ and $\mathbf{W}_n = \epsilon_n$. The parameters of the model are the standard deviations σ, $\sigma^{(T)}$ and $\sigma^{(S)}$ of the noise components introduced above, as well as the parameter ϕ. The data is not very noisy, the challenge lies in the fact that the time series is short for the filter to settle down and have a chance to self-correct possible errors in parameters and initializations. For the purposes of the present experiment we chose $\sigma = 0.0001$, $\sigma^{(T)} = 0.0001$ and $\sigma^{(S)} = 0.00001$, and $\phi = 1.03$ corresponding to an annual growth of about 3 %. As explained earlier, even more problematic is the choice of initial estimates for the state and its measurement-error matrix. To obtain the results we report below we chose $\widehat{\mathbf{X}_0} = [0.05, -0.01, 0.03, -0.01]^t$ and 0.02 times the 4×4 identity matrix for the error matrix. Figure 7.6 shows a couple of years of quarterly earnings, together with the plot of the predictions for a 2-years horizon computed at the end of the third quarter of 2000. We chose this date to be able to compare the predictions with the actual data which are superimposed on the same plot. We also give upper and lower "two-standard-deviations" error bounds for the predictions.

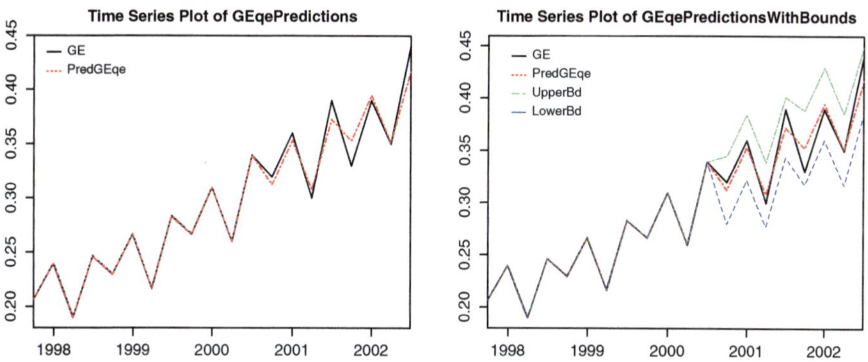

Fig. 7.6. The *left pane* gives the plot of the GE quarterly earnings together with the predictions for the next 2 years of earnings as computed using Kalman filtering at the end of the third quarter of 2000. The *right pane* includes the *upper* and *lower* 2 standard deviations error bands

For the sake of completeness, we give and comment the R code used to produce these results. We first set up the parameters of the state space model with partial observations as explained above.

```
SIGMA <- .0001; SIGMAT <- .0001;
SIGMAS <- .00001; PHI <- 1.03
FF <- matrix(c(PHI,0,0,0,0,-1,-1,-1,0,1,0,0,0,0,1,0),
                 ncol=4,byrow=T)
SigV <- diag(c(SIGMAT,SIGMAS,0,0))
GG <- matrix(c(1,1,0,0),ncol=4,byrow=T)
SigW <- matrix(SIGMA,ncol=1,byrow=T)
X0hat <- matrix(c(.05,-.01,.03,-.01),ncol=1,byrow=T)
Omega0 <- diag(rep(.02,4))
```

The computations of the one-step-ahead predictions are stored in a matrix `Xhat` as follows:

```
Xhat <- matrix(rep(0,284),ncol=71,byrow=T)
Xhat[,1] <- X0hat
Omega <- Omega0
for (n in 1:70)
{
    KALMAN <- kalman(FF=FF, SigV=SigV, GG=GG, SigW=SigW,
     Xhat=Xhat[,n], Omega=Omega, Y=seriesData(GEqeps.ts)[n])
    Xhat[,n+1] <- KALMAN$xpred
    Omega <- KALMAN$error
}
```

Finally, the actual predictions of the observations, together with the upper and lower bounds, are computed and stored in the `timeSeries` objects `GEpred.ts`, `GEpredUB.ts`, and `GEpredLB.ts` respectively.

```
XPRED <- matrix(rep(0,36),ncol=9,byrow=T)
YPRED <- rep(0,8)
YVAR <- rep(0,8)
XPRED[,1] <- Xhat[,71]
MAT <- GG
for (I in 1:8)
{
    YPRED[I] <- GG %*% XPRED[,I]
    XPRED[,I+1] <- FF %*% XPRED[,I]
    YVAR[I] <- MAT %*% Omega %*% t(MAT)
    MAT <- MAT %*% FF
}
```

The plots of Fig. 7.6 were produced with the following commands.

```
GEpred.ts <- timeSeries(pos=seriesPositions(GEqeps.ts),data=
    c(seriesData(GEqeps.ts)[1:70],YPRED),units="PredGEqe")
GEpredUB.ts <- timeSeries(pos=seriesPositions(GEqeps.ts),data=
 c(seriesData(GEqeps.ts)[1:70],YPRED+2*sqrt(YVAR)),units="UpperBd")
GEpredLB.ts <- timeSeries(pos=seriesPositions(GEqeps.ts),data=
 c(seriesData(GEqeps.ts)[1:70],YPRED-2*sqrt(YVAR)),units="LowerBd")
GEqePredictions <- merge(GEqeps.ts[59:78],GEpred.ts[59:78])
```

```
plot.timeSeries(GEqePredictions)

GEqePredictions <- merge(GEqePredictions,GEpredUB.ts[59:78])
GEqePredictionsWithBounds <- merge(GEqePredictions,
                                   GEpredLB.ts[59:78])
plot.timeSeries(GEqePredictionsWithBounds)
```

PROBLEMS

Ⓔ Ⓢ **Problem 7.1** *The purpose of this problem is to examine the various ways the components of a multivariate time series can be dependent. We consider the following AR(1) model for a bivariate time series* $\{\mathbf{X}_t\}_{t=0,1,\cdots}$*:*

$$\mathbf{X}_t = A\mathbf{X}_{t-1} + \mathbf{W}_t, \qquad t = 1, 2, \cdots$$

where:

$$\mathbf{X} = \begin{bmatrix} X_t^{(1)} \\ X_t^{(2)} \end{bmatrix}, \qquad \text{and} \qquad A = \begin{bmatrix} 0.2 & 0 \\ 0 & 0.4 \end{bmatrix},$$

and where $\{\mathbf{W}_t\}_t$ *is a bivariate Gaussian white noise with variance/covariance matrix*

$$\Sigma = \begin{bmatrix} 0.2 & 0 \\ 0 & 1 \end{bmatrix}.$$

1. *Give a condition on the initial random variables* $X_0^{(1)}$ *and* $X_0^{(2)}$ *which guarantees that all the random variables* $X_t^{(1)}$ *and* $X_s^{(2)}$ *are independent for the various values of t and* s*. Explain why. From now on we assume that this condition is satisfied.*
2. *Explain how one can change some of the entries of the matrix* A *to make sure that the time series* $\{X_t^{(1)}\}_t$ *and* $\{X_t^{(2)}\}_t$ *are dependent.*
3. *Setting back the entries of the matrix* A *to their original values, explain how one can change some of the entries of the matrix* Σ *to make sure that the time series* $\{X_t^{(1)}\}_t$ *and* $\{X_t^{(2)}\}_t$ *are dependent.*
4. *Can you identify the differences in the statistical properties resulting from these two procedures? You can use simulations if you cannot quantify these differences theoretically.*

The goal of Problems 7.2–7.5 is to quantify various risk exposures associated with temperature basket options similar to the one analyzed in the text. The data needed for these problems are contained in the `timeSeries` objects `TEMPS1.ts` and `TEMPS2.ts` included in the library `Rsafd`.

Ⓔ **Problem 7.2** *This problem uses the data contained in the* `timeSeries` *object* `TEMPS1.ts`*. For each day of the period* P *starting from 09/06/1948 (the first row) and ending 11/08/2001 (the last row) it gives the daily average temperatures recorded at the meteorological stations of the international airports of Reno (first column), Las Vegas (second column) and Sacramento (third column). We consider two time periods.*

- *Period* P *is the period of the measurements reported in* `TEMPS1.ts`

- *Period P' is the period of the option for which one needs to predict the number of heating degree days. It starts on 12/1/2001 and ends 2/28/2002 inclusive.*

We consider a basket option written on the average number of heating degree days over these three cities. On each given day one computes the average of the three following numbers: the number of heating degree days in Reno, the number of heating degree days in Las Vegas, and the number of heating degree days in Sacramento. One then computes the sum of these daily averages over the period P', and one compares this sum to the strike of the option, using the same rules as usual to compute the pay-off of the option. The purpose of this problem is to perform a simple regression analysis of such a basket option.

1. *For each year in the period from 1961 to 2000 inclusive, compute the yearly index of the option, compute the average of these yearly indexes over the last 15 years of this period, and call this average K.*
   From now on we consider a European call option at the money (i.e. with strike K computed above), with tick α equal to US\$5,000 per degree day, and cap US\$1,000,000, written on the cumulative average number of HDD's in Reno, Las Vegas and Sacramento from December 1, 2001 to February 28, 2002.
2. *Use the linear regression approach introduced in Problem 6.15 to give an estimate of the underlying index on which the basket option is written, i.e. of the average of the three numbers of HDD in the cities of Reno, Las Vegas and Sacramento over the period P', and give a 95 % confidence or prediction interval for this prediction.*
3. *Answer questions 2 and 3 of Problem 6.15 in the present situation.*

Ⓔ **Problem 7.3** *Follow the steps of the joint analysis of the daily average temperatures at the Des Moines, Portland and Cincinnati given in Sect. 7.1.4 of the text, and use the data of the period P to fit a trivariate model to the daily average temperatures in Reno, Las Vegas and Sacramento of the* `timeSeries` *object* `TEMPS1.ts`. *Use this fitted model and follow the prescriptions given in the text to derive an estimate of the value of the average of the three numbers of HDD's in the cities of Reno, Las Vegas and Sacramento over the period P', and to provide new answers to the questions of Problem 7.2 above.*

Ⓔ **Problem 7.4** *This problem provides another numerical illustration of the points addressed in Problems 6.15, 7.2 and 7.3. The data needed for this problem are contained in the* `timeSeries` *object* `TEMPS2.ts` *giving the daily average temperatures, starting from 1/1/1972 (the first row) ending 12/31/2001 (the last row) at the meteorological stations of Newark, La Guardia and Philadelphia. The data have been cleaned in the sense that all the February 29th's of the leap years of this period have been removed. For the purpose of the questions below, we shall use two time periods.*

- *Period P is the period of the time stamps of the timeSeries object.*
- *Period P' is the period for which you will need to predict the number of cooling degree days (CDD's for short). It starts on 5/1/2002 and ends 9/30/2002 inclusive.*

1. *For the location Newark, and for each year in the period P, compute the yearly CDD index, i.e. the total number of CDD's between May 1, and September 30 of that year (these 2 days being included). Bundle these yearly aggregates into a vector which we shall call* `INDEX`.
2. *Using only the numerical values contained in the vector* `INDEX`
   2.1. *Compute a prediction for the value of the yearly cumulative number of CDD's in Newark for the period P'.*

	Newark	La Guardia	Philadelphia
Number of CDD's during P'	1,325	1,355	1,463.5

Table 7.2. Actual numbers of CDD's over the period P'

2.2. *Give a 95 % confidence interval for this prediction*
2.3. *Let us assume that on January 1, 2002, you buy a call option on the number of CDD's in Newark over the period P', with strike K being equal to the average of the yearly numbers of CDD's over the last* 15 *years, nominal pay-off rate (also called the tick) $5,000, and cap $1,000,000 for a premium of $400,000. Use the linear model so fitted to give an estimate of the probability that you will loose money on this deal?*
2.4. *Compute your profit/loss, should the prediction you computed in part 2.1 be perfect in the sense that it is equal to the actual number of CDDs which accumulated over the period P'.*

3. *Use the daily data in the period P to fit a model to the average daily temperature in Newark, and use this estimated model to propose new answers to the questions (a) through (d) above. Compare the results for each of these questions.*
4. *Give an estimate and a 95 % confidence interval for the average of the three numbers of CDD's in the cities of Newark, La Guardia and Philadelphia over the period P',*
5. *Let us assume that on January 1, 2002, you want to buy a "basket" call option at the money, on the average of the numbers of CDD's in Newark, La Guardia and Philadelphia over the period P', with the same tick and the same cap as above. Say how much you would be willing to buy this option for, and explain why.*

FYI: *Even though you should not use these numbers in the solutions of the questions above, the actual numbers of CDD's over the period P' in these three cities are given in Table 7.2.*

Ⓔ **Problem 7.5** *The data needed for this problem are contained in the* `timeSeries` *object* `TEMPS3.ts`*. The latter gives the daily temperatures, starting from 1/1/1981 (the first row), and ending 1/11/2010 (the last row) at the meteorological stations of Saint Etienne and Aix en Provence in France. The data have been cleaned in the sense that all the February 29th's of the leap years of this period have been removed. For the purpose of the questions below, we shall use two time periods.*

- *Period P is the period of the time stamps of the* `timeSeries` *object.*
- *Period P' is the period for which you will need to predict the temperature in these cities. It starts on 1/15/2010 and ends 1/31/2010 inclusive.*

1. *For each of the two locations, and for each year in the period P, compute the average temperature over the period Jan. 15–31, and the number of HDDs over the same period. Bundle these yearly statistics into two separate matrices, each with two columns and as many rows as years in the period P: a matrix* `TEMP` *for the average temperature, and a matrix* `HDD` *for the number of HDDs.*
2. *The questions of this first part rely on linear regression.*
2.1. *Using only the numerical values contained in* `TEMP`*, fit a linear model to the yearly average temperature in Saint Etienne over the period Jan. 15–31, compute a prediction for the value of this variable in 2010, and give a 95 % prediction interval for this prediction.*

2.2. *Using only the numerical values contained in* `TEMP`, *fit a linear model to the yearly average temperature in Aix en Provence over the period Jan. 15–31, compute a prediction for the value of this variable in 2010, and give a 95 %* prediction *interval for this prediction.*
2.3. *Using only the numerical values contained in* `HDD`, *fit a linear model to the yearly number of HDDs in Saint Etienne over the period Jan. 15–31, compute a prediction for the value of this variable in 2010, and give a 95 %* prediction *interval for this prediction.*
2.4. *Using only the numerical values contained in* `HDD`, *fit a linear model to the yearly number of HDDs in Aix en Provence over the period Jan. 15–31, compute a prediction for the value of this variable in 2010, and give a 95 %* prediction *interval for this prediction.*

3. *Use the daily data originally given in* `TEMPS3.lts` *to fit a model to the daily temperature in both locations, and use your fitted model to propose new answers to the questions 2.1 through 2.4 above. Compare the results for each of these questions. We now consider a call option (without cap) on the sum over the period Jan. 15–31, 2010, of the numbers of HDDs in the two cities, with strike given by the mean of this quantity over the last* 15 *years, and tick* 1,000 *EUR per HDD.*
4. *Using* **ALL** *the models constructed above which can provide sensible answers, give estimates of*
   4.1. *The probability that the option is exercised;*
   4.2. *The expected pay-off of the option at maturity.*
5. *Compute for each day of the period P, the sum of the average temperatures in Saint Etienne and in Aix en Provence, call Y the numeric vector so obtained, and assume that instead of actually observing the temperatures in the two cities separately, one has access to the value of Y only, up to a Gaussian noise with variance* 2.6 *independent each day. Assuming that the model you fitted in question 3 above is correct, how would you go about answering the two questions 4.1 and 4.2 if the option was written on the average temperature in Aix en Provence instead? Assume that the strike is given by the mean of this quantity over the last* 15 *years and that the tick is* 1,000 *EUR per HDD.*

Ⓔ **Problem 7.6** *The goal of this problem is to quantify the risks and rewards of some hybrid derivatives written on temperature and energy commodities. The data are contained in the data set* `TempGasNorthGate.ts`, *a bivariate* `timeSeries` *object whose first column gives the daily average temperature at NorthGate and the second column, the price of the month ahead contract of natural gas for delivery at NorthGate.*

1. *We first study the temperature data.*
   1.1. *For each year $j \in \mathcal{J} = \{1980, 1981, \cdots, 2007, 2008\}$ compute the yearly index $HDD[j]$ as the sum of the daily HDD's over the period P comprising the months of January and February of year j. Compute the average of these yearly indexes over the last* 15 *years of this period, and call this average KT.*
   1.2. *Decompose the time series of daily average temperatures at NorthGate into a trend, a seasonal component, and a remainder which you call* `TREND`, `SEA`, *and* `REM` *respectively. Fit an auto-regressive model to* `REM` *and report the order and the estimates of the coefficients. Fit a distribution to the residuals.*
   1.3. *Use $N = 1{,}000$ Monte Carlo simulations to compute an approximation of the probability that the cumulative number of HDDs over the period P in* 2009 *will exceed KT, in other words, provide a Monte Carlo approximation of the probability*

$$\mathbb{P}\{HDD[2009] > KT\}.$$

2. *We now analyze the natural gas prices at NorthGate. For the sake of convenience we let* $G[t]$ *denote the price of the contract of natural gas on day t. The number* KG *will denote the average daily natural gas price over the last* 5 *years. Compute its value.*
   2.1. *Compute and plot the time series* LR *of daily log-returns of the natural gas contract. Fit an auto-regressive model to* LR *and report the order, the estimates of the coefficients, and fit a distribution to the residuals.*
   2.2. *Use* $N = 1{,}000$ *Monte Carlo simulations to compute an approximation of the expectation*

$$\mathbb{E}\{\sum_{t\in P}(G[t] - KG)^{+}\},$$

   namely the expected cumulative amount by which the daily price overshoots the level KG *over the period* P *in* 2009. *The sum appearing in the expectation will be called the overshoot and we will denote it by* $OVER$.
3. *Assuming that we need to buy gas every day of the period* P *in 2009, and that we do not want to pay more than* KG *on any given day, we enter before December 31, 2008 into a contract which guarantees the pay-off*

$$PAYOFF[2009] = \mathbf{1}_{\{HDD[2009]>KT\}}\sum_{t\in P}(G[t] - KG)^{+}$$

   at the end of the period P. *In other words, assuming that we purchase one unit of gas everyday, this hybrid derivative guarantees that, if at the end of the month of February 2009, the cumulative number of HDDs over the period* P *exceeds the strike* KT, *then we are paid the overshoot for all the days we had to buy gas at a price higher than the strike* KG.
   Use $N = 1{,}000$ *Monte Carlo simulations and compute approximations of the probability that the pay-off (at the end of the period* P *in 2009) is non zero, and the expected pay-off.*
4. *We now assume that we will need to buy (one unit of) gas every day of period* P *in 2009 and every day of period* P *in 2010 as well, and we buy an option which guarantees the pay-off* $PAYOFF[2009]$ *if we decide to exercise our option at the end of February 2009, or the pay-off* $PAYOFF[2010]$ *if we decide not to exercise our option at the end of February 2009. Clearly, the decision to exercise or not to exercise the option at the end of February 2009 should be made at that time, i.e. on the basis of information available at that time. Also,* $PAYOFF[2010]$ *is defined similarly to* $PAYOFF[2009]$, *using the HDDs and the daily prices during the period* P *of* 2010.
   Generate $N = 1{,}000$ *Monte Carlo samples of the temperature and the price of natural gas at NorthGate over the period ranging from January 1, 2009 to the last day of February 2010. For each scenario* $n = 1, 2, \cdots, 1{,}000$, *compute* $HDD[2009, n]$ *the number of HDDs over the period* P *in 2009,* $HDD[2010, n]$ *the number of HDDs over the period* P *in 2010,* $OVER[2009, n]$ *the overshoot of the gas price over the period* P *in 2009, and* $OVER[2010, n]$ *the overshoot over the period* P *in 2010, Next, compute the payoffs* $PAYOFF[2009, n]$ *and* $PAYOFF[2010, n]$ *using the defining formula given above.*

Ⓢ **Problem 7.7** *The purpose of this problem is to illustrate by simulation the cointegration properties stated in the example given at the beginning of Sect. 7.1.6 of the text.*

1. *Construct a sample* $\{W_n\}_{n=1,\ldots,N}$ *of size* $N = 1{,}024$ *from the normal distribution* $N(0,1)$, *construct the random walk* $S_n = S_0 + W_1 + \ldots + W_n$ *starting from* $S_0 = 0$, *and construct two independent white noise sequences* $\{\epsilon_n^{(1)}\}_{n=1,\ldots,N}$ *and* $\{\epsilon_n^{(2)}\}_{n=1,\ldots,N}$ *from the distribution* $N(0, 0.16)$ *which are independent of* $\{W_n\}_{n=1,\ldots,N}$. *Give on the*

same figure the plots of the graphs of the time series $\{X_n^{(1)}\}_{n=1,\dots,N}$ *and* $\{X_n^{(2)}\}_{n=1,\dots,N}$ *defined by:*

$$X_n^{(1)} = S_n + \epsilon_n^{(1)} \quad \text{and} \quad X_n^{(2)} = S_n + \epsilon_n^{(2)}.$$

2. *Give the plot of the linear combination of* $\{X_n^{(1)}\}_{n=1,\dots,N}$ *and* $\{X_n^{(2)}\}_{n=1,\dots,N}$ *given by the cointegration vector identified in the text, compare to the previous plot and explain why it confirms cointegration.*
3. *Construct two independent samples* $\{W_n^{(1)}\}_{n=1,\dots,N}$ *and* $\{W_n^{(2)}\}_{n=1,\dots,N}$ *of size* $N = 1{,}024$ *from the normal distribution* $N(0,1)$*, construct the random walks* $S_n^{(1)} = S_0^{(1)} + W_1^{(1)} + \cdots + W_n^{(1)}$ *and* $S_n^{(2)} = S_0^{(2)} + W_1^{(2)} + \cdots + W_n^{(2)}$ *starting from* $S_0^{(1)} = 0$ *and* $S_0^{(2)} = 0$ *respectively, and construct as before two independent white noise sequences* $\{\epsilon_n^{(1)}\}_{n=1,\dots,N}$ *and* $\{\epsilon_n^{(2)}\}_{n=1,\dots,N}$ *from the distribution* $N(0, 0.16)$ *which are independent of* $\{W_n^{(1)}\}_{n=1,\dots,N}$ *and* $\{W_n^{(2)}\}_{n=1,\dots,N}$*. Give on the same figure the plots of the graphs of the time series* $\{X_n^{(1)}\}_{n=1,\dots,N}$ *and* $\{X_n^{(2)}\}_{n=1,\dots,N}$ *defined by:*

$$X_n^{(1)} = S_n^{(1)} + \epsilon_n^{(1)} \quad \text{and} \quad X_n^{(2)} = S_n^{(2)} + \epsilon_n^{(2)}.$$

4. *Give the plot of the same linear combination of* $\{X_n^{(1)}\}_{n=1,\dots,N}$ *and* $\{X_n^{(2)}\}_{n=1,\dots,N}$ *as before, compare to the plot obtained in part 2, and explain why there is no cointegration this time.*
5. *Give the scatterplot of* $X_n^{(2)}$ *against* $X_n^{(1)}$*, add the least squares regression line, test if the regression is significant, and give a time series plot of the residuals as well as their acf. Comment.*

Ⓢ **Problem 7.8** *We consider the state-space model given by:*

$$\begin{cases} X_{t+1} = FX_t + V_t \\ \quad\; Y_t = GX_t + W_t \end{cases}$$

where the covariance matrices of the white noises $\{V_t\}_t$ *and* $\{W_t\}_t$ *are the identity matrices and where the other parameters are given by:*

$$F = \begin{bmatrix} 0.2 & -0.1 & 3 \\ 1 & 0 & 0 \\ 0 & 1 & 0 \end{bmatrix}, \qquad \text{and} \qquad G = \begin{bmatrix} 1 & 1 & 0 \\ 0 & 1 & 1 \end{bmatrix}.$$

We assume that the values of the one step ahead estimates $\hat{X}_{t_0}$ *of the state vector, and* Ω_{t_0} *of its covariance matrix are given by:*

$$\hat{X}_{t_0} = \begin{bmatrix} 1.2 \\ 0.3 \\ 0.45 \end{bmatrix}, \qquad \text{and} \qquad \Omega_{t_0} = \begin{bmatrix} 1.25 & 1 & 1 \\ 1 & 1.25 & 1 \\ 1 & 1 & 1 \end{bmatrix}.$$

We also assume that the next five values of the observation vector Y *are given by:*

$$Y_{t_0} = \begin{bmatrix} -0.1 \\ 1 \end{bmatrix}, \quad Y_{t_0+1} = \begin{bmatrix} 0.3 \\ 0.9 \end{bmatrix}, \quad Y_{t_0+2} = \begin{bmatrix} 0.47 \\ -0.8 \end{bmatrix}, \quad Y_{t_0+3} = \begin{bmatrix} 0.85 \\ -1.0 \end{bmatrix}, \quad Y_{t_0+4} = \begin{bmatrix} 0.32 \\ 0.9 \end{bmatrix}.$$

For each time $t = t_0 + 1$, $t = t_0 + 2$, $t = t_0 + 3$, $t = t_0 + 4$ *and* $t = t_0 + 5$*, use Kalman filtering to compute the one step ahead estimates* $\hat{X}_t$*, its prediction quadratic error* Ω_t*, and of the next observation vector* $\hat{Y}_{t|t-1}$*, and its prediction quadratic error* $\Delta_{t-1}^{(1)}$*.*

Ⓢ **Problem 7.9** *The goal of this problem is to write* R*-functions to simulate and visualize the time evolution of the state and the observation of a linear state-space system of the form:*

$$\begin{cases} \mathbf{X}_{t+1} = F\mathbf{X}_t + \boldsymbol{V}_t \\ \mathbf{Y}_t = G\mathbf{X}_t + \mathbf{W}_t \end{cases} \tag{7.37}$$

where both the state vector $\mathbf{X}_t$ *and the observation vector* $\mathbf{Y}_t$ *are of dimension* 2*, and where* $\{\boldsymbol{V}_t\}_t$ *and* $\{\mathbf{W}_t\}_t$ *are independent Gaussian white noises with variance/covariance matrices* Σ_V *and* Σ_W *respectively. As usual we assume that* $\boldsymbol{V}_t$ *and* $\mathbf{W}_t$ *are independent of* $\mathbf{X}_t$, $\mathbf{X}_{t-1}$, $\mathbf{X}_{t-2}, \ldots$. *We shall also visualize the performance of the forecasts derived from the Kalman filter theory.*

1. *Write an* R*-function, say* `ksim`*, with parameters F, G, SigV, SigW, X0, Omega0, N and SEED which:*
   - *Initializes the seed of the random generator of* R *to SEED;*
   - *Creates a* $N \times 2$ *array of type numeric containing realizations of the* N *values* $\{\boldsymbol{V}_t;\, t = 0, 1, \ldots, N-1\}$ *of the white noise and a* $2 \times (N+1)$ *array of type numeric containing realizations of the* $N+1$ *values* $\{\mathbf{X}_t;\, t = 0, 1, \ldots, N\}$ *of the state vector* $\mathbf{X}_t$ *of the linear system given by Eq. (7.37) starting with* $\mathbf{X}_0 =$ `X0`
   - *Creates an* $(N+1) \times 2$ *array of type numeric containing realizations of the* $N+1$ *values* $\{\mathbf{W}_t;\, t = 0, 1, \ldots, N\}$ *of the observation white noise and uses the values of the state vector created above to produce an* $(N+1) \times 2$ *array of type numeric containing realizations of the* $N+1$ *values* $\{\mathbf{Y}_t;\, t = 0, 1, \ldots, N\}$ *of the observations* $\mathbf{Y}_t$ *satisfying the observation equation;*
   - *Assuming that* X_0 *is perfectly known at time* $t = 0$ *(this is obviously an unrealistic assumption but please bear with me, this is merely homework stuff) computes for* $t = 1, \ldots, N$ *the value of* $\hat{\boldsymbol{X}}_{t|t} = \mathbb{E}\{\boldsymbol{X}_t | \boldsymbol{Y}_{\leq t}\}$, $\hat{\boldsymbol{X}}_t = \mathbb{E}\{\boldsymbol{X}_{t+1} | \boldsymbol{Y}_{\leq t}\}$ *and* $\hat{\mathbf{Y}}_t = \mathbb{E}\{\mathbf{Y}_{t+1} | \boldsymbol{Y}_{\leq t}\}$ *and return the list of the five* $N \times 2$ *arrays of type numeric say* `$xt`, `$yt`, `$hxtt`, `$hxt` *and* `$hyt` *containing the values of* $\mathbf{X}_t$, $\mathbf{Y}_t$, $\hat{\boldsymbol{X}}_{t|t}$, $\hat{\mathbf{X}}_t$ *and* $\hat{\mathbf{Y}}_t$ *for* $t = 1, \ldots, N$.

     We use the notation $\mathbf{Y}_{\leq t} = \{Y_t, Y_{t-1}, \ldots, Y_0\}$ *for the set of all the observations prior to time t. Recall that the notation* $E_t(\mathbf{X})$ *was used for what is now denoted by* $\mathbf{X}_{t|t}$.

2. *Run the* R *command*

```
> KTST <- ksim(F, G, SigV, SigW, X0, Omega0, N, SEED)
```

with:

$$\mathtt{F} = \begin{bmatrix} 0.2 & -0.1 \\ 1 & 0.3 \end{bmatrix}, \qquad \textit{and} \qquad \mathtt{G} = \begin{bmatrix} 1 & 1 \\ 0 & 1 \end{bmatrix},$$

$$\mathtt{SigV} = \begin{bmatrix} 0.2 & 0 \\ 0 & 1 \end{bmatrix}, \quad \mathtt{SigW} = \begin{bmatrix} 1 & 0 \\ 0 & 0.4 \end{bmatrix}, \quad \mathtt{X0} = \begin{bmatrix} 1.2 \\ 0.45 \end{bmatrix} \quad \textit{and} \quad \mathtt{Omega0} = \begin{bmatrix} 1.25 & 0 \\ 0 & 1.25 \end{bmatrix},$$

and finally with $N = 125$ *and* $SEED = 14$.

3. *Write an* R *function, say* `kanim` *which, with the appropriate parameters taken from the list* `KTST` *obtained in the previous question, will produce the following three animations.*
   - *Plots on the same graphic window of the successive values of* `KTST$xt` *and* `KTST$hxtt`*;*
   - *Plots on the same graphic window of the successive values of* `KTST$xt` *and* `KTST$hxt`*;*

- *Plots on the same graphic window of the successive values of* `KTST$yt` *and* `KTST$hyt`.

Ⓣ **Problem 7.10** *Find the state-space representation for the following time series models:*

1. $(1-\phi_1 B)Y_t = (1-\theta_1 B)\epsilon_t$
2. $Y_t = \phi_1 Y_{t-1} + \epsilon_t - \theta_1\epsilon_{t-1} - \theta_2\epsilon_{t-2}$

where in both cases, $\{\epsilon_t\}_t$ *denotes a white noise.*

Ⓣ **Problem 7.11** *Let* $\{X_t\}_t$ *be the univariate AR(1) time series:*

$$X_t = \phi X_{t-1} + V_t$$

and let Y_t *be the noisy observation:*

$$Y_t = X_t + W_t$$

where we assume that $\{V_t\}_t$ *and* $\{W_t\}_t$ *are independent one dimensional Gaussian white noises with variances* σ_V^2 *and* σ_W^2 *respectively. Give the best predictions (in the sense of the expected squared error) for* X_{t+1}, Y_{t+1} *and their variances given the past observations* Y_t, $Y_{t-1}, \ldots$.

Ⓣ **Problem 7.12** *Derive a state space representation for the univariate ARIMA(1,1,1) model:*

$$(1-\phi B)(1-B)X_t = (1-\theta B)W_t$$

when $\{W_t\}_t$ *is a* $N(0,\sigma^2)$ *white noise, and identify all the parameters of the model in terms of* ϕ, θ, *and* σ^2. *Recall that* $(1-B)$ *merely stands for the operation of differentiation.*

Ⓔ **Problem 7.13** *Use the data of the weekly excess returns of AEP contained in the* `timeSeries` *object* `AEP.wer.ts` *to produce the same results as those reported in Sect. 4.5.5 for DUKE, by following all the steps of the filtering analysis of the time varying extension of the CAPM theory described in the text. Compare the beta estimates so obtained to those given in the text for DUKE. Use the* `timeSeries` *object* `MARKET.wer.ts` *for the weekly excess returns of the market portfolio. These excess returns were computed from the weekly returns of the S&P500 index.*

Ⓔ **Problem 7.14** *The data needed for this problem are contained in the* `timeSeries` *objects* `PEPqeps.ts` *and* `IBMqeps.ts` *giving the quarterly earnings per share of Pepsi Co. and IBM.*

1. *Use the Pepsi data and reproduce the filtering analysis done in the text for the DUKE Energy quarterly earnings.*
2. *The goal of this second question is to perform the same analysis for IBM.*
   - 2.1. *Fit a state space model and use it to predict the last 2 years of quarterly earnings using only the data up to the last quarter of 2000.*
   - 2.2. *Fit a state space model to the data of the quarterly earnings over the period starting with the first quarter of 1996, and ending with the third quarter of 2000. Use this model and Kalman filtering as above to predict the last 2 years of quarterly earnings using only the data used to fit the model.*
   - 2.3. *Fit a model to the data up to the last quarter of 1990, and use it to predict the quarterly earnings of the period 1991–1995 using only the data up to the last quarter of 1990.*

2.4. *Comment on the performance of the prediction in each case.*

Ⓣ **Problem 7.15** *The goal of this problem is to use Monte Carlo simulations to quantify the performance and the risk of some hybrid portfolios. The data are contained in the four dimensional* `timeSeries` *object* `ENERGY.ts` *whose first column gives the daily closing price of Constellation Energy Group stock (CEG), second column the daily closing price, say NG, of the nearby Henry Hub natural gas contract, third column the daily close, say SPX, of the S&P500 index, and finally whose fourth column gives the rate of return on a 3 month Treasury Bill, which we will use as proxy for the short interest rate. Notice that this last figure is given in percent. These data span the period from Jan. 2006 to Dec. 2010.*

1. *Compute the times series* `LR1` *and* `LR2` *of the daily log returns of CEG and NG respectively. Fit a bivariate auto-regressive model to* `LR1` *and* `LR2`. *Report the order and the estimates of the parameters of the model. Call the two unidimensional vectors of residuals* `RES1` *and* `RES2` *respectively.*
2. *Fit a joint distribution based on a Gaussian copula to* `RES1` *and* `RES2`*: estimate the marginal distributions of* `RES1` *and* `RES2`, *and the parameter of the Gaussian copula.* *Having holdings in CEG, and being interested in investing in NG contracts, you decide to apply the results of the above analysis to quantify the potential risks, profits, and losses of your portfolio. More precisely, assume that you decided to invest* $10,000 *in CEG and* $10,000 *in NG on Dec. 31, 2010, so that at the start, the proportions of your portfolio invested in CEG and NG are* 50 *and* 50 %. *Imagine that each day, you trade without incurring any transaction costs and rebalance your portfolio so that the proportions invested in CEG and NG remain the same, and you hold the same proportions invested in the two assets throughout the entire year. Also, you should ignore the issues related to weekends, holidays, ... The goal is to compute, at the end to the year 2011, the expected value, the median and the* 95 %*-tile of your portfolio.*
3. *Simulate* `NbSim=1000` *Monte Carlo samples of daily CEG and NG in 2011 from the joint distribution estimated in question 2, and compute the value of your portfolio on Dec. 31, 2011 for each scenario, if the proportions invested in the two assets remain constant. More specifically, do the following,*
   3.1. *Generate* `NbSim=1000` *Monte Carlo samples of* $RES1$ *and* $RES2$ *from the joint distribution based on the Gaussian copula and the marginal distributions you just fitted. For each residual such a path should have length* `LL=365`.
   3.2. *Take the samples you just generated as innovations and compute the corresponding returns of CEG and NG using the fitted bivariate AR model.*
   3.3. *Use these returns to compute the final values of the samples of your portfolio.*
   3.4. *Compute the required statistics.*
4. *Suppose now that in order to diversify your portfolio you invest in the S&P 500 (SPX) index as well.*
   4.1. *Fit a multivariate auto-regressive model to the log returns of CEG, NG and SPX. Report the order and the estimates of the parameters. Call the three residual vectors* $R\hat{E}S1$, $R\hat{E}S2$ *and* $R\hat{E}S3$.
   4.2. *In the spirit of question 2. above, fit distributions to* $R\hat{E}S1$, $R\hat{E}S2$ *and* $R\hat{E}S3$ *separately, and propose estimates for the parameter(s) of a Gaussian copula which we propose to use as their copula. Explain your choices.*
   4.3. *In the spirit of question 3 above, generate* `NbSim=1000` *Monte Carlo samples of CEG, NG and SPX using the tri-variate distribution you just fitted, and compute Monte Carlo estimates of the same statistics relative to the value of your portfolio on Dec. 31,*

2011, if you invest \$10,000 *in CEG,* \$10,000 *in NG, and* \$10,000 *in SPX on Dec. 31, 2010, and hold the proportions* $1/3$, $1/3$ *and* $1/3$ *invested in the three assets constant throughout.*

Ⓔ **Problem 7.16** *The goal of this problem is to use Monte Carlo simulations to quantify the performance of the standard Black-Scholes Delta-hedging of an option on the S&P 500 index. The data needed for this problem are contained in the two dimensional* `timeSeries` *object* `SP.ts` *whose first column gives the daily closing prices of the S&P500 index, and second column the daily closing values of the VIX index between January 2, 2008 and January 31, 2012. The VIX index is an index computed from the implied volatilities of options on the S&P500. Its goal is to give a sense of the market implied volatility.*

1. *Compute the numeric vectors* `LR1` *and* `LR2` *of the daily log returns of the daily S&P500 and the VIX index respectively. We shall freely talk about log-return of the VIX, ignoring the fact that the VIX is only indirectly traded. Fit a bivariate auto-regressive model of order* 2 *to* `LR1` *and* `LR2`. *Report the estimates of the parameters of the model. Call the two unidimensional vectors of residuals* `RES1` *and* `RES2` *respectively, and compute their (Pearson) correlation coefficient.*
2. *The purpose of this question is to find an estimate of the joint distribution of* `RES1` *and* `RES2` *in a form which will allow for the generation of Monte Carlo samples from this joint distribution. Estimate the marginal distributions of* `RES1` *and* `RES2`, *and capture their dependence with a Gumbel copula.*

NB: *The Gumbel copula (whose parameter you are expected to determine) does not need to be* ***the copula of*** `RES1` *and* `RES2`*! Recall the value you found for their correlation coefficient.*

Having purchased an ***at the money*** *call option on the S&P500 on January 31, 2012, with maturity* 90 *days at implied volatility* 15.2 %, *we try to estimate by means of Monte Carlo computations, the fair value of this option and the cost of hedging due to transaction costs. For the sake of simplicity, we shall assume that the short interest rate is zero, i.e.* $r = 0$. *By fair value, we mean the expectation of the terminal value of the Delta hedge as given by Black-Scholes theory when we use the value of the VIX index instead of the constant volatility prescribed by the theory. By the cost of hedging, we mean the sum of the daily transaction costs paid to rebalance the hedge: each time we buy or sell a unit of the index, we incur a cost equal to the so-call spread between the best bid and ask prices, and we shall use a multiple of the volatility (in our case the value of the VIX index) as a proxy for this spread.*

3. *The purpose of this question is to set up the Monte Carlo simulations. More specifically, do the following,*
   3.1. *Generate* `NbSim=1000` *Monte Carlo samples of* `RES1` *and* `RES2` *from the joint distribution determined by the copula and the marginal distributions you identified earlier. For each residual such a path should have length* `LL=90`.
   3.2. *Take the samples you just generated as innovations and generate corresponding scenarios for the log-returns of the S&P500 and VIX indexes using the fitted bivariate AR model.*
   3.3. *Use these log-returns to generate corresponding samples for the actual indexes over the same period of* 90 *days ahead of the last day for which we have data.*
4. *We now compute approximations for the fair value of the option as given by the Monte Carlo approximation of the expected value of the cost of the hedging portfolio, and for the transaction costs as given by the Monte Carlo approximation of the expected value of the cost of* crossing the spread *each time the hedge is rebalanced.*

4.1. For each Monte Carlo scenario, and for each day t between today (January 31, 2012) and the maturity T of the option (90 days later) compute the Black-Scholes Delta

$$\Delta_t = \Phi\left(\frac{\log(S_t/K)}{\sigma_t\sqrt{T-t}} + \frac{1}{2}\sigma_t\sqrt{T-t}\right)$$

where as usual Φ stands for the cumulative distribution function of the standard Gaussian distribution, S_t is the price of the underlying (the value of the S&P500 index in our case) on day t, and σ_t is the volatility. For each scenario, we use for σ_t, 0.12 times the value of the VIX index given by the scenario for that day.

4.2. Compute the Monte Carlo approximation of the expectation

$$\mathbb{E}\left[\sum_{t=1}^{90} \Delta_t(S_t - S_{t-1})\right]$$

where S_0 is the value of the S&P500 index on January 31, 2012.

4.3. Compute the Monte Carlo approximation of the expectation

$$\mathbb{E}\left[\frac{1}{500}\sum_{t=1}^{90} |\Delta_t - \Delta_{t-1}|\sigma_t\right]$$

Note that the quantity Δ_0 can be computed from the same formula as above on January 31, 2012. It is obviously the same for all the Monte Carlo scenarios.

Notes & Complements

Despite serious technical complications, the linear theory of time series presented in the previous chapter in the univariate case can be extended to multivariate time series models. A thorough account can be found in Hamilton's exhaustive exposé [44]. Cointegration is a concept of great theoretical significance, and it is now an integral part of most econometric theory textbooks. It was introduced in the seminal work [32] of Engle and Granger. This fundamental contribution was cited as the main reason to grant the 2003 Nobel prize to these authors. The interested reader can also consult Johansen's book [52] for the general statistical theory of cointegrated time series. Cointegration appears in economic theories as a way to imply equilibrium relationships between time series. These equilibria are the result of relaxation of the dynamics toward *steady states,* and identifying them usually requires long spans of low frequency data, for some form of ergodicity to take effect. In financial applications, cointegration appears as a way to identify arbitrage opportunities. See for example Sect. 8.6 entitled *Threshold cointegration and arbitrage* of Tsay's book [92]. A complete exposé of cointegration theory would be beyond the scope of this book. Nevertheless, we mention that its importance is now recognized by the financial engineering community as well as the econometric community, and new research programs have been started to understand and control better pricing and risk management issues in the presence of cointegrated price series. Spread options in the energy markets, and basket options in the equity markets are two of the many examples of instruments naturally written on cointegrated price series. Our interest in these financial products drove our attempt to introduce the concept of cointegration, even if we knew that our short discussion could not do it justice. The textbooks of Chan [22], Tsay [92], and especially Zivot and Wang [99] can be consulted for examples, error correction forms, and statistical tests. `R` does not have core functions to test for cointegration. The implementation of the Phillips-Ouliaris co-integration test given by the function `count` of the library `Rsafd` is merely a wrapper over the function `po.test` of the package `tseries`. Alternatively, the package `urca` also include functions to test for co-integration in multivariate time series.

Even after so many years, the best way to learn about the classical theory of CUSUM tests is still from the original paper of Brown, Durbin and Evans [11].

The use of linear state-space models in the analysis of time series is now part of the folklore. Most of the modern textbooks on time series do include at least one chapter on the state-space models and the powerful results of the filtering theory of partially observed systems. See for example [10, 44] or [22]. Their use in financial applications has experienced a similar growth. See for example [55]. A discussion in the delicate estimation of the parameters of the model, including an application of the EM algorithm, can also be found in the book of Shumway and Stoffer [87].

The shortcomings of the CAPM model and its failure to pass the empirical tests of its validity (recall Sect. 4.5.5 of Chap. 4) have been studied extensively. The discussion of this chapter, especially the use of filtering theory to track the values of a potentially time varying beta, are borrowed from the book of Gençay, Selçuk and Whitcher, where references to the extensive literature devoted to the *fixing* of the CAPM model can be found. The idea of the application of recursive filtering equations to the prediction of the quarterly earnings of a company was borrowed from Shumway and Stoffer [87].

Problem 7.16 is a scaled down version of a Monte Carlo study aimed at quantifying the cost of transactions, typically *crossing the spread* between the best ask and bid prices when rebalancing at discrete time intervals a portfolio aimed at replicating the payoff of an option. In the world of Black-Scholes theory, a portfolio replicating perfectly the pay-off of the option exists: this hedging strategy consists in holding at each time, a given amount (the so-call Delta of the option) of the underlying interest. This theoretical result does not hold in practice for many reasons, not the least the fact that rebalancing the hedging portfolio can only be done at finitely many (discrete) times. Moreover, Black-Scholes hedging is based on a constant volatility model, so the first goal of the problem is to use at each rebalancing step, the *current sentiment* of the market regarding implied volatility which, according to its very definition, is best given by the value of the VIX index. Another important source of slippage is the fact that financial assets cannot be bought and sold at their theoretical price. The frictions caused by imbalances between supply and demand are captured by an *order book* and the minimal cost of a transaction is proportional to the difference between the best ask and bid prices.. For the purpose of the problem, we do not model the order book, and use the common belief that this difference (the so-called spread) is proportional to the implied volatility for which we use the proxy given by the VIX index.

8

NONLINEAR TIME SERIES: MODELS AND SIMULATION

Most financial time series exhibit nonlinear features which cannot be captured by the linear models seen in the previous two chapters. In this last chapter, we present the elements of a theory of nonlinear time series adapted to financial applications. We review a set of standard econometric models which were first introduced in the discrete time setting. They include the famous, ARCH, GARCH, ... models, but we also discuss stochastic volatility models and we emphasize the differences between these concepts which are too often confused. However, because of the growing influence of the theoretical developments of continuous time finance in the everyday practice, we spend quite a significant part of the chapter analyzing the time series models derived from the discretization of continuous time stochastic differential equations. As ordinary differential equations can be used as framework for time evolution modeling, their stochastic extensions are adapted to the requirements of modeling uncertainty, and powerful intuition from analyses of physical and mechanical systems can be brought to bear. We examine its implications at the level of simulation. The last part of the chapter is devoted to a new set of algorithms for the filtering of nonlinear state space systems. We depart from the time honored tradition of the extended Kalman filter, and we work instead with discrete approximations called particle filters. This modern approach is consistent with our strong bias in favor of Monte Carlo simulations. We illustrate the versatility of these filtering algorithms with the example of price volatility tracking.

8.1 First Nonlinear Time Series Models

This introductory section builds on some of the ideas of the linear theory presented in Chap. 6. It is devoted to the discussion of a couple of nonlinear time series models based on natural generalizations of classical linear time series models introduced in that chapter. The first of these two models was inspired by the desire to develop a time series analog of the so-called fractional Brownian motion whose modeling potential for financial data was popularized by Benoit Mandelbrot. The second model is a straightforward generalization of the notion of auto-regressive model.

R. Carmona, *Statistical Analysis of Financial Data in R*, Springer Texts in Statistics,
DOI 10.1007/978-1-4614-8788-3_8,

8.1.1 Fractional Time Series

The process of fractional Brownian motion has a certain number of desirable properties which are present quite often in financial data. Even though self-similarity has limited use outside continuous time models, long range dependence is a feature which most of the linear models analyzed in Chap. 6 do not share. The time series model introduced here is an attempt to capture this feature.

Definition 1. *If p and q are integers and* $d \in (0,1)$, *we say that the time series* $\{X_t\}_t$ *is an ARIMA(p,d,q) series if* $(I-B)^d X_t$ *is a stationary ARMA(p,q) time series where the fractional difference operator* $(I-B)^d$ *is defined by the infinite sum:*

$$(I-B)^d = I + \sum_{j=1}^{\infty} \frac{d(d-1)\cdots(d-j+1)}{j!}(-1)^j B^j. \tag{8.1}$$

As usual we use the notation B for the backward shift operator. The time series of the type defined above are called fractional processes or fractional time series. This definition calls for a few remarks.

- The infinite series in the right hand side of formula (8.1) mimics the Taylor expansion of the fractional powers:
$$(1-z)^d = 1 + \sum_{j=1}^{\infty} \frac{d(d-1)\cdots(d-j+1)}{j!}(-1)^j z^j$$
which converges for $|z| < 1$, hence the terminology of fractional differentiation.
- The cases $d = 0$ and $d = 1$ appear as limiting cases of the above definition. The case $d = 0$ corresponds to the usual stationary ARMA(p,q) model, while the case $d = 1$ corresponds to the classical ARIMA(p,1,q) model. Fractional processes provide a continuum of models interpolating between these extreme cases.
- Fractional time series are asymptotically stationary when $d < 1/2$. This means that even though they are not technically speaking stationary, they behave as if they were in the regime $t \to \infty$. In fact, it is possible to prove that in this case (i.e. when $d < 1/2$) the auto-correlation function $\rho_X(h)$ converges toward zero like a power for large lags (i.e. when $h \to \infty$). More precisely:
$$\rho_X(h) \sim h^{2d-1} \qquad \text{as} \quad h \to \infty. \tag{8.2}$$
Notice that $2d - 1 < 0$ since $d < 1/2$. This slow polynomial decay of the acf is in contrast with what we found in the case of the classical ARMA models for which the acf either vanishes after some lag (like in the case of the MA series) or decays at an exponential rate (like in the case of the AR series). The long range dependence (also called long range memory) is the reason why these models were introduced. They exhibit persistence while most linear models don't. The slow decay of the empirical estimate of the acf is a strong indication that a fractional model may be relevant.

- Formula (8.2) can be turned into a method of estimation of the exponent d. Indeed, at least asymptotically, one should have:

$$\log \rho_X(h) \sim \beta_0 + (2d - 1)\log h$$

for some constant β_0. Consequently, after estimating the sample acf $\hat{\rho}_X(h)$ in the usual way, a simple linear regression of $\log \hat{\rho}_X(h)$ against the logarithm $\log h$ of the lag should give a slope equal to $2d - 1$, from which one can derive an estimate of the value of d. Unfortunately, this estimate is very poor in the case of non-Gaussian time series.

The core `R` distribution does not have a special function for the estimation and simulation of the fractional time series defined above. Computation and simulation of fractional time series can be done by including the package `fracdiff`. See also the discussion below of the fractionally integrated GARCH models.

8.1.2 Nonlinear Auto-Regressive Series

The first examples of genuinely nonlinear time series models are provided by the nonlinear auto-regressive models. Like their linear counterparts, they generalize the simple model:

$$X_t = \mu + \sigma W_t$$

to the case where the mean is a function of the past values of the series itself. But instead of assuming that this function is linear as in the case of the classical AR(p) models, we shall now assume that it can be any nonlinear function of X_{t-1}, X_{t-2}, ..., X_{t-p}. In other words, we shall assume the existence of a deterministic function $\mu : \mathbb{R}^p \hookrightarrow \mathbb{R}$ such that:

$$X_t = \mu(X_{t-1}, X_{t-2}, \ldots, X_{t-p}) + \sigma W_t.$$

However, we shall not limit the nonlinear dependence on the past lags of the series to the mean term. We shall also allow it in the variance of the series. More precisely:

Definition 2. *If p is an integer, we say that the time series $\{X_t\}_t$ is a nonlinear AR(p) series if there exist a white noise $\{W_t\}_t$ and two deterministic functions:*

$$\mu : \mathbb{R}^p \ni (x_1, \ldots, x_p) \hookrightarrow \mu(x_1, \ldots, x_p) \in \mathbb{R}$$

and

$$\sigma : \mathbb{R}^p \ni (x_1, \ldots, x_p) \hookrightarrow \sigma(x_1, \ldots, x_p) \in \mathbb{R}_+$$

such that:

$$X_t = \mu(X_{t-1}, X_{t-2}, \ldots, X_{t-p}) + \sigma(X_{t-1}, X_{t-2}, \ldots, X_{t-p})W_t. \tag{8.3}$$

Let us consider the simplest case of a nonlinear AR(1) for the sake of illustration. Such a series satisfies an equation of the form:

$$X_t = \mu(X_{t-1}) + \sigma(X_{t-1})W_t$$

Moreover, if we further assume that the white noise $\{W_t\}_t$ is an $N(0,1)$ i.i.d. sequence, then we see that conditioned on its past, X_t is still normally distributed. More precisely:

$$X_t|X_{t-1} = N(\mu(X_{t-1}), \sigma(X_{t-1})^2)$$

which shows that the marginal distribution of X_t is a mixture of normal distributions, and as such, is likely to have heavy tails and excess kurtosis as proved in Sect. 8.3 below and Problem 8.1.

8.1.3 **Statistical Estimation**

As before we limit the scope of our discussion to the particular case $p = 1$. The general case can be treated exactly in the same way, but the notation become so cumbersome that we refrain from discussing the general case all together. There are two ways to approach the statistical estimation of a nonlinear AR model. Either by parametrization of the unknown functions $\mu(x)$ and $\sigma(x)$ or by appealing directly to nonparametric techniques.

- ***Parametric Approach*** Let us assume that the functions μ and σ are known up to a parameter θ. This parameter θ can be estimated by the maximum likelihood method when the white noise is Gaussian. Indeed, in this case, the likelihood function is the product of normal densities, and its maximization is not more difficult than in the case of the estimation of the mean and the variance of a normal sample. The maximum likelihood estimate (MLE for short) is much more difficult to find in the general case. A reasonable approximation can be obtained by *acting as if* the white noise was still Gaussian. The estimate so obtained is usually called the quasi-MLE or the pseudo-MLE. So given a sample $x_1, x_2, \ldots, x_{T-1}, x_T$ of size T, the maximum likelihood estimator $\hat{\theta}_T$ is given by:

$$\begin{aligned}\hat{\theta}_T &= \arg\max_{\theta} \log L(x_1, \ldots, x_T|\theta) \\ &= \arg\max_{\theta} \sum_{t=1}^{T} \log f(x_t|x_{t-1}, \theta)\end{aligned}$$

  if we use the notation $f(x|x', \theta)$ for the conditional density of X_t given that $X_{t-1} = x'$. Since in general we cannot compute it, we compute instead the quasi-maximum likelihood estimator $\hat{\theta}_T$ is given by:

$$\hat{\theta}_T = \arg\min_{\theta} \sum_{t=1}^{T} \left(\frac{1}{2}\log 2\pi + \frac{1}{2}\log \sigma^2(x_{t-1}, \theta) + \frac{[x_t - \mu(x_{t-1}, \theta)]^2}{2\sigma^2(x_{t-1}, \theta)} \right) \tag{8.4}$$

which is obtained by assuming that the white noise is Gaussian. How difficult this optimization problem is depends upon the explicit form of the functions $\mu(\,\cdot\,,\theta)$ and $\sigma(\,\cdot\,,\theta)$. But what is remarkable is that, asymptotically, the resulting estimator has essentially the same desirable properties as in the case of Gaussian white noise. Indeed, it can be proven that, if $\{W_t\}_t$ is an i.i.d. sequence of mean-zero variance-one random variables, then the quasi-MLE $\hat{\theta}_T$ is consistent in the sense that whatever the true value θ of the parameter is, we have:

$$\lim_{T\to\infty} \hat{\theta}_T = \theta,$$

and it is asymptotically normal in the sense that the distribution of $\sqrt{T}(\hat{\theta}_T - \theta)$ converges toward a normal distribution with mean zero and a variance given by the Fisher information matrices which we shall not make explicit here. This last property makes it possible to derive (asymptotically correct) tests of significance and confidence intervals for the parameter.

- ***Nonparametric Approach*** If we do not know enough about the functions $\mu(x)$ and $\sigma(x)$, and if we cannot reduce their estimation to estimating one or a small number of scalar parameters, there is always the possibility of appealing to nonparametric estimation techniques. In particular, building on the expertise we developed in Chap. 5, we can solve the estimation problem by choosing a kernel function K, a bandwidth $b_T > 0$, and for each value of x, computing the estimates:

$$\hat{\mu}_T(x) = \frac{\sum_{t=2}^T x_t K\left(\frac{x-x_{t-1}}{b_T}\right)}{\sum_{t=2}^T K\left(\frac{x-x_{t-1}}{b_T}\right)} \tag{8.5}$$

and

$$\hat{\sigma}_T^2(x) = \frac{\sum_{t=2}^T x_t^2 K\left(\frac{x-x_{t-1}}{b_T}\right)}{\sum_{t=2}^T K\left(\frac{x-x_{t-1}}{b_T}\right)} - \hat{\mu}_T(x)^2. \tag{8.6}$$

Notice that we applied the procedure learned in Chap. 5 to the couples (x_{t-1}, x_t) with x_{t-1} playing the role of the explanatory variable and x_t playing the role of the response variable. Because we are well versed in the theory of kernel estimation, we know that the properties of these estimators depend upon the value of the bandwidth b_T. Hence, we should not be surprised to learn that the consistency and the asymptotic normality of these estimates will only hold for specific forms of this dependence. To be specific, the estimates are consistent, i.e.

$$\lim_{T\to\infty} \hat{\mu}_T(x) = \mu(x) \qquad \text{and} \qquad \lim_{T\to\infty} \hat{\sigma}_T(x) = \sigma(x),$$

whenever $T \to \infty$ in such a way that $\lim_{T\to\infty} Tb_T = \infty$. So as the sample size grows (i.e. as $T \to \infty$) we can take smaller and smaller a bandwidth b_T (which is desirable) but the latter cannot go to zero too fast. Under the same condition the bivariate distribution of the vector

$$\sqrt{Tb_T}\left(\begin{bmatrix}\hat{\mu}_T(x)\\ \hat{\sigma}_T^2(x)\end{bmatrix}-\begin{bmatrix}\mu(x)\\ \sigma^2(x)\end{bmatrix}\right)$$

converges toward a bivariate normal distribution with zero mean. This result shows that even though the kernel estimate converges toward the true value, the rate of convergence is smaller since it is given by $\sqrt{Tb_T}$ instead of the usual $\sqrt{T}$. This is the price to pay for not having to make assumptions on the particular forms of the functions $\mu(x)$ and $\sigma(x)$. In simple terms, this illustrates a statement we made several times already: the kernel method can be used even when we do not know much about the functions to estimate, but in order to get the same precision as parametric methods, it needs more data.

8.2 More Nonlinear Models: ARCH, GARCH & All That

It is now time to investigate the most popular of the nonlinear time series models: the famous ARCH and GARCH models.

8.2.1 **Motivation**

As a matter of illustration, we revisit the example of the daily log-returns of the Brazilian coffee which we considered earlier in Chap. 3. However, the reader should be aware of the fact that most of what we are about to do or say applies to most financial time series as well. The plot of the auto-correlation function of these log-returns is given in the left pane of Fig. 8.1. Obviously, this acf looks pretty much like the auto-correlation function of a white noise. If, ignoring this warning, we try to fit an AR model anyway, we get:

```
> BLRet.ar <- ar(BLRet)
> BLRet.ar$order
[1] 0
```

which confirms our fear that the series may not carry more information than a white noise would. We saw in Chap. 3 that the marginal distribution of these daily log-returns was not Gaussian, and that it had heavy tails. Recall the normal Q-Q plot comparing the distribution of these daily log-returns to the normal distribution in the left pane of Fig. 3.7. This remark is crucial at this stage: indeed it is only for Gaussian time series that the absence of correlation implies independence. So, even though we have an uncorrelated sequence, it is still possible that significant forms of dependence between the successive terms of the series exist. Indeed, if the terms of a time series are not jointly Gaussian, the lack of correlation does not imply independence. The existence of dependencies is confirmed by the plot of the auto-correlation function of the square of the time series which we reproduce in the right pane of Fig. 8.1. Indeed, this plot does not look like the acf of a white noise. Since squares of random variables are independent whenever the original random variables are independent, this proves that the original series of daily log-returns was not an independent series. So, there is still some hope for us to be able to untangle these dependencies.

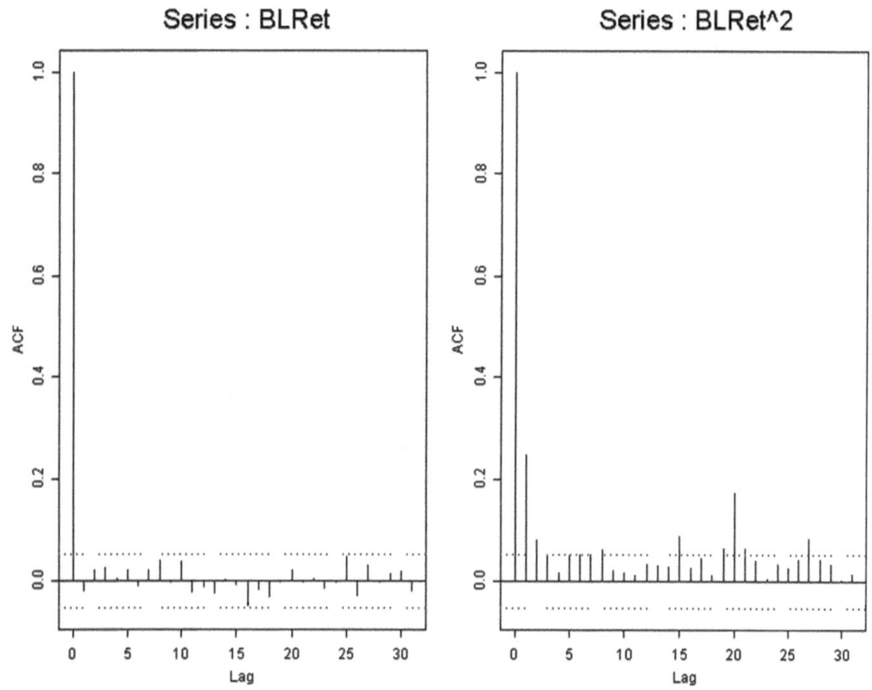

Fig. 8.1. Plot of the auto-correlation function of the Brazilian coffee daily log-returns (*left*) and of their squares (*right*)

8.2.2 **ARCH Models**

The next step is to model the volatility (i.e. the instantaneous standard deviation) as a random process of its own. There are two main reasons for that. The first one is that time series plots of these log-returns show that the variance seems to change over time. This goes under the name of heteroskedasticity. The second is that, as we already saw, one of the easiest ways to create distributions with heavy tails, is to mix together Gaussian distributions with different variances. See Problem 8.1 at the end of this chapter for example. By making sure that the instantaneous standard deviation is random, we may force the distribution of the log-returns to be a mixture of Gaussian distributions, and hence to have heavy tails. These two basic ideas are at the core of the approach taken in the next several sections. First, we introduce ARCH(p) models. Not surprisingly, ARCH stands for:

Auto-**R**egressive **C**onditional **H**eteroskedasticity

A formal definition is as follows.

Definition 3. *A time series* $\{X_t\}_t$ *is said to be of type ARCH(p) if:*

$$X_t = \sigma_t W_t$$

where $\{W_t\}_t$ is a strong Gaussian white noise (i.e. an i.i.d. sequence of $N(0,1)$ random variables), and where σ_t is a (positive) function of X_{t-1}, X_{t-2}, ... determined by:

$$\sigma_t^2 = \alpha_0 + \sum_{j=1}^{p} \alpha_j X_{t-j}^2$$

with $\alpha_0 > 0$ and $\alpha_j \geq 0$ for $j = 1, 2, \ldots, p$.

In most of the textbook definitions of the ARCH(p) models, it is not required that the white noise $\{W_t\}_t$ is Gaussian. However, for the sake of simplicity, we do make this restrictive assumption. As a warm up, we first consider the simple case of an ARCH(1) model. In this case we have:

$$X_t = \sigma_t W_t \qquad \text{and} \qquad \sigma_t^2 = \alpha_0 + \alpha_1 X_{t-1}^2 \tag{8.7}$$

which gives inductively:

$$\begin{aligned} X_t^2 &= \sigma_t^2 W_t^2 \\ &= \alpha_0 W_t^2 + \alpha_1 X_{t-1}^2 W_t^2 \\ &= \alpha_0 W_t^2 + \alpha_1 \alpha_0 W_{t-1}^2 W_t^2 + \alpha_1^2 X_{t-2}^2 W_t^2 W_{t-1}^2 \\ &= \alpha_0 W_t^2 + \alpha_0 \alpha_1 W_{t-1}^2 W_t^2 + \alpha_0 \alpha_1 W_t^2 W_{t-1}^2 W_{t-2}^2 + \cdots + \alpha_0 \alpha_1^n W_t^2 W_{t-1}^2 \cdots W_{t-n}^2 \\ &\qquad \alpha_1^{n+1} W_t^2 W_{t-1}^2 \cdots W_{t-n}^2 X_{t-n-1}^2 \end{aligned}$$

by successive substitutions using the definition formula (8.7). This series converges if

- $\{X_t\}_t$ is causal (so X_{t-n-1} depends on W_s for $s < t - n$)
- $\{X_t\}_t$ is stationary (so that $\mathbb{E}\{X_{t-n-1}\}$ is a constant and hence bounded)
- $|\alpha_1| < 1$

In this case:

$$X_t^2 = \alpha_0 \sum_{j=0}^{\infty} \alpha_1^j W_t^2 W_{t-1}^2 W_{t-2}^2 \cdots W_{t-j}^2$$

from which we get:

$$\mathbb{E}\{X_t\} = 0 \qquad\qquad \mathbb{E}\{X_t^2\} = \frac{\alpha_0}{1 - \alpha_1}$$

and the final formula:

$$X_t = \sigma_t W_t \quad \text{with} \quad \sigma_t = \sqrt{\alpha_0 \left(\sum_{j=0}^{\infty} \alpha_1^j W_{t-1}^2 W_{t-2}^2 \cdots W_{t-j}^2 \right)}$$

It is important to notice that:

$$\mathbb{E}\{X_{t+h}X_t\} = \mathbb{E}\{\mathbb{E}\{X_{t+h}X_t|W_{t+h-1}, W_{t+h-2}, \ldots\}\} = 0$$

so that, in the sense of "weak white noise" introduced in Chap. 6, the time series $\{X_t\}_t$ is a white noise, not a white noise in the strong sense of i.i.d. sequences, but STILL A WHITE NOISE!!!

8.2.3 GARCH Models

The above discussion of ARCH models can be regarded as a warm up for the introduction of models with a broader appeal in the financial econometric community.

Definition 4. *We will say that the time series $\{X_t\}_t$ is GARCH(p,q) if:*

$$X_t = \mu_t + \sigma_t W_t$$

where as before, we assume that the noise sequence $\{W_t\}_t$ is i.i.d. $N(0,1)$, so that the conditional distribution of $\tilde{X}_t = X_t - \mu_t$ given $\tilde{X}_{t-1}$, $\tilde{X}_{t-2}$, … is $N(0, \sigma_t^2)$ with:

$$\sigma_t^2 = \sigma^2 + \sum_{j=1}^{p} \phi_j \sigma_{t-j}^2 + \sum_{j=1}^{q} \theta_j \tilde{X}_{t-j}^2. \tag{8.8}$$

Obviously this definition generalizes the notion of ARCH(p) models which can be recovered by setting $q = 0$. The term μ_t should be understood as a mean. It could be zero (as assumed in the previous subsection) or it could be the result of the fit of an ARMA model, in which case $\tilde{X}_t$ should be viewed as the residual time series after the fit of such a linear time series model. It is sometimes called the regression term.

The case $p = q = 1$ of GARCH(1,1) models is by far the most frequently used model. In such a case, we try to model the time series $\tilde{X}_t$ as:

$$\tilde{X}_t = \sigma_t W_t$$

in such a way that the conditional distribution of $\tilde{X}_t$ given $\tilde{X}_{t-1}$, $\tilde{X}_{t-2}$, … is $N(0, \sigma_t^2)$ with:

$$\sigma_t^2 = \sigma^2 + \phi_1 \sigma_{t-1}^2 + \theta_1 \tilde{X}_{t-1}^2$$

> This formula for the instantaneous variance is screaming for an interpretation in terms of ARMA(1,1) models!

We address this question next.

8.2.4 Summary

Even though we stated the definitions of ARCH and GARCH models with variance one strong white noise series, it is easy to see that the main properties of the models remain true if we only assume that the innovations are only white noise in the weak sense. In this subsection, we shall only consider weak sense white noise series, and instead of working in the general case of ARCH(p) and GARCH(p,q) models, we state results only for the case $p = 1$ and $q = 1$.

The formal way to identify an ARCH(1) time series is to write it first in terms of its conditional standard deviation, i.e. in the form $X_t = \sigma_t W_t$ where $\{W_t\}_t$ is a (stationary) weak white noise with variance one. Then the ARCH(1) prescription becomes:

$$\sigma_t^2 = \alpha_0 + \alpha_1 X_{t-1}^2.$$

which can be rewritten in the form:

$$X_t^2 = \alpha_0 + \alpha_1 X_{t-1}^2 + \tilde{W}_t$$

for some new weak white noise series $\{\tilde{W}_t\}_t$. This shows that (provided we generalize the definitions to accept weak white noise series)

$$\{X_t\}_t \sim \text{ARCH(1)} \quad \Longleftrightarrow \{X_t^2\}_t \sim \text{AR(1)}.$$

We now consider the case of GARCH(1,1) series. In this case the determining condition on the conditional variance reads:

$$\sigma_t^2 = \alpha_0 + \alpha_1 X_{t-1}^2 + \beta_1 \sigma_{t-1}^2.$$

which can be rewritten in the form:

$$X_t^2 = \alpha_0 + (\alpha_1 + \beta_1) X_{t-1}^2 + \tilde{W}_t - \beta_1 \tilde{W}_{t-1}$$

for some new weak white noise series $\{\tilde{W}_t\}_t$. Under the same proviso as before, this shows that

$$\{X_t\}_t \sim \text{GARCH(1,1)} \quad \Longleftrightarrow \{X_t^2\}_t \sim \text{ARMA(1,1)}.$$

8.2.5 `R` Commands

The function `garch` assumes that the argument is a numeric vector or a univariate `timeSeries` object with mean zero. In other words, it assumes that the mean component μ_t in the definition formula (8.8) has already been identified and subtracted. So typically, we will run the function `garch` on mean zero time series, often times on series of residuals. The order of the model is specified by the command `order=c(p,q)`. We choose to fit a GARCH(1,1) model to the log-returns of the Brazilian coffee futures contracts (recall Chap. 3) for the sake of illustration.

```
> BLRet.g <- garch(BLRet,order =c(1,1))
```

The prescription `order = c(1,1)` is not needed in the above command as `c(1,1)` is the default value for the argument `order`. The object `BLRet.g` returned by this function has the form:

```
> BLRet.g
Call:
garch(x = BLRet)
Coefficient(s):
        a0          a1          b1
0.0000271   0.1261165   0.8707195
```

The correspondence between the coefficients `a0, a1, .... , aq, b1, ..., bp` appearing in the output and the parameters appearing in the definition formula (8.8) is given by:

◇ `a0` stands for the mean variance σ^2
◇ `aj` stands for the moving average coefficient θ_j
◇ `bi` stands for ϕ_i

The R `garch` objects can be summarized in still another way. One can use the command `summary(BLRet.g)` which produces on the top of the information given above, an entire set of tests and p-values which we shall not reproduce here.

8.2.6 Fitting a GARCH Model to Real Data

It is now time to consider a first practical example to see why, when and how one fits a ARCH and/or a GARCH model to data. We use the example of the Dow Jones Enron index to demonstrate how the concepts introduced so far can be implemented on real data. At the risk of being accused of nostalgia for the exciting period of the seemingly endless expansion in the energy trading boom, we chose to restrict ourselves to the period pre-bankruptcy. The data we used is plotted in Fig. 8.2.

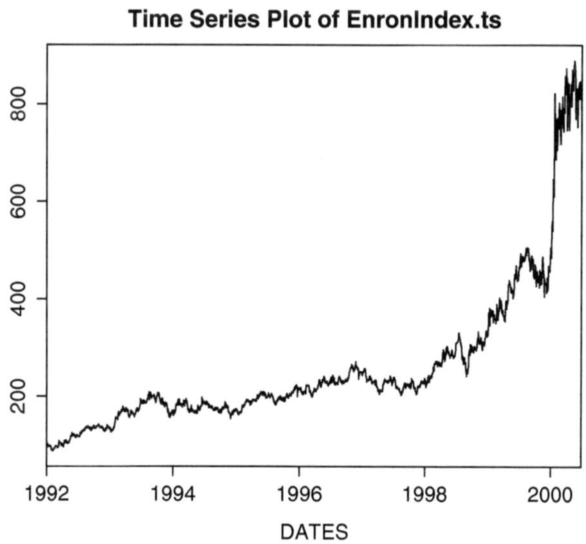

Fig. 8.2. Time series plot of the index created from the values of Enron's stock

Indexes are normalized aggregates of stock prices. Even when the index is computed from a single stock, it differs from the actual stock price in many ways. However, we should view them as providing essentially the same returns as the stock price would since they are merely normalized to have a specific value (say 100) on a given date (January 1, 1992 in this particular instance. The `timeSeries` object `EnronIndex.ts` which we use is included in the library `Rsafd`.

We use the following `R` commands to compute the log-returns from the Enron index, to compute and plot their partial auto-correlation function, fit an AR model, and plot the AIC criterion. The graphical results are reproduced in Fig. 8.3.

```
> LRET <- diff(log(EnronIndex.ts))
> acf(LRET,type="partial")
> LRET.ar <- ar(LRET)
> LRET.ar$order
[1] 4
> plot(LRET.ar$aic, type="l")
> title( "AIC criterion for ENRON LRET")
```

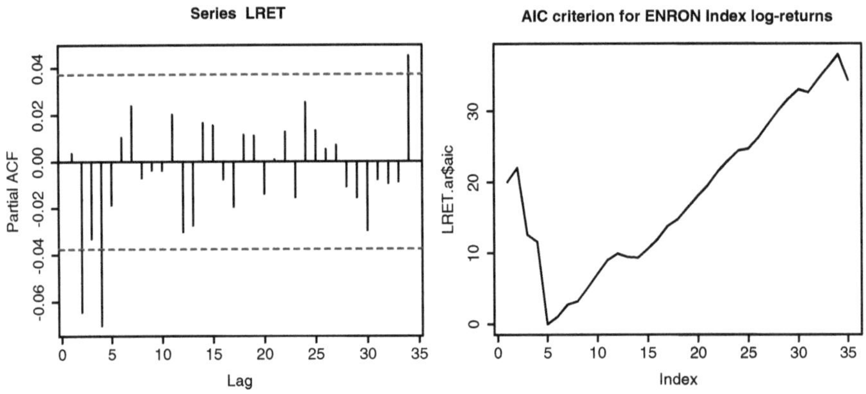

Fig. 8.3. Auto-correlation function of the log returns computed from the Enron index (*left*) and AIC criterion from the fit of an AR model to these log returns (*right*)

Except for the value at lag 34 (which is much too large to be considered as a possible order of the model), the partial auto-correlation function is not significantly different from zero for lags of 5 and above. The AIC has a clear minimum for lag 5. Both plots points to a model of order 4, which is what was fitted by the function `ar`.

Remark 1. The fact that we found an AR(4) for a financial log-return time series could appear as a contradiction with the intuition developed throughout the book according to which we expect those log-returns to be *white noise* like, and hence lead to AR(0) models. Knowing what we know in hindsight of what was actually driving Enron's stock price, we should not be surprised by the fact that it behaves differently than typical stock prices more aligned with an efficient market.

The time series plot of the residuals of this AR model, together with their Q-Q norm plot are obtained with the commands:

```
> plot(LRET.ar$resid)
> RES <- seriesData(LRET.ar$resid)
> qqnorm(RES)
```

Notice that we had to perform the Q-Q plot on the data of the time series since the function `qqnorm` does not accept `timeSeries` objects as parameters. The results are shown in Fig. 8.4. If lack of serial correlation is a possibility in light of what

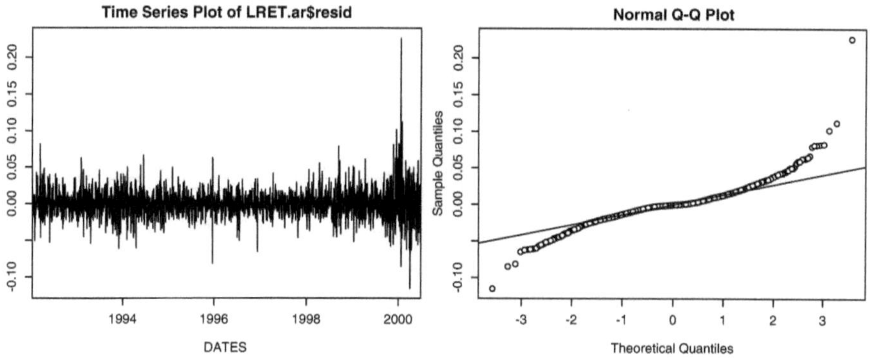

Fig. 8.4. Time series plot of the residuals of the AR model as fitted to the log returns computed from the Enron index (*left*) and their Q-Q plot against the Gaussian distribution (*right*)

we see in the left pane of Fig. 8.4, it is clear that too many measurements end up several standard deviation away from the mean. The marginal distribution of these residuals is presumably not normal. This is confirmed by the normal Q-Q plot of these residuals reproduced in the right pane of this figure. This plot shows that the distribution of the residuals has heavy tails. So even if the AR model was able to capture the serial correlation contained in the log returns, we cannot be sure that the residuals are independent, and that there are no serial dependencies left. We settle this question by computing the acf of the squares of these residuals.

```
> acf(LRET.ar$resid)
> acf(LRET.ar$resid^2)
```

The results are shown in Fig. 8.5. We are now in a rather delicate situation. Indeed, after transforming the data to stationarity (by first taking their logarithms and then the difference of the latter), and fitting a time series model (AR(4) to be specific), we end up with a residual time series with an `acf` of the white noise type, i.e. of the form:

$$1, 0, 0, \cdots .$$

Should we consider ourselves done? as we did in Chap. 6 when we were restricted to linear time series models? Or should we try to further analyze the data by digging into the residual series? Our discussion of the ARCH/GARCH models indicates that this is a reasonable option at this stage. For this reason, we proceed with the fitting of a GARCH(1,1) to the residuals of the AR model fitted earlier to the log-returns of the Enron Index data.

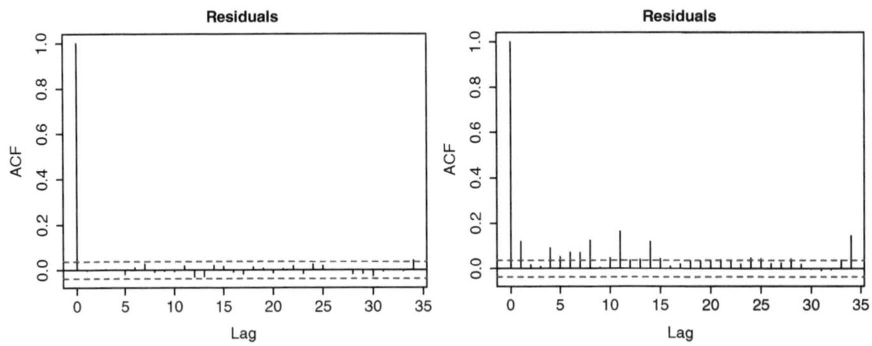

Fig. 8.5. Auto-correlation function of the raw residuals from the fit of an AR model to the Enron log-returns (*left*) and of the squares of these residuals (*right*)

8.2.6.1 ***Fitting a GARCH(1,1) Model to ENRON Index Log Return***

The function `garch` does not handle `NA`s, but the function `ar` of `Rsafd`, when applied to a `timeSeries` object, produces residuals in the form of a `timeSeries` object without `NA`s, the first `order` entries, which are bound to be `NA`s since they cannot be computed, being removed. As explained in the previous subsection, we fit a GARCH(1,1) model to the residual series with the following command:

```
  > LRES.g <- garch(LRET.ar$resid)
  > LRES.g
Call:
garch(x = LRET.ar$resid)
Coefficient(s):
       a0          a1          b1
2.636e-06  4.209e-02  9.494e-01
```

We do not reproduce here the output of the function `garch` which gives a trace of the convergence process of the algorithm. Recall the meaning of the parameters `a0`, `a1` and `b1` given in Sect. 8.2.5. Checking if the serial correlation present in the AR residuals has been removed by the GARCH model is a reasonable step to take at this stage. This can be done with the commands:

```
  > acf(LRES.g$resid)
  > acf(LRES.g$resid^2)
```

The results are reproduced in Fig. 8.6. These plots of the auto-correlation functions of the residuals and squared residuals of the fitted GARCH(1,1) model show that the serial correlation has practically been completely removed. In dealing with ARCH

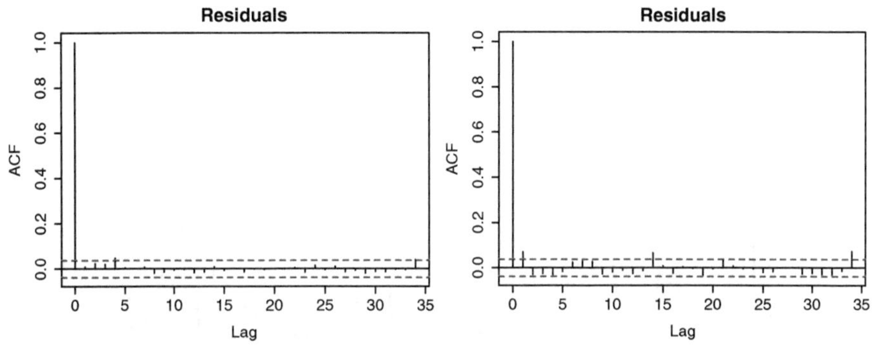

Fig. 8.6. Plots of the auto-correlation functions of the residuals of the GARCH(1,1) model fitted to the AR residuals from the Enron log-returns (*left*), and of their squares (*right*)

and GARCH models, part of the fitting procedure involves the estimation of the instantaneous conditional variance, and it is always instructive to take a look at such an estimate. In `R`, the fitted values and the estimates of the conditional variances are returned by the function `garch` and can be retrieved with the extensions $\cdots$`$fitted.values` and $\cdots$`$sigma.t` at the end of the name of the GARCH object produced by the fitting procedure. These objects are numeric vectors when the data input is itself a numeric vector, and a `timeSeries` object when the input is a `timeSeries` object. Note that these objects are shorter than the original data since the first `p` entries cannot be computed because of the auto-regressive part of the GARCH model.

The plot of the `timeSeries` object `LRES.g$sigma.t` giving the estimate of the instantaneous conditional standard deviation of the fitted series appears in Fig. 8.7 as the dark continuous line. This estimator is quite realistic and quite accurate. We see a large peak indicative of the huge growth and the subsequent highly volatile period corresponding to the crisis of 2000, and regularly spaced smaller peaks reminiscent of the seasonal nature of the business. So the estimation of the conditional standard deviation seems to be a powerful tool. It is interesting to compare this estimate to the result of a *pedestrian* approach which, on any given day, compute the empirical standard deviation of the entries of the series in the window containing the entries of the last 60 days. Notice that this computation is non-anticipative because the sliding window ends at the time of the computation, i.e. at any given time, we use only past observations to compute the standard deviation estimate. The plot of this naive sliding window estimate appears in Fig. 8.7 as the lighter dashed line. Notice that the results are very similar to the plot of `LRES.g$sigma.t` obtained by fitting the GARCH model. The naive estimator is smoother (the peaks are not as sharp) and lags the GARCH estimator. Both features should be expected given the fact that the window in which the empirical estimate of the standard deviation is computed is

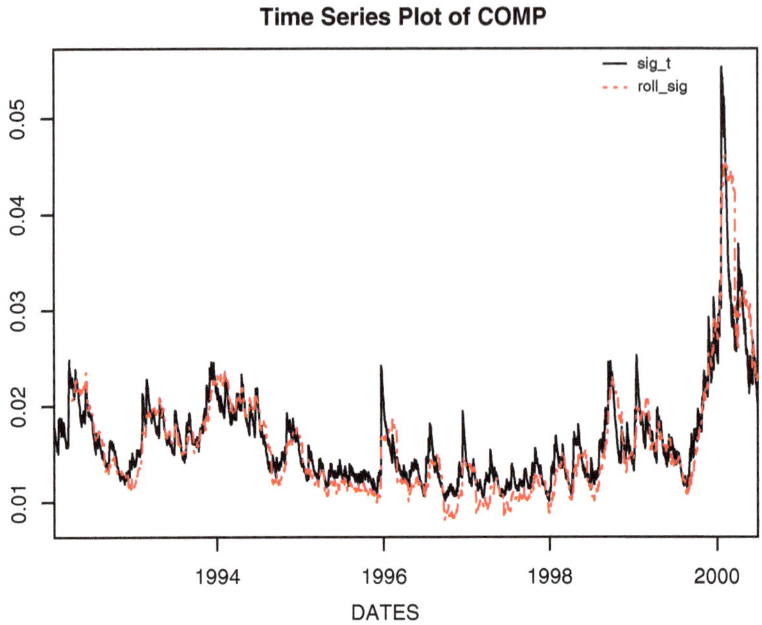

Fig. 8.7. Conditional standard deviation as estimated in the fitting of a GARCH(1,1) model to the AR residuals from the Enron index log-returns together with the estimate of the standard deviation in a 60 days tailing sliding window

trailing. But given the complexity of the GARCH fitting procedure, this comparison is rather anti-climatic. Figure 8.7 was produced with the commands

```
> SIG <- rep(0,LL-59)
> for (I in 60:LL) SIG[I-59] <- sd(RES[(I-59):I])
> SIG.ts <- timeSeries(positions =
seriesPositions(LRET.ar$resid)[60:LL],data=SIG,units=
  "roll_sig")
> COMP <- merge(LRES.g$sigma.t,SIG.ts)
> plot(COMP)
```

One can easily imagine from looking at the left pane of Fig. 8.4 that it is rather difficult to assess graphically the quality of the fit of the GARCH model. Figure 8.8 shows the scatter plot of the fitted values against the actual values to which the GARCH model was fitted. It was produced with the commands

```
> RES <- seriesData(LRET.ar$resid)
> LL <- length(RES)
> FITTED <- seriesData(LRES.g$fitted.values)
> plot(RES[2:LL],FITTED,main="Scatterplot of GARCH
      fitted values against Enron AR residuals")
```

The fact that the points are found on or near the diagonal is a good indication of the quality of the fit.

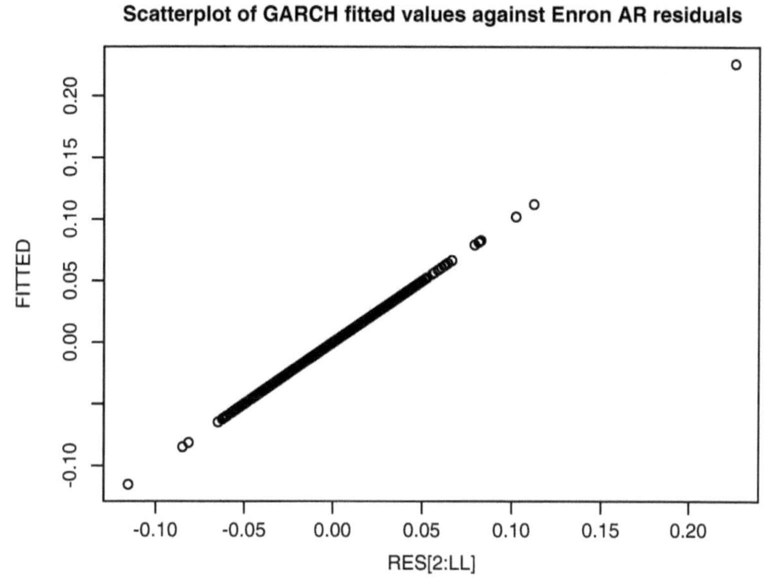

Fig. 8.8. Scatterplot of the fitted values against the actual values to which the GARCH model was fitted

8.2.6.2 *Monte Carlo Simulations*

The library Rsafd provides the function sim.garch to generate Monte Carlo samples from a GARCH model. Here is an example of its use.

```
> MODEL <- list(sig2=LRES.g$coef[1],ma=LRES.g$coef[2],
                      ar=LRES.g$coef[3])
> PRESIG <- seriesData(LRES.g$sigma.t)[LL-1]
> PREX <- FITTED[LL-1]
> SIM <- sim.garch(model=MODEL,n.ahead = 1024, n.start = 0,
        n.sim = 1, presigmat=PRESIG, preXt=PREX )
> plot(SIM$sigma.t,type="l")
```

The result is reproduced in Fig. 8.9. Several remarks are in order at this point. The parameter model specifies the GARCH model to be simulated. It should be a list whose first element is a number sig2 giving the mean variance σ^2 of the model (the number a0 returned by the function garch), a vector ar of the p coefficients ϕ_1, $\cdots$, ϕ_p (the numbers b1, $\cdots$, bp returned by the function garch), and a vector ma of the q coefficients θ_1, $\cdots$, θ_q (the numbers a1, $\cdots$, aq returned by the function garch). Here we use the model fitted to the residuals of the AR model fitted to the log-returns of the Enron index. The parameter n.ahead gives the length of the desired simulated scenarios, while n.sim (which is equal to 1 by default) gives the number of desired scenarios. As in most simulations of time series models, the parameter n.start represents the number of samples which are disregarded. They are

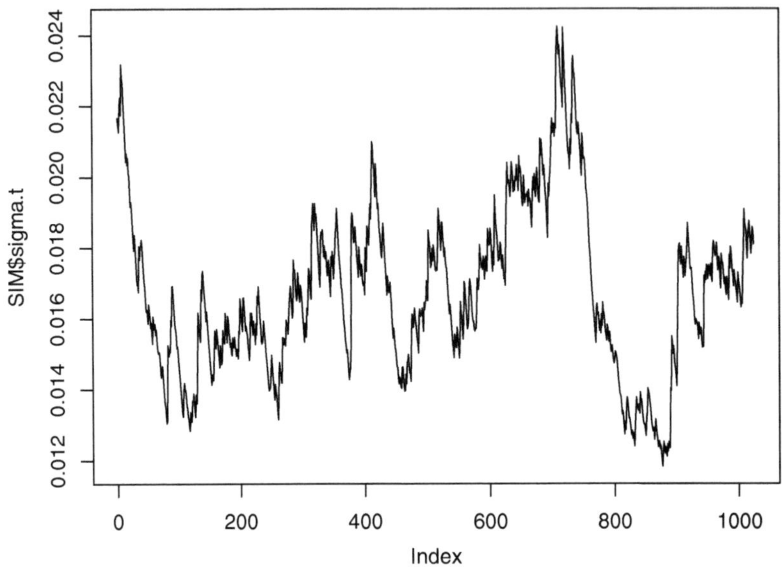

Fig. 8.9. Simulated stretch of length 1,024 of the conditional standard deviation of the GARCH(1,1) model fitted to the AR residuals from the Enron log-returns, as an extension of the in-sample estimate

generated in order for the series to reach a steady state regime in which the dynamics are stationary/stable. One should experiment and try several values of this parameter to get a feeling for its effect. In particular, too small a value of this parameter will produce a sample from a non-stationary time series. This is usually detected because of significant differences between the first part of the simulation which exhibits non-stationary behavior, and the later part where the stationary regime is reached, and the effects of the initial values have been wiped out of the statistics. Obviously, this parameter is set to 0 when the simulation tries to produce scenarios extending the time series, in which case one sets the parameters `presigmat` and `preXt` to the vector of the last p values of the fitted conditional standard deviation, and the vector of the last q fitted values of the time series. This is what we did in the above example.

The output of the function `sim.garch` is a list of two matrices each with `n.sim` rows (one for each Monte Carlo scenario) and `n.ahead` columns (one for each time in the future for which we desire to have a scenario. The first matrix is called `X.t`. It gives the values of the time series. The second matrix is called `sigma.t`. It gives the values of the instantaneous conditional standard deviation.

It is always rather difficult to assess the quality of Monte Carlo simulations. In particular, there is no reason to believe that the simulation appearing in Fig. 8.9 is not appropriate. However, we may be suspicious of its quality because it does not have any of the deterministic features (in particular it does not have a significant seasonal component) which we identified in the data. Next, we expand on this problem.

8.2.6.3 *Simulation Can Be a Touchy Business*

A wrong value for the `n.start` parameter is not the only possible misuse of the Monte Carlo simulation function `sim.garch`. Indeed, quite unrealistic simulated samples can be produced when a deterministic seasonal pattern exists, and the GARCH model simply believes that it is part of the typical fluctuations of the random variance. We illustrate this phenomenon in the case of high frequency trading intra-day data, and we refer the reader to Problem 8.4 for a similar example in the case of temperature data.

We use high-frequency data of all the quotes on the stock of IBM throughout the month of June 1999. We pre-processed the data to produce regularly spaced quotes to reduce the bid-ask spread and end up with only one single price every 30 s. There were 22 trading days on that month, so the data became a 22×779 matrix `IBM`. For the sake of the present illustration, we shall only use the first row (i.e. the first trading day). We fit a GARCH(1,1) model to the log-returns and we plot the estimate of the conditional standard deviation in the way described above.

```
> I <- 1
> plot(IBM[I,],type = "l", main="IBM High Frequency (30 s).
                   on First Day of Trading in June 1999")
> IBMLR <- diff(log(IBM[I,]))
> IBM.g <- garch(IBMLR)
> plot(IBM.g$sigma.t,type="l", main="Conditional
                                  Volatility on that Day")
```

Figure 8.10 gives the plots produced by these commands. The instantaneous volatility shows an interesting pattern: high levels in the morning (around 9:30 or 10:00 a.m.), and in the later part of the trading day, with a smaller surge before lunch time. This pattern is pretty typical, and it can be identified most days, so we would like to treat it as a deterministic seasonal component. Whether or not it is a deterministic component, the GARCH fitting procedure, picked it up. So this looks pretty nice, and it is indeed. The problems arise when we try to use the model fitted in this way, in order to simulate Monte Carlo samples to be used as possible scenarios of a trading day. As before, we can use commands of the following type to produce such scenarios:

```
> MODEL <- list(sig2=IBM.g$coef[1],ma=IBM.g$coef[2],
                    ar=IBM.g$coef[3])
> IBM.sim <- sim.garch(model=MODEL,n.ahead = 779,
                    n.start = 1000)
> plot(IBM.sim$X.t, type="l", main="Simulated GARCH(1,1)
                        with same parameters as in IBM.g")
> plot(IBM.sim$sigma.t, type="l", main="Conditional
                   Volatility of the Simulated GARCH(1,1)")
```

The results are reproduced in Fig. 8.11. There one cannot find the special pattern of high activity levels at specific times of the day. The simulation algorithm uses

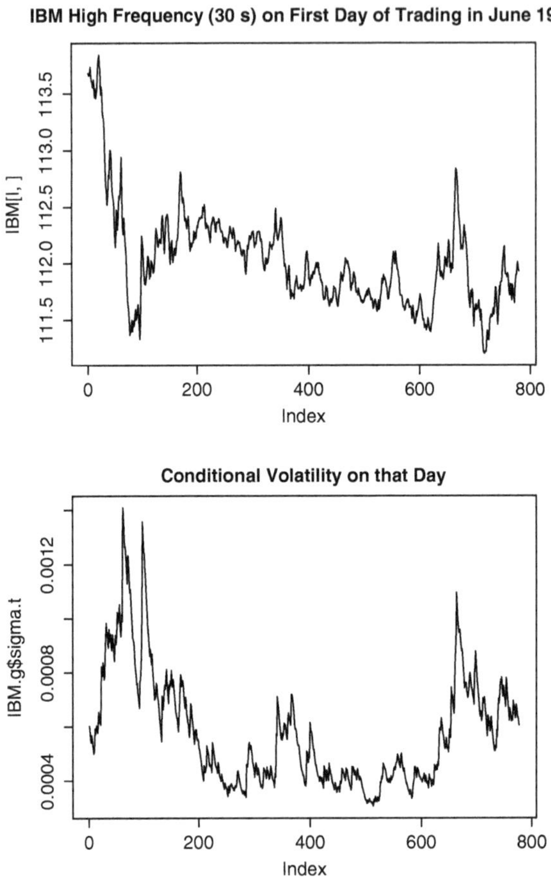

Fig. 8.10. IBM Quotes Regularly Spaced by 30 s on the first trading day of June 1999 (*top*) and estimate of the conditional volatility (*bottom*) given by a GARCH(1,1) model fitted to the log-returns

the coefficients estimated for the GARCH(1,1) model, and using them, it produces a sample which is as stationary as possible. In particular, the simulation algorithm makes sure that the relative frequencies of the periods with high and low levels of volatility come with their expected frequencies, say two or three periods of high volatility per day, for example. However, there is absolutely no reason for the algorithm to try to set these periods of high volatility in the early morning and/or in the later part of the trading day. Because of stationarity, they appear any time throughout the day. These deterministic features of the term structure of volatility are not part of the model which is simulated. This may not be a problem if we are interested in statistics involving the whole day, but if we care about statistics depending on the time of the day, the scenarios produced by this particular use of the function `sim.garch` become misleading, and possibly useless. Worse than that, they may end up being very dangerous since they may lead to gross errors.

8.2.7 Generalizations

ARCH and GARCH models have been generalized in many different directions, and the menagerie of models available to the data analyst is amazing. This great variety can be intimidating to the non-specialist. Moreover, because of the anxiety resulting from this lack of "one size fits all", and of the difficulties encountered in interpreting and using the results of the fits, many potential users have shied away from the GARCH methodology in favor of more robust and stable alternatives.

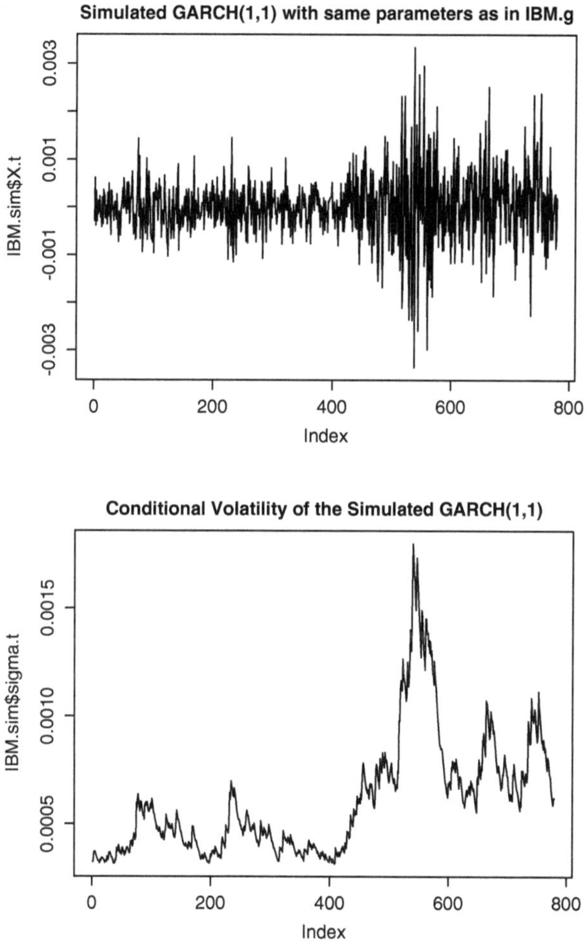

Fig. 8.11. Simulation of one trading day for IBM high-frequency data (*top*), and corresponding instantaneous volatility (*bottom*)

8.2.7.1 More Univariate GARCH Models

Because of the high level of complexity of the technicalities of these generalizations, we shall refrain from discussing here the various forms of GARCH which have been proposed to accommodate the features missed by the standard ARCH and GARCH models introduced in this chapter. The following is a very short list of some of the most popular of these variations on the GARCH theme.

- **LGARCH** Leverage GARCH
- **PGARCH** Power GARCH
- **EGARCH** Exponential GARCH
- **TGARCH** Threshold GARCH
- **CGARCH** Component GARCH
- **GARCH-M** GARCH in the Mean
- **FIGARCH** Fractionally Integrated GARCH

We refer the interested (or desperate) reader to the Notes & Complements at the end of the chapter for precise references to textbooks in which these generalizations are described.

8.2.7.2 Multivariate Models

In the same way AR models were effortlessly generalized to the multivariate setting, GARCH models can be extended as well, and the fitting methods keep the same structure. A multivariate time series $\{\mathbf{X}_t\}_t$ is said to be a GARCH series if it is of the form:

$$\mathbf{X}_t = \mu + \mathbf{W}_t$$

where

- μ and $\mathbf{W}_t$ are $d \times 1$ vectors
- The distribution of $\mathbf{W}_t$ conditioned on $\mathbf{W}_{t-1}, \mathbf{W}_{t-2}, \ldots$, can be:
  - d-variate normal;
  - d-variate Student;
- With $d \times d$ variance/covariance matrix $\mathbf{V}_t$ satisfying

$$\mathbf{V}_t = A + \sum_{k=1}^{p} A_k * [\mathbf{W}_{t-k}\mathbf{W}_{t-k}^t] + \sum_{h=1}^{q} \mathbf{B}_h * \mathbf{V}_{t-k}$$

  where
  - $\mathbf{A}$, $\mathbf{A}_k$ and $\mathbf{B}_h$ are symmetric $d \times d$ matrices
  - $*$ stands for the Hadamard product (entry by entry)
  - As usual ^{t} stands for the transpose of a matrix/vector

  Finally, note that one can replace μ by the result of a regression on exogenous variables.

8.3 Stochastic Volatility Models

A stochastic volatility (SV) model is based on a couple of independent variance one white noise series $\{W_t\}_t$ and $\{u_t\}_t$ and the defining dynamical equations:

$$\begin{cases} X_t = \sigma_t W_t \\ \log \sigma_t = \phi_0 + \phi_1 \log \sigma_{t-1} + \gamma u_t \end{cases} \tag{8.9}$$

where ϕ_0, ϕ_1 and γ are constants. We shall assume that $|\phi| < 1$ to guarantee that the second equation gives a stationary AR(1) time series for the logarithm of the conditional variance σ_t^2. This implies that the time series $\{X_t\}_t$ is stationary since both series $\{W_t\}_t$ and $\{\sigma_t\}_t$ are. Notice that $\{X_t\}_t$ is a (weak) white noise, and that σ_t^2 is its conditional variance. So this SV model has a lot in common with the ARCH and GARCH models discussed in this chapter. But there is a fundamental difference: *the model is driven by two sources of randomness.* This innocent looking difference has dramatic consequences, and this model will behave quite differently from the ARCH and GARCH models discussed in the previous section.

Remark 2. As stated above, we assume that the two sources of randomness (i.e. the white noise series $\{W_t\}_t$ and $\{u_t\}_t$) are independent. This assumption is for convenience as it makes the mathematical analysis of stochastic volatility models more tractable. However, as we shall see, it is often desirable to assume that they are negatively correlated to account for the so-called leverage effect discussed later in this section.

8.3.1 **Information Structure**

The first problem to consider is related to the presence of two driving white noise terms, and for that reason, it is specific to the SV models: which is the right notion of past information. At any given time t, should the past information be the information contained in the past values of the series as encapsulated in $\underline{X}_{\leq t-1} = \{X_{t-1}, X_{t-2}, \ldots, X_1\}$, or in the past values of the series together with the past values of the volatility (or equivalently the noise driving it) as captured by:

$$(\underline{X}_{\leq t-1}, \underline{u}_{\leq t-1}) = \{(X_{t-1}, u_{t-1}), (X_{t-2}, u_{t-2}), \ldots, (X_1, u_1)\}.$$

This question is not purely academic, it has very practical consequences. For example, if one uses the first notion of past information, the corresponding notion of conditional variance should be:

$$h_t = \text{var}\{X_t | \underline{X}_{t-1}\} = \text{var}\{X_t | X_{t-1}, X_{t-2}, \ldots, X_1\} \tag{8.10}$$

like in the case of the ARCH and GARCH models. On the other hand, if we use the second notion of past information, then the natural conditional variance to use is given by:

$$\begin{aligned}\sigma_t^2 &= \text{var}\{X_t|(\underline{X}_{\le t-1}, \underline{u}_{\le t-1})\} \\ &= \text{var}\{X_t|(X_{t-1}, u_{t-1}), (X_{t-2}, u_{t-2}), \ldots, (X_1, u_1)\}.\end{aligned}$$

Notice that, because of the tower property of conditional expectations we have:

$$h_t = \mathbb{E}\{\sigma_t^2 | \underline{X}_{\le t-1}\}.$$

8.3.2 State Space Formulation

Mostly because of the fact that they are driven by two different white noise series, the stochastic volatility systems fit very naturally in the framework of state space systems. Indeed the second equation in (8.9) can be written in the form:

$$x_t = 2\phi_0 + \phi_1 x_{t-1} + v_t$$

if we set $x_t = \log \sigma_t^2$ and $v_t = 2\gamma u_t$, which is exactly the form of a state equation (with linear dynamics in the present situation). Moreover, the first equation in (8.9) can be written in the form:

$$\log |X_t| = \log \sigma_t + \log |W_t|$$

which can be rewritten in the form:

$$y_t = a_0 + a_1 x_t + w_t$$

which is a linear observation equation of the state x_t, provided we set $y_t = \log |X_t|$, $a_0 = \mathbb{E}\{\log |W_t|\}$, $a_1 = 1/2$ and $w_t = \log |W_t| - \mathbb{E}\{\log |W_t|\}$. So the stochastic volatility model (8.9) rewrites as a linear state space model, and all the estimation/filtering techniques studied in Chap. 7 can be applied. We shall come back to this important remark later in this chapter when we implement nonlinear filtering techniques to the continuous time analog of the stochastic volatility models of this section.

The analysis of AR(1) models done in Chap. 6 implies that:

$$\mu_x = \mathbb{E}\{x_t\} = \frac{\phi_0}{1 - \phi_1} \qquad \text{and} \qquad \sigma_x^2 = \text{var}\{x_t\} = \frac{\sigma^2}{1 - \phi_1^2}$$

where σ^2 is the variance of the white noise of the AR(1) dynamics of x_t, i.e. $\sigma^2 = 4\gamma^2$ if we come back to the parameters of the original dynamical equation (8.9).

8.3.3 Excess Kurtosis

In this subsection we assume that both noise terms are Gaussian. Notice that, because we assume that the two white noise series are independent, for each integer $k \ge 1$ we have:

$$\mathbb{E}\{X_t^k\} = \mathbb{E}\{\sigma_t^k\}\mathbb{E}\{W_t^k\}$$

for all the integers k. This k-th moment will be zero for all odd power k because the distribution of the noise W_t is symmetric. When $k = 2h$ is even, using the expressions for the Laplace transform (moment generating function) and the moments of the Gaussian distributions, we find that:

$$\mathbb{E}\{X_t^{2h}\} = \mathbb{E}\{e^{hx_t}\}\mathbb{E}\{W_t^{2h}\} = e^{h\mu_x + h^2\sigma_x^2/2}\frac{(2h)!}{h!2^h}$$

In particular, the kurtosis of X_t is given by:

$$k\{X_t\} = \frac{\mathbb{E}\{X_t^4\}}{\mathbb{E}\{X_t^2\}^2} = \frac{e^{2\mu_x + 2\sigma_x^2}\frac{4!}{2!2^2}}{e^{2\mu_x + \sigma_x^2}\frac{2!}{1!2}} = 3e^{\sigma_x^2}$$

which is strictly greater than 3, proving that SV models exhibit excess kurtosis and heavy tail distributions.

8.3.4 **Leverage Effect**

Both GARCH and SV models have been shown to give an account of excess kurtosis and persistence in financial data. There is still another stylized fact which needs to be checked against these models: the so-called leverage effect. The latter is attributed to the fact that large up-moves in the value of a stock are usually accompanied by decreases in volatility while at the contrary down-moves in the value of a stock are usually accompanied by surge in volatility. This effect appears as a form of negative correlation between the changes in prices and the changes in volatility. This is the way we shall detect the leverage effect in the mathematical models.

◇ *The Case of ARCH Models*

Let us assume for example that the time series $\{X_t\}_t$ is ARCH(1), and let us try to compute the sign of the correlation coefficient between the changes in X_t and the changes in its conditional variance. At any given time t, we have the information of the past values of the series, so the probabilities, expectations, variances, correlations, ... are computed conditionally on the knowledge of $\underline{X}_{\leq t-1} = \{X_{t-1}, X_{t-2}, \ldots, X_1\}$. We shall emphasize that conditioning by adding a subscript $t-1$ to all the expectations, variances and covariances which we compute. Our ARCH(1) assumption can be written in the form

$$X_t = \sqrt{\alpha_0 + \alpha_1 X_{t-1}^2} W_t$$

for some positive coefficients α_0 and α_1, and we have:

$$\text{cov}_{t-1}\{X_t - X_{t-1}, \sigma_{t+1}^2 - \sigma_t^2\} = \text{cov}_{t-1}\{X_t, \sigma_{t+1}^2\}$$

because X_{t-1} and σ_t^2 are known at time $t-1$. Consequently:

$$\begin{aligned}
\mathrm{cov}_{t-1}\{X_t - X_{t-1}, \sigma_{t+1}^2 - \sigma_t^2\} &= \mathrm{cov}_{t-1}\{X_t, \alpha_0 + \alpha_1 X_t^2\} \\
&= \alpha_1 \mathrm{cov}_{t-1}\{X_t, X_t^2\} \\
&= \alpha_1 \mathrm{cov}_{t-1}\{\sqrt{\alpha_0 + \alpha_1 X_{t-1}^2} W_t, \\
&\qquad\qquad (\alpha_0 + \alpha_1 X_{t-1}^2) W_t^2\} \\
&= \alpha_1 (\alpha_0 + \alpha_1 X_{t-1}^2)^{3/2} \mathrm{cov}_{t-1}\{W_t, W_t^2\} \\
&= \alpha_1 (\alpha_0 + \alpha_1 X_{t-1}^2)^{3/2} \mathbb{E}\{W_t^3\}.
\end{aligned}$$

The conclusion is that the leverage effect is present in the model only when the noise distribution is skewed to the left, i.e. when $\mathbb{E}\{W_t^3\} < 0$.

◇ *The Case of the SV Models*

The leverage effect is more difficult to pinpoint in the case of the stochastic volatility models. As explained earlier, the main difficulty lies in the notion of information available at time t, and this difficulty is rooted in the presence of several sources of randomness. Notice that, because we assume that the two noise terms are independent, we have:

$$\mathrm{cov}\{X_t - X_{t-1}, \sigma_t^2 - \sigma_{t-1}^2 | \underline{X}_{\le t-1}, \underline{u}_{\le t-1}\} = 0$$

$$\mathrm{cov}\{X_t - X_{t-1}, \sigma_{t+1}^2 - \sigma_t^2 | \underline{X}_{\le t-1}, \underline{u}_{\le t-1}\} = 0.$$

On the other hand, the conditional covariance:

$$\mathrm{cov}\{X_t - X_{t-1}, \sigma_{t+1}^2 - \sigma_t^2 | \underline{X}_{\le t-1}, \underline{u}_{\le t-1}\}$$

can be different from zero. Here we use the notation $h_t = \mathrm{var}\{X_t | \underline{X}_{\le t-1}\}$ already defined in (8.10) for the conditional variance of X_t given its own past values (excluding the knowledge of the past values of the noise driving the volatility). The use of h_t as a conditional variance is very natural since we observe the values of X, while we have no way to guess the values of the noise u in general. Unfortunately, it is very difficult to handle this quantity mathematically, and proving rigorously the presence of a negative correlation in SV models is usually very difficult.

8.3.5 **Comparison with ARCH and GARCH Models**

A SV time series $\{X_t\}_t$ is a martingale difference in the sense that:

$$\mathbb{E}\{X_t | \underline{X}_{t-1}, \underline{u}_{t-1}\} = 0.$$

It is a weak white noise (recall that we assume $|\phi_1| < 1$ to guarantee stationarity). Concentrating on the series of squares, and computing their auto-covariance function we get:

$$\begin{aligned}\operatorname{cov}\{X_t^2, X_{t-s}^2\} &= \mathbb{E}\{e^{x_t + x_{t-s}} W_t^2 W_{t-s}^2\} - \mathbb{E}\{X_t^2\}^2 \\ &= e^{2\mu_x + (1+\phi_1^s)\sigma_x^2} - \mathbb{E}\{e^{x_t}\}^2 \\ &= e^{2\mu_x + \sigma_x^2}(e^{\phi_1^s \sigma_x^2} - 1)\end{aligned}$$

and the auto-correlation function is given by:

$$\rho_{X^2}(s) = \frac{\operatorname{cov}\{X_t^2, X_{t-s}^2\}}{\operatorname{var}\{X_t^2\}} = \frac{e^{\phi_1^s \sigma_x^2} - 1}{3e^{\sigma_x^2} - 1} \sim \frac{e^{\sigma_x^2} - 1}{3e^{\sigma_x^2} - 1}\phi_1^s$$

which prompts the following remarks:

Remarks.

1. This acf changes signs when $\phi_1 < 0$, which is not the case for ARCH(1) models.

2. This acf looks very much like the acf of an ARMA(1,1) model. For this reason, one should think that a SV model is closer to a GARCH(1,1) model than to a ARCH(1) model!

8.3.6 The Smile Effect

The smile effect is next on the list of stylized facts which models of financial prices should capture. This effect is explained in detail in Appendix 9.2, justifying the important role of implied volatility in our discussion of the nonparametric approach to option pricing in Chap. 5. We mentioned the analog role of implied correlations in our discussion of CDOs in Chap. 3.

Because of the existence of smiles in option prices, and the tremendous impact this discovery had on the everyday trading practices, we think it is important to review their theoretical underpinnings. But before we can define them rigorously, we need to introduce the notion of implied parameter. which is based on a pricing formula (a pricing algorithm would do as well) providing a one-to-one correspondence between the price of a financial instrument and certain parameters (short interest rate, volatility, time to maturity, etc.). This one-to-one correspondence makes it possible to infer the value of a parameter which needs to be fed to such a formula or algorithm in order to recover the value of a price actually quoted on the market. Let us consider for example the case of implied volatility, assuming that we are using Black-Scholes formula to price options. For each option price quoted on the market, one can take note of the strike price, the time to maturity, ..., and we can find which value of the volatility parameter σ we have to use in the Black-Scholes formula to get the price actually quoted. This value is called the *implied volatility.* As explained in Appendix 9.2, it is not a statistical estimate of the parameter σ, it is a value implied by a transaction (or a set of transactions).

If on a given day and for a given time to maturity one can observe the prices of several European call options for different strike prices, the Black-Scholes theory tells us that the corresponding implied volatilities should be equal to each other.

But there is strong empirical evidence to the contrary. The implied volatilities of these options are different. Quite often, if one plots them against the strike prices, these implied volatilities form a convex curve having a minimum when the option is *at the money*, i.e. when the strike price is exactly equal to the current price of the underlying interest. This curve is called the volatility smile. The implied volatility is higher when the strike price is above the current price (in this case we say that the option is out of the money) and when it is below the current price (in this case we say that the option is in the money).

This empirical fact has prompted analysts and traders to revise the simultaneous use of a pricing model and a pricing formula (the Samuelson's log-normal model and the Black-Scholes formula in the discussion above) when they lead to a contradiction of the type we just described. One of the great successes of the stochastic volatility models was the discovery that they can account for the volatility smile. Such a derivation would be far beyond the scope of this book, but we could not resist giving one of the main reasons for the popularity of these models. The interested reader should consult the Notes & Complements section at the end of the chapter for references.

8.4 Discretization of Stochastic Differential Equations

Continuous time finance has seen a tremendous growth over the last 40 years. Among the many abstract concepts brought to bear in the analysis of financial models, martingales and Ito's stochastic calculus are the most obscure to the non-mathematically inclined financial analysts. Any attempt to present Ito's theory of stochastic integration, and stochastic differential equations would take us beyond the scope of this book. In this section, we consider discrete time analogs of the continuous time dynamic models most often used in practice and the theoretical literature.

We already encountered one example of Ito stochastic differential equation (SDE, for short). It is the most famous of all the SDE's used by the financial community:

$$dS_t = S_t[\mu dt + \sigma dW_t]. \tag{8.11}$$

It describes the stochastic dynamics of the so-called geometric Brownian motion introduced by Samuelson as a model for the time evolution of stock prices. More generally, stochastic differential equations appear in the form:

$$dX_t = \mu(t, X_t)dt + \sigma(t, X_t)dW_t \tag{8.12}$$

where $(t, x) \hookrightarrow \mu(t, x)$ and $(t, x) \hookrightarrow \sigma(t, x)$ are functions of the time variable t, and the state variable x. These functions will be deterministic and real valued in the applications discussed below. But they could very well be vector valued, or matrix valued, and random as well. Essentially the same theory would apply. For the sake of the present discussion, one should think of X_t as describing the state at time t of a set of economic factors. The "dt" – term has the usual interpretation of an infinitesimal change in the time variable t. The $''dW_t''$ – term tries to play an analogous role for an infinitesimal random change. Its rigorous definition is very delicate, and far beyond the scope of this book. The intuitive interpretation of dW_t should be that of an infinitesimal random shock of the white noise type, but since the notion of continuous time white noise is very intricate, this term has to be understood as the (stochastic) differential of its antiderivative. Indeed, the latter can be defined more easily, as a stochastic process with independent increments. This process $\{W_t\}_{t\geq 0}$ is usually called a Wiener process, or a process of Brownian motion since one of its early use was to model the motion of particles in suspension, investigated by Brown. Even though Einstein is usually credited for the first development of the theory of Brownian motion, a growing part of the scientific community is now making a case that Bachelier should be getting this credit because of his earlier work on the theory of speculation.

Equation (8.12) is a concise way to describe the dynamics (time evolution) of the (possibly random) quantity X. The coefficient $\mu(t, X_t)$ gives the instantaneous mean of the infinitesimal increment dX_t. The coefficient $\sigma(t, X_t)$ represents its instantaneous standard deviation. So, Eq. (8.12) is not random when $\sigma \equiv 0$. In this case, it reduces to an ordinary differential equation, and it can be analyzed using classical calculus. Things are much more complicated when σ is not identically zero. However, we shall not need to get involved in the meanders of Ito's theory of stochastic calculus, we shall limit ourselves to discretized versions of these equations, and this will give us a chance to bridge continuous time finance with the time series analysis of financial econometrics.

8.4.1 **Discretization Schemes**

Instead of working directly with a continuous time model, we assume that snapshots of the system are taken at discrete time intervals. We assume that measurements take place at regular times $t_j = t_0 + j\Delta t$, and we use the notation $X_j^{(\Delta t)}$ for the value X_{t_j} of the (random) variable X at time t_j. We sometimes drop the superscript (Δt) specifying the length of the sampling interval from the notation for the sake of easier typesetting. This abuse of notation should not create confusion. The sampling frequency is usually defined as the inverse of the length Δt of the time interval separating two successive measurements.

8.4.1.1 ***The Euler Scheme***

A natural question is now to identify a discrete time dynamical equation for $X_j^{(\Delta t)}$ which would be consistent with the continuous time dynamics given by Eq. (8.12). This is usually done in the following manner, dropping the superscript (Δt) for notational convenience. We consider the evolution given by the recursive equation:

$$X_j - X_{j-1} = \mu(t_{j-1}, X_{j-1})\Delta t + \sigma(t_{j-1}, X_{j-1})\sqrt{\Delta t}\epsilon_j \qquad (8.13)$$

where $\{\epsilon_j\}_{j\geq 1}$ is an $N(0,1)$ white noise, and where the initial condition X_0 is assumed to be given. A few important remarks are granted at this stage.

Remarks.

1. The (stochastic) differential dX_t appearing in the left hand side of (8.12) was discretized as $X_{t_j} - X_{t_{j-1}}$, while the X_t's appearing in the right hand side were discretized as $X_{t_{j-1}}$. This is specific to the Ito's stochastic integration theory. The stochastic increments dW_t should be taken ahead of the instant of the discretization. This non-anticipative idiosyncratic feature is crucial in financial applications (unless you own a crystal ball, in which case it should not apply to you!)
2. The stochastic differential dW_t appearing in (8.12) was discretized as $\sqrt{\Delta t}\epsilon_j$. This is a consequence of the fact that the increments $W_t - W_s$ of a Wiener process are mean-zero Gaussian random variables with standard deviations $\sqrt{t-s}$, and because they are independent of each other when computed over non-overlapping intervals.
3. If σ is constant, i.e. if $\sigma(t,x) \equiv \sigma$, and μ is linear, say $\mu(t,x) = \phi_0 + \phi_1 x$, then Eq. (8.12) is called an equation of the Ornstein-Uhlenbeck type, and it is plain to see that its discretized form (8.13) defines an AR(1) process. In general equation (8.13) defines a (possibly nonlinear) auto regressive process of order 1 as defined in Sect. 8.1.2. So we are still in known territory.
4. It should be understood that, even if they both start from the same initial values, say X_0, the solution X_t of the continuous time stochastic differential equation (8.12), and the solution $X_j^{(\Delta t)}$ of the recursive equation (8.13) have no reason to coincide, *not even at the sampling times* $t_j = t_0 + j\Delta t$. In other words, we should not expect that $X_{t_j} = X_j^{(\Delta t)}$. But there is a justice, and whenever the coefficients μ and σ are smooth, and whenever the discretization step Δt tends to zero, then the difference between these values tends to zero in a controlled manner. We shall not state a precise mathematical theorem, but obviously, this result is a clear justification of the extensive use of discrete models to simulate and approximate continuous time models. In fact, it is possible to show that continuous time stochastic volatility models can also appear as diffusion limits of appropriately set up ARCH and GARCH models.
5. Finally, we notice that the recursive form of equation (8.13) is perfectly suited for Monte Carlo simulations. Indeed, it is very easy to simulate a white noise, and from white noise samples, it is plain to implement formula (8.13) to generate samples of X_j.

The discretization given by Eq. (8.13) is known as the Euler scheme. It not the only way to derive a discrete time approximation to a continuous time evolution. Other procedures have been introduced to speed up the convergence toward the true dynamics. But it is nevertheless the simplest one, and we shall use it for that reason.

8.4.2 Monte Carlo Simulations: A First Example

In this subsection, we give examples of direct random simulations based on explicit formulae for the solutions of SDE's. In particular, these simulations do not use the Euler's scheme described earlier. However, due to the fact that they depend upon the existence of exact formulae, their realm of application remains limited.

As we explain in Appendix 9.2, the form of equation (8.11) is so simple that an explicit formula can be found for the solution. We restate the form of the solution given in (9.6):

$$S_t = S_0 e^{(\mu - \sigma^2/2)t + \sigma W_t} \tag{8.14}$$

or more generally as:

$$S_t = S_s e^{[\mu - \sigma^2/2](t-s) + \sigma[W_t - W_s]} \tag{8.15}$$

if we know the value S_s of the solution at time $s < t$ instead of 0. As earlier, we emphasize the presence of the unexpected term $-\sigma^2/2$ which should not have appeared according to the rules of classical calculus. Its presence is forced on us by the special rules of Itô's calculus developed in order to accommodate integrals and differentials with respect to the Wiener process $\{W_t\}_t$. It is often referred to the *Itô's correction*.

8.4.2.1 Simulation of a Random Variable

In many instances, one is interested in the value of the index at a given time T in the future. This is for example the case if one is interested in a contingent claim with European exercise and maturity T. In such a case, formula (8.15) can be used with $s = 0$ and $t = T$. It shows that the random variable S_T is log normal with mean $\log S_0 + T(\mu - \sigma^2/2)$ and variance $\sigma^2 T$, and simulation is plain: generating samples from this distribution can be done with simple commands. For example, the R commands:

```
> N <- 1024; S0 <- 845; MU <- .05; SIG <- .2; TT <- .9
> MUT <- log(S0) + TT*(MU-SIG^2/2)
> SIGT <- SIG*sqrt(TT)
> SAMPLE <- rlnorm(N, meanlog=MUT,sdlog=SIGT)
```

produce a sample of size $N = 1{,}024$ for S_T with $T = .9$, when $S_0 = 845$, $\mu = 0.05$, and $\sigma = 20\,\%$. The risk manager will be able to use such a sample to compute means, probabilities, quantiles (such as VaR's), conditional expectations (such as expected shortfall) $\cdots$ involving $S(T)$, as we did in the first chapters of the book.

8.4.2.2 Simulation of a Time Series

In other circumstances, for example in the analysis of American options, simulation of the entire series may be needed. In such a case, instead of relying directly on the

Euler scheme, one may still use the exact formula (8.15) before calling on random number generators. This is best achieved at the level of the price logarithms rather than the level of the prices themselves. Indeed, it is more convenient to simulate first a sample sequence for the log-prices, and then compute the exponentials of the numbers so obtained. This is based on the fact that:

$$\log S_t = \log S_s + [\mu - \frac{\sigma^2}{2}](t-s) + \sigma[W_t - W_s], \tag{8.16}$$

showing that if the value of the process $\{\log S_t\}_t$ is known at a given time s, the knowledge of its value at a later time t is equivalent to the knowledge of the increment $W_t - W_s$ of the Wiener process. So if we want to generate samples $X_0, X_1, \dots, X_N$ of log prices $X_t = \log S_t$ at times t_j separated by the time interval Δt, we set $X_0 = 0$, we generate an $N(0,1)$ white noise $\epsilon_1, \dots, \epsilon_N$, and we use the recursive formula:

$$X_{j+1} = X_j + [\mu - \frac{\sigma^2}{2}]\Delta t + \sigma\sqrt{\Delta t}\epsilon_{j+1}.$$

As before, we use the notation X_j with j integer, as a short for $X_j^{(\Delta t)}$ which intends to represent X_t for $t = t_j = t_0 + j\Delta t$. This is a plain consequence of the formula (8.16) if we set $s = t_0 + j\Delta t$ and $t = t_0 + (j+1)\Delta t$, and if we use the fact that the increments of $\{W_t\}_t$ over disjoint intervals are independent Gaussian random variables, and that $W_t - W_s \sim N(0, t-s)$. The following R code was used to produce the plot in Fig. 8.12.

```
> N <- 1024; DELTAT <- 1/365; MU <- .05; SIG <- .2
> DELTAX <- rnorm(1024, mean=DELTAT*(MU-SIG^2/2),
          sd=SIG*sqrt(DELTAT))
> GBMDATA <- c(1,exp(cumsum(DELTAX)))
> GBM.ts <- timeSeries(positions=timeSequence(
    from="2001-01-02",by="days",length=N+1), data=GBMDATA)
> plot(GBM.ts)
```

After setting the length of the simulated time series and the values of the parameters Δt, μ and σ, we generate the sequence of the increments $X_{j+1} - X_j$ in a vector which we called DELTAX. This uses the fact that these increments are independent and normally distributed with mean $(\mu - \sigma^2/2)\Delta t$ and with variance $\Delta t\ \sigma^2$. Then, we use the R function cumsum to sum these increments in order to recover the X_j from their increments, and finally, we compute the exponentials giving the desired prices $S_{j\Delta t}$. For plotting purposes, we chose to turn the geometric Brownian motion numeric vector GBMDATA into a daily time series starting from January 2, 2001.

We shall revisit the problem of the simulation of geometric Brownian motion in Sect. 8.5.1. There we use an alternative approach based on Euler's discretization scheme.

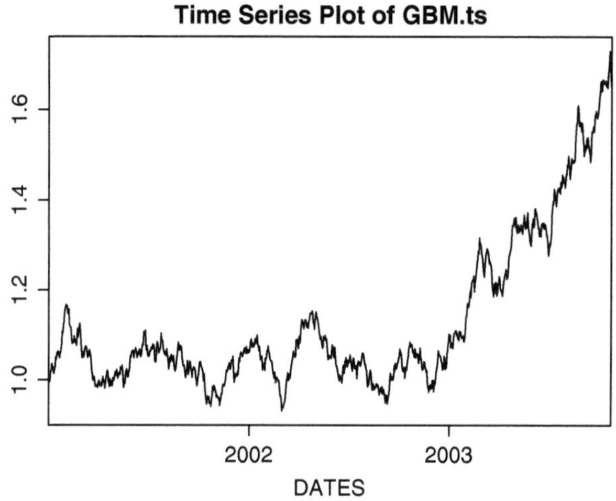

Fig. 8.12. Result of the simulation of a geometric Brownian motion

8.5 Random Simulation and Scenario Generation

In this section, we use our newly acquired expertise in discretization of continuous time stochastic differential equations to prepare a set of scenarios for the purpose of portfolio risk management even though we do not intend to address the delicate problem of active portfolio risk management. Because of the pervasive use of *scenarios* in the finance and insurance industries, we show how Monte-Carlo scenarios can be generated for the purpose of stress analysis, but we leave risk assessment and decision making under uncertainty untouched.

We assume that the investment portfolio is based on three economic factors which are monitored on a monthly basis. These factors are the cost of short term borrowing, the cost of long term borrowing and a stock indicator. The data which we use for the purpose of estimation are contained in the data set `MONTHLY` included in the library `Rsafd`. It provides monthly quotes (between May 1986 and November 1999) of the 1 year Treasury Bill yield (which we shall subsequently call the short interest rate), the 30 years US Government Bond yield (which we shall call the long interest rate) and the value of the S&P 500 composite index. Note that the 30 years Treasury Bonds have been retired since we first performed this experiment. These data are further analyzed in Problems 8.5 and 8.6.

8.5.1 A Simple Model for the S&P 500 Index

As we explained in our discussion of the nonparametric pricing of options in Appendix 9.2, geometric Brownian motion is the time-honored model for stock price and financial index dynamics. We follow this tradition in this section. In other words, denoting by S_t the value of the index at time t, we assume that its time evolution is given by the geometric Brownian motion model proposed by Samuelson, i.e. by the solution of the stochastic differential equation (8.11) where the constant μ has the interpretation of a mean rate of growth, while the constant $\sigma > 0$ plays the role of the volatility of the index. In this equation, W_t is a Wiener process, (as introduced to model the physical process of Brownian motion).

Contrary to the approach followed in Sect. 8.4.2, we ignore the fact that we have an explicit form for the distribution of S_t at any given time, and we use a simple Euler scheme to produce Monte Carlo samples of the time evolution of the index. This gives us the opportunity to illustrate the use of this discretization scheme in a setup where we already used direct simulation, allowing for a comparison of the two methods. But most importantly, the same scheme can be used for much more general models of the dynamics of the underlying index. So we discretize the (stochastic) differential equation (8.11) directly, and construct solutions of the discretized forms of this equation. Equation (8.11) rewrites:

$$S_{t_{j+1}} - S_{t_j} = S_{t_j}[\mu \Delta t + \sigma\sqrt{\Delta t}\epsilon_{j+1}] \tag{8.17}$$

where $\{\epsilon_j\}_{j\geq 1}$ is an $N(0,1)$ i.i.d. white noise. Once more, we see one of the main curiosities associated with the process of Brownian motion. The increment dW_t is essentially proportional to $\sqrt{\Delta t}$ instead of being proportional to Δt! Dividing both sides of equation (8.17) by S_{tj}, we get an expression for the raw return RR_j over the period $[t_j, t_{j+1}]$:

$$1 + RR_j = \frac{S_{t_{j+1}}}{S_{t_j}} = (1 + \mu\Delta t) + \sigma\sqrt{\Delta t}\epsilon_{t+1} \tag{8.18}$$

which shows that, according to this discretization procedure, the raw return RR_j over one time period $[t_j, t_{j+1}]$ is a normal random variable with mean $\mu\Delta t$ and standard deviation $\sigma\sqrt{\Delta t}$.

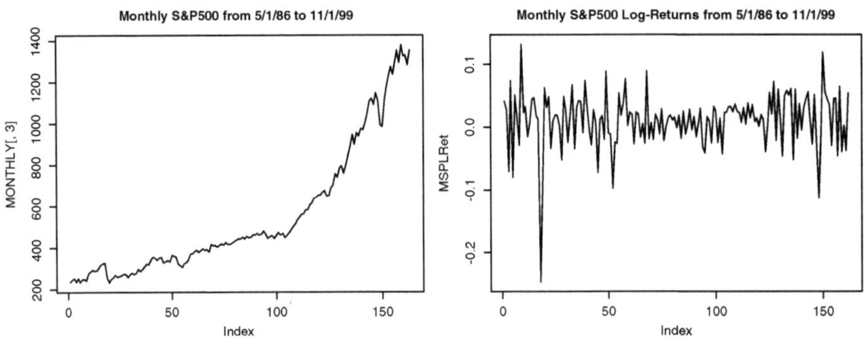

Fig. 8.13. Sequential plot of the monthly values of the S&P 500 composite index (*left*), and of the corresponding log-returns for the same period (*right*)

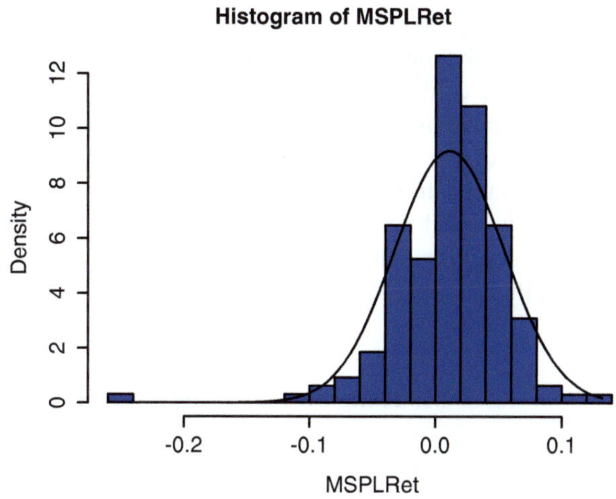

Fig. 8.14. Histogram of the log-return values and density of the normal distribution having the same mean and the same variance

The plot of the monthly values of the S&P 500 index is given in the left pane of Fig. 8.13. This time series is obviously non-stationary. In the terminology of Sect. 6.3.2, it is a random walk with drift, and as explained earlier, it is integrated of order 1, i.e. of type $I(1)$. So it will be more convenient to work with the stationary time series of raw returns $\{RR_j = S_{t_{t+1}}/S_{t_j}\}_{j\geq 1}$, or with the stationary time series $\{\log(1 + RR_j)\}_{j\geq 1}$ of log-returns. The plot of the log-returns is given in the right pane of Fig. 8.13. This series looks definitely more stationary, and as such, it is more amenable to statistical inference.

Figure 8.13 was produced with the following R-commands:

```
> plot(MONTHLY[,3],type="l")
> title("Monthly S&P500 from 5/1/86 to 11/1/99")
> MSPLRet <- diff(log(MONTHLY[,3]))
> plot(MSPLRet,type="l")
> title("Monthly S&P500 Log-Returns from 5/1/86 to 11/1/99")
```

while Fig. 8.14 was produced with the following commands:

```
> mean(MSPLRet)
[1]  0.01080645
> sqrt(var(MSPLRet))
[1]  0.04348982
> hist(MSPLRet,nclass=25,probability=T,col="blue")
> LX <- seq(from=min(MSPLRet),to=max(MSPLRet),length=1000)
> lines(LX,dnorm(LX,mean=.010806,sd=0.04349))
```

We already argued in Chaps. 1 and 3 that the log-normal model is inconsistent with some of the empirical statistics computed from daily stock returns. Monthly data do not behave much differently. Indeed, the time evolutions given by Eqs. (8.11) and (8.18) imply that the distribution of the monthly log-return is normal, but unfortunately, this fact cannot be confirmed by even the simplest of the exploratory data analysis tools such as the histogram. Figure 8.14 shows that the distribution of the log-returns cannot be Gaussian. This graphical evidence could be complemented by Q-Q plots against the Gaussian distribution (with the function `qqnorm`) and tests of goodness of fit, but we shall not delve on that at this stage, especially since our intention is to keep the log-normal model given by Eq. (8.11) for the purposes of the simulation. See nevertheless the discussion of possible extensions at the end of this section.

8.5.2 Modeling the Short Interest Rate

We now work with the data contained in the first column of the data set `MONTHLY`. Figure 8.15 was produced with the following `S`-commands:

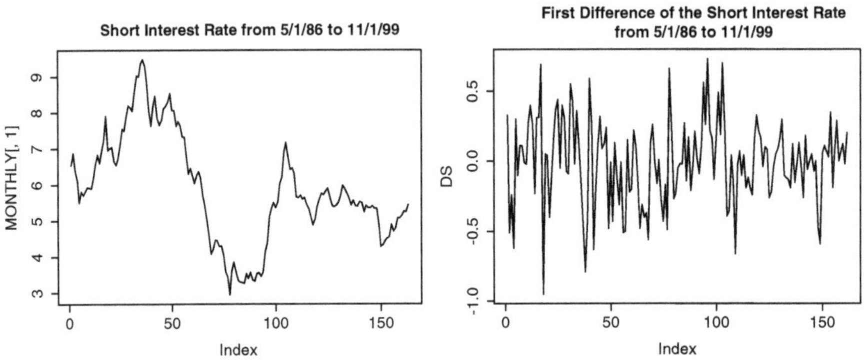

Fig. 8.15. Time series plot of the monthly values of the short interest rate (*left*) contained in the first column of the data set `MONTHLY`, and of its first difference for the same period (*right*)

```
> plot(MONTHLY[,1],type="l")
> title("Short Interest Rate from 5/1/86 to 11/1/99")
> DS <- diff(MONTHLY[,1])
> plot(DS,type="l")
> title("First Difference of the Short Interest Rate
                                  from 5/1/86 to 11/1/99")
```

One of the advantages of modeling the stock index via its logarithm is that, the resulting S_t is always positive, and obviously, this is a desirable feature for a stock index. Unfortunately, the short interest rate r_t cannot be modeled in the same way with the same success. Instead, we shall model its (stochastic) time evolution by a mean reverting Ornstein Uhlenbeck process. This model was introduced in the remark following the definition of the Euler discretization scheme. In the financial community, this model for the dynamics of the short interest rate is known as the Vasicek model. It is given by an equation of the form:

$$dr_t = -\lambda_r(r_t - \overline{r})dt + \sigma_r dW^{(r)}(t) \tag{8.19}$$

where λ_r, $\overline{r}$ and σ_r are positive constants and where $W^{(r)}(t)$ is another Wiener process (or process of Brownian motion). If it weren't for the presence of this Brownian motion, the equation would read:

$$dr_t = -\lambda_r(r_t - \overline{r})dt,$$

and its solution would be given by an exponential function converging toward $\overline{r}$ when $t \to \infty$. This is the relaxation property given by Hooke's law in physics. In the present situation, it is perturbed by the (random) kicks $\sigma_r dW_t^{(r)}$, but the mean reverting tendency remains, its strength being given by the constant λ_r.

Equation (8.19) has three parameters, λ_r, $\overline{r}$, and σ_r which have to be estimated from the data. This is the object of Problems 8.5 and 8.6 at the end of the chapter. As before, we use Euler's scheme, leading to the following finite difference equation:

$$r_{t_{j+1}} = r_{t_j} - \lambda_r(r_{t_j} - \overline{r})\Delta t + \sigma_r\sqrt{\Delta t}\,\epsilon_{j+1}^{(r)} \tag{8.20}$$

for an i.i.d. N(0,1) strong white noise $\{\epsilon_j^{(r)}\}_{j\geq 1}$ which we assume independent of the white noise $\{\epsilon_j\}_{j\geq 1}$ driving the stochastic time evolution in our model for the S&P 500 index.

Remark. As already mentioned, the mathematical model for the short term interest rate given by Eq. (8.19) is known in the financial literature as the Vasicek model. According to this model, the short interest rates at different times are jointly Gaussian random variables. This is very convenient because it leads to closed form formulae for the prices of many fixed income derivatives, including the forward and yield curve manipulated in Problem 4.12. This fact is the main reason for the extreme popularity of the model. Nevertheless, this model has annoying shortcomings. One of the most frequently voiced complains is that, since a normal random variable can take values ranging from $-\infty$ to $+\infty$, it is quite possible that $r_{t_{j+1}}$ becomes negative in any time period. Most practitioners regard this possibility as heretic. However, some have argued that r_t should be viewed as a *real* interest rate as defined by the difference between the short rate and the inflation rate, and as such, it could become negative. But more realistically, the reason why the Vasicek model is reasonable is the following: normal random variables can take arbitrarily large values indeed, but with overwhelming probability they remain within three standard deviations from their means. So if the constants λ_r giving the rate of mean reversion to the long term relaxation level, and this relaxation level $\overline{r}$ are large enough compared to σ_r, then the random quantity given by Eq. (8.19) will essentially never be negative.

8.5.3 Modeling the Spread

We now work with the data contained in the second column of the data set `MONTHLY`. The plots of Fig. 8.16 were produced with the following `R`-commands:

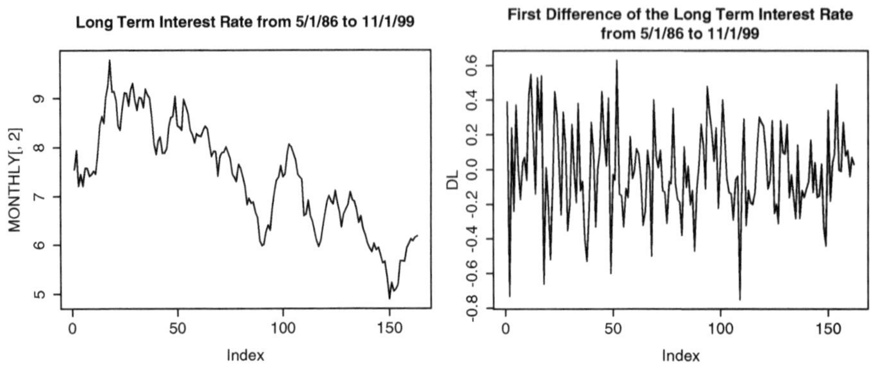

Fig. 8.16. Time series plot of the monthly values of the long interest rate (*left*) and of its first difference for the same period (*right*)

```
> plot(MONTHLY[,2], type="l")
> title("Long Term Interest Rate from 5/1/86 to 11/1/99")
> DL <- diff(MONTHLY[,2])
> plot(DL,type="l")
> title("First Difference of the Long Term Interest Rate
                                    from 5/1/86 to 11/1/99")
```

Obviously, the long interest rate could be modeled in exactly the same way as the short rate. Indeed, Fig. 8.16 shows that most of the features of the short rate r_t are shared by the long rate ℓ_t. By looking at the simultaneous plots of the long and the short interest rates given in Fig. 8.17, we see that they cross rarely, and that the long interest rate is essentially always higher than the short interest rate. In other words, their difference (which is usually called the spread in interest rate and which quantifies the time value of money) is almost always positive.

At this stage of our discussion, it is important to recall the discussion of Sect. 7.1.6. There, we argued that the time evolutions of the short and long interest rates r_t and ℓ_t, could be modeled by I(1) time series, i.e. by non-stationary time series integrated of order one. However, we insisted that their difference was stationary, i.e. integrated of order zero, and we used that example as an illustration for the notion of cointegration. In this section, we still assume that the difference between the long and short interest rates is stationary. Nevertheless, to make our life easier, we also assume that both the short and long interest rates are stationary, i.e. I(0) by assuming that $\lambda_r > 0$. This is a slight departure from the assumptions of Sect. 7.1.6.

For reasons to be discussed later, we decide to model the spread $s_t = \ell_t - r_t$ instead of the long interest rate ℓ_t. As before, we model the (stochastic) time evolution by a mean reverting Ornstein Uhlenbeck process.

$$ds_t = -\lambda_s(s_t - \overline{s})dt + \sigma_s dW_t^{(s)} \tag{8.21}$$

where λ_s is a positive number measuring the speed with which s_t reverts to its long term average $\overline{s}$, and $W_t^{(s)}$ is still another Wiener process. As above, this equation has

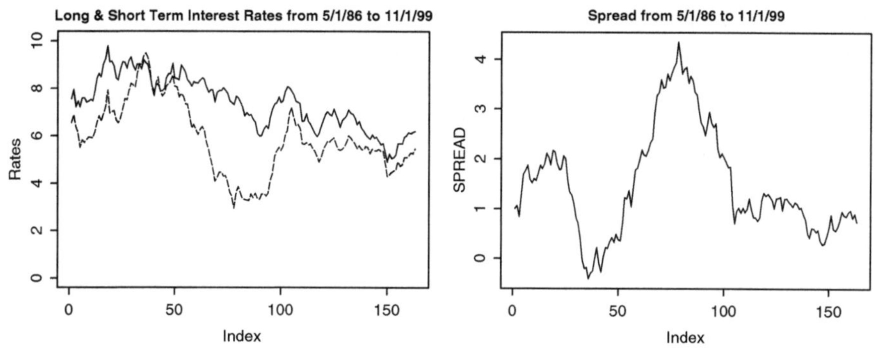

Fig. 8.17. Simultaneous plots of the long and short interest rates (*left*) and time series plot of the spread (*right*) for the same period. Notice that the scales of the vertical axes are different

three parameters, λ_s, $\overline{s}$ and σ_s, which need to be estimated from the data. And again as before, we shall discretize this equation in the following form:

$$s_{t_{j+1}} = s_{t_j} - \lambda_s(s_{t_j} - \overline{s})\Delta t + \sigma_s\sqrt{\Delta t}\;\epsilon^{(s)}_{j+1} \tag{8.22}$$

for an i.i.d. N(0,1) random white noise $\{\epsilon^{(s)}_j\}_{j\geq 1}$ which we shall assume independent of the other white noises $\{\epsilon_j\}_{j\geq 1}$ and $\{\epsilon^{(r)}_j\}_{j\geq 1}$.

8.5.4 **Putting Everything Together**

According to the models presented above, the three economic factors r_t, s_t and S_t are independent of each other. Indeed the three equations (8.11), (8.19), and (8.21) are un-coupled and driven by independent Wiener processes. Obviously, this statement applies as well to their time-discretizations (8.20), (8.22), and (8.11) as the white noise sequences are also assumed to be independent.

Because of our discussion in Sect. 7.1.6, using a model without a strong dependence between the short and long interest rate dynamics would not have been acceptable. Notice nevertheless that cointegration does not necessarily preclude the independence of r_t and s_t as we have it so far in our model. In any case, the lack of dependence between the three time evolutions is a serious shortcoming of our model as it stands, and the remaining part of this section is devoted to possible improvements.

Minor changes in the equations can remedy this lack of correlation. In order to model the fact that the spread has a tendency to increase when the short interest rate decreases to unusually low levels, and that it has a tendency to decrease when the short interest rate is high, one could for example replace (8.21) by an equation of the form:

$$ds_t = [-\lambda_s(s_t - s_0) + \mu_s(r_t - r_0)]dt + \sigma_s dW^{(s)}(t) \tag{8.23}$$

and (8.22) by the corresponding discretization of (8.23). Using this equation introduces a coupling between the dynamics of r_t and of s_t, and consequently, between the dynamics of r_t and ℓ_t. Unfortunately, this modification of the dynamics of s_t creates a couple of problems. First, it introduces a new parameter, μ_s, which may be difficult to estimate from the data. Second, this monotonic dependence between s_t and r_t may not hold all the time, and as a consequence, it may lead to erroneous forecasts in regimes in which this specific relationship between r_t and s_t does not hold. Thus, we refrain from implementing it. Problem 8.6 is concerned with still another way to build dependencies between the three stochastic differential equations giving the dynamics of r_t, s_t and ℓ_t.

For each integer $j \geq 1$ (corresponding to time $t_j = t_0 + j\Delta t$), we define the (random) vector $\mathbf{X}_j$ by:

$$\mathbf{X}_j = \begin{bmatrix} RR_j \\ r_{t_j} \\ s_{t_j} \end{bmatrix}.$$

With this notation, the stochastic dynamics given above can be rewritten as:

$$\mathbf{X}_{j+1} = F\mathbf{X}_j + B + \Sigma\mathbf{W}_j \tag{8.24}$$

where the constant matrix F is given by:

$$F = \begin{bmatrix} 0 & 0 & 0 \\ 0 & 1 - \lambda_r \Delta t & 0 \\ 0 & 0 & 1 - \lambda_s \Delta t \end{bmatrix},$$

the constant vector B is given by:

$$B = \begin{bmatrix} \mu \Delta t \\ \lambda_r \overline{r} \Delta t \\ \lambda_s \overline{s} \Delta t \end{bmatrix},$$

and the matrix Σ is given by:

$$\Sigma = \begin{bmatrix} \sigma\sqrt{\Delta t} & 0 & 0 \\ 0 & \sigma_r\sqrt{\Delta t} & 0 \\ 0 & 0 & \sigma_s\sqrt{\Delta t} \end{bmatrix},$$

and finally, where the random noise vector $\underline{\mathrm{W}}_t$ is defined by:

$$\mathbf{W}_t = \begin{bmatrix} \epsilon_t \\ \epsilon_t^{(r)} \\ \epsilon_t^{(s)} \end{bmatrix}.$$

The fact that the matrices F and Σ are diagonal, is the reason why the three equations are not coupled. So if the three components of the noise term $\mathbf{W}$ are independent, the time evolutions of the three components are statistically independent. As explained earlier, including non-zero off diagonal terms in the matrices F and/or Σ couples the equations and creates dependencies between the time evolutions. Obviously, another possibility is to assume that the components of the noise are dependent. See Problem 8.5 for details.

The decision to work with Gaussian white noise terms offers a very convenient way to *inject* dependencies between the various white noise terms by changing the off-diagonal terms of the matrix Σ, whose square gives the variance/covariance matrix of the innovations ϵ_j, $\epsilon_j^{(r)}$ and $\epsilon_j^{(s)}$ for j fixed. However, as we mentioned earlier, we know by experience that heavy tails are present, at least in the distribution of ϵ_j. So another avenue for a possible improvement of the performance of Monte Carlo scenarios would be to keep formula (8.24) for the dynamics of the system, but produce samples from noise terms generated from a GPD fitted to the residuals of the S&P 500 index. This is suggested in Problem 8.6 where other possible alternatives are implemented.

8.6 Filtering of Nonlinear Systems

The state space models analyzed in the previous chapter are linear. Our goal is now to extend the recursive filtering procedures to nonlinear models. With this in mind, we introduce the terminology of hidden Markov models (HMM for short). The latter form the most popular class of models for partially observed stochastic systems, and being able to filter them efficiently is of great practical importance. Rather than attempting to develop the mathematical theory, which would take us far beyond the scope of this book, we restrict ourselves to the discussion of general features, and to a thorough presentation of a recently discovered implementation of great computational usefulness. Like the Kalman filter of the linear models, it is based on a recursive scheme. But since the nonlinearities take us out of the domain of Gaussian distributions, it is not possible to restrict ourselves to tracking conditional means and conditional variances. Tracking the full conditional distribution of the partially observed state is done via particle approximations requiring Monte Carlo simulations of the time evolution of the state, and clever resampling.

8.6.1 **Hidden Markov Models**

Let us first recall the notation (7.6) and (7.8) introduced earlier for the analysis of general state space models. We assume that at each time n, the state of a system is described by a state vector $\mathbf{X}_n$ in a state space $\mathcal{X}$. In most applications, the state space is a subset of a Euclidean space $\mathbb{R}^{dx}$ where d_X stands for the dimension of the state $\mathbf{X}$, and there is little loss of generality in assuming $\mathcal{X} = \mathbb{R}^{dx}$. It is assumed that the dynamics of the state vector are given by a Markov chain. Rather than being bugged down by the intricacies of the general theory of Markov chains, we shall assume the following convenient form for the dynamics. We assume that the transition from state $\mathbf{X}_n$ to $\mathbf{X}_{n+1}$ is given by an explicit equation:

$$\mathbf{X}_{n+1} = F_n(\mathbf{X}_n, \mathbf{V}_{n+1}) \tag{8.25}$$

where $\mathbf{V} = \{\mathbf{V}_n\}_{n\geq 1}$ is a white noise i.i.d. sequence in $\mathbb{R}^{d'}$ where the dimension d' could be different from the dimension d_X of the state vector. The functions $F_n : \mathbb{R}^{d_X} \times \mathbb{R}^{d'} \hookrightarrow \mathbb{R}^{d_X}$ will be independent of n in all the applications considered in this chapter. However, it is important to emphasize that we do not assume that they are linear, as we did in the previous chapter. Nevertheless, we assume that these functions are known, and that we have a way to compute $F_n(\mathbf{x}, \mathbf{v})$ for any couple $(\mathbf{x}, \mathbf{v}) \in \mathbb{R}^{d_X} \times \mathbb{R}^{d'}$. The states $\mathbf{X}_n$ of the system are not directly observable, and the data with which we have to work, are given in the form of partial observations $\mathbf{Y}_n$ derived from the states $\mathbf{X}_n$ via a formula of the form:

$$\mathbf{Y}_n = G_n(\mathbf{X}_n, \mathbf{W}_n). \tag{8.26}$$

Even though the functions G_n are linear (as in (7.9) of the previous chapter) in many applications, we shall only assume that the noise is additive. In other words, we assume that the observations are of the form:

$$\mathbf{Y}_n = G_n(\mathbf{X}_n) + \mathbf{W}_n \tag{8.27}$$

for some (possibly nonlinear) function $G_n : \mathbb{R}^{d_X} \hookrightarrow \mathbb{R}^{d_Y}$ and a sequence $\mathbf{W} = \{\mathbf{W}_n\}_{n\geq 1}$ of mean zero independent random vectors in $\mathbb{R}^{d_Y}$ with densities ψ which we assume to be known.

The challenge is framed in the same way as in the linear case: given an integer $n' \geq 1$, and observations $\mathbf{Y}_1, \ldots, \mathbf{Y}_n$, we want to *guess* $\mathbf{X}_{n'}$. For the sake of convenience we shall use the notation $\mathbf{Y}_{\leq n}$ for the set of n d_Y-dimensional vectors $\mathbf{Y}_1, \ldots, \mathbf{Y}_n$.

8.6.2 General Filtering Approach

As explained in the previous chapter, the three main problems of state space models are (1) data smoothing ($n' < n$), (2) filtering ($n' = n$), and (3) prediction ($n' > n$), and as before, we concentrate on the filtering and prediction problems.

All the information about the state $\mathbf{X}_n$ which one can hope to derive from the knowledge of the observations $\mathbf{Y}_{\leq n}$ up to (and including) time n, is contained in the conditional distribution of $\mathbf{X}_n$ given $\mathbf{Y}_{\leq n}$ which we denote by:

$$\pi_n(\,\cdot\,, \mathbf{Y}_{\leq n}) = \mathbb{P}\{\mathbf{X}_n \in \cdot\,|\mathbf{Y}_{\leq n}\}. \tag{8.28}$$

This conditional distribution obviously depends upon the past observations $\mathbf{Y}_{\leq n}$. We assume that a given sequence $\mathbf{Y}_1 = \mathbf{y}_1, \ldots, \mathbf{Y}_n = \mathbf{y}_n$ of observations has been acquired, and we drop the dependence of π_n upon $\mathbf{Y}_{\leq n}$ in the notation. This abuse of notation should not hinder the understanding of the arguments.

The main result of the theory of stochastic filtering is that given a sequence of observations $\mathbf{y}_1, \ldots \mathbf{y}_n, \ldots$, the conditional distributions π_n can be computed recursively. In other words, there exists an algorithm:

$$\pi_{n+1} = \rho_n(\pi_n, \mathbf{y}_{n+1}) \tag{8.29}$$

which gives π_{n+1} in terms of π_n and the *new observation* $\mathbf{y}_{n+1}$. This says that the dependence of π_{n+1} upon the past observations $\mathbf{Y}_1 = \mathbf{y}_1, \ldots, \mathbf{Y}_n = \mathbf{y}_n$ is already contained in the conditional distribution π_n, so that, in order to compute π_{n+1}, it is enough to *update* the previous conditional distribution π_n using the algorithm (8.29) and the new observation $\mathbf{y}_{n+1}$.

The proof of this beautiful theoretical result is the conclusion of 30 years of intensive research following the original works of Kalman and Bucy on linear systems. In the case of these linear systems (studied in the previous chapter), if the state white noise $\mathbf{W}$ and the observation white noise $\mathbf{V}$ are Gaussian, then the random vectors $\mathbf{X}_n$ and $\mathbf{Y}_{\le n}$ are jointly Gaussian, and the conditional distribution of $\mathbf{X}_n$ given $\mathbf{Y}_{\le n}$ is also Gaussian. π_n being Gaussian, it is entirely determined by its mean vector, say $\hat{\mathbf{X}}_n$, and its variance/covariance matrix, say Ω_n. In this case, the algorithm (8.29) can be rewritten as an update algorithm for $\hat{\mathbf{X}}_n$ and Ω_n:

$$\begin{bmatrix} \hat{\mathbf{X}}_{n+1} \\ \Omega_{n+1} \end{bmatrix} = \rho_n\left(\begin{bmatrix} \hat{\mathbf{X}}_n \\ \Omega_n \end{bmatrix}, \mathbf{y}_{n+1}\right). \tag{8.30}$$

Working out the details of this update algorithm, we would recover the update formulae derived in the previous chapter. So what could be viewed as a miracle, can now be explained. It is only because a Gaussian distribution is completely determined by its mean vector and its variance/covariance matrix, that the optimal filter could be given by a recursive update of the mean and the variance. In general, the conditional distributions π_n are not Gaussian, and the update algorithm cannot be reduced to such a simple form as (8.30).

8.6.3 **Particle Filter Approximations**

As suggested by the fact that the update algorithm involves the entire conditional distribution, any practical implementation of the solution to the nonlinear filtering problem has to be based on an approximation. The idea of the particle approximation to nonlinear filtering is simple and natural. The basic principle of the particle method is to approximate the desired distribution π_n by the empirical distribution of a sample of randomly chosen states. There would not be anything special to this idea if it weren't for the fact that these approximations can be updated in a very natural way without affecting the quality of the approximation.

Nothing can be so simple without being deep and powerful! What is remarkable is the fact that it can be implemented in a recursive manner, as in the case of the Kalman filter for linear systems.

At each time n we consider a set $\{x_n^j\}_{j=1,\ldots,m}$ of m elements of the state space in which $\mathbf{X}_n$ takes its values, and we consider the approximation to the optimal filter:

$$\pi_n(dx) = \mathbb{P}\{\mathbf{X}_n \in dx | \mathbf{Y}_{\le n}\} \approx \hat{\pi}_n^{(m)}(dx) = \frac{1}{m}\sum_{j=1}^{m} \delta_{x_n^j}(dx)$$

given by the empirical distribution of the sample $\{x_n^j\}_{j=1,\ldots,m}$. Choosing m large enough, and choosing the x_n^j appropriately, one can make sure that the approximation is as good as desired. But the main question remains: do we have to recompute the entire approximation each time we have a new observation, or could it be possible to update the approximation of π_n in an efficient way, and get a reasonable approximation of π_{n+1}. In other words, can we:

- Compute $\hat{\pi}_{n+1}^{(m)}$ as $\rho_n(\hat{\pi}_n^{(m)}, \mathbf{y}_{n+1})$?
- And still have: $\lim_{m\to\infty} \hat{\pi}_n^{(m)} = \pi_n$ for each n.

The algorithm which we present now, does just that. To implement it we need two kinds of particles:

- Those used to simulate $\mathbb{P}\{\mathbf{X}_n | \mathbf{Y}_{\leq n}\}$

$$x_n^1, \ldots, x_n^m$$

- Those used to simulate $\mathbb{P}\{\mathbf{X}_{n+1} | \mathbf{Y}_{\leq n}\}$

$$p_{n+1}^1, \ldots, p_{n+1}^m$$

for the one-step-ahead distribution. At this stage it is important to remember the fundamental role played by the one-step-ahead prediction in the linear case.

8.6.3.1 *One Step Ahead Prediction*

To describe and justify the algorithm we assume temporarily that

$$x_n^1, \ldots, x_n^m$$

form a random sample from the distribution $\pi_n = \mathbb{P}\{\mathbf{X}_n | \mathbf{Y}_{\leq n}\}$. If we assume that $v_n^1, \ldots, v_n^m$ are m independent realizations of the noise $\mathbf{V}_n$ (remember that we assume that we know the distribution of the state equation noise terms), and if we define the new particles $p_{n+1}^1, \ldots, p_{n+1}^m$ by:

$$p_{n+1}^j = F_n(x_n^j, v_n^j),$$

in other words by simulating the dynamics given by the state equation, then it is clear that we have a random sample

$$p_{n+1}^1, \ldots, p_{n+1}^m$$

from the conditional distribution $\mathbb{P}\{\mathbf{X}_{n+1} | \mathbf{Y}_{\leq n}\}$.

8.6.3.2 *Filtering, or* **Updating**

The second step of the algorithm is less obvious. Assuming that we have a sample

$$p^1_{n+1}, \dots, p^m_{n+1}$$

from the conditional distribution $\mathbb{P}\{\mathbf{X}_{n+1}|\mathbf{Y}_{\leq n} = \mathbf{y}_{\leq n}\}$, and assuming that a new observation $\mathbf{y}_{n+1}$ is made available, we compute the likelihoods

$$\alpha^j_{n+1} = \mathbb{P}\{\mathbf{Y}_{n+1} = \mathbf{y}_{n+1}|p^j_n\}$$

of the particles p^j_{n+1} given this new observation. These likelihoods can be computed because we assume that we know the distribution of the observation noise, and we also assume that we can invert the observation equation. More precisely, since $\mathbf{Y}_{n+1} = G_{n+1}(\mathbf{X}_{n+1}, \mathbf{V}_{n+1})$, we must have:

$$\alpha^j_{n+1} = \psi(H_{n+1}(p^j_{n+1}, y_{n+1}))|\frac{\partial H_{n+1}}{\partial y}(H_{n+1}(p^j_{n+1}, y_{n+1}))|$$

provided we denote by ψ the density of the observation white noise, and by H_{n+1} the inverse of the function G_{n+1} when the state variable is held fixed. In other words, for $\mathbf{X}_{n+1}$ given, $\mathbf{Y}_{n+1} = G_{n+1}(\mathbf{X}_{n+1}, \mathbf{V}_{n+1}) \Leftrightarrow \mathbf{V}_{n+1} = H_{n+1}(\mathbf{X}_{n+1}, \mathbf{Y}_{n+1})$. The second step of the algorithm is justified by the following important computation. By definition of conditional probabilities we have:

$$\begin{aligned}\mathbb{P}\{\mathbf{X}_{n+1} &= p^i_{n+1}|\mathbf{Y}_{\leq n+1} = \mathbf{y}_{\leq n+1}\} \\ &= \frac{\mathbb{P}\{\mathbf{X}_{n+1} = p^i_{n+1}, \mathbf{Y}_{n+1} = \mathbf{y}_{n+1}|\mathbf{Y}_{\leq n} = \mathbf{y}_{\leq n}\}}{\mathbb{P}\{\mathbf{Y}_{n+1} = \mathbf{y}_{n+1}|\mathbf{Y}_{\leq n} = \mathbf{y}_n\}}\end{aligned} \tag{8.31}$$

For the sake of notation, we compute separately the numerator and the denominator of the right hand side. Using again the very definition of conditional probabilities, and mimicking the classical derivation of Bayes rule, we get:

$$\begin{aligned}&\mathbb{P}\{\mathbf{X}_{n+1}{=}p^i_{n+1}, \mathbf{Y}_{n+1}{=}\mathbf{y}_{n+1}|\mathbf{Y}_{\leq n}{=}\mathbf{y}_{\leq n}\} \\ &{=}\mathbb{P}\{\mathbf{Y}_{n+1}{=}\mathbf{y}_{n+1}|\mathbf{X}_{n+1}{=}p^i_{n+1}, \mathbf{Y}_{\leq n}{=}\mathbf{y}_{\leq n}\}\mathbb{P}\{\mathbf{X}_{n+1}{=}p^i_{n+1}|\mathbf{Y}_{\leq n}{=}\mathbf{y}_{\leq n}\}.\end{aligned}$$

Given the knowledge of $\mathbf{X}_{n+1}$, formula (8.26) says that the observation $\mathbf{Y}_{n+1}$ depends only upon the observation noise $\mathbf{W}_{n+1}$ which is independent of all the previous observations $\mathbf{Y}_{\leq n}$. Consequently, the conditioning by the fact that $\mathbf{Y}_{\leq n} = \mathbf{y}_{\leq n}$ can be removed from the first probability in the above right hand side, showing that this probability is in fact equal to α^i_{n+1}. Moreover, according to the construction of the one-step ahead candidates p^i_{n+1}, we have

$$\mathbb{P}\{\mathbf{X}_{n+1} = p^i_{n+1}|\mathbf{Y}_{\leq n} = \mathbf{y}_{\leq n}\} = \mathbb{P}\{\mathbf{X}_n = x^i_n|\mathbf{Y}_{\leq n} = \mathbf{y}_{\leq n}\} = \frac{1}{m}$$

by definition of the fact that $x^1_n, \cdots, x^m_n$ is a random sample from the conditional distribution $\pi_n = \mathbb{P}\{\mathbf{X}_n|\mathbf{Y}_{\leq n} = \mathbf{y}_{\leq n}\}$ of $\mathbf{X}_n$ given the knowledge of the past observations. Putting together these two facts, we get the following expression

$$\mathbb{P}\{\mathbf{X}_{n+1} = p^i_{n+1}, \mathbf{Y}_{n+1} = \mathbf{y}_{n+1} | \mathbf{Y}_{\leq n} = \mathbf{y}_{\leq n}\} = \frac{1}{m}\alpha^i_{n+1}. \tag{8.32}$$

for the numerator of the right hand side of (8.31). Using similar arguments, we can develop the denominator:

$$\begin{aligned}
&\mathbb{P}\{\mathbf{Y}_{n+1} = \mathbf{y}_{n+1} | \mathbf{Y}_{\leq n} = \mathbf{y}_{\leq n}\} \\
&= \sum_{j=1}^{m} \mathbb{P}\{\mathbf{Y}_{n+1} = \mathbf{y}_{n+1} | \mathbf{X}_{n+1} = p^j_{n+1}, \mathbf{Y}_{\leq n} = \mathbf{y}_n\} \\
&\qquad\qquad\qquad \mathbb{P}\{\mathbf{X}_{n+1} = p^j_{n+1} | \mathbf{Y}_{\leq n} = \mathbf{y}_n\} \\
&= \sum_{j=1}^{m} \mathbb{P}\{\mathbf{Y}_{n+1} = \mathbf{y}_{n+1} | \mathbf{X}_{n+1} = p^j_{n+1}\} \mathbb{P}\{\mathbf{X}_{n+1} = p^j_{n+1} | \mathbf{Y}_{\leq n} = \mathbf{y}_{\leq n}\} \\
&= \left(\sum_{j=1}^{m} \alpha^j_{n+1}\right)\frac{1}{m}.
\end{aligned} \tag{8.33}$$

Now, putting together (8.31), (8.32) and (8.33), we get:

$$\mathbb{P}\{\mathbf{X}_{n+1} = p^i_{n+1} | \mathbf{Y}_{\leq n+1} = \mathbf{y}_{\leq n+1}\} = \frac{\alpha^i_{n+1}}{\sum_{j=1}^{m} \alpha^j_{n+1}}. \tag{8.34}$$

This computation suggests that we define x^j_{n+1} by:

$$x^j_{n+1} = \begin{cases} p^1_{n+1} \text{ with probability} \frac{\alpha^1_{n+1}}{\alpha^1_{n+1}+\ldots+\alpha^m_{n+1}} \\ \quad\vdots \\ p^m_{n+1} \text{ with probability} \frac{\alpha^m_{n+1}}{\alpha^1_{n+1}+\ldots+\alpha^m_{n+1}} \end{cases}$$

In words, this means that for each $j = 1, \cdots, m$, the value of the state x^j_{n+1} is obtained by *sampling/drawing with replacement* from the set $\{p^1_{n+1}, \cdots, p^m_{n+1}\}$ with probability $\alpha^j_{n+1}/(\alpha^1_{n+1} + \ldots + \alpha^m_{n+1})$. In doing so, we can be sure, because of formula (8.34), that $x^1_{n+1}, \ldots, x^m_{n+1}$ appear as particles which form a random sample from the conditional distribution $\pi_{n+1} = \mathbb{P}\{\mathbf{X}_{n+1} | \mathbf{Y}_{\leq n+1} = \mathbf{y}_{\leq n+1}\}$, putting us at time $n+1$, in the situation we started from at time n. This second step completes at the level of the random samples, the implementation of the update algorithm (8.29) for the conditional distributions.

8.6.3.3 ***Algorithm Summary***

We can summarize the algorithm as follows:

1. *Initialization:* generate an initial random sample of m particles $x^1_0, \ldots, x^m_0$
2. *Iteration:* for each time step n, repeat the following:

- Generate independent samples v_n^j from the distribution of the state noise;
- Compute the values p_{n+1}^j from the formula $p_{n+1}^j = F(x_n^j, v_n^j)$;
- Given the value y_{n+1} of the new observation, compute the likelihoods α_{n+1}^j;
- Resample the p_{n+1}^js to produce the x_{n+1}^js.

The remainder of this chapter is devoted to the discussion of possible implementations of this particle filtering algorithm. The applications that we have in mind share the following features.

- Nonlinear dynamics for the unobserved state;
- Nonlinear observation equation;
- Parameters as components of the state vector;
- Regime switching which can be incorporated in the Markov dynamics of the partially observed state.

8.6.4 **Filtering in Finance? Statistical Issues**

Some of the financial applications of filtering can be motivated by a quick look at the plots in Fig. 8.18 where we show the time evolution of the indexes of a few energy companies before the crash following the California energy crisis and Enron's bankruptcy. One of the time series (obviously Enron) clearly stands out. Is it because of a change in the mean rate of return? Is it because of a sudden change in the volatility? Could it be because of the issuance of debt? A change in rating of some of its debts? A sudden change in other economic factors? Tracking these important parameters is a natural task that most econometricians, economists, analysts, financial engineers and traders do on a day-by-day basis. We propose to illustrate the use of nonlinear filtering techniques to do just that.

The tracking motivation outlined above fits perfectly the statistical theory of so-called change point problems. Nevertheless, several important differences need to be pointed out.

- As a general rule, real time computations are not an issue for economic/econometric applications for which computations can be performed *ex ante* (i.e. after the fact, overnight for example). Only highly speculative operations (such as program trading or energy spot price tracking) require real time computations.
- The filtering approach is very natural when it comes to non-anticipative estimation of some of the parameters needed for scenario generation and risk management. Nevertheless, some insight into these issues can easily be gained without having to introduce the heavy machinery of filtering theory. As we pointed out in Sect. 8.2.6, standard estimation can be performed in trailing sliding windows, and classical multivariate analysis tests can be run inside these windows to test for the equality of rates of mean return or volatility. See the Notes & Complements at the end of the chapter for a discussion of an application to the subindexes of the Dow Jones Total Market Index.
- Even when the problem is naturally framed as a problem of parameter estimation, or detection of change in parameters (as opposed to a filtering problem),

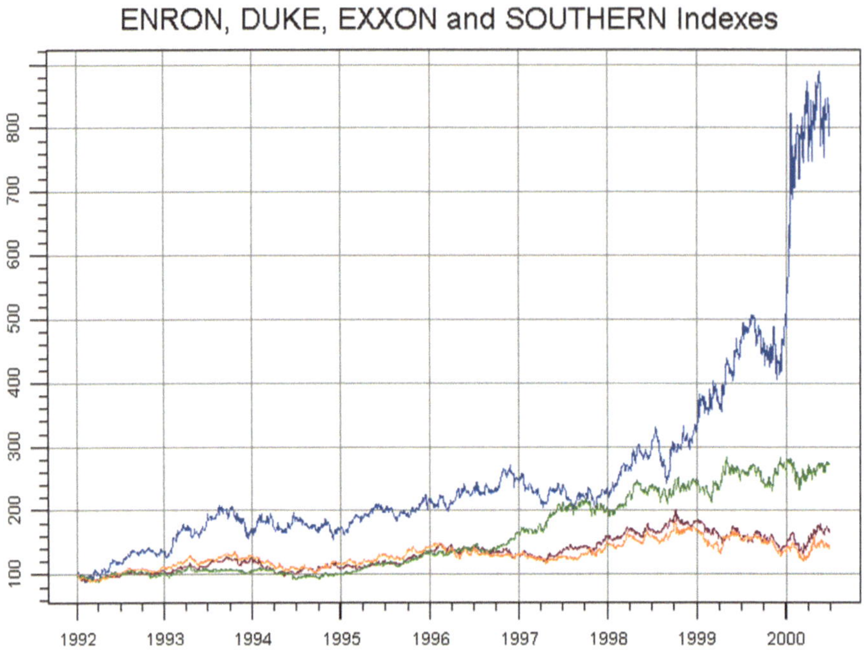

Fig. 8.18. Time series plots of the Enron, Exxon, Duke and Mirant (formally SouthWest) indexes

particle approximations are still of great value. Indeed, parameter estimation and hypotheses testing are very often based on the computation of the likelihood function of the model. This task is unfortunately too difficult in many practical applications. But according to our discussion of Sect. 7.4.5, it is often possible to express the likelihood function in a recursive fashion, in terms of the optimal filter. It is not a surprise to learn that many active researchers have based their computation of the likelihood (and consequently their estimations and test procedures) on the particle approximation of the optimal filter. This recursive computation of the likelihood approximation is currently investigated as a leading candidate for efficient estimation procedures and test statistic computations.

The next subsection is devoted to the detailed analysis of a specific application chosen for the sake of illustration of the versatility of particle filters.

8.6.5 Application: Tracking Volatility

Stochastic volatility models of continuous time finance are factor models, and most of them are two-factor models, one of these factors being the volatility in question. Recall the presentation of the discrete time stochastic volatility models of Sect. 8.3. The dynamics of the two factors are given by stochastic differential equations of the

Itô's type driven by two different Wiener processes. In this section, we consider the discretizations of these models obtained by Euler's scheme.

As already explained when we discussed discrete time stochastic volatility models in Sect. 8.3, the main motivation for the introduction of these models is the smile effect. The popularity of the stochastic volatility models is deeply rooted in their intuitive rationale, and the fact that they are capable of producing "smiles" like the ones encountered in practice.

8.6.5.1 ***The Model***

In analogy with the Samuelson's framework used for Black-Scholes' theory, we assume that the dynamics of the underlying asset price are given by a stochastic differential equation of the form:

$$dS_t = S_t(\mu dt + \sigma_t dW_t),$$

but instead of assuming that σ_t is a positive constant, we assume that σ_t is in fact another random quantity which changes with time, and the model is based on the choice of another stochastic differential equation for the dynamics of this stochastic quantity. For the sake of definiteness we shall assume that σ_t satisfies:

$$d\sigma_t = -\lambda(\sigma_t - \overline{\sigma})dt + \gamma d\tilde{W}_t \tag{8.35}$$

for some positive constants λ, $\overline{\sigma}$ and γ, and for another Wiener process $\tilde{W}_t$ providing the source of the random kicks driving the time evolution of the volatility. If we recall the discussions of the stochastic models used for the short interest rate and the interest rate spread in Sects. 8.5.2 and 8.5.3, Eq. (8.35) says that the stochastic volatility evolves as an Ornstein – Uhlenbeck process, and we can reuse the discussion of the interpretations of the parameters of the model. $\overline{\sigma}$ is a mean level of volatility toward which σ_t tries to revert. In fact $\overline{\sigma}$ would be the deterministic limit of σ_t for large values of the time t if it were not for the random kicks given by the Wiener process $\tilde{W}_t$. The constant λ is the rate of mean reversion toward $\overline{\sigma}$, while γ is the volatility of the volatility σ_t (volvol in the jargon of the street). So the resulting force $d\sigma_t$ acting on the volatility is an aggregate of a restoring force $-\lambda(\sigma_t - \overline{\sigma})dt$ and a random kick $\gamma d\tilde{W}_t$. The balance between these two terms determines the actual dynamics of the stochastic volatility model chosen here.

8.6.5.2 ***Discretization***

In order to make the above model amenable to implementation and analysis by the filtering techniques introduced in this chapter, we discretize the above equations in order to set up a state space system in the form of a hidden Markov model.

We first consider the equation for the asset price. We choose a sampling interval Δt, times $t_j = t_0 + j\Delta t$ forming a regular time grid (we shall set $t_0 = 0$ for convenience), and we write the dynamical equation given by Euler's scheme. Mimicking the procedure followed in Sect. 8.5.1, we introduce the variable RR for the raw return over one time period. The Euler discretization scheme gives:

$$RR_j = \frac{S_{t_{j+1}}}{S_{t_j}} - 1 = \mu \Delta t + \sigma_t \sqrt{\Delta t} \epsilon_{j+1}. \tag{8.36}$$

Next, we consider Eq. (8.35). Using Euler's scheme again, we get:

$$\sigma_{t_{j+1}} = \lambda \overline{\sigma} \Delta t + (1 - \lambda \Delta t) \sigma_{t_j} + \gamma \sqrt{\Delta t} \tilde{\epsilon}_{j+1}. \tag{8.37}$$

Here $\{\epsilon_j\}_{j \geq 1}$ and $\{\tilde{\epsilon}_j\}_{j \geq 1}$ are $N(0,1)$ strong white noise time series independent of each other.

Remarks.

1. Equation (8.37) is not the only way to discretize the dynamical equation (8.35). Indeed the latter can be solved *explicitely* and, as we explained in Sect. 8.4.2, it is then preferable to discretize the exact solution (as opposed to solving the discretized equation). If we do that in the present situation, we end up with the discretization:

$$\sigma_{t_{j+1}} = \overline{\sigma} + e^{-\lambda \Delta t}(\sigma_t - \overline{\sigma}) + \sqrt{\frac{\gamma^2}{2\lambda}(1 - e^{-2\lambda \Delta t})} \tilde{\epsilon}_{t+\Delta t}. \tag{8.38}$$

2. The independence of the white noise terms is not a realistic assumption. Indeed, the leverage effect discussed earlier would suggest that the two white noise time series should be negatively correlated. It is not difficult to force this condition and still develop the filtering apparatus presented below. We refrain from doing it because this requires adding noise terms to the dynamical equations, and as a consequence, rather cumbersome notation, and more involved formulae.

8.6.5.3 Setting Up a Hidden Markov Model (HMM)

In the present set up, we can assume that the asset prices are observed, hence so are the raw returns. However, it is clear that the volatility σ_t cannot be observed directly.

1. Using the experience gained in rewriting ARMA models as state space models, one is tempted to choose the observation equation

$$\mathbf{Y}_j = \begin{bmatrix} 1 \; 0 \end{bmatrix} \begin{bmatrix} RR_{j-1} \\ \sigma_{t_j} \end{bmatrix} \tag{8.39}$$

which would lead us to choose the state vector $\mathbf{X}_j$ and the state noise vector $\mathbf{V}_{j+1}$ as:

$$\mathbf{X}_j = \begin{bmatrix} RR_{j-1} \\ \sigma_{t_j} \end{bmatrix}, \quad \text{and} \quad \mathbf{V}_{tj+1} = \begin{bmatrix} \epsilon_{j+1} \\ \tilde{\epsilon}_{j+1} \end{bmatrix}, \tag{8.40}$$

with state equation:

$$\mathbf{X}_{j+1} = F(\mathbf{X}_j, \mathbf{V}_{j+1}) \tag{8.41}$$

where the function F is defined by:

$$F(\mathbf{x},\mathbf{v})=\begin{bmatrix}\mu\Delta t+x_2\sqrt{\Delta t}v_1\\ \lambda\overline{\sigma}\Delta t+(1-\lambda\Delta t)x_2+\gamma\sqrt{\Delta t}v_2\end{bmatrix},\quad \text{if}\quad \mathbf{x}=\begin{bmatrix}x_1\\ x_2\end{bmatrix},\ \mathbf{v}=\begin{bmatrix}v_1\\ v_2\end{bmatrix}.$$

Clearly, the observation is partial in the sense that not all the components of the state vector can be observed. However, the observation is perfect in the sense that there is no noise coming to perturb the accuracy of the observation. Strangely enough, this can be a serious handicap, and filtering methods have a harder time handling perfect observations. As a result, the practitioner adds an artificial (small) noise to the right hand side of the observation equation (8.39), just to regularize the problem. We shall not follow this approach here.

2. Another possibility is to choose $X_j = \sigma_{t_j}$ for the state of the system, and $Y_j = RR_{j-1}$ for the observation. Note that we do not use bold characters as both the state and the observations are univariate, which is a simplification. In any case, the dynamics of the state are given by

$$X_{j+1} = F(X_j, V_{j+1}) \tag{8.42}$$

with

$$F(x,v) = \lambda\overline{\sigma}\Delta t + (1-\lambda\Delta t)x + \gamma\sqrt{\Delta t}v$$

if we choose the system noise V_{j+1} to be $\tilde{\epsilon}_{j+1}$. Accordingly, the observation equation reads:

$$Y_j = G(X_j, W_j)$$

with

$$G(x,w) = \mu\Delta t + x\sqrt{\Delta t}w$$

provided the observation noise W_j is chosen to be ϵ_j. Notice that the noise is not additive, i.e. the observation function $G(x,w)$ is not of the form $G(x)+w$. However, the function G is still invertible, and as we saw when we derived the particle filtering algorithm, that's all we need!

3. The above formulation is quite appropriate for the implementation of the particle filtering algorithm described in the previous section, as long as we have estimates of the parameters μ, $\overline{\sigma}$, λ and $c = \gamma^2/2\lambda$ of the problem. Rather than trying to estimate them first, and then run the filtering algorithm, we include them in the state of the system, letting the filter estimate their values dynamically. This idea is very appealing, and this practice is pretty common in some circles, despite the fact that it still lacks rigorous mathematical justification. For the sake of simplicity, we only include the parameters λ and c in the state, estimating directly the parameters μ and $\overline{\sigma}$ off line. So the final form of the state space model which we use to track (stochastic) volatility is given by the following state dynamical equation:

$$\begin{bmatrix}\sigma_{t_{j+1}}\\ \lambda_{j+1}\\ c_{j+1}\end{bmatrix} = \begin{bmatrix}\overline{\sigma}+e^{-\lambda_j\Delta t}(\sigma_{t_j}-\overline{\sigma})+\sqrt{c_j(1-e^{-\lambda_j\Delta t})}\tilde{\epsilon}_{j+1}\\ \lambda_j\\ c_j\end{bmatrix} \tag{8.43}$$

coupled with the observation equation:

$$R_{j-1} = \mu \Delta t + \sigma_{t_j} \sqrt{\Delta t} \epsilon_j.$$

Notice that the dynamics of the parameters are trivial since their true values do not change over time. In a particle filtering implementation, the success of the procedure depends upon the initial random samples $\lambda_0^1, \cdots, \lambda_0^m$ and $c_0^1, \cdots, c_0^m$. Indeed, the only changes in λ_j and c_j can occur during the resampling step as we update their conditional distributions when we compute our new best guess for the state of the system given a new set of observations. Obviously, the convergence of these estimates toward their true values is not guaranteed, but it seems to be happening in many applications, giving an empirical justification for this practice. But to be on the safe side, practical implementations often include extra noise terms to allow the update steps of the algorithm more freedom in the search for the true values of these parameters. In such cases the dynamics of the second and third components of the state are given by equations of the form:

$$\lambda_{j+1} = \lambda_j + \epsilon_{j+1}^{(\lambda)} \quad \text{and} \quad c_{j+1} = c_j + \epsilon_{j+1}^{(c)}$$

where $\{\epsilon_j^{(\lambda)}\}_{j \geq 1}$ and $\{\epsilon_j^{(c)}\}_{j \geq 1}$ are independent $N(0, \sigma^{(\lambda)2})$ and $N(0, \sigma^{(c)2})$ i.i.d. noise series with small variances $\sigma^{(\lambda)2}$ and $\sigma^{(c)2}$ respectively, instead of the trivial dynamics of equation (8.43). See also the Notes & Complements at the end of the chapter for references on this approach to adaptive parameter estimation.

Remarks.

1. Positivity of σ_t is not guaranteed in the above model. This prompts us to search for other stochastic differential equations for the dynamics of σ_t. We could use some of the models proposed for the short interest rate because of their positivity. This is for example the case of the square root model also called CIR model. Unfortunately, the direct approach (analogous to (8.38)) require random number generators from the conditional distribution of $\sigma_{t_{j+1}}$ given σ_{t_j} whose technical description would take us beyond the scope of the book.
2. Because of these positivity problems, geometric Ornstein-Uhlenbeck processes (i.e. exponentials of Ornstein-Uhlenbeck processes) are often used as models for stochastic volatility.

8.6.5.4 *Empirical Results*

We implemented the particle filtering algorithm on the third state space systems described above. The results of some Monte Carlo experiments are reproduced graphically in Figs. 8.20 and 8.21. Figure 8.19 gives the plot of the price series we generated to run the filter. We used the values $\Delta t = 0.004$, $\mu = 0.006$, $\lambda = 2$, $c = 0.5$, $\sigma_0 = 0.1$ and $\overline{\sigma} = 0.2$ for the parameters.

Figure 8.20 gives the plot of the volatility which was actually used to generate the price shown in Fig. 8.19 together with two estimates of this instantaneous volatility. The first one is obtained by the particle filter described above. It tracks the true volatility reasonably well. The other estimate is the result of the empirical estimation of the standard deviation in a sliding non-anticipative window of size 30. As we can see, this naive estimate is too slow to react to changes in the volatility.

Finally, Fig. 8.21 gives another instance of the superiority of the particle filter over the sliding window naive estimate. In fact the particle filter estimate is always better for simulated data. It is not clear how to compare the two on real data. Indeed, we do not know the exact value of the *true* volatility which drove the price dynamics.

From the many experiments which we ran on this specific application, we found that the particle filter approximation to the (true) optimal filter

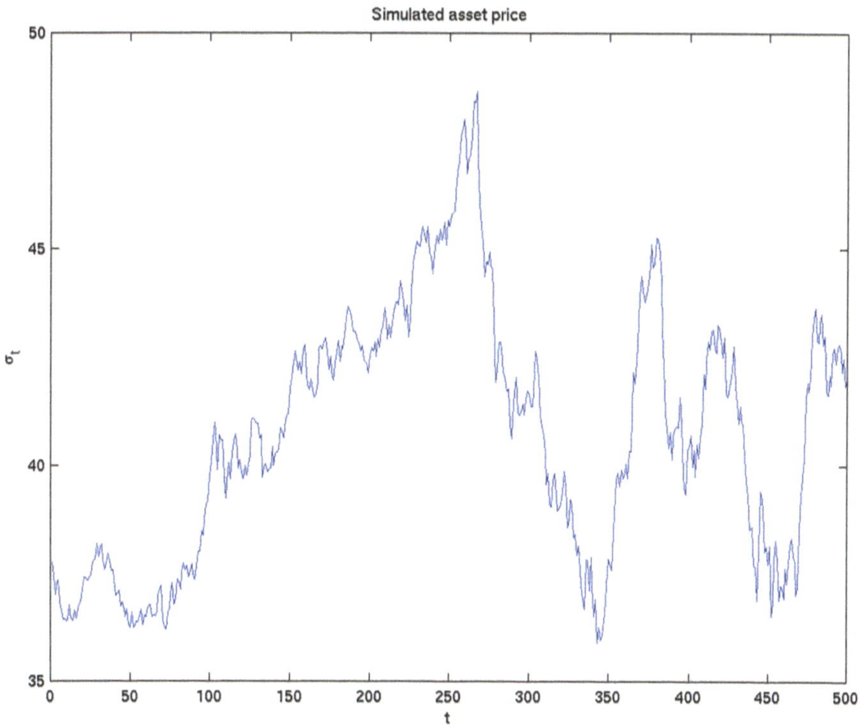

Fig. 8.19. Simulated asset price

- Provides exceptionally good tracking if initial parameters (mean reverting rates, vol of vol, etc.) in right range;
- Lacks robustness (poor estimation) if initial parameters too far from the true values;
- Reacts and relaxes faster (in all cases) than the sliding window estimate of the historical volatility.

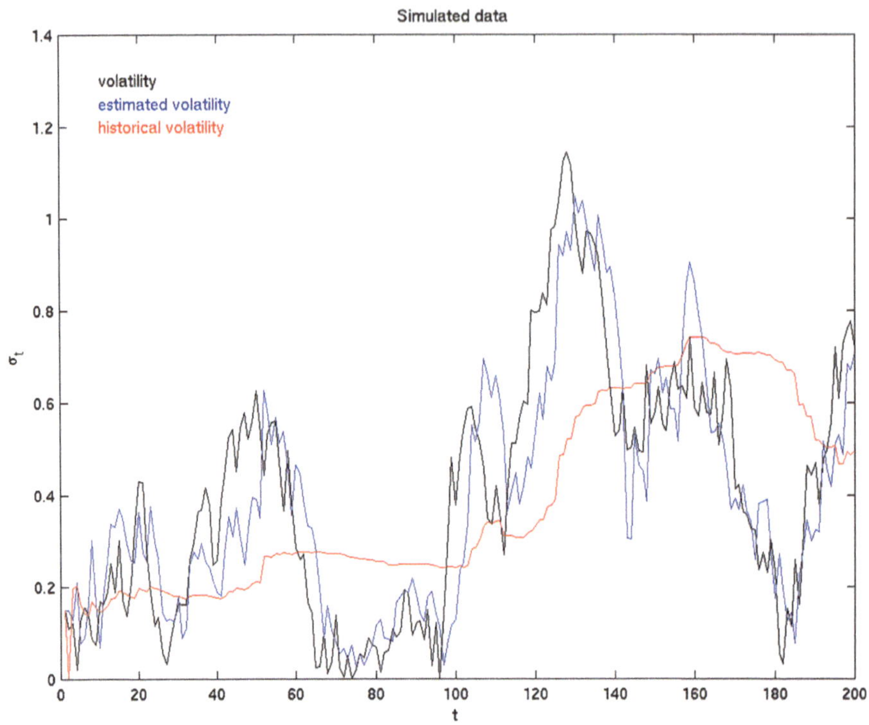

Fig. 8.20. Another example of volatility tracking by particle filtering

8.6.5.5 ***Possible Extensions***

The application considered above can be made more realistic by extending the model to include more features of real data. As a first possible extension, we should mention that the particle filters can be used without much change to the case of a regime switching model in which the volatility is a function of a factor changing value over time in a limited and restricted way, though remaining Markovian. Another possible generalization is to adjust the filtering model to volatility processes with several time scales. These multiscale volatility models are very fashionable, and as long as filtering is concerned, they can be handled without much significant change to what we presented above.

PROBLEMS

Ⓣ **Problem 8.1** *This problem shows once more that a conditionally Gaussian random variable has excess kurtosis. Let us assume that* X *and* σ^2 *are two random variables and that* $X|\sigma^2 \sim N(0, \sigma^2)$*, i.e. that conditioned on the value of* σ^2*,* X *is a mean-zero Gaussian random variable with variance* σ^2*. Prove that:*

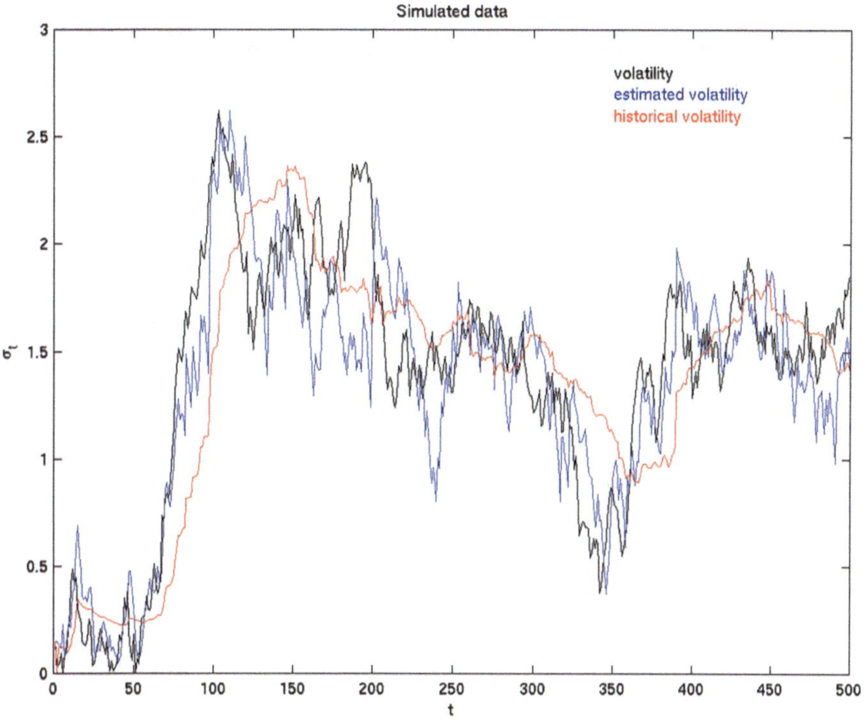

Fig. 8.21. Still another example of volatility tracking by particle filtering

$$\frac{\mathbb{E}\{X^4\}}{var\{X\}^2} = 3\left[1 + \frac{var\{\sigma^2\}}{\mathbb{E}\{\sigma^2\}^2}\right]$$

proving the claim of excess kurtosis when σ^2 is not deterministic.

Ⓣ **Problem 8.2** *This problem shows that (linear) AR(p) time series can lead to (nonlinear) ARCH models when they have random coefficients.*

Let $\{\epsilon_t\}_t$ be a strong univariate white noise with $\epsilon_t \sim N(0,1)$, and let $\{\phi_t\}_t$ be a p-variate time series independent of $\{\epsilon_t\}_t$, and such that all the ϕ_t are independent of each other, and for each time t, the vector $\phi_t = (\phi_{t,1}, \phi_{t,2}, \dots, \phi_{t,p})$ is a vector of jointly Gaussian random variables with mean zero and variance/covariance matrix Σ. We study the time series $\{Y_t\}_t$ defined by:

$$Y_t = \phi_{t,1}Y_{t-1} + \phi_{t,2}Y_{t-2} + \cdots + \phi_{t,p}Y_{t-p} + \epsilon_t.$$

1. *Determine the conditional distribution of Y_t given $Y_{\leq t-1}$ by integrating out the ϕ-random variables.*
2. *Assume that the components $\phi_{t,1}$, $\phi_{t,2}$, ..., $\phi_{t,p}$ of ϕ_t are independent, and show that at least in this case, $\{Y_t\}_t$ has an ARCH representation.*

Ⓣ **Problem 8.3** This problem addresses the very important issue of the stability of models under change of sampling frequency. Here is a typical example: assuming that a GARCH(1,1) model was found appropriate for some daily financial time series, and assuming that we need now to look at the end of the week data entries only, could we assume that this time series will also follow a GARCH(1,1) model, or should we start the model fitting process from scratch directly on weekly data?

We assume that a series $\{Y_t\}_t$ of log-returns has the GARCH(1,1) representation of the form:

$$Y_t = \sigma_t \tilde{\epsilon}_t, \qquad \sigma_t^2 = c + b\sigma_{t-1}^2 + aY_{t-1}^2$$

where we assume that $\{\tilde{\epsilon}_t\}_t$ is a strong $N(0,1)$ white noise, and where the coefficients a, b and c are such that σ_t^2 is stationary.

1. *Show that the squared log-returns have an ARMA(1,1) representation in the sense that:*

$$Y_t^2 = c + (b+a)Y_{t-1}^2 + \epsilon_t - b\epsilon_{t-1}$$

for some weak white noise $\{\epsilon_t\}$ which you should identify. Show that the number $\alpha = \mathbb{E}\{\epsilon_t^2\}$ is independent of t and that:

$$Y_t^2 = c(1+b+a) + (b+a)^2 Y_{t-2}^2 + u_t$$

with $u_t = \epsilon_t + a\epsilon_{t-1} - b(a+b)\epsilon_{t-2}$.

2. *Compute the number $\beta = \mathbb{E}\{u_t^2\}$ in terms of the numbers a, b and c, and check that it is independent of t. Show that $\mathbb{E}\{u_t u_{t-2k}\} = 0$ for all integers $k > 1$, and that:*

$$\frac{\mathbb{E}\{u_t u_{t-2}\}}{\mathbb{E}\{u_t^2\}} = -\frac{b(a+b)}{1+a^2+b^2(a+b)^2}.$$

3. *We now use the process $\{v_t\}_t$ defined by:*

$$v_t = \frac{1}{1-\lambda B^2} u_t$$

for some $\lambda \in (0,1)$ where B stands for the backward shift operator. Show that:

$$\mathbb{E}\{v_{2t}v_{2t-2s}\} = \frac{\lambda^{s-1}}{1-\lambda^2}[\lambda \mathbb{E}\{u_t^2\} + (1+\lambda^2)\mathbb{E}\{u_t u_{t-2}\}], \qquad s \geq 1.$$

4. *Show that if λ is chosen as the root of the equation:*

$$\frac{\lambda}{1+\lambda^2} = \frac{b(a+b)}{1+a^2+b^2(a+b)^2}$$

which belongs to the interval $(0,1)$, then the v_t with even indices are uncorrelated, i.e.

$$\mathbb{E}\{v_{2t}v_{2t+2s}\} = 0, \qquad s \geq 1$$

and that:

$$\mathbb{E}\{v_{2t}^2\} = \frac{\beta}{1+\lambda^2}.$$

5. *Show that the square* $\overline{Y}_t^2$ *of the series* $\{\overline{Y}_t\}_t$ *defined by:*

$$\overline{Y}_t = Y_{2t}$$

has an ARMA(1,1) representation, and conclude that the series $\{\overline{Y}_t\}_t$ *admits a GARCH(1,1) representation similar to the original GARCH(1,1) representation of* $\{Y_t\}_t$ *which we started from.*

Ⓔ **Problem 8.4** The goal of this problem is to give another example of the misuse of the simulation of GARCH models fitted to data with a deterministic seasonal pattern in the instantaneous variance.

1. *We first use the daily average temperatures in La Guardia as encapsulated in the* `timeSeries` *object* `LaGuardia.ts`. *Identify the trend and seasonal components, fit a GARCH(1,1) model to the remainder, and plot the estimate of the instantaneous conditional variance.*
2. *In each CDD season, identify the periods of high volatility and the periods of low volatility. Same question for the HDD seasons.*
3. *Using the fitted model, simulate a Monte Carlo sample for* 5 *years worth of temperature in La Guardia. Do you find the periods of high and low volatility identified in question 2 above? Explain.*
4. *Redo questions 1, 2, and 3 for the temperature in Las Vegas contained in the time series* `LasVegas.ts`.

Ⓢ **Problem 8.5** *This problem is based on the data* `MONTHLY` *described in Sect. 8.5. Each row corresponds to a month between May 1986 and November 1999. The first column represents the values of the short interest rate as given by the 1 year T-bill between, the second column gives the long interest rate as given by a 30 years US Government Bonds, and the third column gives the values of the S&P 500 index.*

1. *Use the equation:*

$$S_{t+\Delta t} - S_t = S_t[\mu\Delta t + \sigma\sqrt{\Delta t}\epsilon_t]$$

proposed in the text as Eq. (8.22) to model the monthly S&P 500 index values, and estimate the parameters μ *and* σ *from the data. Remember,* $\Delta t = 1/12$ *stands for 1 month. Explain your work.*

2. *Similarly, use the data and the equation:*

$$r_{t+\Delta t} = r_t - \lambda_r(r_t - \overline{r})\Delta t + \sigma_r\sqrt{\Delta t}\ \epsilon_t^{(r)}$$

used in the text as Eq. (8.20) to model the monthly values of the short interest rate, and estimate the parameters λ_r, $\overline{r}$, *and* σ_r. *Be imaginative and explain your work.*

3. *Finally, use the data and the equation:*

$$s_{t+\Delta t} = s_t - \lambda_s(s_t - \overline{s})\Delta t + \sigma_s\sqrt{\Delta t}\ \epsilon_t^{(s)}$$

used in the text as Eq. (8.22) to model the monthly values of the interest rate spread, to estimate the parameters λ_s, $\overline{s}$, *and* σ_s. *Again, explain your work.*

4. *Assuming that the three white noise sequences* $\{\epsilon_t\}_t$, $\{\epsilon_t^{(r)}\}_t$, *and* $\{\epsilon_t^{(s)}\}_t$ *are independent, use the above equations and parameter estimates to generate a* $100 \times 120 \times 3$ *array containing* 100 *scenarios of* 120 *possible monthly values of the short interest rate, the long interest rate and the S&P 500 index, starting from the last entries of the data in the file* `monthly.asc`. *Call these three sets of* 100 *scenarios* `IndepShort`, `IndepLong` *and* `IndepSP500` *respectively.*

5. *From the data, compute the variance/covariance matrix of the log-return of the S&P 500, the first difference of the short interest rate, and the first difference of the spread. Assuming now that the variance covariance matrix of the three white noise sequences* $\{\epsilon_t\}_t$, $\{\epsilon_t^{(r)}\}_t$, *and* $\{\epsilon_t^{(s)}\}_t$ *is the same as the empirical variance/covariance matrix you just computed, generate a new set of* $100 \times 120 \times 3$ *array containing* 100 *scenarios of* 120 *possible monthly values of the short interest rate, the long interest rate and the S&P 500 index. Call these* 3 *sets of scenarios* `CorrelShort`, `CorrelLong` *and* `CorrelSP500` *this time.*
6. *From the data, fit a bivariate distribution to the log-returns of the S&P 500, and the first difference of the short interest rate (use the techniques developed in Chap. 3), and assuming that the first difference of the long interest rate is independent of the log-returns of the S&P 500 and the first difference of the short interest rate, generate a new set of* $100 \times 120 \times 3$ *array containing* 100 *scenarios of* 120 *possible monthly values of the short interest rate, the long interest rate and the S&P 500 index. Call these* 3 *sets of scenarios* `CopShort`, `CopLong` *and* `CopSP500` *this time.*
7. *Choose two statistics (numerical function of the outcomes of the S&P 500, the short and the long interest rates), compute them for the three sets of scenarios generated in questions 4, 5, and 6, to illustrate the impact of the different notions of dependence included in the three Monte Carlo simulation procedures.*

Ⓣ **Problem 8.6** *In this problem, we assume that the dynamics of a stock are given by Samuelson's geometric Brownian motion model, and we assume that the rate of growth* μ *and the volatility* σ *are known.*

1. *Use formula (8.16) to derive the distribution of the raw returns* RR_t *over a 1-month period* Δt.
2. *Compare with the distribution derived from (8.18) given by a blind application of Euler's scheme. Comment.*

Ⓢ **Problem 8.7** *The goal of this problem is to implement a volatility tracking algorithm based on the particle filtering of a stochastic volatility model where the dynamics of the volatility are given by a geometric Ornstein Uhlenbeck process.*

1. *For the purpose of this question we consider the price of an asset evolving with time according to a geometric Brownian motion with a stochastic volatility satisfying equation (8.35) used in the text. Such a process is usually called an Ornstein-Uhlenbeck process. Explain why, at any given time t, the volatility* σ_t *has a positive probability of being negative. In doing so, you can choose to work directly with the continuous time equation (8.35) if you feel comfortable, or with its discretized version if you find it more intuitive.*
2. *Explain why this shortcoming is not shared by the geometric Ornstein-Uhlenbeck model in which the volatility evolves according to the equation:*

$$d\sigma_t = \sigma_t[-\lambda(\log \sigma_t - \mu)dt + \gamma d\tilde{W}_t] \tag{8.44}$$

where λ, μ *and* γ *are strictly positive constants. Again, feel free to choose to work with the continuous time equation (8.44) or its discretized form.*

Ⓔ **Problem 8.8** *You will need to use the* `timeSeries` *objects* `enelret.ts` *and dynlret.ts included in the library* `Rsafd`. *Each file contains the daily log return of a large energy company.*

1. *Fit an AR model to the part of the time series* `dynlret.ts` *corresponding to the period of the* 500 *trading days ending January 1, 2001. Use this model to predict the values of the log return for all the remaining days after January 1, 2001, and compute the sum of square errors.*
2. *Same question for the time series* `enelret.ts`
3. *Concatenate the two time series in a bivariate time series, fit an AR model to the bivariate series so obtained for the same period as before, and use this model to predict the values of the two series following January 1, 2001. Compute the sum of the squares errors and compare to the previous results. Comment your results.*

NOTES & COMPLEMENTS

The data and the proprietary algorithm used to compute the indexes used in this chapter and throughout the book were graciously provided by Dow Jones.

Mandelbrot was presumably the first one to warn the scientific community of the effects of the presence of heavy tails and long range dependence in financial time series. A good account of his work on the subject can be found in [64, 65] and [66]. Nonlinear time series models are studied with great care and lucidity in the recent graduate textbook [35] by Fan and Yao, where the interested reader will find many examples of analyzes relying on nonparametric regression techniques.

Financial time series differ from typical time series in many different ways, but most noticeably, in the properties of the conditional standard deviation, i.e the so-called volatility. Persistence (tendency of large changes to be followed by more large changes), are at the root of the introduction of the ARCH and GARCH models discussed in this chapter. The first ARCH models were introduced in their simplest form by Engle in 1982, see [30]. This introduction had far-reaching implications on subsequent developments in financial economics, hence the award to Engle of the 2003 Nobel prize in economics. ARCH models were later generalized by Engle and Bollerslev in [31], and Bollerslev [6] who introduced the GARCH models. The review article [7] is a comprehensive introduction to the early developments of the subject. Since then, the literature on the subject exploded, and the analysis of these models is now an integral part of financial econometrics. For more recent developments on these models we refer the interested reader to the more modern account given in the monograph [41].

One of the major difficulties with the stochastic volatility models is their statistical estimation. Many methods have been proposed and tested: generalized method of moments (GMM for short), quasi-likelihood method based on Monte Carlo Markov Chain (MCMC for short) computations and importance sampling, and even the simulated EM (expectation maximization algorithm). We chose to recast the SV models in the framework of linear state space models, and refer to the statistical estimation methods used for these models.

In the second half of the nineties, many books on the applications of Ito's stochastic calculus to finance were published. One of the earliest ones, and presumably the best one, was the short book of Lamberton and Lapeyre [58]. Many followed but they were often technically difficult and rarely self-contained. We shall only refer to the introductory texts of Michael Steele [90]. Discretization schemes for stochastic differential equations have been studied in details. It would have taken us far afield to attempt to review this literature. We limited ourselves to the Euler scheme because of its simplicity and its intuitive appeal. Further properties of these schemes and their relations to the ARCH and GARCH and SV models discussed in the text can be found in the recent text by C. Gourieroux and J. Jasiak [42].

The continuous time stochastic volatility models have been studied extensively by means of the properties of the Ito's stochastic calculus. The names of Heston, and Hull and White are often attached to these models. See for example the account given in the book [50]. The analysis of the volatility smile for these models was done rigorously in Touzi and Renault [78]. For a complete exposé of the facts of continuous time stochastic volatility models from the point of view of Ito's stochastic calculus and asymptotic expansions we refer the interested reader to the excellent monograph [38] by Fouque, Papanicolaou and Sircar.

The abstract form of the equation giving the optimal filter update in the form of a dynamical system is due to Stettner. The update algorithm (8.29) can be viewed as a discrete time dynamical system in the space of probability distributions. Unfortunately, this space is infinite dimensional and the analysis of such a dynamical system is very difficult. This dynamical system can also be viewed as a discrete time version of the stochastic partial differential equation derived for the optimal filter of partially observed systems driven by Ito's stochastic differential equations. This stochastic partial differential equation is known under the name of Kushner equation, or Zakai equation, depending on our looking at the update of the (normalized) density of the optimal filter, or the density of the unnormalized filter. Both are regarded as natural generalizations of the Ricatti equation giving the update of the Kalman filter in the linear case. Indeed, even when we look at conditional density functions for the conditional distributions, the update algorithm appears (because of the infinite dimensionality of the functional space) as a partial differential equation. Because the coefficients of this partial differential equation depend upon the observations, they can be viewed as random. These equations were some of the first stochastic partial differential equations studied, and solving the general stochastic filtering problem was one of the main impetus in the development of the field of stochastic partial differential equations.

Unfortunately, the numerical solution of these equations is extremely computer intensive and, as of today, there is essentially no example of practical problem whose solution can be computed in real time. The particle method is the first method which stands a real chance at cracking this nagging problem.

The particle approximation algorithm presented in the text is due to Kitagawa [56]. The mathematical theory of this approximation (and several variations on this approximation) was subsequently developed by several probabilists including Del Moral, Guillonnet, Lyons, Crisan, Le Gland, , Cérou, . . . and many others. Among other things, they proved that at each fixed time n, the particle approximation $\hat{\pi}_n^{(m)}$ converges toward π_n as $m \to \infty$.

It is interesting to track the volatility of indexes and sub-indexes when the latter are computed by sectors, and even by subdividing the sectors according to the credit ratings of the companies comprising the sectors in question. In his 2001 Princeton PhD, M. Sales considered amon other things the sector "Materials / Mid-Cap's" of DOW JONES TMI (Total Market Index) subindexes. He noticed a general increase in volatility in 1999. But if after breaking the sector into two sub-sectors, say the companies whose bonds are rated "investment grade", and the companies whose bonds are rated "non-investment grade" (using for example Moody's ratings), then as expected, he noticed an increase in volatility in the non-investment grade sub-sector, not the investment grade. However, it should be emphasized that the reverse though less natural, does occur as well. `R` can be used to obtain simultaneous Likelihood Ratio Tests for the significance of the differences in mean returns. Unfortunately, similar tests for the comparison of the variances of correlated samples are not available, and part of Sales' PhD was devoted to the development of such tests.

The application of the particle filter to stochastic volatility tracking was done by A. Papavasiliou's as part of her Princeton PhD. The classical (linear) Kalman filter has been applied to the construction of commodity forward curves. The estimation of these curves is of crucial importance in energy risk management. The reader interested in energy derivatives may consult the book of Clewlow and Strickland [24] and the recent review [20] by Carmona and Coulon. The desire to present the details of one specific form of the particle filtering algorithm, together with a financial application was motivated by their lack of availability in book form, and the many successful implementations the author developed for the purpose of commodity data analysis (e.g. estimation of the convenience yield of crude oil and natural gas. The reader interested in applications of particle methods in finance is referred to the survey article [19] by Carmona, Del Moral, Hu and Oudjane.

Part IV

BACKGROUND MATERIAL

9

APPENDICES

9.1 APPENDIX A: A QUICK INTRODUCTION TO R

This appendix gives a streamlined introduction to the basics of R. Following the prescriptions given in this appendix will take you through an introductory session intended for readers who are first time users of R. We do not expect that such a session will turn R-novices into experts. However, it should help beginners feel comfortable enough with the language to start practicing with the examples given in the book.

9.1.1 Starting R Under Mac OS, Windows and Under UNIX

The instructions given below are restricted to versions 2.0 and higher of R. One of the main attraction of R is its being platform independent. However, the GUI (Graphic User Interface, pronounced "gooohee") of a typical R distribution can still change from one platform to another, and in any case, it is not as sophisticated as for most commercial applications. But since most every task can be described independently of the platform, we will limit ourselves to the use of commands typed in the command window of the application. We take advantage of this universal way to input and execute commands, and ignore the possible alternatives relying on menu items and the use of a mouse.

Despite the merits of this discourse on the virtues of platform independence, I will presumably err on the side of Windows and Mac users since after all, this book was prepared on a laptops running various versions of their operating systems.

Finally, even though this textbook does not require prior knowledge of R, because its previous incarnation was written in conjunction with a S-Plus library, given the significant number of S users switching to R, we shall sprinkle the text with small remarks emphasizing the similarities and the differences between the two languages when appropriate.

R. Carmona, *Statistical Analysis of Financial Data in R*, Springer Texts in Statistics, DOI 10.1007/978-1-4614-8788-3_9, © Springer Science+Business Media New York 2014

9.1.2 Creating R Objects

Typing the command

```
> X <- 1:16
```

in the Command Window, creates a vector of length 16 containing the first 16 integers in increasing order. As in S, the arrow "<-" which is typed by using first the "less than" key "<" and then the minus sign "-", is used to assign the instance appearing on its right hand side to the object on its left. It reads "gets" to avoid the possible confusion with left arrow key. Whatever is on the left of this "gets" is an object created by R with the result of the command on the right of the "gets" sign. If an object with this name already exists, it is automatically replaced. Note that the same result would be obtained with the command:

```
> X <- seq(from=1,to=16,by=1)
```

To understand what the function seq does, we can use the online help by typing:

```
> help(seq)
```

Now that we know what the function seq can do for us, we can type:

```
> help
```

and notice that, instead of getting a help file of some kind, we get something which looks more like code. This is a general rule: if one types the name of an R function or method without parentheses, the system returns the R-code of the function in question. Only if we add the parentheses, will we see the command be executed (or an error message returned telling us that we did not provide the parameters expected by the function). We can experiment further by typing:

```
> objects()
```

which lists all the R objects contained in the current workspace, and

```
> objects
```

which merely gives its code. Notice that the effect is the same as if we were to use the command ls which I believe, was kept for compatibility with S. R organizes objects into workspaces. They are ordered in a hierarchy which can be displayed by the command search(). The command:

```
> dim(X) <- c(8,2)
```

reshapes the one dimensional vector X into a 8 by 2 matrix. The right hand side c(8,2) creates a vector of length 2 with entries 8 and 2. The function c which stands for concatenation, is one of the most commonly used core functions. So in order to save typing and memory space, it was given a name with only one letter. Typing the command

```
> class(X)
```

will return "matrix". All objects created by R have a class attribute, and this attribute can be retrieved with the function class. The command:

```
> Y <- X*X
```

creates a new R-object, named Y, which is also a 8 by 2 matrix. Its entries are obtained by multiplying X by itself *entry by entry*. It is important to remember that the symbol * does not give the usual matrix product. The latter is obtained by sandwiching the * symbol in between % 's as in:

```
> Z <- X %*% X
```

This command gives an error message

```
> Z <- X %*% X
Error in X %*% X : non-conformable arguments
```

for one cannot multiply a matrix with 2 columns by a matrix with 8 rows. However, the commands

```
> Z1 <- X %*% t(X)
> Z2 <- t(X) %*% X
```

will produce an 8 by 8 matrix Z1, and a 2 by 2 matrix Z2. Indeed, we used the R command t(X) which computes the transpose of the matrix X. Notice that both matrices are *non-negative definite*. This property is characteristic of variance/covariance matrices of multivariate random vectors, and it plays an important role in multivariate random simulation.

9.1.2.1 ***Tidying Up, Saving Objects, and Exiting the Workspace***

Exiting R can be done at any time with the menu item **File** ▷ **Exit**, or by typing the command q(). In either case, you will be asked if you want to save the contents of the workspace. Answering YES is equivalent to the command save.image(). In fact, you do not have to wait for the end of a session to save the workspace. It is possible to save the workspace to a file at any time with the command save.image(). An image of the workspace will be saved in the current directory/folder under the name .RData or _RData. This image will automatically be restored when the program R is launched in this directory/folder. It is also possible to give an alternative name to this image by specifying a filename (enclosed in double quotes) for the argument of save.image. Individual objects can be saved with the command save. This command writes the objects whose names are passed as parameters to a specified file. The objects can be read back from the file at a later date by using the function load.

The number of objects created during an R session can be very large, and can quickly clutter your disk or at least your workspace. In order to save memory, it is good practice to remove the objects which you will not need again by using the command rm. For example

```
> rm(X,Y)
```

will delete the objects X and Y if they are in the active workspace.

All the objects saved with the methods described above are written to binary files which cannot be opened and read by other programs: only R can open and re-use them. The function dump is a convenient way to produce text representations of the objects. The function dump needs two parameters; a vector of names of objects to be found in the workspace, and a file name to dump the representations to. A typical usage reads:

```
> dump(c("Z1","Z2"),"mymatrices")
```

When needed, a dump file like mymatrices can be sourced with the command source("mymatrices") into another R session.

9.1.3 Random Generation and White Noise

The commands

```
> WN <- rnorm(1024)
> plot (WN,type="l")
```

create a vector WN of length 1,024, and produce a sequential plot of the entries of this vector. The entries of WN are realizations of independent Gaussian (i.e. *normal*) random variables with the same distribution $N(0, 1)$. We specified the length of the random vector but we did not specify the values of the parameters of the distribution. R uses the default values 0 and 1 for the mean and the standard deviation of the normal distribution. The function plot gives a sequential plot of the vector WN, by plotting its values against the variable which takes the values 1, 2, ..., 1,024. Without providing other arguments, the function plot outputs a plain scatterplot., i.e. draws a discrete set of isolated points on the plane of the graph. Adding type = "l" forces the plotting routine to join the successive points of the scatterplot by a line segment. This form of the plot is often more instructive than a plain scatterplot.

NB: A command WN <- rnorm(1024,1.2,4.0) would have created a sample of 1,024 of realizations of independent identically distributed (i.i.d. for short) variates from the Gaussian distribution with mean 1.2 and standard deviation 4, in other words the distribution $N(1.2, 16)$ if we use the standard notation used in the text. The commands:

```
> par(mfrow=c(2,1))
> plot(WN,type = "l")
> plot(WN[1:64],type = "l")
> par(mfrow=c(2,1))
```

divide the graphics window into two subplot areas, one on the top of the other, give a time series plot of the full WN vector on top, a plot of the first 64 entries of the vector WN, and reset the graphic window to a single plotting area. The results are shown in Fig. 9.1. But before we comment on the output, notice how we extracted a subvector from a vector, just by providing the range `1:64` of the values of the indices we wanted to keep in the subvector. In the above command, we did subscript the vector `WN` to only keep the set of its first 64 entries. To do so, we specified the range of the indices we wanted to keep. Subscripting is also conveniently used to extract sub-matrices of matrices. For example, if `X` is a matrix, say with 5 rows and 6 columns, the command `X[1:3,1:3]` will produce the 3 by 3 matrix extracted from the top-left corner of `X`. A similar form of subscripting is frequently used to extract a column or a row, or a set of columns or a set of rows of a matrix. For example, `X[,2]` produces a vector equal to the second column of `X`, `X[2:4,]` produces a matrix with three rows and as many columns as `X`, the three rows of the new matrix being the rows numbers 2, 3 and 4 of the original matrix `X`.

We will see that *subscripting* is a very powerful tool to manipulate objects, and we will make full use of its versatility in our data analyzes. Coming back to Fig. 9.1, the top plot shows what we should expect from a white noise: very erratic because of the lack of correlation of the successive entries of the series. Identifying a white noise is of crucial importance to a statistician. Indeed, most model fitting efforts are devoted to the identification and the extraction of organized structures from the data. This is usually done up until the remainder terms (which we call the residuals) form a white noise. From that point on, it is unreasonable to try to extract any more information, and statistical inference should take place at that time, and definitely before we start trying to fit the noise.

Strangely enough, the bottom plot of Fig. 9.1 looks smoother. The reason is simple: it is viewed on a very different scale. We plotted one 16th of the data on a plot

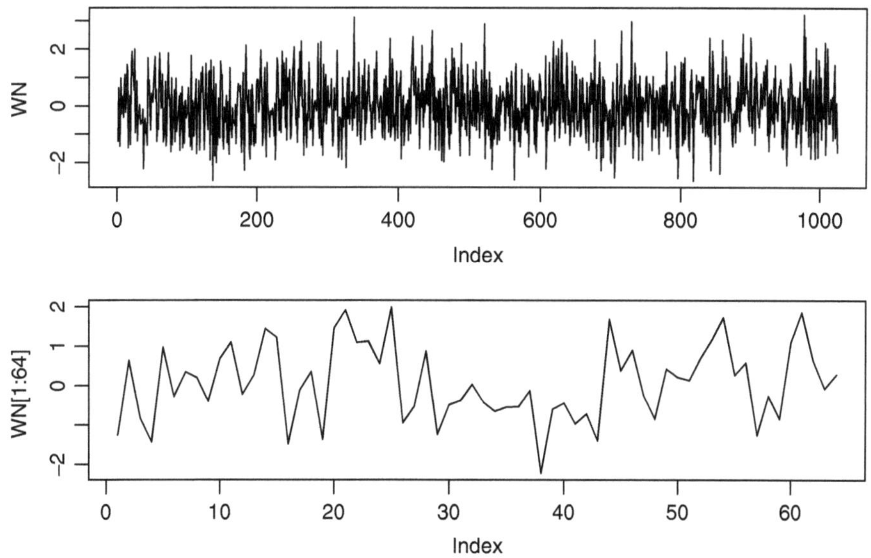

Fig. 9.1. Sequential plot of the full white noise series `WN` (*top*) and of the series of its first 64 entries (*bottom*)

with the same width! The *white noise look* which was identified earlier has practically disappeared. It is important to realize that, looking at the same data at different scales will leave different visual impressions. Don't be fooled by this artifact of the graphic settings used, and always look at the ticks appearing on the axes before forming an opinion on the characteristics of the data actually plotted.

9.1.4 More Functions and `for` Loops

The function `diff` should be viewed as a discrete analog of the operation of differentiation for functions of continuous variables. We learned in calculus that the inverse operation is integration. The discrete analog is given by the function `cumsum`. If we apply it to the white noise vector `WN`:

```
> RW <- cumsum(WN)
```

we get a numeric vector `RW` which represents a sample of length 1,024 from a random walk. The sequential plot of `RW` (i.e. the plot of the values of `RW` against the successive integers) is given in the left pane of Fig. 9.2. Let us now try to create an object according to the Samuelson's model for stocks. According to this model, the time evolution of a stock is represented by the exponential of a random walk (with a given volatility) with drift given by the rate of appreciation of the stock. More precisely, we define the objects `SIG`, `MU`, `TIME` and `STOCK` by:

```
> DELTAT <- 1/252
> SIG <- .2*sqrt(DELTAT)
> MU <- .15*DELTAT
> TIME <- (1:1024)/252
> STOCK <- rep(0,1024)
```

for the volatility, the rate of return, the time (expressed in years) and an initial zero vector for the stock (by repeating 0 as many as 1,024 times with the command `rep`). With these objects at hand, one should be able to fill in the entries of the vector `STOCK` with the commands:

```
> for (I in 1:1024) STOCK[I] <- exp(SIG*RW[I]+MU*TIME[I])
```

This command is typical of the use of the `for` loops in `R`. Unfortunately, in `R` like in all other implementations of interpreted languages, computations with loops are very slow, and they should be avoided whenever possible. There is a very simple way to avoid the above loop. Indeed, most numerical functions in `R` when applied to a vector (resp. matrix, array, ...) will create a vector (resp. matrix, array, ...) with entries given by the computations of the function on the entries of the vector (resp. matrix, array, ...) passed as argument to the function. So in the above example, the command:

```
> STOCK <- exp(SIG*RW+MU*TIME)
```

will give the same result, and the computations will be much faster. The plot of the entries of this vector is given in the bottom pane of Fig. 9.2. The latter was produced with the commands:

```
> plot(RW,type="l")
> title("Sample of Size 1024 from a Random Walk")
> plot(TIME,STOCK,type="l")
> title("Corresponding Geometric Random Walk with Drift")
```

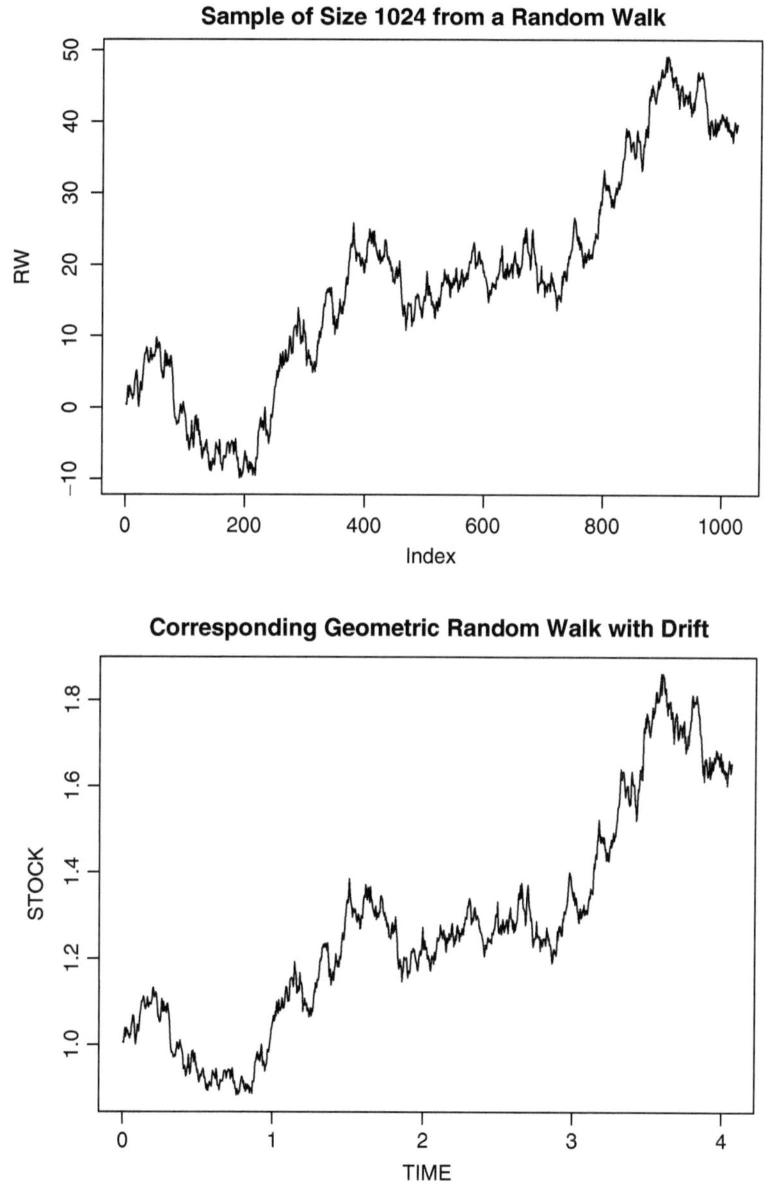

Fig. 9.2. Sequential plot of random walk sample RW (*top*) and of the corresponding geometric random walk with drift (*bottom*)

Notice that we use the command `plot` with two arguments. We need to pass two vectors to the command `plot`. The first one gives the values of the first coordinates of the points to plot while the second vector gives the second coordinates of these points. We use the option `type="l"` to joint the points by straight line segments to get the visual impression of a continuous curve instead of isolated points. The plots look very similar, until we notice the difference in scale on the vertical axes.

9.1.5 **Importing Data**

In most data analysis applications, the data are not created within the statistical package used for the analysis: they have to be imported. We give several examples as illustrations. These examples can be used as templates for further applications and to work out the problems (homework assignments) given in the text.

9.1.5.1 Using Packages

Libraries of *packages* have been developed to add functionality to the core distribution of `R`. Packages are sets of data and `R` functions bundled together with the intent to be used as "add-on"s to a session. For the purpose of this book, you will only need the core distribution of `R` and the package `Rsafd` which was developed for use with this book. The package is loaded into `R` using the `library` command:

```
> library(Rsafd)
```

In this way, all the functions used in the book, as well as all the data sets used in the applications, illustrations, and problems become available to your `R` session. Libraries are loaded for the life of a session. In other words, you do not need to *unload* them when you exit, but you need to reload them each time you start the program and you want to use its contents. This shortcoming can be remedied with the use of the function `.First`.

9.1.5.2 The Function `.First`

Each time the program `R` is started, and even before any command from the user can be executed, the program looks for a function `.First`, and if it exists and is found, `R` runs it first. So any instruction or command included in such a function will be executed at the start of the program. This is especially convenient for repetitive loading of libraries and data sets which are needed on a regular basis. The following example is clearly self-serving. Typing the command

```
> .First
```

at the prompt will let us know if such a function already exists on our system. If it does, its code is displayed and we can see what it does or does not do. Assuming for the moment that such a function does not exist, typing the command

```
> fix(.First)
```

opens a window on the desktop. A template for the function `.First` is proposed. We only need to type the code in between the two curly brackets. See below for a more detailed discussion of how to write functions. Let us type

```
library(Rsafd)
cat("\n Welcome to my world\n\n")
```

save the content of the window and close it. From now on, each time a new R session is started, the command `library(Rsafd)` is executed, followed by a carriage return, the display of the text "Welcome to my world" followed by two carriage returns.

9.1.5.3 ***Attaching a*** `data.frame`

```
> attach(hills)
```

Doing so makes it possible to use all the variables in this frame by simply referring to their names. For example, the following commands produce the results shown in Fig. 9.3.

```
> plot(climb,time)
> plot(dist,time)
```

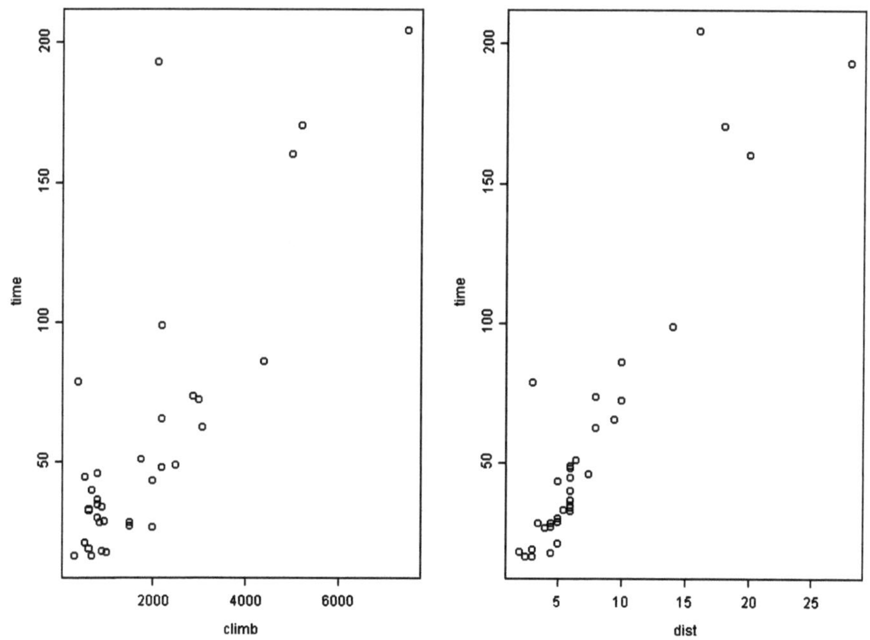

Fig. 9.3. Scatterplot of the record `time` versus `climb` (*left*) from the `hills` data frame, and of the `time` versus the distance variable `dist` (*right*)

9.1.5.4 *Importing the Contents of Text Files in* R

As an example, we show how to import the text (ASCII) file containing numerical data into your R session. We assume that the file is called HOWAREYOU.txt. This data vector is used in Chap. 6 as a mere illustration. For the sake of definiteness we assume that the file is located in the directory

C:\My Documents\ORF405\book\data.

We give explicit commands as if we were working under Windows, similar commands would work under Unix or Mac OS, obviously after replacing the backward slashes by forward ones in the file names. We could open the file with a text editor to check its contents. We show the first seven rows of the file for the sake of illustration.

```
0.0000000000000000
0.0359678408504965
0.0635196843758394
0.0750752080020106
0.0628095887777292
0.031177878417794
-0.00520235015344319
-0.0294474884975495
```

Under Windows, we import this file into R with the dialog created by the **File** ▷ **Import Data** ▷ **From File**, and clicking on the button BROWSE to navigate to the location where the file is, and selecting it. The full pathname of the file HOWAREYOU appears, its being an ASCII file also shows, and OK'ing all that allows R to create an R object called HOWAREYOU in the current data base. The same result could be obtained with the command:

```
> HOWAREYOU <- scan("C:\\My Documents\\......\\book
                                        \\data\\HOWAREYOU")
```

which could also be used under Unix or Mac OS provided the double backward slashes \\ are replaced by single forward ones /. As the result of the next command shows, it is a vector of length 5,151 and we can print its first entries using the command head:

```
> length(HOWAREYOU)
[1] 5151
> head(HOWAREYOU)
[1] 0.000000 0.035968 0.063520 0.075075 0.062810 0.031178
```

9.1.5.5 *Importing R Dumps*

The previous example was concerned with plain text files. Next, we consider the case of text files produced by R, and containing information about the R-objects they were issued from. The main feature of this procedure is to be able to port files from one version of R to another and from one platform to another. For example, files created and manipulated under Unix, Linux or Mac OS can be easily read and manipulated under Windows. R objects can be dumped to a text file with the dump command (see the help file for details). The format of these text files is such that they can be read by any other version of R irrespective of the platform. Under Windows, a text file resulting from dumping R objects should be open from the dialog created by **File** ▷ **Open**, or by double-clicking the open folder icon in the task bar. R creates a script window, and displays the contents of the text file in this window. Hitting the **F10** key, or running the script using the menu item **Script** ▷ **Run F10**, is what is needed to create the desired object. These dumps are text files and as such, they can be opened and edited with a text editor. We suggest to use the extension .asc or .R to distinguish these dumps from regular text files for which the extension .txt is commonly used. Note that all of the data sets used for illustration purposes in the text, or needed in the problem sets are included in the library Rsafd.

9.1.5.6 ***Getting Data from the Web via Excel***

We now show how to import data into an R session when they were originally processed and saved within Excel. For the sake of definiteness we assume that the data were saved in the form of a .csv file, and for the purpose of illustration we choose to work with the data of Calpine, an independent power producer whose headquarters are in San Jose CA.

We import the data in an R object by running the command

```
> CPN.tab <- read.csv("calpine.csv")
```

One may have to give the full path name of the file "calpine.csv" depending upon its location on the disk. With this command, the content on the table saved from the Excel spreadsheet are imported into a data frame named CPN.tab.

Indeed, typing the command

```
> class(CPN.tab)
 "data.frame"
```

confirms that this is indeed the format in which it was imported into R. We can check the dimensions of this data frame and verify that the data have been imported properly as follows.

```
> dim(CPN.tab)
[1] 2101    7
> head(CPN.tab)
         Date Open High  Low Close   Volume Adj..Close.
1   27-Jan-05 3.38 3.42 3.33  3.36  5469600        3.36
2   26-Jan-05 3.45 3.45 3.38  3.40 10899700        3.40
3   25-Jan-05 3.30 3.40 3.30  3.37 12769200        3.37
4   24-Jan-05 3.12 3.27 3.10  3.25  9242600        3.25
5   21-Jan-05 3.23 3.23 3.07  3.13 13690600        3.13
6   20-Jan-05 3.01 3.31 3.01  3.24 25324300        3.24
```

which displays the top six rows of the data frame. Using the command tail (CPN.tab) would display the last six rows. Notice that the small row indexes correspond to the most recent quotes. We shall come back soon to this strange convention. We can extract and plot the daily close prices of Calpine stock.

```
> CPN.close <- CPN.tab[,5]
> plot(CPN.close,type="l")
```

The plot is reproduced in the right pane of Fig. 9.4. The plot function produces a plot of the entries of the vector CPN.close against the integers in increasing order, the parameter type="l" stating that straight lines should connect the points of the plot to give the impression of a continuous graph. The remark made above is confirmed. The increasing dates are running from right to left which is quite unusual. To conform with the standard convention, we need to reverse the order of the data using the function rev provided by R,. The data file CPN is included in the library Rsafd.

```
> CPN <- rev(CPN.close)
```

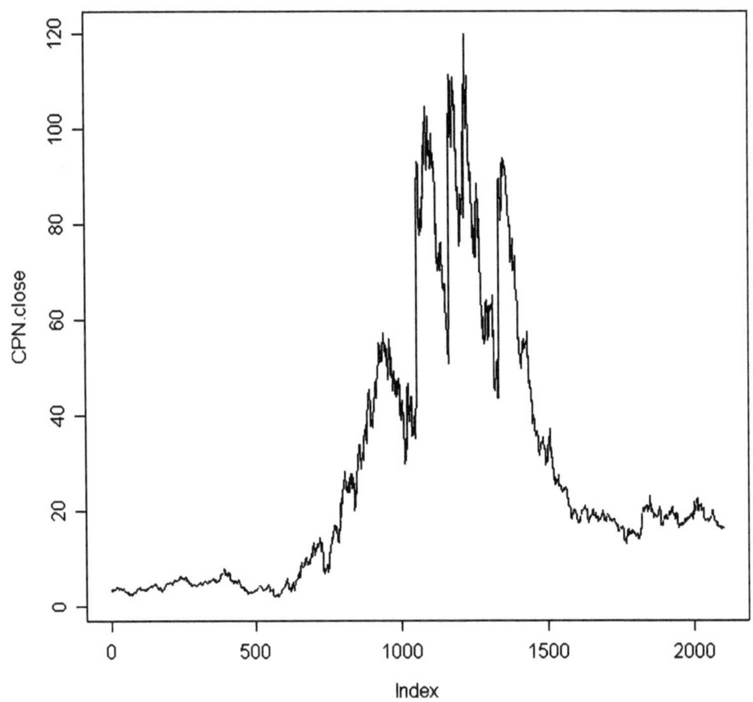

Fig. 9.4. Daily Calpine closing prices as imported from the internet. The data does not appear in the natural order

9.1.6 Programming in R: A First Function

When we suspect that a set of R commands will be needed frequently, it is important to save them, either separately or as part of a workspace. Moreover, it is also possible to encapsulate these commands in a script-like function which can be saved and re-used. This is especially true when we can structure the set of commands in order to take parameters as arguments. We illustrate this point with the example of the Black-Scholes formula for the price of a European call option. See next appendix for a thorough discussion. This formula states that the price C at time t when the underlying asset value is S, of a European call option with maturity T and strike K is given by the formula:

$$C = S\Phi(d_1) - Ke^{-r(T-t)}\Phi(d_2) \tag{9.1}$$

where we used the notation

$$d_1 = \frac{\log(S/K) + (r + \sigma^2/2)(T-t)}{\sigma\sqrt{T-t}} \quad \text{and} \quad d_2 = d_1 - \sigma\sqrt{T-t}. \tag{9.2}$$

The remaining parameters appearing in these formulae have the following meanings. r is a numerical constant which stands for the short interest rate, and σ is the annualized volatility at which the option is priced. Also recall that throughout the book, we use the notation Φ for the cumulative distribution function of the standard normal distribution. As explained in Chap. 1 computing values of this function in R can be done with the command pnorm.

Notice that, instead of depending separately upon the date t at which the option is priced and the date of maturity T, the actual price of the option depends only upon the difference $\tau = T - t$. The interpretation of this difference is very simple: it is the *time to maturity* of the option.

For the sake of illustration, let us compute in R, the price of an option at the money which matures in 90 trading days when the underlying asset is valued at 100, the short interest rate is 10% and the option is priced at 20% annualized volatility. We set up the parameters of our computation by typing the following in the command window.

```
> S <- 100
> K <- S
> R <- .1
> SIG <- .2
> TAU <- 90/252
```

The time spans need to be expresses in years, so we chose to divide 90 by the (approximate) number of trading days in 1 year in order to assign to the variable TAU the time to maturity in years. We can now compute the values of the constants d_1 and d_2 defined above. We choose to do so with the following commands

```
d1 <- log(S/K) + TAU * (R + SIG^2/2)
d1 <- d1/(SIG * sqrt(TAU))
d2 <- d1 - SIG * sqrt(TAU)
```

Finally, we assign the price of the option to a variable C by typing the command

```
> C <- S * pnorm(d1) - K * exp( - R * TAU) * pnorm(d2)
```

and typing C at the prompt will return the value of the option price as given by the Black-Scholes formula.

```
> C
[1] 6.643238
```

Very likely, we will want to compute many more prices of options. Also, we may want or need to compute values of the parameters which guarantee a given option price. For this reason it is convenient to encapsulate the commands used above in a single script which can be run more efficiently. R has this capability, allowing us to define home made functions. We can do just that by typing the following in the command window, or by using the command fix(bscall) which opens a window in which we can type and edit the code of the function, or even by saving the text below in a file, say myfunction.txt and then sourcing the file with the command source(myfunction.txt).

```
> bscall <- function(TAU=90/252, K=100, S=100, R=.1, SIG=.2)
{
# Parameters:    TAU time to maturity in yrs
#                K strike
#                S current value of the underlying
#                R yearly interest rate
#                SIG annualized volatility
# Return         Black-Scholes call price

    d1 <- log(S/K) + TAU * (R + SIG^2/2)
    d1 <- d1/(SIG * sqrt(TAU))
    d2 <- d1 - SIG * sqrt(TAU)
    C <- S * pnorm(d1) - K * exp( - R * TAU) * pnorm(d2)
    C
}
```

The following remarks are in order, especially if this is your first R function.

- As defined, the function bscall can be used without any parameters. Indeed, we chose to give default values to all the parameters, so each time a value is not specified for a parameter, the default value contained in the definition of bscall() is used. In particular the call bscall() produces the same result as bscall(90/252.100.100..1,.2);
- Whatever is found on the same line and to the right of a symbol # is ignored at run time. This is used to include comments which explain what the function is doing and what the parameters mean. This is extremely useful when you re-use a function long after it was originally written;
- Typically, a function in R is a set of commands included in between two curly brackets. The value of the object present on the last line of the code of the function is what is returned by a call to the function: hence the last line being C in the

present situation. A safer practice would be to specify the object returned by the function with the R function `return`, and replace the last line of the code of the function by `return(C)`.

We make extensive use of this function in the examples given in this book.

It is always a good idea to name your functions in such a way that their names will hint at what the functions are actually doing. Moreover, if you use a name already used by R, you will (at least temporarily) remove access to this R function. For example, if you were to create a function c with the command

```
> c <- function(x){x+1}
```

you would automatically replace the original concatenation core function c by your function. Running the command `Y <- c(12)` would assign the value 13 to Y, and the command `dim(X) <- c(4,4)` would now produce an error message. Fortunately, the concatenation core function c would only be *temporarily* masked by your function, and the system would restore it the next time you launch the program. There was no danger of masking the concatenation function when we computed the price of an option because we used an upper case letter C and R is case sensitive.

9.2 Appendix B: A Crash Course on Black-Scholes Option Pricing

This appendix presents the background material which we did not want to include in the part of the chapter on nonparametric regression devoted to the pricing of options. This material can be found in any basic textbook on financial mathematics. It is reproduced here for the sake of completeness.

9.2.1 **Generalities on Option Pricing**

For the sake of simplicity the following discussion is restricted to plain vanilla options on an underlying asset which can be thought of as a stock or an index. Moreover, to make our life easier, we use the terminology *option* when we actually mean *European call option.* The problem of pricing *European put options* can be addressed in exactly the same way, or can be reduced to the pricing of European call options because of a parity argument. We leave to the interested reader the easy task of adjusting the following discussion to this case. Also, we shall not discuss the pricing of options with American exercise or the so-called exotic options, such as barrier options.

An option is a contract which gives the buyer (of the option) the right (but not the obligation) to purchase at time T (called the expiration or maturity date of the option) one unit of the underlying asset or instrument at an agreed upon price K (called the strike price of the option). Such an option is a contingent claim and the contract unambiguously defines the payoff of the claim. In the situation at hand the payoff is a simple function of the price of the underlying asset at expiration. If we use the notation S_t to denote the price of this underlying asset at time t, then the payoff of our option is given by $f(S_T)$ where the function f is defined by $f(x) = \max\{x-K, 0\}$. Indeed, if at time of expiration the price S_T of the asset is smaller than K, then the option is worth nothing. Why would we buy one unit of the underlying for K if we can get it for less on the open market! Moreover, if the price S_T is greater than K, then it is clear that one should exercise the option, buy the unit of the underlying for K and re-sell it immediately on the open market for a profit of $S_T - K$. The graph of the payoff function is given in Fig. 9.5.

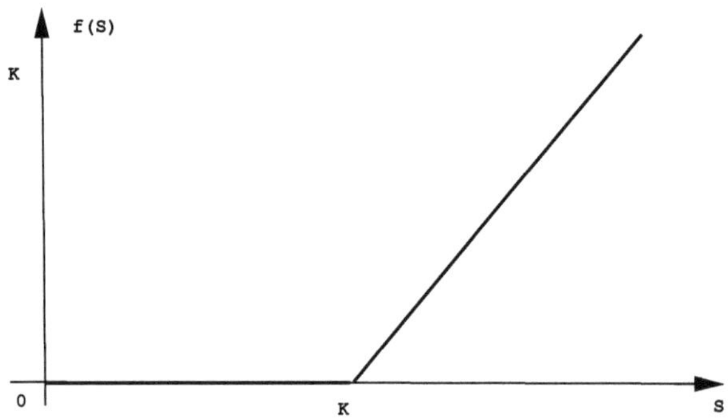

Fig. 9.5. Graph of the "Hockey Stick" function giving the payoff of European call options

We shall use the notation $C_{T,K}(t, S)$ for the price of such an option at time t if the price (at this time) of the underlying asset is S.

9.2.1.1 ***Stochastic Models for Asset Prices***

Obviously, the value S_T of the asset price at maturity cannot be predicted with certainty, and hence, it is reasonable to model it as the outcome of a random variable. Under this assumption the payoff is also a random variable, and the value of the option now depends on the probability that the price S_T at expiration is lower than the strike K, and on the various probabilities of the different ways this price could end up being greater than the strike. Because of this inherent randomness, it is reasonable to price the option by expectation, as an average over all the possible future scenarios weighted by their respective probabilities. At any given time t one can observe the current value of the underlying asset, say S, and consider the expectation of the payoff (which takes place precisely at time T since we are only considering options with an European exercise). It is quite clear that, because of the time value of money (from our discussion of the fixed income markets, we agree that 1 dollar later does not sound as valuable as 1 dollar now) this expectation should be discounted to give the current value of this expected payoff. Consequently, one is naturally lead to the following formula for the price of the option:

$$C_{K,T}(t,S) = e^{-r(T-t)}\mathbb{E}\{f_K(S_T)|S_t = S\}. \tag{9.3}$$

The positive number $r > 0$ appearing in the exponential in the right hand side stands for the interest rate. It is assumed to be constant over the life of the option. Finally the expectation is the expectation of the pay-off $f_K(S_T)$ conditioned by the knowledge that at time t the price of the underlying is S (i.e. $S_t = S$). This conditional expectation (like any other expectation) can be a sum when the random variables are discrete, but in general, it can be computed as an integral if one knows the density of the random variables in question. In the present situation, the expectation is in fact a conditional expectation, but this does not change the fact that the expectation can be computed as an integral: we just have to use the conditional density instead of the a-priori density. So, if we denote by $p(\,\cdot\,|S)$ the density of the random variable S_T of the asset price at time T knowing that $S_t = S$, then the expectation in (9.3) can be computed as:

$$\mathbb{E}\{f_K(S_T)|S_t = S\} = \int f_K(s')\,p(s'|S)\,ds'. \tag{9.4}$$

Intuitively, $p(s'|S)$ gives the probability that $S_T = s'$ given the fact that $S_t = S$. This probability density is often called the *objective density* or the *historical density* because it describes the statistics of the price of the asset at time T, as predicted by historical data. It depends upon many factors which need to be specified.

9.2.1.2 *Aside on Dynamical Models*

The static models of the statistical distribution of the underlying asset at a *fixed* given time are enough to price European call and put options, but they cannot suffice for the pricing of options whose settlements depend upon the entire trajectories of the underlying instruments. Even though we will not need the following material in this chapter, we include it as preparation for the forthcoming discussions in the last chapters of the book. Obviously, this paragraph can be skipped in a first reading.

One of the most popular dynamical models for stock prices is certainly the so-called geometric Brownian motion model proposed by Samuelson and used by Black, Scholes and Merton in their ground-breaking work on option pricing in the early 1970s. The widespread use of their pricing formulae earned Merton and Scholes the Bank of Sweden Prize in Economics in Memory of Alfred Nobel, usually called the Nobel prize in economics, in 1997. Fisher Black passed away in 1995. According to this model, the dynamics of the underlying asset price are given by the *stochastic differential* equation:

$$dS_t = S_t[\mu dt + \sigma dW_t], \tag{9.5}$$

where the term dW_t is a white noise in continuous time (some sort of a mathematical monstrosity to which one has to give a rigorous meaning). The deterministic constant $\mu \in \mathbb{R}$ represents the rate of growth of the (logarithmic) return of the stock, while $\sigma > 0$ represents the volatility of the stock (log) return. Because the coefficients are constant, the stochastic differential equation (9.5) can be solved explicitly. The solution is given by:

$$S_t = S_0 e^{[\mu - \frac{\sigma^2}{2}]t + \sigma W_t} \tag{9.6}$$

and the unexpected term $-t\sigma^2/2$ appearing in the exponential is called the Ito correction. This model is called the geometric Brownian motion model for stock prices. These technical facts are quoted here as preparation for Sects. 8.4, 8.4.2 and 8.5.1 in Sect. 8.5.

We now come back to the derivation of a pricing formula for the option. The important fact is that, conditioned on being equal to S at time t, the asset price at time T is log-normally distributed with mean $\log S_t + (\mu - \sigma^2/2)\tau$ and variance $\sigma^2\tau$ where $\tau = T - t$. In other words:

$$\log S_T \sim N(\log S_t + (\mu - \sigma^2/2)\tau, \sigma^2\tau). \tag{9.7}$$

This fact is a consequence of the above choice for a dynamical model, but it can also be postulated independently. In any case, under these conditions, the expectation in (9.4) can be computed explicitly (see details below for a similar computation).

9.2.1.3 **Risk Neutral Pricing**

Even though the expectation of formula (9.3) is a very natural candidate for the price of the option, even in Samuelson's geometric Brownian motion model, it is not the correct one! Indeed, if options were priced in this way, market makers would have arbitrage opportunities making it possible to make a profit starting from nothing. So since we all know that there is no such a thing as a *free lunch,* something must be wrong with the pricing formula (9.3) and the distribution (9.7). In the absence of arbitrage, and for reasons well beyond the scope of the book, the option can indeed be priced by discounting an expectation, as long as the expectation is computed with respect to another log-normal distribution. This new density is called the *state price density,* and in the case of the Black-Scholes theory, it is obtained by merely replacing μ by r. In other words, in order to price an option on an underlying asset, we act as if the rate of growth of the (log) returns was the short interest rate. This justifies the name *risk neutral* probability for this new log-normal density.

9.2.1.4 **Black-Scholes Formula**

For the sake of completeness, we give the details of the computations leading to the Black-Scholes formula for the price of the option. This formula will be used in one of the nonparametric approaches presented below.

Using the explicit form of the density of the $N(0,1)$ random variable and formula (9.7), we derive the value of $C_{T,K}(t,S)$ as follows:

$$C_{T,K}(t,S) = \frac{e^{-r(T-t)}}{\sqrt{2\pi}} \int_{-\infty}^{+\infty} f\left(Se^{(r-\sigma^2/2)(T-t)} e^{\sigma\sqrt{T-t}z}\right) e^{-z^2/2}\, dz. \tag{9.8}$$

Now, using the fact that $f(x) = \max\{x - K, 0\} = (x - K)^+$ in the case of a European call option and using the notation:

$$d_1 = \frac{\log(S/K) + (r + \sigma^2/2)(T - t)}{\sigma\sqrt{T - t}} \quad \text{and} \quad d_2 = d_1 - \sigma\sqrt{T - t},$$

we get:

$$\begin{aligned}
C_{T,K}(t, S) &= \frac{e^{-r(T-t)}}{\sqrt{2\pi}} \int_{-\infty}^{+\infty} \left(Se^{(r-\sigma^2/2)(T-t)}e^{\sigma\sqrt{T-t}z} - K\right)^+ e^{-z^2/2}\,dz \\
&= \frac{1}{\sqrt{2\pi}} \int_{-\infty}^{+\infty} \left(Se^{-\sigma^2(T-t)/2+\sigma\sqrt{T-t}z} - Ke^{-r(T-t)}\right)^+ e^{-z^2/2}\,dz \\
&= \frac{1}{\sqrt{2\pi}} \int_{-d_2}^{+\infty} \left(Se^{-\sigma^2(T-t)/2+\sigma\sqrt{T-t}z} - Ke^{-r(T-t)}\right) e^{-z^2/2}\,dz \\
&= \frac{S}{\sqrt{2\pi}} \int_{-d_2}^{+\infty} e^{-\sigma^2(T-t)/2+\sigma\sqrt{T-t}z}e^{-z^2/2}\,dz \\
&\qquad - Ke^{-r(T-t)}\frac{1}{\sqrt{2\pi}} \int_{-d_2}^{+\infty} e^{-z^2/2}\,dz \\
&= S\Phi(d_1) - Ke^{-r(T-t)}\Phi(d_2),
\end{aligned}$$

where we performed the substitution $y = z + \sigma\sqrt{T - t}$ in the first of the two integrals. Recall that we use the notation Φ for the cumulative distribution function of the standard normal distribution. We summarize this computation in the following box:

> The price at time t of a European call with strike K and maturity T is given by the formula:
>
> $$C_{T,K}(t, S) = S\Phi(d_1) - Ke^{-r(T-t)}\Phi(d_2) \tag{9.9}$$
>
> if the price of the underlying risky asset is S at time t.

Notice that, instead of depending separately upon t and T, the conditional density and the actual price of the option depend only upon the difference $\tau = T - t$. The interpretation of this difference is very simple: it is the time to maturity of the option.

Figure 9.6 shows plots of the prices $C_{T,K}(t, S)$ (together with the prices $P_{T,K}(t, S)$ of the corresponding European put options) for various times to expiration $T - t$ when the strike price K, the volatility σ and the interest rate r are fixed. We notice that when regarded as functions of the underlying price S, the European call and put values appear as smoothed forms of the original hockey stick pay-off functions. Moreover, $C_{T,K}(t, S)$ vanishes near small values of S while $P_{T,K}(t, S)$ vanishes for large values of S.

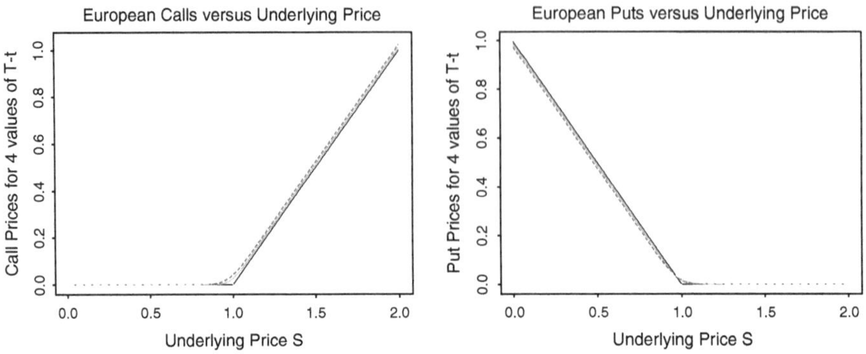

Fig. 9.6. Plots of the prices of European call (*left*) and European put (*right*) vanilla options as functions of the underlying price S when all the other parameters are held fixed. We used the values $K = 1$, $\sigma = 15\%$, $r = 10\%$ and $T - t = 0, 30/252 = 0.1190476, 60/252 = 0.2380952, 90/252 = 0.3571429$

9.2.1.5 *Historical Volatility, Implied Volatility and the Smile Effect*

Since the rate of growth μ of the stock was evacuated from the formula for the price of the contingent claims, the only parameter from the dynamics of the risky asset that is entering these formulae is the volatility σ and the problem we wish to discuss now is the estimation of its value.

In theory, the observation of a single trajectory (realization) of the price should be enough to completely determine the value of this parameter. This would be true if the price process S_t could be observed continuously! Unfortunately this cannot be the case in practice and we have to settle for an approximation. Given observations S_{t_j} of past values of the risky asset (usually the times t_j are of the form $t_j = t - j\delta t$), we use the fact that in the Black-Scholes model the random variables $\log(S_{t_j}/S_{t_{j+1}})$ are independent and normally distributed with mean $(\mu - \sigma^2/2)\delta t$ and variance $\sigma^2\delta t$. Consequently, the volatility can be estimated by the formula:

$$\hat{\sigma} = \frac{1}{(N-1)\sqrt{\delta t}} \sum_{j=0}^{N-1} [\log \frac{S_{t_j}}{S_{t_{j+1}}} - \overline{LS}]^2, \tag{9.10}$$

where the sample mean $\overline{LS}$ is defined by the formula:

$$\overline{LS} = \frac{1}{N} \sum_{j=0}^{N-1} \log \frac{S_{t_j}}{S_{t_{j+1}}}.$$

The volatility estimate provided by formula (9.10) is called the historical volatility. Even though $\overline{LS}$ provides (at least up to the multiplicative factor δt) an unbiased estimator of the growth rate μ, this fact is not used since the growth rate μ does not appear in the Black-Scholes formula of the arbitrage pricing of the claim.

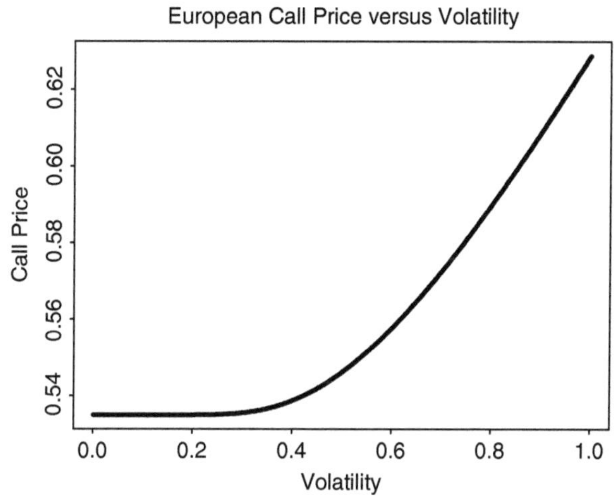

Fig. 9.7. Values of the price of a European call option as a function of the volatility when all the other parameters are held fixed. We used the values, $T - t = 90/252$, $r = 10\%$, $S = 1.5$ and $K = 1$

Everything seems to be fine except for the fact that we are now about to encounter our first shocking evidence that the market does not have the good taste to follow the Black-Scholes model. This will be the first of a series of rude awakenings.

A quick look at Fig. 9.7, which gives the plot of $C = C_{T.K}(t, S)$ as a function of σ when all the other parameters (namely r, T, K and S) are held fixed, shows that C is an increasing function of σ, and consequently, that there is a one-to-one correspondence between the price of the option, and the volatility parameter σ. For each value of C, the unique value of σ which, once injected into the Black-Scholes formula, gives the option price C, is called the implied volatility of the option. This one-to-one correspondence is so entrenched in the practice of the markets that prices of options are most of the time quoted in percentage points (indicating a value for the *implied volatility*) rather than in dollars (for a value of C).

We now demonstrate the fact that the assumptions of the Black-Scholes model are in contradiction with market reality. We use quotes from January 6, 1993 (when the S&P index was at the $S = 435.6258$ level) on European calls on the S&P 500 index with maturity February 19, 1993, but with different strike prices. In general, when everything else is fixed, for a given set of strike prices K_j we have the corresponding call prices C_j quoted by the market. If the Black-Scholes model was in force we should expect that these quotes had been computed using formula (9.1) with a single volatility value for σ, the differences in the quotes being due only to the differences in the strikes (since the expiration is the same for all these options). Since given a strike price there is a one-to-one correspondence between option price and volatility, one should be able to compute the (implied) volatility used to price all these options. Moreover, plotting the values of these implied volatilities versus the corresponding strike prices one should expect a flat graph indicative of the unique-

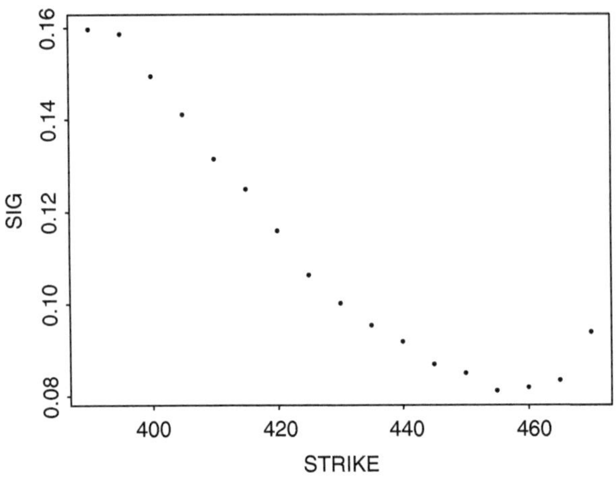

Fig. 9.8. Plot of the implied volatility values versus the strike prices of call options with the same maturity on the S&P 500 index. A convex (upward) curve is found instead of the horizontal line predicted by the Black-Scholes theory. This curve is called the smile

ness of the volatility used to price the options. Figure 9.8 shows that this is certainly not the case! The volatility plot forms a convex curve which has been called the volatility smile. This volatility smile is striking empirical evidence of the inadequacy of the Black-Scholes option pricing formula.

NOTES & COMPLEMENTS

The internet is a great source of information on the open source code project, the language R, and the many contributed libraries which you can download and use freely to enhance the working capabilities of the core language.

References

1. S. Amari. *Differential Geometrical Methods in Statistics*, volume 28 of *Lecture Notes in Statistics*. Springer Verlag, New York, NY, 1985.
2. N. Anderson, F. Breedon, M. Deacon, A. Derry, and G. Murphy. *Estimating and Interpreting the Yield Curve*. Wiley, Chichester, 1996.
3. A. Antoniadis, J. Berruyer, and R. Carmona. *Régression Non-linéaire et Applications*. Economica, 1992.
4. P. Artzner, F. Delbaen, J.M. Eber, and D. Heath. Coherent measures of risk. *Mathematical Finance*, 9(3):203–228, 1999.
5. E. Banks, editor. *Weather Risk Management*, New York, NY, 2002. Palgrave.
6. T. Bollerslev. Generalized Auto Regressive Heteroskedasticity. *Journal of Econometrics*, 31:307–327, 1986.
7. T. Bollerslev, R.F Engle, and J.M. Wooldridge. ARCH models. In *Handbook of Econometrics, IV*, pages 2959–3038. 1994.
8. G.E.P. Box and G.M. Jenkins. *Time Series Analysis: Forecasting and Control*. Holden Day, San Francisco, revised edition, 1976.
9. L. Breiman, J.H. Friedman, R.A. Olshen, and C.I. Stone. *Classification And Regression Trees*. Wadsworth and Brooks Cole, Monterey, CA, 1984.
10. P.J. Brockwell and R.A. Davis. *Introduction to Time Series and Forecasting*. Springer Texts in Statistics. Springer Verlag, New York, NY, 1996.
11. R.L. Brown, J. Durbin, and J.M. Evans. Techniques for testing the constancy of regression relationship over time (with comments). *Journal of the Royal Statistical Society*, 37:149–192, 1975.
12. A. Bruce and H.Y. Gao. *Applied Wavelet Analysis with S-Plus*. Springer Verlag, New York, N.Y, 1996.
13. J.Y. Campbell, A.W. Lo, and A.C. MacKinlay. *The Econometrics of Financial Markets*. Princeton University Press, Princeton, N.J., 1997.
14. R. Carmona, W. Hwang, and B. Torresani. *Time Frequency Analysis: Continuous Wavelet and Gabor Transform, with an implementation in* `S-Plus`. Academic Press, New York, N.Y., 1998.
15. R. Carmona and J. Morrisson. EVANESCE, an `S-Plus` library for heavy tail distributions and copulas. Technical report, Dept. of Operations Research & Financial Engineering, Princeton University, 2000. `http://www.princeton.edu/~rcarmona`.

R. Carmona, *Statistical Analysis of Financial Data in R*, Springer Texts in Statistics, DOI 10.1007/978-1-4614-8788-3,

16. R. Carmona, J.P. Fouque, and D. Vestal. Interacting Particle Systems for the Computation of CDO Tranche Spreads with Rare Defaults. *Finance and Stochastics*, 13:613–633, 2009.
17. R. Carmona and S. Crepey. Importance Sampling and Interacting Particle Systems for the Estimation of Markovian Credit Portfolio Loss Distributions. *Intern. J. of Theoretical and Applied Finance*, 13:577–602, 2010.
18. R. Carmona and M. Tehranchi. *Interest Rate Models: an Infinite Dimensional Stochastic Analysis Perspective*. Springer Verlag, New York N.Y., 2010.
19. R. Carmona, P. Del Moral, P. Hu and N. Oudjane. An introduction to particle methods in Finance. in *Numerical Methods in Finance* eds *R. Carmona, P. Del Moral, P. Hu and N. Oudjane*, pp. 1–45, Springer Verlag, 2012.
20. R. Carmona and M. Croulon. A Survey of Commodity Markets and Structural Approaches to Modeling Electricity. In *Energy Markets, Proceedings of the WPI Special Year* eds. *F. Benth*, Springer Verlag, pp. 1–42, 2012.
21. J.M. Chambers. *Programming with Data: A Guide to the S Language*. MathSoft, Seattle WA, 1998.
22. N.H. Chan. *Time Series: Applications to Finance*. John Wiley & Sons, New York, N.Y., 2002.
23. W.S. Cleveland. *The Elements of Graphing Data*. Hobart Press, Summit, N.J., 1994.
24. L. Clewlow and C. Strickland. *Energy Derivatives, Pricing and Risk Management*. Lacima Publications, London, 2000.
25. P. Dalgaard. *Introductory Statistics with R*. Springer-Verlag, New York, 2002.
26. C. de Boor. *A Practical Guide to Splines*. Springer Verlag, New York, N.Y., 1978.
27. D. Duffie and J. Singleton. *Credit Risk: Pricing, Measement, and Management*. 2003.
28. P. Embrechts, R. Frey, and A. McNeil. *Quantitative Risk Management: Concepts, Techniques and Tools*. Princeton University Press, Princeton, NJ, 2005.
29. P. Embrechts, C. Klüppelberg, and T. Mikosch. *Modelling Extremal Events for Insurance and Finance*. Springer-Verlag, New York, 2000.
30. R.F. Engle. Autoregressive conditional heteroskedasticity with estimates of the variance of the United Kingdom inflation. *Econometrica*, pages 987–1006, 1982.
31. R.F. Engle and T. Bollerslev. Modeling the persistence of conditional variances. *Econometric Reviews*, 5:1–50, 1986.
32. R.F. Engle and C. Granger. Cointegration and error-correction: Representation, estimation and testing. *Econometrica*, 55:251–276, 1987.
33. J. A. Greenwood et al. Extreme Value Theory based on the r largest annual events: a robust approach. *Water Resources Research*, 15:1049–1054, 1979.
34. E.F. Fama and R.R. Biss. The information in long-maturity forward rates. *American Economic Review*, 77:680–692, 1987.
35. J. Fan and Q. Yao. *Nonlinear Time Series: Nonparametric and Parametric Methods*. Springer Verlag, New York, N.Y., 2003.
36. H. Foellmer and A. Schied. *Stochastic Finance: An Introduction in Discrete Time*. De Gruyter, Berlin, 2002.
37. Bank for International Settlements. Zero-coupon yield curves. Technical report, March 1999.
38. J.P Fouque, G. Papanicolaou, and R. Sircar. *Derivatives in Financial Markets with Stochastic Volatility*. Cambridge University Press, London, 2000.
39. R. Gençay, F. Selçuk, and B. Whitcher. *An Introduction to Wavelets and Other Filtering Methods in Finance and Economics*. Academic Press, New York, N.Y., 2002.

40. P. Glasserman. *Monte Carlo Methods in Financial Engineering*. Springer-Verlag, New York, N.Y., 2004.
41. C. Gouriéroux. *ARCH Models and Financial Applications*. Springer Verlag, New York, N.Y., 1997.
42. C. Gourieroux and J. Jasiak. *Financial Eeconometrics: Problems, Models and Methods*. Princeton University Press, Princeton, NJ, 2001.
43. W. Haerdle. *Non-parametric Regression*. Springer Verlag, New York, N.Y., 1994.
44. J.D. Hamilton. *Time Series Analysis*. Princeton University Press, Princeton, N.J., 1994.
45. T. Hastie, Tibshirani, and J. Friedman. *The Elements of Statistical Learning, Data mining, Inference and Prediction*. Springer Verlag, New York, N.Y., 2001.
46. J. R. M. Hosking. L-moments: Analysis and estimation of distributions using linear combinations of order statistics. *Journal of the Royal Statistical Society, Series B*, 52(1):105–124, 1990.
47. J. R. M. Hosking and J. R. Wallis. Parameter and quantile estimation for the generalized Pareto distribution. *Technometrics*, 29(3):339–349, 1987.
48. J. R. M. Hosking, J. R. Wallis, and E. F. Wood. Estimation of the generalized extreme value distribution by the Method of Probability-Weighted Moments. *Technometrics*, 27(3):251–261, 1985.
49. P.J. Huber. *Robust Statistics*. John Wiley & Sons, New York, N.Y., 1981.
50. J.C. Hul. *Options, Futures, and Other Derivatives*. Prentice Hall, New York, N.Y., 6th edition, 2006.
51. J.M. Hutchinson, A.W. Lo, and T. Poggio. A nonparametric approach to the pricing and hedging of derivative securities via learning networks. *Journal of Finance*, 49, 1994.
52. S. Johansen. *Likelihood-Based Inference in Cointegrated Vector Autoregressive Models*. Oxford University Press, Oxford, 1995.
53. P. Jorion. *Value at Risk: The New Benchmark for Managing Financial Risk*. McGraw Hill, New York, N.Y., 2nd edition, 2000.
54. J. Rice. *Mathematical Statistics and Data Analysis*. Duxbury Press, Belmont, CA, 2nd edition, 1995.
55. C.J. Kim and C.R. Nelson. *State-Space Models with Regime Switching*. Cambridge, MA, 1999.
56. G. Kitagawa. Monte Carlo filter and smoother for non-Gaussian nonlinear state space models. *Journal of Computational and Graphical Statistics*, 5:1–25, 1996.
57. T. Kohonen. *Self-organizing Maps*. Springer Verlag, New York, N.Y., 1995.
58. D. Lamberton and B. Lapeyre. *Introduction to Stochastic Calculus Applied to Finance*. CRC Press, 1996.
59. D. Lando. *Credit Risk Modeling: Theory and Aplications*. 2004.
60. R. Litterman and J. Scheinkman. Common factors affecting bond returns. *Journal of Fixed Income*, 1:49–53.
61. F.M. Longin. From value at risk to stress testing: The extreme value approach. *Journal of Banking and Finance*, 24:10971130, 2000.
62. F. A. Longstaff and R. S. Schwartz. Valuing American options by simulation : A simple least-square approach. *Review of Financial Studies*, 14:113–147, 2001.
63. G. Lindren M. R. Leadbetter and H. Rootzén. *Extremes and Related Properties of Random Sequences and Processes*. Springer-Verlang, New York, 1983.
64. B.B. Mandelbrot. New methods in statistical economics. *Journal of Political Economy*, 71:421–440, 1963.
65. B.B. Mandelbrot. The variation of certain speculative prices. *Journal of Business*, 36:394–419, 1963.

66. B.B. Mandelbrot. *Fractal, Form Dimension and Chance*. W.H. Freeman & Co., San Francisco, CA, 1977.
67. K.V. Mardia, J.T. Kent, and J.M. Bibby. *Multivariate Analysis*. Academic Press, New York, N.Y., 1979.
68. D. Drouet Mari and S. Kotz. *Correlation and Dependence*. Imperial College Press, London, 2001.
69. D.C. Montgomery and E.A. Peck. *Introduction to Linear Regression Analysis*. John Wiley & Sons, New York, N.Y., 2nd edition, 1992.
70. S. Menendez N. Perez and L. Seco. New families of distributions fitting l-moments for modelling financial data, 2004.
71. G.P. Nason. *Wavelet Methods in Statistics with R*. Springer Verlag, New York, NY, 2008.
72. A. Mc Neil, R. Frey, and P. Embrechts. *Quantitative Risk Management: Concepts, Techniques and Tools*. Princeton University Press, Princeton, NJ, 2005.
73. R. B. Nelsen. *An Introduction to Copulas*. Springer-Verlag, New York, 1999.
74. C.R. Nelson and A.F. Siegel. Parsimonious modeling of yield curves.
75. J. Pickands. Multivariate extreme value distributions. pages 229–231, 1981.
76. M.B. Priestley. *Nonlinear and Non-stationary Time Series Analysis*. Academic Press, New York, N.Y., 1988.
77. R. Rebonato. *Interest-Rate Option Models: Understanding, Analyzing and Using Models for Exotic Interest-Rate Options*. Wiley, New York, N.Y., 1996.
78. E. Renault and N. Touzi. Option hedging and implied volatility in a stochastic volatility model. *Mathematical Finance*, 6:215–236, 1996.
79. S.I. Resnick. *Extreme Values, Regular Variation and Point Processes*. Springer Verlag, New York, 1987.
80. R. Roll. A critique of the asset pricint theory's test, part one: Past and potential testability of the theory. *Journal of Financial Economics*, 4, 1977.
81. M. Rosenblatt. *Gaussian and Non-Gaussian Linear Time Series and Random Fields*. Springer Verlag, New York N.Y., 2000.
82. P.J. Rousseeuw and A.M. Leroy. *Robust Regression and Outlier Detection*. New York, N.Y., 1984.
83. B. Van Roy and J. N. Tsitsiklis. Regression methods for pricing complex American-style options. *IEEE Trans. on Neural Networks*, 2000.
84. P.A. Ruud. *An Introduction to Classical Econometric Theory*. Oxford University Press, New York, N.Y., 2000.
85. Y. Ait Sahalia and A.W. Lo. Nonparametric estimation of state price densities implicit in financial asset prices. *The Journal of Finance*, 53:499–547, 1998.
86. P. Schönbucher. *Credit Dderivatives Pricing Models: Model, Pricing and Implementation*. John Wiley & Sons, New York, N.Y., 2003.
87. R.H. Shumway and D.S. Stoffer. *Time Series Analysis and Its Applications*. Springer Verlag, Newy York, N.Y., 2000.
88. B.W. Silverman. *Density Estimation for Statistics and Data Analysis*. Chapman & Hall, London, 1986.
89. R. L. Smith. Threshold methods in statistics. In J. Tiago de Olivera, editor, *Statistical Extremes and Applications*, volume 131 of *NATO ASI Series*, pages 621–638. Reidel, 1984.
90. J.M. Steele. *Stochastic Calculus and Financial Applications*. Springer Verlag, New York N.Y., 2000.
91. L.E.O. Svensson. Estimating and interpreting forward interest rates: Sweden 1992–94. 4871, 1994.

92. R.S. Tsay. *Analysis of Financial Time Series*. John Wiley & Sons, New York, N.Y., 2002.
93. O. Vasicek and G. Fong. Term structure estimation using exponential splines. *Journal of Finance*, 38:339–348, 1982.
94. W. N. Venables and B. D. Ripley, *Modern Applied Statistics with S-PLUS*. Springer-Verlag, New York, 1997.
95. H. von Storch and F.W. Zwiers. *Statistical Analysis in Climate Research*. Cambridge University Press, New York, N.Y., 1999.
96. L. De Vroye. *Non-uniform Variates*. Springer Verlag, 1984.
97. M.V. Wickerhauser. *Adapted Wavelet Analysis: from Theory to Software*. A.K.Peters LTd, Wellesley, MA, 1994.
98. I.H. Witten and E. Frank. *Data Mining*. Academic Press, San Diego, CA, 2000.
99. E. Zivot and J. Wang. *Modeling Financial Time Series with S-PLUS*. Springer Verlag, New York, N.Y., 2003.

Notation Index

R. Carmona, *Statistical Analysis of Financial Data in R*, Springer Texts in Statistics,
DOI 10.1007/978-1-4614-8788-3,

Data Set Index

R. Carmona, *Statistical Analysis of Financial Data in R*, Springer Texts in Statistics,
DOI 10.1007/978-1-4614-8788-3, © Springer Science+Business Media New York 2014

R Index

R. Carmona, *Statistical Analysis of Financial Data in R*, Springer Texts in Statistics,
DOI 10.1007/978-1-4614-8788-3, © Springer Science+Business Media New York 2014

Author Index

R. Carmona, *Statistical Analysis of Financial Data in R*, Springer Texts in Statistics,
DOI 10.1007/978-1-4614-8788-3,

SUBJECT INDEX

R. Carmona, *Statistical Analysis of Financial Data in R*, Springer Texts in Statistics,
DOI 10.1007/978-1-4614-8788-3, © Springer Science+Business Media New York 2014

D

MIX
Papier aus verantwortungsvollen Quellen
Paper from responsible sources
FSC® C105338

If you have any concerns about our products,
you can contact us on
ProductSafety@springernature.com

In case Publisher is established outside the EU,
the EU authorized representative is:
Springer Nature Customer Service Center GmbH
Europaplatz 3, 69115 Heidelberg, Germany

Printed by Libri Plureos GmbH
in Hamburg, Germany